69 Springer Series in Solid-State Sciences

Edited by Peter Fulde

Springer Series in Solid-State Sciences

Editors: M. Cardona P. Fulde K. von Klitzing H.-J. Queisser

40 **Semiconductor Physics** An Introduction
3rd Edition By K. Seeger

41 **The LMTO Method**
Muffin-Tin Orbitals and Electronic Structure
By H. L. Skriver

42 **Crystal Optics with Spatial Dispersion, and Excitons**
By V. M. Agranovich and V. L. Ginzburg

43 **Resonant Nonlinear Interactions of Light with Matter**
By V. S. Butylkin, A. E. Kaplan, Yu. G. Khronopulo, and E. I. Yakubovich

44 **Elastic Media with Microstructure II**
Three-Dimensional Models
By I. A. Kunin

45 **Electronic Properties of Doped Semiconductors**
By B. I. Shklovskii and A. L. Efros

46 **Topological Disorder in Condensed Matter**
Editors: F. Yonezawa and T. Ninomiya

47 **Statics and Dynamics of Nonlinear Systems**
Editors: G. Benedek, H. Bilz, and R. Zeyher

48 **Magnetic Phase Transitions**
Editors: M. Ausloos and R. J. Elliott

49 **Organic Molecular Aggregates,** Electronic Excitation and Interaction Processes
Editors: P. Reineker, H. Haken, and H. C. Wolf

50 **Multiple Diffraction of X-Rays in Crystals**
By Shih-Lin Chang

51 **Phonon Scattering in Condensed Matter**
Editors: W. Eisenmenger, K. Laßmann, and S. Döttinger

52 **Superconductivity in Magnetic and Exotic Materials**
Editors: T. Matsubara and A. Kotani

53 **Two-Dimensional Systems, Heterostructures, and Superlattices**
Editors: G. Bauer, F. Kuchar, and H. Heinrich

54 **Magnetic Excitations and Fluctuations**
Editors: S. Lovesey, U. Balucani, F. Borsa, and V. Tognetti

55 **The Theory of Magnetism II**
Thermodynamics and Statistical Mechanics
By D. C. Mattis

56 **Spin Fluctuations in Itinerant Electron Magnetism** By T. Moriya

57 **Polycrystalline Semiconductors,**
Physical Properties and Applications
Editor: G. Harbeke

58 **The Recursion Method and Its Applications**
Editors: D. Pettifor and D. Weaire

59 **Dynamical Processes and Ordering on Solid Surfaces**
Editors: A. Yoshimori and M. Tsukada

60 **Excitonic Processes in Solids**
By M. Ueta, H. Kanzaki, K. Kobayashi, Y. Toyozawa, and E. Hanamura

61 **Localization, Interaction, and Transport Phenomena**
Editors: B. Kramer, G. Bergmann, and Y. Bruynseraede

62 **Theory of Heavy Fermions and Valence Fluctuations**
Editors: T. Kasuya and T. Saso

63 **Electronic Properties of Polymers and Related Compounds**
Editors: H. Kuzmany, M. Mehring, and S. Roth

64 **Symmetries in Physics, Group Theory Applied to Physical Problems**
By W. Ludwig and G. Falter

65 **Phonons: Theory and Experiments II**
Experiments and Interpretation of Experimental Results By P. Brüesch

66 **Phonons: Theory and Experiments III**
Phenomena Related with Phonons
By P. Brüesch

67 **Two-Dimensional Systems: Physics and New Devices**
Editors: G. Bauer, F. Kuchar, and H. Heinrich

68 **Phonon Scattering in Condensed Matter V**
Editors: A. C. Anderson and J. P. Wolfe

69 **Nonlinearity in Condensed Matter**
Editors: A. R. Bishop, D. K. Campbell, P. Kumar and S. E. Trullinger

70 **From Hamiltonians to Phase Diagrams**
The Electronic and Statistical-Mechanical Theory of sp-Bonded Metals and Alloys
By J. Hafner

Volumes 1–39 are listed on the back inside cover

Nonlinearity in Condensed Matter

Proceedings of the Sixth Annual Conference,
Center for Nonlinear Studies,
Los Alamos, New Mexico, 5–9 May, 1986

Editors: A. R. Bishop, D. K. Campbell,
P. Kumar and S. E. Trullinger

With 163 Figures

Springer-Verlag Berlin Heidelberg New York
London Paris Tokyo

Dr. Alan R. Bishop
Dr. David K. Campbell
Dr. Pradeep Kumar
Dr. Steven E. Trullinger

Center of Nonlinear Studies, Los Alamos National Laboratory,
Los Alamos, NM 87545, USA

Series Editors:

Professor Dr., Dr. h. c. Manuel Cardona
Professor Dr., Dr. h. c. Peter Fulde
Professor Dr. Klaus von Klitzing
Professor Dr. Hans-Joachim Queisser

Max-Planck-Institut für Festkörperforschung, Heisenbergstrasse 1
D-7000 Stuttgart 80, Fed. Rep. of Germany

ISBN-13:978-3-642-83035-8 e-ISBN-13:978-3-642-83033-4
DOI: 10.1007/978-3-642-83033-4

Softcover reprint of the hardcover 1st edition 1987

2153/3150-543210

Preface

The Sixth Annual Conference of the Center for Nonlinear Studies at the Los Alamos National Laboratory was held May 5–9, 1986, on the topic "Nonlinearity in Condensed Matter: Lessons from the Past and Prospects for the Future." As conference organizers, we felt that the study of nonlinear phenomena in condensed matter had matured to the point where it made sense to take stock of the numerous lessons to be learned from a variety of contexts where nonlinearity plays a fundamental role and to evaluate the prospects for the growth of this general discipline.

The successful 1978 Oxford Symposium on nonlinear (soliton) structure and dynamics in condensed matter (Springer Ser. Solid-State Sci., Vol. 8) was held at a time when the ubiquity of solitons was just beginning to be appreciated by the condensed matter community; in subsequent years the soliton paradigm has provided a rather useful framework for investigating a large number of phenomena, particularly in low-dimensional systems. Nevertheless, we felt that the importance of nonlinearity in wider arenas than "solitonics" merited a significant expansion in the scope of the conference over that of the 1978 symposium. Indeed, many of the lessons are quite general and their potential for cross-fertilization of otherwise poorly connected disciplines was certainly one of the prime motivations for this conference. Thus, while these proceedings contain many contributions pertaining to soliton behavior in different contexts, the reader will find much more as well, particularly in the later chapters.

The introductory review (Part I) by R. Landauer provides a historical and technological perspective which we feel is of particular importance in learning some of the lessons from the past. We wish to thank Dr. Landauer for his very thoughtful views which have relieved us of the duty of writing an introduction! The next two parts survey fields which have received a great deal of attention in the last several years, namely low-dimensional magnetism (Part II) and conducting polymers (Part III). As with the following parts, each of these contains a mixture of invited and contributed papers. Part IV contains contributions on some experimental techniques for probing nonlinearity. We hasten to add that several contributions in other parts also describe valuable experimental methods such as optical probes and transport measurements. Part V deals with theoretical issues and techniques and again we note that contributions appearing in other parts could also have been collected there. The traditional arena of structural phase transitions now presents some new and interesting nonlinear phenomena which are discussed in Part VI. Spatial (and temporal) complexity resulting from "competing" space (and/or time) scales are rapidly becoming a paradigm in their own right. Several illustrations are collected in Parts VII and VIII. In Part VII one finds articles on recent experimental and theoretical developments in the areas of spin-glasses and random field systems, while in Part VIII we have collected several articles in the general areas of frustrated, incommensurate and nonequilibrium systems. Finally, in the summary (Part IX) by Gordon Baym, the reader will find much delight in a perceptive summary of the essence of the conference.

Sponsorship of the conference was generously provided by the Program in Applied Mathematical Sciences, Office of Basic Energy Sciences, U.S. Department of Energy. We are also grateful for expert secretarial support from the Center for Nonlinear Studies – especially from Frankie Gomez, Marian Martinez and Valerie Ortiz. Finally, we thank Los Alamos National Laboratory for use of their excellent conference facilities and organizational staff.

The prospect of learning new and powerful techniques beckons the "linear" traditionalist to cross the threshold into the remarkably rich world of nonlinear phenomena, where forays into unfamiliar territory may be rewarded by much improved perspectives on his own field of endeavor. We hope that this collection of articles lights the way.

Los Alamos, December 1986

A.R. Bishop
D.K. Campbell
P. Kumar
S.E. Trullinger

Contents

Part IV Experimental Techniques

Part V Theoretical Techniques

Part VI Structural Phase Transitions

Part VII Spin-Glasses and Random Field Systems

Part VIII **Frustrated, Incommensurate, and Nonequilibrium Systems**

Part IX **Summary**

Part I

Historical Perspective

Nonlinearity: Historical and Technological View

R. Landauer

IBM Thomas J. Watson Research Center, P.O. Box 218,
Yorktown Heights, NY 10598, USA

1. INTRODUCTION

The title of this conference causes us to ask about the meaning of Condensed Matter Physics and of nonlinearity. The ambiguities in Condensed Matter Physics are trivial, and need no belaboring. Every highly developed branch of science has borderline subjects which can be excluded, or included, according to taste. The ambiguity in *nonlinear* is deeper, and reflects the mode of description. We tend to consider classical motion in a constant, linear, or quadratic potential as *linear* and refer to anharmonicity when we go beyond that. But the Hamilton-Jacobi equation, even for a free particle, is already nonlinear. On the other hand the Schrödinger equation is always linear, no matter how complex the potential, and as a limiting case can always describe the classical Hamilton-Jacobi equation. The classical Liouville equation is also linear and represents *the* universal trick for making things look linear, in the case of deterministic problems. We populate the space, in which the solutions evolve, with a density, and all the real complexities are shifted to the description of the motion of a point in this solution space. The kind of linearity obtained via the Liouville equation does not require deterministic behavior, the master equation is also linear, and treats stochastic problems. The low field Hall current, in crossed electric and magnetic fields, represents a linear conductance, if we consider the magnetic field to be a parameter. If, however, we consider electric and magnetic fields as simultaneous components of a total electromagnetic field, then it is clearly nonlinear. Chaos is taken as a symptom of nonlinearity. We can, however, ask about the deviations from a given chaotic solution and return to linearity [1]. It is possible that there is some kind of rational path through this perplexity, and that there is a broadly applicable definition of *nonlinearity*. I do not know about it, and interpret *nonlinear* in a more phenomenological sense. Nonlinearity has to do with thresholds, with multi-stability, with hysteresis, with phenomena which change *qualitatively* as we change excitation.

Nonlinearity is ubiquitous. We cannot *superpose* two timetables to get a valid timetable. Sexual reproduction does depend on the scrambling of two genetic structures, but it is hardly a linear superposition. Most of our technical machinery is nonlinear. Nonlinearity in condensed matter physics is not new, nonlinear phenomena were studied decades ago. Schottky barrier rectifiers, ferromagnetic hysteresis loops [2], the plastic flow theory of dislocations, or PEIERLS's [3] view of lattice anharmonicity and Umklapp processes in heat transport, are just a few of a myriad of examples. Harmonic lattice vibrations cannot explain thermal expansion; it was obvious long ago, that one had to go beyond that. What was missing, until about a decade ago, was not a sensitivity to nonlinear phenomena, but only an appreciation for the *remarkable diversity* of behavior available in nonlinear systems. It is the willingness to *celebrate* nonlinearity that is new. Undoubtedly, solid state physicists were too preoccupied, for some decades, with the implications of quantum mechanics [4], and this raised the threshold for attention to questions not directly related to that. Eventually we learned, as pointed out by ANDERSON [5], that we face a hierarchy of phenomena. The intellectual wonders of the higher levels have little to do with the lowest level. The Schrödinger equation does not really help us to understand macro-economics. Note, however, that this point is still not generally understood. One of our leading physicists [6] states: *Important new theories* do *emerge in other sciences: Cluster chemistry, superconductivity and molecular biology are examples. How truly fundamental are they? Do they not result from a complex interplay among many atoms, about which Heisenberg and his friends taught us all we need to know long ago?*

To illustrate the hierarchy, quantum mechanics is needed to give the transistor inventor and designer a few initial concepts, but once those are available quantum theory provides little further help. The computer architect, in turn, must learn how transistors can give him useful switching circuits; then he can go off on his own. Freed from their earlier preoccupation with quantum mechanics, the physicist and the physical chemist have rediscovered that interesting technical and biological machinery is related to nonlinearity, very often to multistable systems, and to systems very far from thermal equilibrium. The engineer and the biologist, of course, always knew that, and may be a little surprised by the fuss accompanying this rediscovery [7].

This author's connection to nonlinearity stems from early exposures to electronics, and a subsequent connection to computers. Switching systems are drastically nonlinear; a logic stage provides an output which reflects the nonlinear interaction of its inputs. Storage and bistability are almost synonyms. Furthermore, the computer can be considered to be the ultimate exemplification of chaos; changing one bit out of many, can change the total significance of a program. A computer oriented environment makes it easy to appreciate nonlinearity, it does not require deep perception. This paper inevitably reflects that personal background. The parts of the subject most closely related to computers, however, will be taken up in a companion paper [8]. Computers and electronics are, of course, not the only application fields steeped in nonlinearity. Almost all applications have to confront nonlinearity. Key early insights into chaos, just to cite one example, came from meteorology [9], and statistical ecology [10]. Early books on nonlinear mechanics, such as MINORSKY [11] attest to the sophistication that was developed in some applications, decades ago. In electrical engineering a remarkably detailed understanding of stability, instability and multistability developed early in this century, as demonstrated by a book published in 1913 [12]. Some appreciation of multistability in chemical-engineering reactors existed as far back as 1918 [13].

In the remainder of this paper, we first discuss a few prototypes of interesting nonlinearities in electrical circuits. I then go on to discuss the role of noise in escape from the metastable state. In a final section, we discuss nonlinear electrical behavior in very small metallic samples.

2. DETERMINISTIC CIRCUITS

2.1 Chaos

Chaotic systems and chaotic differential equations, whose solutions are not periodic, or even multiply-periodic, have become a highly fashionable subject, following a lengthy dormant period. Once again, there are long-standing partial anticipations of this notion, in electronics. I quote from a recent note by KENNEDY and CHUA [14]:

> In a paper entitled "Frequency Demultiplication", van der Pol described an experiment in which by tuning the capacitor in a neon bulb R-C relaxation oscillator, driven by a sinusoidal voltage source, "currents and voltages appear in the system which are whole submultiples of the driving frequency". He noted that as the capacitance was increased from that value ... for which the natural frequency of the undriven relaxation oscillator equalled that of the sinusoidal source, the system frequency made "discrete jumps from one whole submultiple of the driving frequency to the next"(detected by means of "a telephone coupled loosely in some way to the system"). van der Pol noted that "often an irregular noise is heard in the telephone receiver before the frequency jumps to the next lower value". Interested primarily in frequency demultiplication, he dismissed the "noise" as "a subsidiary phenomenon."

KENNEDY and CHUA [14] then continue with a modern analysis of the circuit originally studied by VAN DER POL and VAN DER MARK [15]. Another closely related modern discussion can be found in [16]. Serious theoretical analysis of the periodically forced relaxation oscillator probably commenced with the work of CARTWRIGHT and LITTLEWOOD [17]. A detailed modern review of the subject is given by LEVI [18] who also discusses the early history, connecting the Cartwright-Littlewood work with World War II work in England on cavity magnetrons.

The paper by van der Pol and van der Mark examined a relaxation oscillator, using a neon glow lamp as its nonlinear component, and driven by an additional sinusoidal signal. It is a circuit closely related to examples we will discuss. My own first exposure to chaotic circuits, based on neon glow lamps, occurred while I was a trainee in the U.S. Navy's electronic technician's training program, around 1945. We were shown a toy, intended to fascinate the observer, displaying a set of neon bulbs blinking in a time sequence with no apparent underlying rhythm. Its chaotic nature was understood, without the use of fancy labels. CASPERSON and ORCHARD [19] have provided an entertaining account of a number of such early circuits, supplemented by computer simulations.

Before discussing our actual chaotic circuit, which does not use periodic excitation, let us, as introduction, discuss two other circuits. Figure 1 shows a simple relaxation oscillator, and its wave form. The battery supplies a charging current, through a resistor, to a capacitor. The voltage across the capacitor rises toward the battery voltage. When this voltage reaches the firing or threshold voltage, V_T, of the neon bulb, the bulb then carries a (relatively) high current, and quickly discharges the capacitor to a lower holding or extinguishing voltage, V_H, below which the glow discharge of the bulb cannot be sustained. Within the framework of our discussion, the discharge will be taken to be

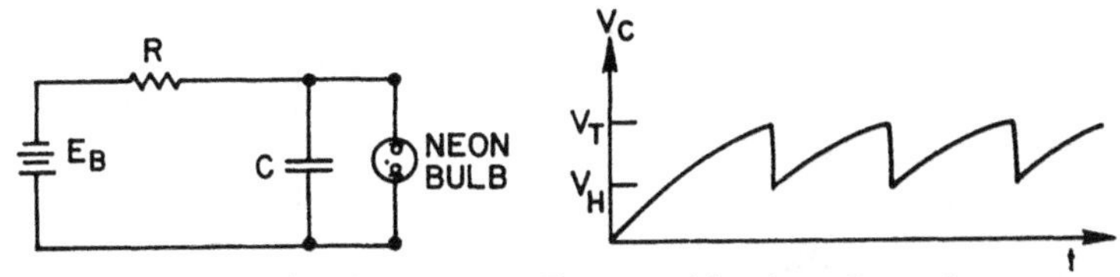

Fig. 1. Neon bulb relaxation oscillator and its time-dependent sawtooth voltage

instantaneous. The charging cycle then starts again, and thus generates periodic relaxation oscillations.

Actually the resistance R must be selected within a limited range. If it is too large the current flow through it can be carried by the leakage conductance of the unexcited neon gas. If the series resistor is too small it can carry enough current to permit the neon bulb to remain continuously in its excited state. (These requirements are equivalent to the condition that the resistive load line intersects only the negative resistance portion of the bulb's i-V characteristic.) The circuit of Fig.1, or minor variations on it, was used widely in early low cost cathode-ray oscilloscopes, to generate the horizontal deflection signal. The waveform rises almost linearly, as needed, and then returns rapidly. The circuit was low in cost, compared to one using vacuum tubes, and furthermore was easily synchronized (to the period of the waveform to be observed, its harmonic, or subharmonic) by the injection of the external signal to be displayed. The noise phenomenon discussed by VAN DER POL and VAN DER MARK [15] will be familiar to anyone who has played with these devices, attempting to establish synchronization. If we want to be precise, we should really distinguish between two phenomena; it is not quite clear that the authors of [15] observed both. When we move between two conditions which both lead to effective synchronization, there can be a long transient when the phases of the relaxation oscillator and the synchronizing signal drift with respect to each other, and have a signal which *temporarily* looks somewhat noisy. Alternatively, there are parameter regions, to be discussed, in which synchronization is really impossible, and the irregularity continues indefinitely.

Let us now go on to the circuit of Fig. 2. A fixed current i is fed to two neon bulbs in series, each shunted by a capacitor. Let us assume $C_1 > C_2$; thus the lower capacitor takes a longer time to be charged to V_T than the upper capacitor. Each capacitor is discharged through its own bulb when the capacitor voltage reaches V_T, but the capacitive discharge current flows only through the adjacent bulb, and leaves the other circuit unaffected. We, thus, achieve two independent relaxation oscillations, and if the

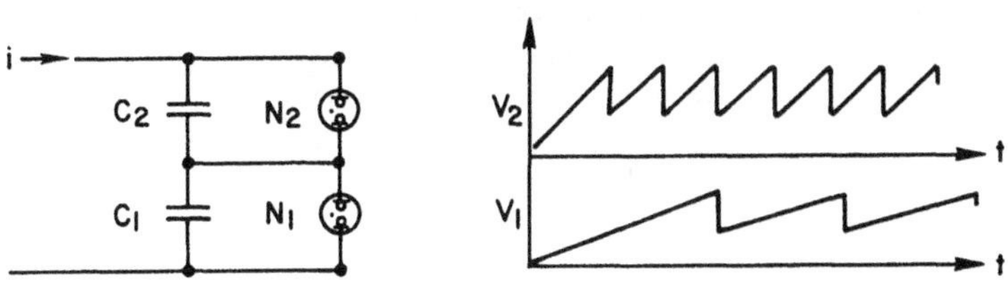

Fig. 2. Fixed current source feeds two neon bulbs and associated capacitors in series. The two exhibit independent relaxation oscillations

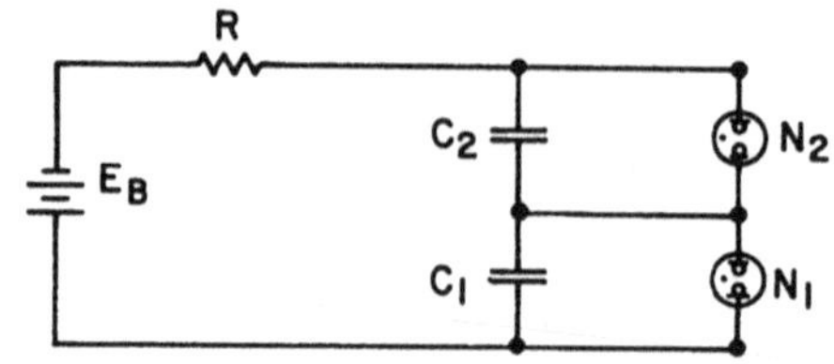

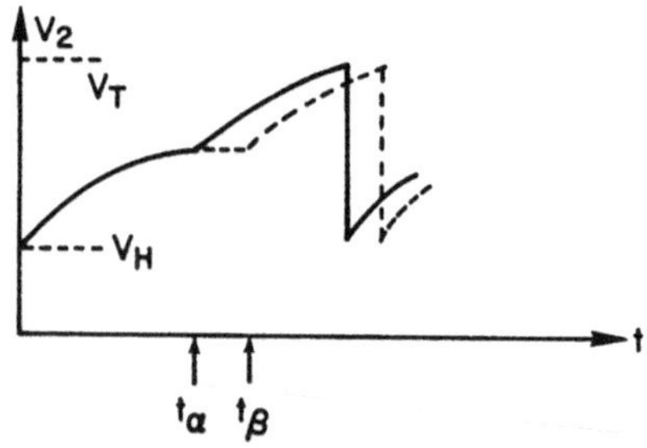

Fig. 3. In contrast to Fig. 2 the current is no longer fixed, and the charge on C_1 controls the current supplied to C_2. The relaxation oscillators are coupled

Fig. 4. Voltage across C_2 in Fig. 3. Solid and dashed line represent differing firing times for C_1

two frequencies are incommensurate the voltage across the two capacitors is multiply periodic. In this case we can still easily predict the exact behavior, at some time many cycles later, without the need to calculate the circuit behavior over the whole intervening period.

Let us now go on to the chaotic circuit shown in Fig.3, once again assuming $C_1 > C_2$. Both capacitors see the same charging current, as long as neither one is in its discharge cycle. The solid wave form, Fig.4, shows V_2 in a typical charge cycle, during which the lower bulb fires. We see that after the firing of circuit 1 more voltage is available for resistor R, and the charging current for C_2 increases. The dotted line in Fig.4 shows the behavior for a charging cycle in which circuit 2 fires slightly later. In that case the transition to the higher charging rate is made later, and V_2 takes slightly longer to reach V_T. Thus, we can see that the charging time for circuit 2 depends on the exact firing time for circuit 1, and almost every charging cycle can have a slightly different history. Let us restate this a little more carefully. We are examining a circuit consisting of two coupled relaxation oscillators. We can think of this as an attempt by one oscillator to synchronize, or phase lock, the other oscillators. We know from experience, common sense, and analysis [20] that if the two oscillators have almost the same natural frequency, a small injection signal (or weak coupling) will be adequate to provide synchronization, but if the frequencies are too far apart, then a weak cross-influence will be inadequate. Similarly a higher harmonic of one oscillator can synchronize some other, very close lying harmonic of the second oscillator, giving rise to complex *periodic* waveforms. We can expect the same behavior, in our circuit, leading either to periodic behavior (perhaps, with very complex waveform), or to a situation in which synchronization is ineffective, and every firing cycle is really different. Thus, we cannot expect to predict the long range behavior of the circuit, without following it in detail, cycle by cycle. All of this has been exhibited very beautifully in experiments by GOLLUB et al. [21], [22]. Reference [21], as indicated in its text, was in fact triggered by an early and unpublished version of this discussion.

Let us now proceed to the slightly more complex circuit shown in Fig. 5. For sufficiently small E_B (say below V_T) neither bulb fires, and after an initial transient period we find a time independent set of currents and voltages. Now go to somewhat larger

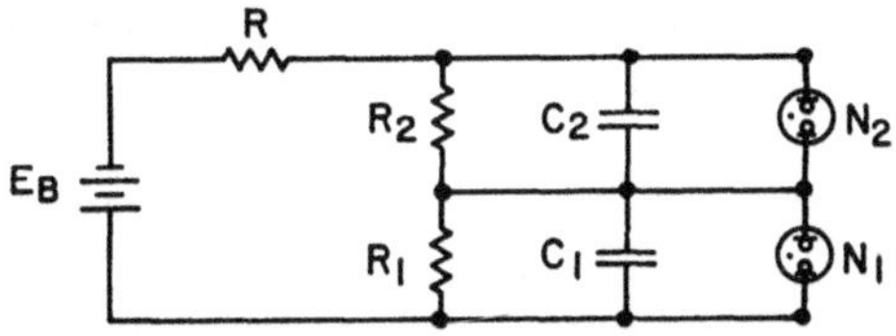

Fig. 5. Additional resistors lead to a more complex behavior

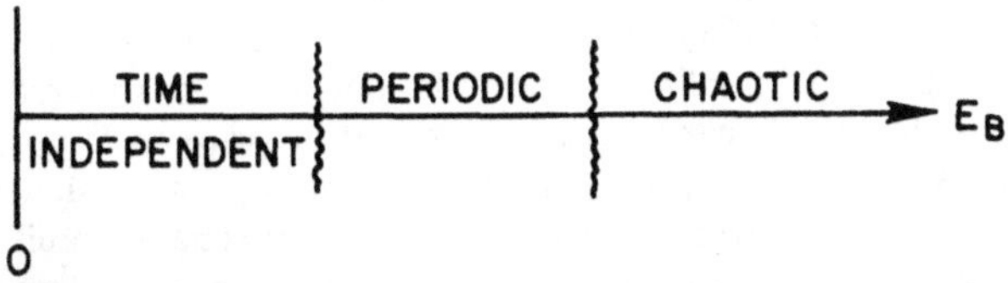

Fig. 6. The circuit of Fig. 5 exhibits different qualitative behavior, for different supply voltages

voltages, but assume that $R_1 << R_2$, and that $E_B(R_1/R) < V_T$. In other words, the maximum current that can possibly flow through R is still not enough to bring the resistor R_1 to the firing voltage V_T. Then, if the circuit parameters are chosen correctly, circuit 2 will fire periodically. At still higher battery voltages we achieve the chaotic behavior described in connection with Fig.3. We thus find a sequence of behavior patterns as illustrated, in Fig.6, much as found in other physical phenomena.

It is well known that chaotic solutions can arise from three simultaneous nonlinear differential equations. Our circuit has only *two* capacitors. Why is this enough? As ROESSLER [23] has pointed out, our circuit invokes neon bulbs, and these are multi-valued resistors. They have memory. At a given voltage, between V_T and V_H, they remember whether the voltage has come up from V_H or down from V_T. Thus, the continuous capacitive degrees of freedom are supplemented by two other digital degrees of freedom, characterizing the existence of a glow discharge, or its absence. Are both of these digital degrees of freedom needed, or will one bulb suffice, if used in the right RC network? This is a question posed by ROESSLER [23]. The appendix gives a physical argument to show that both bulbs are needed, one is inadequate. On the other hand in circuits containing both capacitors and inductors, a single negative resistance is enough to cause chaos [24].

Our discussion of chaos has had a technological flavor. Chaos is not just of historical interest to the technologist. In the design of supposedly well-behaved systems, we must be able to anticipate chaos. One example: HENDRIKS [25] has studied impact printing mechanisms, to learn the conditions under which chaos can be expected. Another example: the people interested in Josephson junction switching circuits have had a long-standing concern with a very sensitive dependence on initial conditions, and discussed this under the name *punchthrough* [26], before chaos achieved its current fashionability. All sufficiently complex systems are likely to exhibit chaos, and the associated difficulty of long range prediction. PIPPARD [27] has pointed out that this includes economic and social systems.

For a discussion of synchronization, chaos, and subharmonics, more closely related to the center of gravity of this conference volume, see [28].

2.2 Parametric Excitation

Parametric effects were already known in a number of forms and understood in RAYLEIGH's [29] time. Parametric effects became a subject for serious practical attention in the middle fifties. The key steps in this direction were independent inventions by VON NEUMANN [30], and by GOTO [31]. For a broader exposition see [32]. These identical proposals invoke the use of nonlinear susceptibilities, for subharmonic generation, in resonant circuits with a nonlinear reactance, excited at twice their approximate resonance frequency. The proposals point out that in the simplest possible case, where the pump signal induces the growth of a signal at the approximate circuit resonance frequency, and at half the pump frequency, the lower frequency can build up with one of *two* possible phases. Bistable circuits, of this sort, are essential to computers. The two inventions proceeded to point out how such parametric circuits could be coupled to each other to carry out all necessary computing functions. Large efforts to apply these inventions came into existence at several laboratories, in the U.S., and in Japan. In Tokyo, actual working "parametron" computers were built. Shortly after the von Neumann and Goto inventions a more general interest in the parametric use of the nonlinear capacitance of reverse biased p-n junctions arose, going beyond computer applications, to general purpose low noise amplification [33].

As stated, the von Neumann and Goto inventions stirred up a good deal of activity and interest, in their time. Despite that, the work and concepts went into complete obscurity, relatively quickly. The work was not remembered when, in the early sixties, interest in optical parametric schemes arose. In recent years when work in chaos and strange attractors has once again brought attention to parametrically excited bifurcation, von Neumann and Goto have also been forgotten. Another related field appeared, circa 1970, when analogies between phase transitions in ordinary thermodynamic systems, and their equivalent in active dissipative systems, became fashionable. The von Neumann and Goto inventions were clear examples of this, but were largely ignored. Young scientists, when their work does not receive its expected attention, are likely to blame their lack of membership in the "establishment." How can von Neumann, one of the most renowned scientists of this century, suffer the same fate? We'll let the reader answer that.

The von Neumann proposal has survived, in one slightly disguised form, as a subject of continuing interest. This is in a field concerned with the fundamental physical limits of the computational process. von Neumann's scheme was adapted to particles in time-dependent potentials in [34] and applied to these fundamental limit questions. In more recent years it was further modified [35] to apply to Josephson junctions.

2.3 Nonlinear Optics

We have already referred to nonlinear optics in connection with parametric excitation. Nonlinear optics is a vast field. While not strictly speaking condensed matter physics, the concern with nonlinearities in optics has a common heritage with that in condensed matter physics, and in electronics. The quantum electronics community seems reluctant, on occasion, to admit some of its precursors, and we will discuss only one, briefly. A more detailed discussion of precursors to nonlinear optics is contained in [36]. Some interesting early history, with particular stress on Bell Laboratory contributions, can be found in [37].

In a remarkable pioneering paper SALINGER [38] analyzed wave propagation along a nonlinear electromagnetic transmission line. Salinger was motivated by telegraph signal propagation, along transoceanic cables, loaded with nonlinear inductors. While incorrect in some of its details, Salinger's paper is entirely correct in its general conceptual points. Salinger realized that different parts of the wave, depending upon their amplitude, will move with different velocities. This will produce wave deformation, which in turn leads to shock waves. Salinger's viewpoint, that different portions of the wave can carry different velocities, was rediscovered in quantum electronics and there led to the suggestions for self-focusing of wave fronts, self-steepening of laser pulses, and self-phase modulation. [39].

3. ESCAPE FROM THE METASTABLE STATE

The theory of noise activated escape out of a well, i.e. the view of escape as Brownian motion in the presence of a restoring force field, started with the pioneering work of KRAMERS [40]. The field has now become a vast industry with many different facets. Table 1 illustrates the current diversity. Within the framework of this discussion we cannot even provide citations for most of the subcategories listed in Table 1, and will comment on only a few of these. A good review paper covering *all* of this does not exist; HANGGI [41] provides some approximation to that. Earlier reviews [42], while authoritative, are more limited in scope.

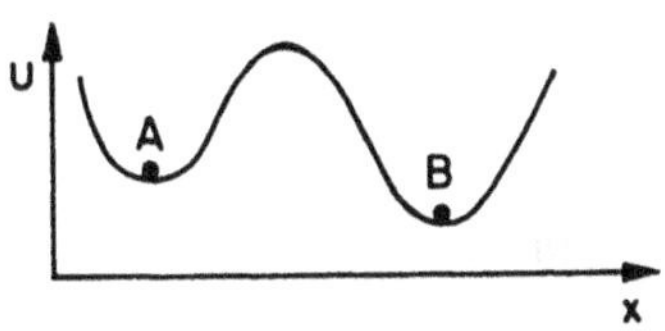

Fig. 7. Damped potential with metastable state A and state of lowest energy at B

Let us start with the simplest possible situation, in which we treat barrier crossing in the one-dimensional situation of Fig. 7 and avoid most of the complexities listed in Table 1. Let us, however, be more general than the situation in Fig. 7 showing motion of a particle in a potential. Consider, instead, a more general law of motion, e.g. an active dissipative system, with two states of local stability separated by a barrier with an intervening state of instability.

In the absence of noise, a system at A will simply stay there; metastability and relative stability are concepts which can be discussed sensibly only in a statistical context.

Table 1

ESCAPE FROM THE METASTABLE STATE

DIMENSIONS AND DEGREES OF FREEDOM

PARTICLE MOVING IN ONE DIMENSION

MANY DEGREES OF FREEDOM

ALL ON SIMILAR TIME SCALE
BORN-OPPENHEIMER SURFACES

INFINITE NUMBER OF PARTICLES

SINE-GORDON CHAIN, ϕ^4 CHAIN, SOLITONS
ONE-DIMENSIONAL ISING

LAW OF MOTION

COMPLETELY HAMILTONIAN (MOLECULAR DYNAMICS, MQT THEORY)

HAMILTONIAN WITH NOISE AND FRICTION

MORE GENERAL DYNAMICS, e.g. MULTISTABLE CIRCUIT, LASER, ECOLOGY

NUMBER OF COMPETING LOCALLY STABLE STATES

SINGLE WELL, ESCAPE INTO UNLIMITED RANGE OF MOTION

BISTABLE WELL

SYMMETRICAL
UNSYMMETRICAL

MANY COMPETING STATES

NO SYMMETRIES (e.g. SPIN GLASS, COMPETING STATES OF ECOLOGY)
ALL ALIKE: PARTICLE IN SINUSOIDAL POTENTIAL
BIASED SINUSOIDAL POTENTIAL

QUANTIZATION

CLASSICAL CONTINUOUS CASE

CLASSICAL DISCRETE CASE

Q.M. CASE

QUANTIZATION IN WELL
ESCAPE BY TUNNELING WITH OR WITHOUT FRICTION

NOISE

THERMAL EQUILIBRIUM

MORE GENERAL, e.g. 1/f NOISE

BIG OCCASIONAL JUMPS (CHEMICAL COLLISION, QUANTIZED RADIATION.)

MARKOVIAN vs. CORRELATED NOISE

NOISE INDEPENDENT OF STATE OF SYSTEM, vs. STATE-DEPENDENT NOISE

TIME DEPENDENCE:

TIME INDEPENDENT LAW OF MOTION

MODULATED BARRIER AND/OR SINUSOIDAL EXCITATION IN WELL.

DAMPING IN WELL, NOISE

STRONG, WEAK, OR VERY WEAK

PURPOSE; ANSWER DESIRED

ESCAPE PROBABILITY

EQUILIBRATION WITHIN A WELL

AVERAGE ESCAPE ENERGY

DWELL TIME NEAR TOP OF BARRIER, OR TUNNELING THROUGH BARRIER

APPROACH TO OPTIMUM WELL (SPIN GLASS ANNEALING)

PUBLICATION OF ONE MORE PAPER

Let $\rho(x)$ denote the distribution function along x. If x followed a deterministic law of motion, the systems would simply be driven toward the states of local stabilities, e.g. A and B in Fig. 7, and accumulate there. In that case we would have a flux of probability $j = \rho v(x)$, where v is the deterministic velocity at x. In the case of the heavily damped potential of Fig. 7, $v = -\mu dU/dx$, where μ is the mobility.

Fluctuations do exist, however, and systems that start at the same value of x do not have to stay together, but can diffuse apart. The noise effects can often be represented by a supplementary diffusive term

$$j = \rho v - D\partial\rho/\partial x. \tag{1}$$

In the steady state $\partial j/\partial x$ must vanish. In many systems, including that of Fig. 7, $\rho(x)$ can be expected to vanish for very large x. Then $j = 0$ in the steady state, and (1) integrates to give:

$$\rho = \exp \int \frac{v}{D} dx. \tag{2}$$

Equation (2) indicates that there is a competition between v, in the numerator of the integral, driving the system toward states of local stability, and the diffusion coefficient D in the denominator, permitting the system to move away from the states of local stability. It is immediately apparent that the relative probability

$$\rho(B)/\rho(A) = \exp \int_A^B (v/D)dx, \tag{3}$$

depends on an integral and, thus, on the behavior of the system in the unlikely intermediate states between A and B, not just on the behavior near A and B. We will return to this point.

Equation (2) is a generalization of the Boltzmann distribution, $\rho \sim \exp(-U/kT)$. To make this clear consider a potential U, exerting a force on an overdamped particle, with mobility μ. The potential U can be written as an integral of the force $F = -\partial U/\partial x$, yielding $\rho \sim \exp(\int F dx/kT)$. Multiply numerator and denominator in the integrand by μ. The numerator μF then becomes the drift velocity v; the denominator, through the Einstein relation, becomes the diffusion constant, giving $\rho \sim \exp \int (v/D)dx$, as in (2).

Equation (3) emphasizes that relative stability, i.e. the relative likelihood of competing states of local stability, depends on the noise, i.e. on D(x) along the whole intervening path between the two states. By playing with the kinetics, even if it only affects relatively rarely occupied parts of the system's space, we can alter the identity of the preferred state. In a number of papers I have argued this point in more detail, by examples, e.g. invoking a heavily damped well as in Fig. 1, but with a temperature which is a function of x. Here, instead, we will use the example of Fig. 8. Laser light is used to alter the kinetics of the transition between two high lying levels. Let the barrier between the wells be sufficient so that the quantized levels, in the two wells, are essentially decoupled. Now apply radiation to induce transitions between the uppermost level in the right-hand well and the level above that. If the level spacing is large compared to kT, these two levels are very unequally occupied, before the laser light is applied. In the

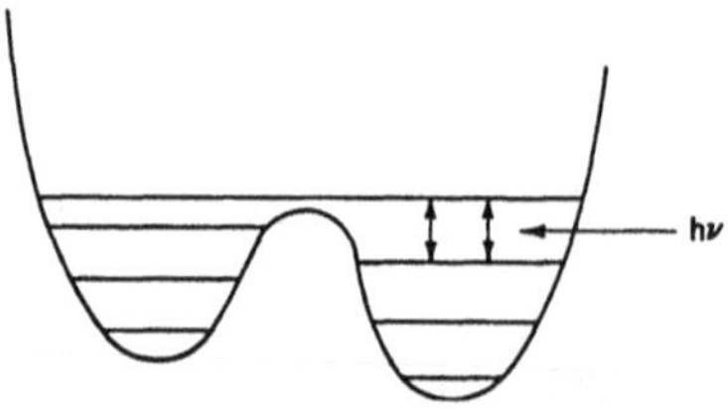

Fig. 8. Biased bistable well. Transition out of highest righ-hand level is pumped with laser light

presence of intense radiation the two level populations can be brought much closer to each other. Note that the light intensity has to be high, but if the lower of the two levels is many kT above the ground state, and sparsely occupied, the net absorption need not be large, and the required optical intensity can be built up by reflection between the mirrors of a cavity wall.

Now let us assume that the interlevel transitions, other than the one controlled by the laser, occur largely between adjacent levels. The *relative* population of all other levels, unaffected by the laser, will then still be controlled by the Boltzmann factor. Now consider the case where the two levels, whose population has been brought to approximate equality by the laser, are further apart than the relative displacement, in energy, of the two well minima. The right-hand well, which was originally the favored well, will then become the more sparsely occupied well. Thus, the relative population of the two wells has been controlled by manipulating the kinetics of sparsely occupied states, without changing the kinetics in the bottom of each well. Any attempt to predict the relative occupation of the two wells that uses local properties characterizing the behavior only near the bottom of each well, cannot be successful. Elsewhere I have discussed the implications of this for evolution and the origin of life, [43] and [44], elaborating on notions first expressed in [45]. These concepts are closely related to those involved in discussions of punctuated equilibrium [46]. This author is not qualified to discuss that subject, but has the impression that fluctuation induced transitions between competing states of relative stability have a role in any reasonable approach to evolution, independent of its finer details [47].

Let us now go on to make a few other comments on the entries of Table 1. The statistical mechanics of solitons in sine-Gordon chains, or in ϕ^4 chains, has received considerable attention, particularly by some of the participants at this conference, and constituted a serious focus for a number of earlier conferences, of which we can cite only a few [48]. This present conference includes a number of additional closely related contributions.

The escape from a very underdamped potential well has also been a subject of considerable recent attention. A particle in a well without damping and fluctuations will just stay at its initial energy, and cannot escape. As the coupling to the reservoir, and therefore the noise and damping, is increased, the chances for a change in energy, and therefore for escape, will improve. At sufficiently high damping the particle will execute diffusive motion over the barrier, and in that range further increases in damping will decrease the escape probability. All this was clear already to KRAMERS [40]. It seems, however, to have escaped the attention of many theoretical physicists, starting with CHANDRASEKHAR's [49] otherwise definitive review paper. The continuing failure to understand the point may, in part, have been an unfortunate and unintended consequence of the very elegant viewpoint, introduced by LANGER [50] in the late sixties, which invokes analytic continuation, and the imaginary part of the free energy to calculate escape rates. That, however, introduces no real kinetics. The physical chemists were more perceptive. After all, molecules in a gas, going through some internal rearrangement which involves barrier crossing, need to be jostled by other molecules, to get the necessary energy. This has been understood, at least, since the early twenties [51]. It is, perhaps, no great surprise that Kramers, a collaborator of CHRISTIANSEN [52] later returned to a more definitive discussion of the subject [40]. For more modern experimental and theoretical explorations of the chemical viewpoint see [53]. In recent years theoretical physics has caught up, though there are also a few older papers in the physics literature which appreciated the point. For a sampling of the recent physics literature on this point see [41], [54] and [55].

Much of the recent progress in this field has had a quantum mechanical flavor. The work on macroscopic quantum tunneling, dealing with tunneling in the presence of friction, is represented at this conference through LEGGETT's [56] paper. See also [57]. Leggett emphasizes the question: Is the phase of a Josephson junction, in a metastable well, a genuinely quantum mechanical object? Reference [58] presents "spectroscopic" microwave measurements on a junction in its anharmonic well. These measurements show that the junction makes transitions between energy levels which correspond to solutions of the Schrödinger equation (assuming no effective alternative classical explanation).

A third recent topic in the quantum mechanical treatment of escape from a metastable state deals with the time that a particle interacts with the barrier, during the escape [59].

Table 1, in its last section, points out that the papers in this field address a variety of questions, not just a variety of physical situations. The reference to "publication of one more paper" is not totally facetious. One recent conference publication had 38 citations; of these 28 were to the group that generated the paper.

In concluding this Section we would like to return to our recurrent technological theme, and note that the treatment of escape from a metastable well, in the heavily damped case, is closely related to the passage of carriers from a transistor emitter, over a base region, into the collector. As discussed elsewhere [44] the viewpoint generated in the semiconductor device field leads to an easy solution for some problems usually treated by more sophisticated techniques in the statistical mechanics literature. In this connection let us make a brief detour to a somewhat different area, that of hopping, along a random chain of sites, or one with a periodic modulation. Diffusion along such a chain, via the Einstein relation, becomes equivalent to the evaluation of the resistance of a chain of resistors in series. *If the resistance variation is bounded*, this is a trivial problem, as pointed out in [60]. The simplicity of the problem does not seem to be widely recognized [61]. An exception: Hopping along a "hierarchical" chain is handled by the series resistance approach in [62]. For the specialist, we note that under the simple series resistance approach discussed in [60] the distinction between symmetrical and asymmetrical hopping probabilities, between adjacent sites, is of no significance.

4. NONLINEAR ELECTRICAL BEHAVIOR OF MESOSCOPIC SAMPLES

Small samples, large compared to atomic dimensions, but small compared to the usual macroscopic dimensions, have received considerable attention in recent years. Electronic device work has, of course, been aimed for decades at smaller structures, and has supplied us with techniques to make these. The more fundamentally oriented physicist has, in a sense, now caught up, and has invoked the word "mesoscopic." The word was, to the best of my knowledge, first used by VAN KAMPEN [63] in connection with noise theory and time-dependent fluctuations; rather than with the sample to sample fluctuations which have been emphasized in the most recent literature. The expression was taken over, and given a broader meaning by IMRY [64], who has an extensive review paper on those parts of the subject which have arisen recently.

One example of the push toward small structures, in transistor development, is given by the recent accomplishments reporting ballistic, i.e. unscattered, transport through the base region of a hot electron transistor structure [65]. There is a complementary potential development in theory; the characterization of devices, not just of linear resistors, by the overall transmission behavior of the structure [66]. A more general point: In a small sample we do not have the equivalent of an ensemble average. A small semiconductor region, say a p type base in a transistor, doped intentionally to a certain acceptor density, may have more acceptors, or fewer, in a particular device. Or the acceptors may be situated preferentially on one side or the other. The *small* device, in its electrical behavior, will reflect its own particular acceptor distribution. In the recently developed jargon in this field, this is called the "fingerprint" of the sample. Recent work, on transport in normal metals, has brought elegant new techniques into play for the calculation of conductance fluctuations [67]. The basic notions, however, are older and were anticipated in the device literature. See, for example, [68]. More recent work on device geometries has emphasized the dramatic effects of one scattering center in a field-effect transistor turning on or off [69], or that of a change in Fermi-level making new states available for hopping through the sample [70]. Resonant tunneling has also been observed [71]. "Fingerprints" are also seen in tunneling structures [72]. Clearly, "fingerprints" are not a desirable aspect of supposedly identical devices, and illustrate a difficulty we face, as we move towards smaller devices.

The recent emphasis on small samples has a number of separate facets. One relates to the effect of magnetic fields on quantum mechanical interferences; solid state versions of the Aharanov-Bohm effect [73]. We will, instead, discuss nonlinear behavior. Small samples are, after all, readily driven into a nonlinear regime.

The recent literature in this subject has emphasized *universal fluctuations*. [67]. These papers point out that conductance fluctuations are of order e^2/h as we change Fermi-level; change magnetic field, move from one ensemble member to another, or move a single scattering center by distances comparable to or larger than the Fermi level wavelength. This is supposed to hold regardless of the dimensionality of the sample, its size, or its exact shape. If the sample is long enough so that the typical conductance $G << e^2/h$, that cannot be true, of course. Similarly, if the sample is short enough to permit an appreciable probability of unscattered transmission, then the fluctuations in G can be larger than e^2/h. The *universal* behavior, therefore, isn't really all that uni-

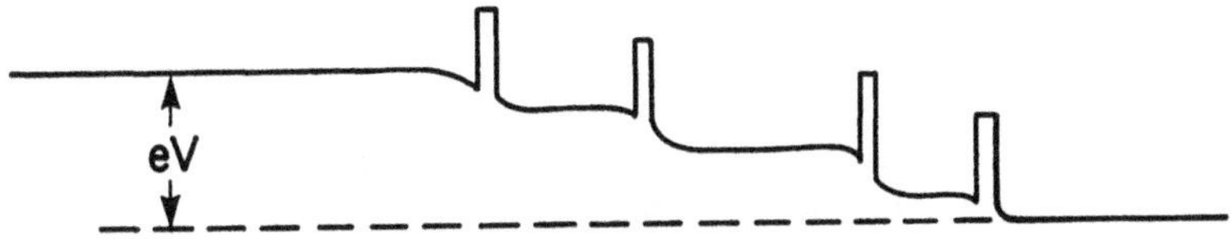

Fig. 9. Random obstacle chain with applied voltage

versal. As pointed out by AL'TSHULER and KHMEL'NITSKII, [74], these fluctuations also determine, directly, the scale of nonlinearity in small samples.

Consider Fig.9, showing a sample with random scatterers, and a voltage eV across the sample. The *universal fluctuation* picture tells us that if we change Fermi-level, and thus the wave-vector, and through that the interference pattern between the various possible multiple-scattering events in the unbiased version (V=0) of Fig. 9, then G changes by order of e^2/h. The theory also tells us that we can expect this change to occur if we change the carrier energy by about E_c, the Thouless energy over which phase correlations, through the sample, persist. E_c is also the energy defined through the uncertainty relation, by the time a carrier needs to diffuse through the sample.

Figure 9 can be used to point to two intrinsic sources of nonlinearity. *Intrinsic* here is used to designate effects related to the basic quantum kinetics as defined by Fig. 9, rather than more remote causes, such as resistance changes induced by Joule heating, or electrostriction. First of all, at T = 0, the carriers are not all exactly at the Fermi-level, and are characterized by a range of parameters, with an energy range which has to grow in proportion to the current flow. But this is a source of symmetric nonlinearity; the resistance will be the same if we reverse the current. We are looking for the lowest possible order of nonlinearity. Now, in Fig. 9, if $eV \sim E_c$, then the phase relations in the second half of the sample will have changed sufficiently to change its contribution to the multiple scattering process. If we allow ourselves some sloppiness about E_c for the right half, vs. E_c for the whole sample, then we find [74]

$$< [G(V) - G(-V)]^2 > \sim \left(\frac{e^2}{h} \frac{eV}{E_c} \right)^2 . \tag{4}$$

In (4) e^2/h represents the scale of the conductance changes, eV/E_c the voltage scale on which the changes occur. This is a nonlinearity of the lowest possible order; if we replace Fig. 9 by one in which carriers come in preferentially from the right, and increase their kinetic energy as they move to the left, we will get an entirely independent effect. In that case, it is a different set of obstacles which are dephased. This is a nonlinearity which will provide second harmonic generation, in contrast to thermal effects which can only provide third harmonics. Nonlinearities are, of course, most easily measured through phased-locked harmonic detection techniques, and will not require $eV \sim E_c$, nor precision i-V curves. Note that the dephasing invoked by [74] present at very small fields *seems* unrelated to the more well known delocalization in strong fields studied elsewhere [75].

Figure 9 also enables us to point to a second and somewhat separate source of nonlinearity which will be manifested even if the sample does not act as a single quantum mechanically coherent scatterer. Figure 9 shows that the voltage drop is not uniform, but localized near the obstacles. Between the obstacles of Fig. 9 we have uniform plane waves. In a self-consistently screened situation we can have departures from neutrality only over distances of the order of a screening length. Thus, the charges giving rise to the localized fields are confined to the vicinity of the scattering centers. These localized fields, called residual resistivity dipoles by this author, in 1957, have been the subject of extensive analysis and debate in electromigration theory [76], [77]. They have managed, however, to elude the perception of the broader transport theory community. AL'TSCHULER [78], for example, tells us: *The Kubo equation or an equivalent equation is always used to calculate the conductivity of metals. This means that the electric field in a sample is assumed to be uniform...* This, *despite* the fact that sophisticated many-body Coulomb effects in disordered samples have received intensive attention [79]. It is remarkable that despite the great technological interest in electromigration, and the fact that electromigration theory is an area in which noted theoreticians, e.g. Friedel, Peierls, Nozières, and Sham have come to differing viewpoints, their work remains largely unknown outside of the small community consisting of the debating partners in the field.

The drawing in Fig. 9 is somewhat idealized, and based on a semi-classical viewpoint. The screening fields are actually accompanied by Friedel oscillations, due to the interference between incident, reflected, and multiply reflected waves. These interference oscillations are not shown in Fig. 9.

Consider, now, the case where there is enough inelastic scattering present in Fig. 9, to make the separate obstacles scatter incoherently. As in the earlier case, involving multiple scattering effects, an obstacle presents a number of sources of nonlinearity. For example, the voltage drop determines a *range* of velocities near the Fermi-surface that are involved; we are no longer dealing with the exact undisturbed original Fermi velocity. Once again, we are particularly interested in the lowest possible order of nonlinearity. If the obstacle is symmetrical, as suggested by Fig. 9, that will not be available, then we can only expect a symmetrical nonlinearity of the form $i = G_oV + G_2V^3$. Let us, therefore, emphasize asymmetrical obstacles, or obstacles in an asymmetrical environment. It is asymmetry with respect to the direction of current flow that counts. Even if the host lattice is symmetrical, a vacancy-interstitial pair, an edge dislocation, or a step at the surface can all have the required lack of inversion symmetry. Can we say something about the scale of expected nonlinearity? We can expect scattering cross-sections to change roughly as $\sigma \sim \sigma_0(1 + \eta eV_D/W_A)$. Here V_D is the localized potential drop and W_A measures the size of the asymmetric part of the scattering potential. η is a coefficient less than, but of order unity, reflecting the fact that σ_0 is, in part, due to a symmetrical potential. η also allows for the fact that the potential drop associated with transport is somewhat more spread out than the scattering center potential, and presumably less effective in scattering. In the one-dimensional version of Fig.9, V_D is simply the applied potential, divided by the number of scattering centers. In three dimensions a line going through a cube of side unit length will encounter roughly $n\sigma$ obstacles, where n is the obstacle density and σ their scattering cross-section. Thus the voltage drop, per obstacle, will be $E/n\sigma$, where E is the space average or "applied" field. We should preface this by a factor 1/3 because the more detailed discussions of the *three dimensional* residual resistivity dipole fields [80] show that only 1/3 of the space integrated three-dimensional dipole field is located between the sources of the localized field, rather than outside of these charges.

Thus, if we assume that the resistance changes are proportional to the changes in scattering cross-section, we can expect

$$R \simeq R_o(1 + \eta e\, E/nW_A), \qquad (1\text{dim})$$

$$R \simeq R_o(1 + \frac{1}{3}\eta e\, E/n\sigma W_A). \qquad (3\text{dim})$$

The above equations assume that E/W_A has the same sign for all obstacles, i.e. their polarizations are aligned. In the more likely case that the obstacles have a random polarization, the effects will tend to offset each other, and will be much smaller. Clearly this is a crude first handwaving approach. The key point, however: Electric fields associated with current transport are not spatially uniform. The non-uniformity will make nonlinearities much more observable.

Small systems also present us with the question: If energy is to be dissipated, e.g. through a conductance, where does it go? Physicists have developed remarkable cleverness in starting from conservative Hamiltonian dynamics, which is easiest to treat, and using it to predict dissipative behavior, which is prevalent. Much of this, however, is based on cheating. A Hamiltonian system with a limited number of degrees of freedom is, of course, just that. It can store energy, it cannot dissipate it. Consider, for example, a one-dimensional metallic ring, of normal metal. *One-dimensional* means that we deal with a chain of atoms; we do not need to be concerned with the transverse variation of the wave function. Instead of an unrealistic chain of atoms we can have a circular FET gate structure, which is narrow enough so that only the lowest transverse state is occupied. The ring can have defects, causing elastic scattering. We have no leads connected to this ring, and can only drive it through flux, by transformer action. At $T = 0$, and in the presence of slowly changing magnetic fluxes, it has been shown [81] that the ring acts just like a superconducting ring with a Josephson junction. The one modification: The charge 2e of a superconducting pair, in the Josephson junction case, is replaced by the single electron charge e, in our case. We then find current oscillations in the presence of a steady accelerating field. The oscillation amplitude is determined by the properties of the ring, and their frequency is proportional to the field. That is strikingly nonlinear. As we go to larger more realistic structures, and to finite temperatures, how does the transition to ordinary resistive behavior take place? Some progress in that direction exists [82], but much remains to be cleared up.

We briefly will mention one other recent set of questions posed by small structures. It has been suggested that Josephson junctions, as well as normal junctions, with small capacitances, fed by constant currents, will exhibit coherent voltage oscillations [83]. We will not try to explain, criticize, or justify these proposals, but only point to them as interesting ideas, deserving further attention.

Small structures pose many other interesting problems, many that we cannot possibly list, and undoubtedly many more that still await recognition.

ACKNOWLEDGMENTS

I am indebted to M. Büttiker and Y. Imry for motivation and help leading to the discussion of Ref. [74]. L. Chua and M. Levi provided help in connection with the early electronic history of chaos. The discussion of chaos in coupled relaxation oscillators stems from my participation in a conference, in 1977, organized by H. Haken.

APPENDIX

Roessler has posed the question: Are two neon bulbs really needed in the sort of relaxation oscillators we discussed in Sec. 2.1. We here argue that they are, and that an arbitrarily complex network consisting of batteries, resistors, and capacitors and *one* gas discharge device has a periodic solution.

Assume that such a periodic solution exists. The neon bulb fires at V_T and extinguishes at V_H. During this firing the voltage across the bulb drops by $V_T - V_H$. The capacitors (possibly more than one) which have at least one terminal connected to the bulb, but do not need to pass their discharge current through a resistor, will change charge during the firing event. Between firing events the bulb is inactive. Therefore, if the bulb fires periodically, it can be replaced by a periodic charge injector, which quickly injects a finite amount of charge into the network, once each period. The total network response of this linear network must then be a superposition of the steady state due to the batteries (and neglecting the bulb), added to the effect of the periodic charge injector, with the batteries short-circuited.

The steady state response due to the batteries must give a voltage across the bulb, V_{SS}, which exceeds V_T, otherwise there is an obvious periodic solution. That would simply be the steady state battery response, without any firing at all. If $V_{SS} > V_T$ then the time sequence of voltages across the bulb will then be as shown in Fig. 10. The injected charge leaks away, monotonically, between firing events; we have a circuit with only resistors and capacitors, and cannot have oscillatory effects. The charge injector is, as already stated, adjusted to give the needed charge discontinuity, and the injected charge relaxes away after the moment of injection. If the injected charge were to leak away completely, without any further firing, then the voltage across the bulb would approach V_{SS} asymptotically. The difference between V_{SS} and V_T arises from the fact that the injected charge disappears incompletely between successive charge injections. Clearly there can be a periodic solution, as sketched in Fig. 10, where after a period the injected charge leaks away, and $V = V_T$ is reached once again, to permit the next firing. The periodic solution is clearly self-consistent: At each firing event the sum of the residual charges left over from the incomplete decay of earlier charge injections amounts to the same, and can be expected to continue its decay, as in the preceding period.

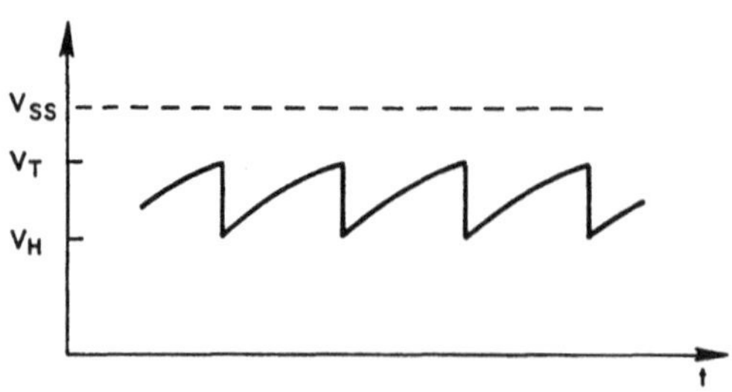

Fig. 10. Time dependence of voltage across bulb in arbitrary RC network with voltage sources

Is this solution stable and unique? We leave that as an exercise to the more mathematically minded, but it would be surprising, in such a simple and heavily damped system, if the proposed solution were not unique and stable. As in any oscillator, of course, the phase of the oscillation is not stable and is easily shifted. If we perturb the charge on some capacitor in the network, it will shift the phase permanently. The phase, once disturbed, sees no effective restoring force, and remembers the change. Other effects of the perturbation, however, can be expected to decay.

References

1. J. Heldstab, H. Thomas, T. Geisel, G. Radons: Z. Phys. B **50**,141 (1983); T. Geisel, J. Heldstab, H. Thomas: Z. Phys. B **55**,165 (1984); S. Grossmann: Z. Phys. B **57**,77 (1984)
2. For a charming review of the very early stages of ferroelectricity, see G. Bush: Helv. Phys. Acta **59**, 1 (1986)
3. R. Peierls: *Bird of Passage*, (Princeton Univ. Press, Princeton, 1985) see p.40-45
4. A. Cottrell: Contemp. Phys. **21**, 85 (1980)
5. P. W. Anderson: Science **177**, 393 (1972)
6. S. L. Glashow: Phys. Today **39**, 11 (Feb., 1986)
7. This paragraph adapted from R. Landauer: Am. J. Physiology **10**, R107 (1981)
8. R. Landauer: In *The Physics of Structure and Complexity,* Physica Scripta, to be published
9. E. N. Lorenz: J. Atmos. Sci. **20**, 130 (1963)
10. R. M. May: Nature **261**,459 (1976)
11. N. Minorsky: *Introduction to Non-Linear Mechanics,* (Edwards, Ann Arbor, 1947)
12. H. Busch: *Stabilität, Labilität und Pendelungen in der Elektrotechnik* (S. Hirzel, Leipzig, 1913)
13. R. A. Schmitz: In *Chemical Reaction Engineering Reviews,* ed. by H. M. Hulburt (Am. Chem. Soc., Washington, D.C., 1975) p. 156; R. Aris: In *Bifurcation Theory and its Applications in Scientific Disciplines*, ed. by O. E. Rössler, O. Gurel (N. Y. Acad. Sci., 1979) p. 314
14. M. P. Kennedy, L. O. Chua: *Van der Pol and Chaos*, IEEE Trans. on Circuits and Systems, to be published
15. B. van der Pol, J. van der Mark: Nature **120**, 333 (1927)
16. M. Van Exter, A. Lagendijk: Phys. Lett. **99A**, 1 (1983)
17. M. L. Cartwright, J. E. Littlewood: J. London Math. Soc. **20**, 180 (1945)
18. M. Levi: *Qualitative analysis of the periodically forced relaxation oscillations* (Am. Math. Soc., Providence, 1981)
19. L. W. Casperson, H. J. Orchard: *Periodic and Chaotic Pulsations in Ring Oscillator Networks,* preprint
20. Y. S. Tang, A. I. Mees, L. O. Chua: IEEE Trans. CAS-**30**, 620 (1983)
21. J. P. Gollub, T. O. Brunner, B. G. Danly: Science **200**, 50 (7 April 1978)

22. J. P. Gollub, E. J. Romer, J. E. Socolar: J. Stat. Phys. **23**, 321 (1980)

23. O. Roessler: Private communication

24. T. Matsumoto, L. O. Chua, M. Komuro: IEEE Trans. CAS-**32**, 798 (1985)

25. F. Hendriks: IBM J. Res. Dev. **27**, 273 (1983)

26. J. A. Blackburn, Z. Yang, S. Vik: Phys. Lett. **114A**, 500 (1986)

27. A. B. Pippard: Interdisciplinary Science Reviews **7**, 92 (1982)

28. S. E. Brown, G. Grüner, L. Mihály: Solid State Comm. **57**, 165 (1986)

29. J. W. S. Rayleigh: Philos. Mag. **15**, 229 (1883)

30. J. von Neumann: U.S. Patent 2,815,488, filed April 28, 1954, issued December 3, 1957

31. E. Goto: J. Elec. Commun. Engrs. Japan **38**, 770 (1955)

32. R. L. Wigington: Proc. IRE **47**, 516 (1959)

33. A. Uhlir, Jr.: Proc. IRE **46**, 1099 (1958)

34. R. W. Keyes, R. Landauer: IBM J. Res. Dev. **14**, 152 (1970)

35. K. K. Likharev: Int. J. Theor. Phys. **21**, 311 (1982); K. K. Likharev, S. V. Rylov, V. K. Semenov: IEEE Trans. MAG-**21**, 947 (1985)

36. R. Landauer: *A Personal View of Nonlinear Electromagnetic Wave Propagation,* February, 1985, unpublished essay

37. J. M. Manley, H. E. Rowe: Proc. IRE **44**, 904 (1956)

38. H. Salinger: Arch. Elektrotech. **12**, 268 (1923)

39. R. Y. Chiao, E. Garmire, C. H. Townes: Phys. Rev. Lett. **13**, 479 (1964); R. J. Joenk, R. Landauer: Phys. Lett. **24A**, 228 (1967); F. DeMartini, C. H. Townes, T. K. Gustafson, P. L. Kelley: Phys. Rev. **164**, 312 (1967)

40. H. A. Kramers: Physica **7**, 284 (1940)

41. P. Hanggi: J. Stat. Phys. **42**, 105 (1986)

42. P. Hänggi, H. Thomas: Phys. Rep. **88**, 207 (1982); N. G. van Kampen: *Stochastic Processes in Physics and Chemistry* (North-Holland, Amsterdam, 1981); C. W. Gardiner: *Handbook of Stochastic Methods* (Springer, Heidelberg, 1983); H. Haken: Rev. Mod. Phys. **47**, 67 (1975)

43. R. Landauer: In *Self-Organizing Systems: The Emergence of Order,* ed. by F. E. Yates, D. O. Walter, G. B. Yates (Plenum, New York, in press)

44. R. Landauer: Helv. Phys. Acta **56**, 847 (1983)

45. R. Landauer: Phys. Rev. A **12**, 636 (1975)

46. M. Kirkpatrick: Am. Nat. **119**, 833 (1982); R. Lande: Proc. Natl. Acad. Sci. **82**, 7641 (1985); R. Lewin: Science **231**, 672 (1986); J. Levinton: Science **231**, 1490 (1986); C. M. Newman, J. E. Cohen, C. Kipnis: Nature **315**, 399 (1985)

47. S. Wright: In *Experimental Results and Evolutionary Deductions* (Univ. Chicago Press, Chicago, 1977) p. 443; B-O. Küppers: In *Der Informationsbegriff in Technik und Wissenschaft* ed. by O. G. Folberth, C. Hackl (Oldenbourg, München, 1986) p. 181; W. Ebeling: In *Strukturbildung Bei Irreversiblen Prozèssen* (Teubner,

Leipzig, 1976) p. 168; H. Kuhn, C. Kuhn: Origins of Life **9**, 137 (1978); P. W. Anderson, D. L. Stein: In [43]; P. W. Anderson: Proc. Nat. Acad. Sci. **80**, 3386 (1983)

48. *Nonlinear Phenomena at Phase Transitions and Instabilities* ed. by T. Riste (Plenum, New York, 1982); *Physics in One Dimension* ed. by J. Bernasconi, T. Schneider (Springer, Heidelberg, 1981); *Nonlinear Problems: Present and Future* ed. by A. Bishop, D. Campbell, B. Nicolaenko (North-Holland, Amsterdam, 1982); Proceedings of Workshop on Structure, Coherence and Chaos in Dynamical Systems, Lyngby, Denmark, August 12-16, 1986
49. S. Chandrasekhar: Rev. Mod. Phys. **15**, 1 (1943)
50. J. S. Langer: In *Systems Far from Equilibrium* ed. by L. Garrido (Springer, Heidelberg, 1980) p. 12
51. F. A. Lindemann: Trans. Faraday Soc. **17**, 598 (1922); J. A. Christiansen: Ph.D. Thesis, Copenhagen, 1921
52. J. A. Christiansen, H. A. Kramers: Z. Phys. Chem. **104**, 451 (1923)
53. D. Chandler: J. Stat. Phys. **42**, 49 (1986); ibid. G. R. Fleming, S. H. Courtney, M. W. Balk: 83; J. T. Hynes: 149; H. Frauenfelder, P. G. Wolynes: Science **229**, 337 (1985); A. Nitzan: *Non-Markovian Theory of Activated Rate Processes VI. Unimolecular Reactions in Condensed Phases*, to be published
54. P. Hänggi, U. Weiss: Phys. Rev. A **29**, 2265 (1984)
55. H. Risken, K. Voigtlaender: J. Stat. Phys. **41**, 825 (1985); F. Marchesoni: Phys. Rev. B **32**, 1827 (1985); H. Dekker: Physica **135A**, 80 (1986); M. H. Devoret, J. M. Martinis, J. Clarke: Phys. Rev. Lett. **55**, 1908 (1985)
56. A. J. Leggett: *Quantum Macroscopic Tunneling,* Conf. on Nonlinearity in Condensed Matter (Los Alamos National Laboratory), May 5-9, 1986
57. H. Grabert: *SQUID '85* ed. by H. D. Hahlbohm, H. Lübbig (Walter de Gruyter, Berlin, 1986) p. 289; U. Weiss, P. Riseborough, P. Hanggi, H. Grabert: Phys. Lett. **104A**, 10 (1984); H. Grabert, P. Olschowski, U. Weiss: Phys. Rev. B **32**, 3348 (1985); S. Washburn, R. A. Webb, R. F. Voss, S. M. Faris: Phys. Rev. Lett. **54**, 2712 (1985)
58. J. M. Martinis, M. H. Devoret, J. Clarke: Phys. Rev. Lett. **55**, 1543 (1985)
59. M. Büttiker, R. Landauer: IBM J. Res. Dev. **30**, No. 5 (September, 1986)
60. Ref. [44], see paragraph preceding Sec. 4.; M. Büttiker, R. Landauer: In *Nonlinear Phenomena at Phase Transitions and Instabilities* ed. by T. Riste (Plenum, New York, 1982) p. 111, see Appendix; M. Büttiker, R. Landauer: Physica Scripta **T9**, 155 (1985), see Sec. 8
61. B. Derrida: J. Stat. Phys. **31**, 433 (1983). For additional citations see T. Schneider, M. P. Soerensen, A. Politi, M. Zannetti: Phys. Rev. Lett. **56**, 2341 (1986)
62. D. Kutasov: M. Sc. Thesis, Weizmann Institute
63. N. G. van Kampen: In *Statistical Physics, Proceedings of the IUPAP International Conference,* ed. by L. Pél, P. Szépfalusy (North-Holland, Amsterdam, 1976) see p. 31; N. G. van Kampen: In *Advances in Chemical Physics,* ed. by I. Prigogine, S. A. Rice (Wiley, New York, 1976) p. 245
64. Y. Imry: In *Memorial Volume in Honor of Prof. Shang-Keng Ma*, ed. by G. Grinstein, G. Mazenko (World Scientific, Singapore, in press)

65. A. L. Robinson: Science **231**, 22 (1986)

66. J. Sinkonnen: In *Proceedings of the Fourth International School on Physical Problems in Microelectronics*, ed. by J. Kassabov (World Scientific, Singapore, in press)

67. Y. Imry: Europhys. Lett. **1**, (5) 249 (1986); B. L. Al'tshuler, B. I. Shklovskii: *Repulsion of Energy Levels and Conductance of Small Metallic Samples,* Preprint 1166 (Leningrad Nuclear Physics Institute, February 1986); H. Fukuyama, P. A. Lee, A. D. Stone: *Universal Conductance Fluctuations in Metals: Effects of Finite Temperature, Interactions and Magnetic Field,* preprint

68. R. Landauer: In *Electrical Transport and Optical Properties of Inhomogeneous Media,* ed. by J. C. Garland, D. B. Tanner (Am. Inst. Phys., New York, 1978) see p. 32, Sec. 11

69. R. E. Howard, L. D. Jackel, P. M. Mankiewich, W. J. Skocpol: Science **231**, 346 (1986)

70. R. A. Webb, A. Hartstein, J. J. Wainer, A. B. Fowler: Phys. Rev. Lett. **54**, 1577 (1985)

71. A. B. Fowler, G. L. Timp, J. J. Wainer, R. A. Webb: *Observation of Resonant Tunneling in Silicon Inversion Layers,* to be published

72. C. T. Rogers, R. A. Buhrman: Phys. Rev. Lett. **55**, 859 (1985); R. H. Koch: In *Noise in Physical Systems and 1/f Noise,* ed. by M. Savelli, G. Lecoy, J-P. Nougier (Elsevier, New York, 1983) p. 377; M. E. Welland, R. H. Koch: Appl. Phys. Lett. **48**, 724 (1986)

73. B. Schwarzschild: Phys. Today **39**, 17 (Jan. 1986); M. Büttiker: In *New Techniques and Ideas in Quantum Measurement Theory,* ed. by D. Greenberger (N. Y. Acad. Sci., 1986) to be published. See also [64], [67]

74. B. L. Al'tshuler, D. E. Khmel'nitskii: JETP Lett. **42**, 359 (1985)

75. C. M. Soukoulis, J. V. Jose, E. N. Economou, P. Sheng: Phys. Rev. Lett. **50**, 764 (1983); E. Cota, J. V. Jose, M. Ya. Azbel: Phys. Rev. B **32**, 6157 (1985)

76. Ref. [68] see Sec. 9; R. S. Sorbello: In *Macroscopic Properties of Disordered Media* ed. by R. Burridge, S. Childress, G. Papanicolaou (Springer, Heidelberg, 1982) p. 251; R. S. Sorbello: Phys. Rev. B **31**, 798 (1985)

77. A. H. Verbruggen: *Electromigration and proton-Hall effect in metalhydrides,* unpublished Ph.D. thesis (Amsterdam, 1985). While I do not agree completely with the theoretical overview presented in Verbruggen's thesis, it is probably the best overall review of the theory that is available

78. B. L. Al'tshuler: JETP Lett. **41**, 648 (1985)

79. P. A. Lee, T. V. Ramakrishnan: Rev. Mod. Phys. **57**, 287 (1985)

80. Ref. [68], Sec. 6, p. 7; R. Landauer: Phys. Rev. B **14**, 1474 (1976)

81. M. Büttiker, Y. Imry, R. Landauer: Phys. Lett. **96A**, 365 (1983)

82. R. Landauer, M. Büttiker: Phys. Rev. Lett. **54**, 2049 (1985); R. Landauer: Phys. Rev. B **33**, 6497 (1986)

83. E. Ben-Jacob, Y. Gefen: Phys. Lett. **108A**, 289 (1985); K. K. Likharev, A. B. Zorin: J. Low Temp. Phys. **59**, 347 (1985); D. V. Averin, A. B. Zorin, K. K. Likharev: Sov. Phys. JETP **61**, 407 (1985); D. V. Averin, K. K. Likharev: J. Low Temp. Phys. **62**, 345 (1986); There are additional preprints from Likharev's group; see also comments in last paragraph of [59]

Part II

Low-Dimensional Magnetism

Soliton Dynamics: Experiments on Magnetic Chains

J.P Boucher[1;+], *L.P. Regnault*[2], *and H. Benner*[3]

[1]Groupe Dynamique de Spin et Propriétés Electroniques,
Centre d'Etudes Nucléaires de Grenoble, Service de Physique, 85X,
F-38041 Grenoble Cedex, France

[2]Groupe Magnétisme et Diffraction Neutronique,
Centre d'Etudes Nucléaires de Grenoble, Service de Physique, 85X,
F-38041 Grenoble Cedex, France

[3]Institut für Festkörperphysik, Technische Hochschule,
D-6100 Darmstadt, Fed. Rep. of Germany

In the last few years, quasi one-dimensional (1D) magnetic compounds have been shown to provide good examples of soliton excitations. Various experimental results obtained recently on the three compounds $CsNiF_3$, TMMC and $CsCoCl_3$ are reviewed in this article. Firstly, for each compound, the "soliton phase diagram" is presented. Secondly, basic questions concerning such soliton gas systems are discussed. Two problems are considered : the nature of the soliton motion, which can be ballistic or diffusive, and the effects of collisions between solitons and between solitons and magnons. Finally experimental perspectives for the future are briefly mentioned.

1. INTRODUCTION

The study of the soliton concept continues to be a very active research field in condensed matter physics. Many physical systems have now been shown to provide good examples of soliton excitations. This is the case for many quasi one-dimensional (1D) magnetic compounds where solitons can be viewed as moving domain walls [1]. These compounds are now used to carry out accurate experimental investigations of the collective behaviour of solitons in real systems.

The discussion presented in this work relies on recent experimental results obtained from various inelastic neutron scattering (INS) and nuclear magnetic and electron spin resonance (NMR and ESR) measurements performed on the three compounds, $CsNiF_3$, TMMC [$(CH_3)_4NMnCl_3$] and $CsCoCl_3$. $CsNiF_3$ gives an example of ferromagnetic (F) chains [2]. TMMC [3] and $CsCoCl_3$ [4] behave antiferromagnetically (AF). While a "classical" approach is used to describe the solitons in $CsNiF_3$ and TMMC, a "quantum" description is proposed for $CsCoCl_3$ [5]. As a result, the soliton picture which can be obtained is different for each compound. In particular, it refers to "broad" solitons in the case of $CsNiF_3$ and TMMC -the number of spins inside the walls is large ($\Delta N_s \simeq 10$-30)- while it refers to "narrow" solitons ($\Delta N_s \simeq 1$) in the case of $CsCoCl_3$ [7].

In section II, the "soliton phase diagram" -temperature versus external field (if any)- will be considered and discussed for each compound, with particular attention to the case of TMMC where two soliton branches can be observed [7].

+ also a member of Equipe de Recherche CNRS 216

"Stability" and "mobility" appear to be the two basic properties of the solitons : while the former is a general feature of solitons, the latter results from the low dimensionality of the medium. The discussion presented in sections III and IV will be focused on the soliton dynamics. The nature of the soliton motion and the effect of impurities will be considered in section III and the possible effects of collisions between solitons and between solitons and magnons, in section IV.

2. SOLITON PHASE DIAGRAMS

For each compound it is essential to define the conditions of field and temperature under which the soliton picture is reliable. A low-temperature condition $T \ll 4JS^2$ (J is the 1D exchange coupling and S the spin value), is definitely required since substantial magnetic ordering must be present along the chains. For $CsNiF_3$ and TMMC, an external magnetic field H must also be applied perpendicular to the chain direction. In $CsCoCl_3$, solitons can be observed in zero field.

2.1 Classical spin chains : $CsNiF_3$ and TMMC

With S = 1 and S = 5/2, $CsNiF_3$ and TMMC are usually treated as classical spin chains. Example of "transverse" (T) solitons in planar ferromagnetic chains (the case of $CsNiF_3$) are shown in fig. 1a. Inside the domain walls, the 2π rotation of the spins is made essentially within the XY plane. Examples of "transverse" (T) and "longitudinal" (L) solitons for planar AF chains (the case of TMMC) are given in figs. 1 b and c. The π rotation of the spins is made within either the XY plane or the YZ plane.

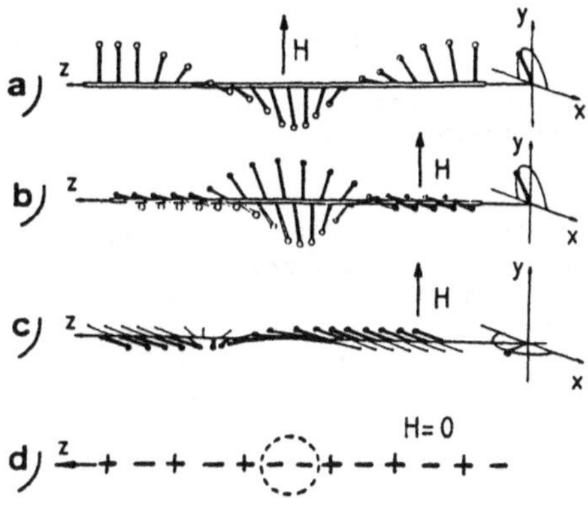

Fig.1 - Transverse (T) soliton in planar ferromagnetic chain (a); T- and longitudinal (L) solitons in planar antiferromagnetic chain, (b) and (c), respectively; quantum soliton in Ising antiferromagnetic chain (d).

Such broad solitons in classical spin chains can be defined with two parameters : the soliton mass $m \simeq 1/\Delta N_s$ and the soliton density $n_s \simeq 1/N_s$ where N_s is the number of spins between the domain walls. Hence, the condition for the existence of solitons is given by $\Delta N_s/N_s \ll 1$. Referring to the sine-Gordon (SG) model, which offers a good starting point for the description of such soliton gas, one obtains

$$\Delta N_s/N_s = 2\pi n_s/m = \sqrt{8\pi}\ \sqrt{E_s^o/T}\ \exp(-E_s^o/T)$$

where E_s^o is the soliton rest energy. For T-solitons $E_s^o \sim \sqrt{H}$ in ferromagnetic chains and $E_s^o \sim H$ in AF chains. Including a renormalisation of about

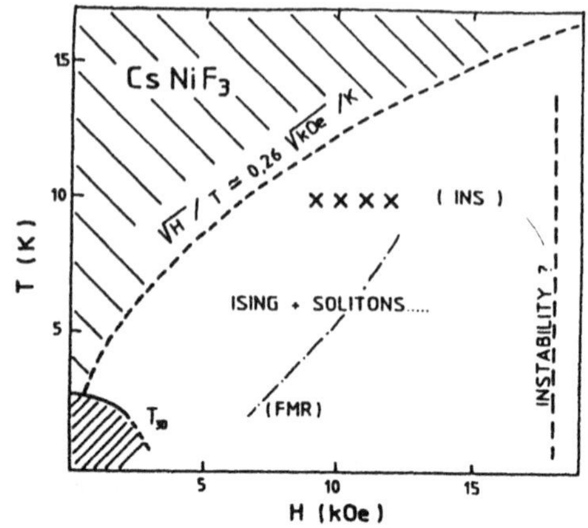

Fig. 2 - Expected soliton phase diagram for $CsNiF_3$. The crosses and the dot-dashed line show where the inelastic neutron scattering (INS) and ferromagnetic resonance (FMR) measurements are performed.

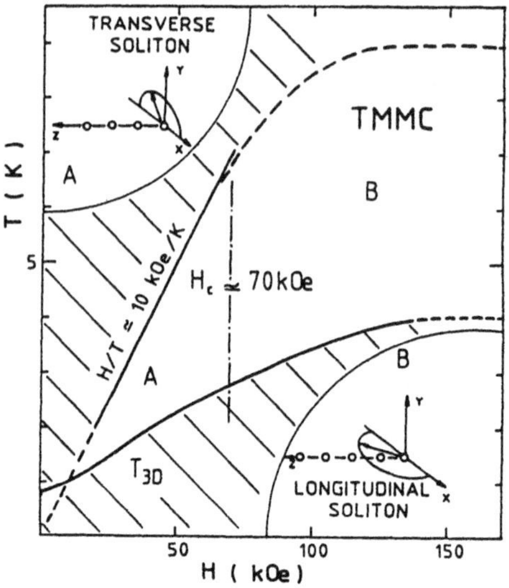

Fig. 3 - Experimental soliton phase diagram for TMMC [5].

30% in the soliton energy with respect to the classical prediction -in agreement with experimental observation [2], [3]- one expects the soliton picture in $CsNiF_3$ and TMMC to be approximately valid for

$$\sqrt{H/T} \gg 0.26\ \sqrt{kOe}/K \text{ and } H/T \gg 10\ kOe/K, \qquad (1)$$

respectively. The corresponding soliton phase diagrams are shown in figures 2 and 3. On these figures, T_{3D} defines the 3-dimensional ordering temperature.

Fig. 2 describes the case of $CsNiF_3$. The dashed line corresponds to condition (1) for ferromagnets. Below this line, one expects the spin system to be essentially in a "Ising" state with the spins parallel to the field direction except in the domain walls. To complete this soliton phase diagram in $CsNiF_3$, it should be mentioned that a soliton instability has been predicted to occur at $H \simeq 18$ kOe [8]. However this instability has not been observed experimentally (see also the article by R. Pynn in this proceedings).

Unlike the case of $CsNiF_3$, the soliton phase diagram for TMMC has been determined experimentally [6]. In particular, condition (1) for antiferromagnets, represented by the full line in fig. 3, agrees well with the experimental observation in low field. However, in the full field range, up to 170 kOe, two different zones can be defined depending on whether $H \gtrless H_c \simeq 70$ KOe. Zone A ($H < H_c$) corresponds essentially to T-solitons while zone B to L-solitons. In fact, T- and L-solitons are two limiting cases of a more complex solution which has been recently investigated by Wysin et al.[7]. In fig. 4 the soliton energy for the general solution is shown as a function of the soliton velocity v for different field values, $H < H_c$, $H = H_c$ and $H > H_c$. On these figures, the "exact" T-soliton solution is represented by the star : it corresponds to static solitons (v = o), with a soliton energy proportional to H. The full lines correspond to T-like domain walls in which the spins are tilted out of the XY plane. For

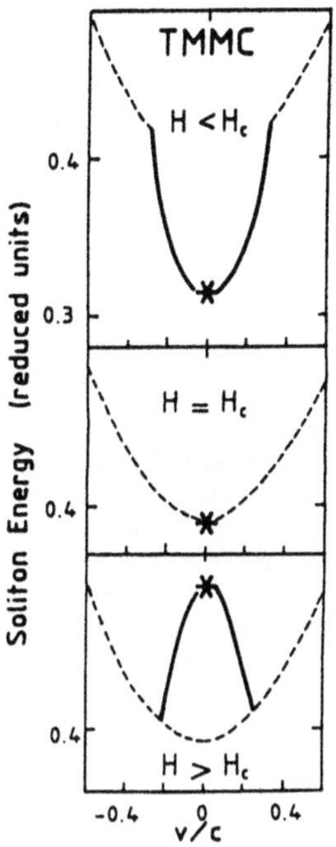

Fig. 4 - Soliton energy as a function of velocity in case of TMMC (from [7]).

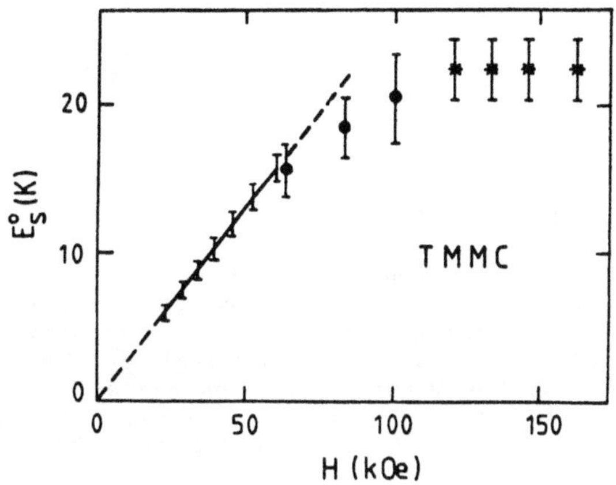

Fig. 5 - Experimental values of the rest soliton energy in TMMC [6]

these solitons, the deviation from the XY plane increases with the velocity. For $H = H_c$, this branch collapses in one point which coïncides with the star : all these solitons become static. The dashed lines describes L-solitons. They can move along the chain and for $H > H_c$ the corresponding rest soliton energy is constant and proportional to H_c.

Such a description corresponds to a one-soliton solution. However, in an experiment, we probe the thermodynamic properties of a soliton gas and the main contribution is given by the lowest energy branch of the excitations. This explains that in TMMC, the soliton energy E_s^o as deduced from the experiments (fig. 5) appears to be proportional to H in low field and field independent in large field.Within the SG model, one evaluates $E_s^o = \alpha H$ with $\alpha = 0.26 \pm 0.02$ K/kOe for T-solitons ($H < Hc$) and $E_s^o \simeq 22 \pm 2$ K for L-solitons ($H > Hc$).

The SG model can account for general features of such soliton gas. However, it has been shown to be insufficient for accurate quantitative comparisons, especially when static properties are concerned [1]. As discussed for the case of TMMC, the out-of-plane spin components are expected to play an essential role in the soliton dynamics [7]. In case of $CsNiF_3$ their contribution to the soliton fluctuations has been established experimentally [9].

2.2 Quantum spin chains : $CsCoCl_3$

$CsCoCl_3$ is a good example of Ising quantum spin ($S = 1/2$) chains. As shown by Villain [5], the first excited state corresponds to the soliton model displayed in fig. 1d. The domain wall extends over one lattice spacing : it defines "narrow" soliton with $\Delta N_s = 1$. The soliton density is given by

$$n_s \simeq \exp(-J/T)$$

The soliton picture becomes reliable for $T < J \simeq 75$ K. In this system the 3D ordering temperature occurs at $T_{N1} \simeq 21$ K. However, below this temperature and down to $T_{N2} \simeq 9$ k it has been shown that one chain over three remains in a paramagnetic state with moving solitons [4]. This peculiar situation results from a frustation effect between the chains due to their mutual AF coupling in an hexagonal structure. In $CsCoCl_3$, the soliton dynamics can therefore be studied in the temperature range : $9 \leqslant T < 75$ K.

3. SOLITON DYNAMICS

In 1D systems, the solitons are expected to move rapidly and it is essential to understand the nature of the soliton motion in real systems. On this crucial question, magnetic chains have given particularly interesting results. Two different dynamical behaviours have been observed : a ballistic motion in pure compounds and a diffusive one in chains containing a small amount of impurities. However, "broad" and "narrow" solitons follow the same ballistic behaviour in pure compounds [6].

3.1 Ballistic behaviour

The ballistic behaviour corresponds to a soliton displacement z proportional to time t : $z = vt$.

In AF chains, such as in TMMC (or $CsCoCl_3$) the solitons, when they travel along the chains, induce rapid spin flips which finally reverse the spin direction in the magnetic sublattices. As a result, the AF short-range order is expected to be limited in time and the transverse correlation function (in the spin direction) $S_\perp(t)$ to relax exponentially as a function of the average number of spin flips $N(t)$ occurring during the time t :

$$S_\perp(t) \simeq \exp\left[-2\, N(t)\right] \tag{2}$$

In the case of ballistic solitons, the number of flips at a given point of the chain is simply given by $N(t) = n_s v_o t$ where $v_o = \sqrt{2/\pi}\; c/\sqrt{E_s^o/T}$ represents the average velocity in a soliton gas and c the characteristic velocity of the medium. One is led to

$$S_\perp(t) \simeq \exp(-\Gamma_\omega\, t) \tag{3}$$

where Γ_ω evaluates the flipping rate given by

$$\Gamma_\omega = \Gamma_q\, v_o \tag{4}$$

with $\Gamma_q = 2\, n_s$. The quantities Γ_ω and Γ_q can be measured independently. For TMMC in low field ($H < H_c$), their values are shown in fig. 6 where they are analysed within the framework of the SG model [3]. The very good agreement between Γ_ω and Γ_q through Eq. (4) proves the ballistic nature of the soliton motion in pure TMMC.

In the case of ferromagnetic chains such as in $CsNiF_3$, there is no such flipping process and the soliton dynamics must be studied by probing directly the fluctuation modes associated with the domain walls. A ballistic behaviour would result in a frequency width of these modes proportional to the wave-vector q [2] :

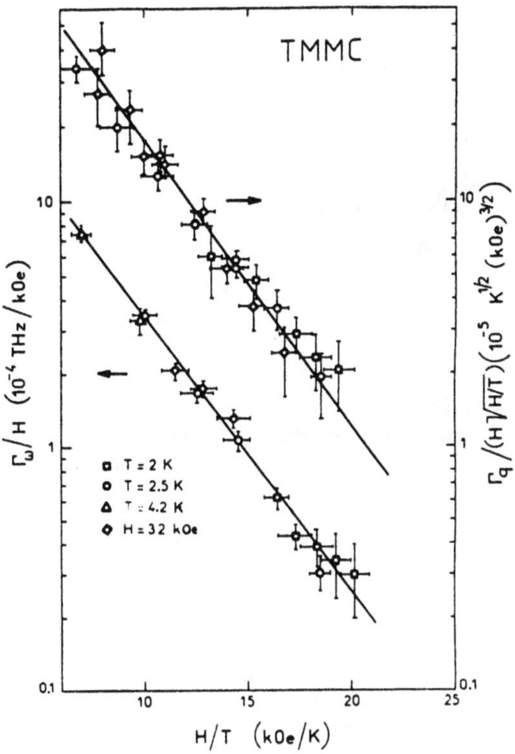

Fig. 6 - Ballistic solitons in TMMC : flipping rate Γ_ω and Γ_q=2n_s as a function of H/T in the T-soliton regime [3].

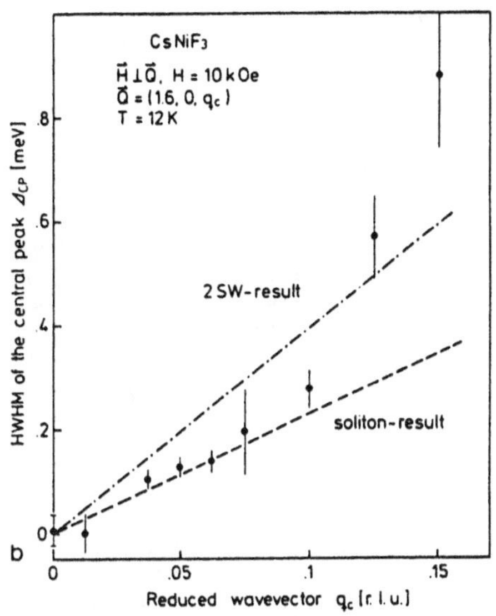

Fig. 7 - Ballistic solitons in $CsNiF_3$: linewidth of the central peak as a function of the wavevector q (from [2]).

$$\Delta_\omega = \sqrt{\pi^3}\, v_o\, q \qquad (5)$$

with q expressed in reciprocal lattice units (r.l.u).

Such behaviour has actually been observed in $CsNiF_3$ for the central peak (CP) associated with fluctuations in the field direction [2]. A linear q dependance is observed at low q, in agreement with the sine-Gordon model (fig.7). It confirms the ballistic behaviour of the solitons in pure $CsNiF_3$.

3.2 Diffusive solitons

A diffusion is characterized by a soliton displacement $z = \sqrt{Dt}$ where D defines the diffusion coefficient. Referring to the flipping process described previously for AF chains, one expects $N(t) = 2\, n_s\sqrt{Dt}$ and the transverse relaxation to behave as

$$S_\perp \simeq \exp(-\sqrt{Bt}) \qquad (6)$$

Such a $\sqrt{t}$ dependance of $S_\perp(t)$ has recently been observed in doped TMMC from Neutron Spin Echo measurements [10].

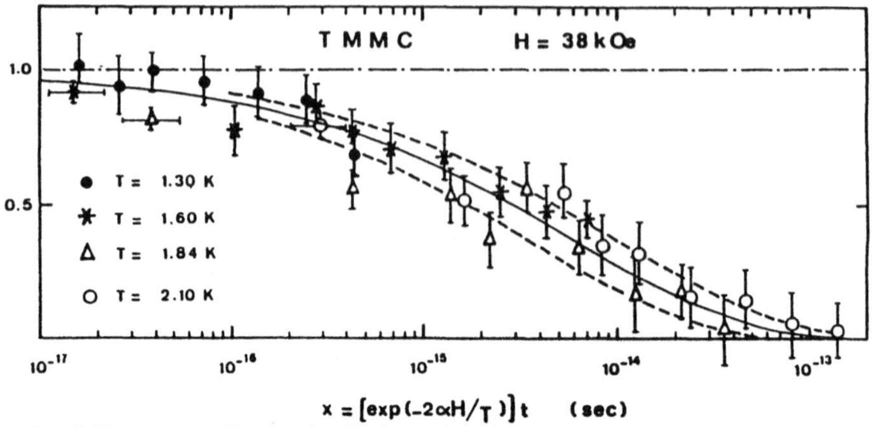

Fig. 8 - Diffusive solitons in TMMC : time decay of the local AF order as a function of the "extended" time scale x, specific to the diffusive model (full line) [10].

Within the previous soliton model, the flipping rate B for a diffusion can be written as

$$B = 4\ n_s^2\ D = 32/\pi\ m^2\ D\ E_s^o/T\ \exp(-\ 2E_s^o/T)$$

A striking representation of the NSE results is shown in fig. 8 where the data for $S_\perp(t)$ are plotted as a function of the "extended time" scale $x = [\exp(-2\alpha H/T)]t$. With this scaling specific to the diffusion model for T-solitons ($E_s^o = \alpha H$), the line-shape analysis covers more than four orders of magnitude : it establishes unambiguously that the motion of solitons is diffusive in doped TMMC.

With TMMC, a unique opportunity is therefore given to study a crossover between two different dynamical regimes. This crossover from ballistic to diffusive behaviours is clearly driven by the impurity concentration. However the unknown origin of the diffusion in such real systems remains to be studied.

This NSE study also provided an answer to a basic question raised a few years ago, concerning the expected time dependance of $S_\perp(t)$ in the case of diffusive solitons. Two different propositions were given. According to Gunther and Imry the $\sqrt{t}$ dependance of $S_\perp(t)$ was expected only for $t < \tau$, where τ is the soliton lifetime [11]. On the contrary, Buttiker and Landauer predicted that Eq.(5) holds for any value of t [12]. Our result agrees with this last proposition, especially if we are in the "strong damping" case where the annihilation process, resulting in the soliton lifetime, occurs each time there is a collision between solitons and anti-solitons. Again, for a definite conclusion on this basic point, a better understanding of the impurity-induced diffusion process is needed.

4. COLLISION EFFECTS

In a soliton gas model, one expects the quasi-particles to undergo frequent collisions, not only with other solitons (the non-linear excitations) but also with the magnons (the linear excitations). A dynamical damping of the soliton or magnon modes can result from these collisions.

In the case of SG ferromagnetic chains, such a prediction has been made for the magnon modes, by Allroth and Mikeska [13]. For the mode $q = 0$, the frequency damping of the magnon is predicted to be

$$\delta = n_s v_o \tag{7}$$

which evaluates the rate of collisions undergone by the magnon with the solitons.

In case of quantum spin chains, similar dynamical damping was also suggested by J. Villain for the soliton modes [5].

4.1 Magnon damping : $CsNiF_3$ and TMMC

An accurate evaluation of the magnon damping in $CsNiF_3$ has recently been obtained from Ferromagnetic Resonance measurements (FMR), which probe the $q = 0$ mode [14]. Expressed in field units Eq. (7) can be written as

$$\delta_o H/H = \pi/16 \exp(-E_s^o/T) \tag{8}$$

with $E_s^o = \alpha \sqrt{H}$ as expected for ferromagnetic chains. The experimental values for the width δH agree very well with this equation which is shown as the full line in fig. 9. The value that one can deduce for α ($\alpha = 11 \pm 1$ K/$\sqrt{\text{kOe}}$) is 20% less than the prediction given by the classical sine-Gordon model, but it is in reasonable agreement with other experimental evaluations [2].

In TMMC, there are two magnon branches corresponding to "transverse" (T) and "longitudinal" (L) fluctuations, as for the solitons. The frequency damping δ_π of the T-magnon mode at $q = 1$ r.l.u. has been evaluated by INS measurements [15]. The data for $H < H_c$ are shown in fig. 10a where the solid line represents the theoretical prediction of a collision model for AF chains :

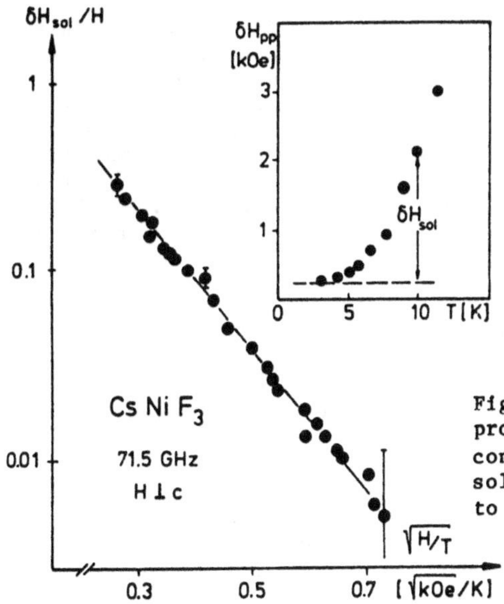

Fig. 9 – Dynamical magnon damping probed at q=0 by FMR in $CsNiF_3$: contribution from collisions with the solitons. The solid line corresponds to Eq. (8) [14].

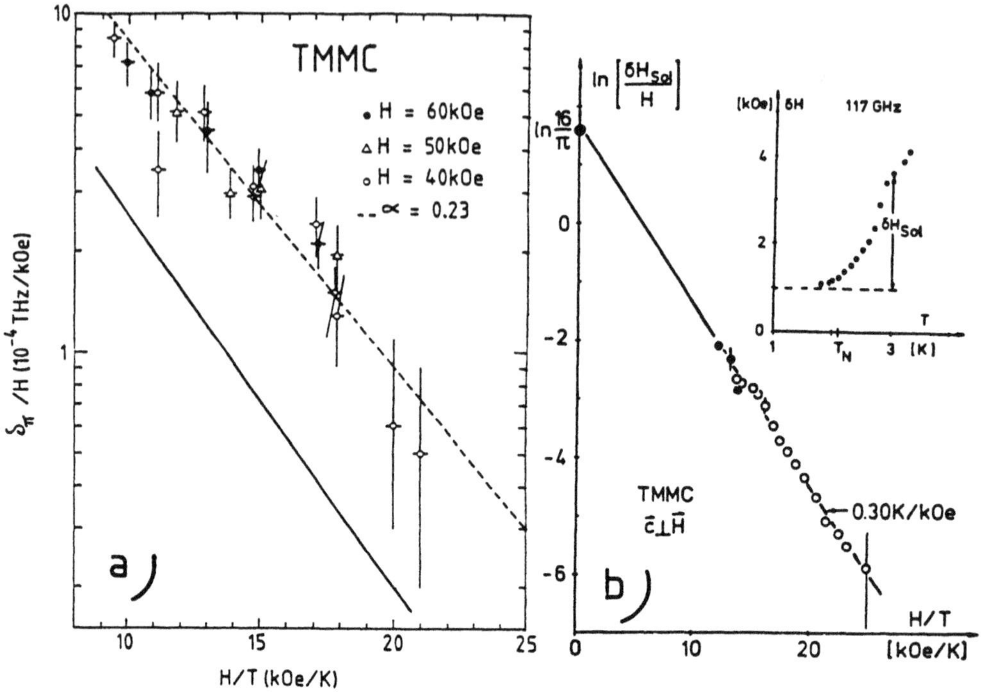

Fig. 10 - Dynamical magnon damping due to collisions with T-solitons in TMMC : a) probed at q=1 r.l.u. from INS [15] and b) probed at q=0 by AFMR [16].

$$\delta_{\pi}/H = \pi/8 \exp(-E_s^o/T) \quad (9)$$

with $E_s^o = \alpha H$ and $\alpha = 0.26$ K/kOe. As discussed by Sasaki and Maki in this proceedings, the damping at q = 1 r.l.u. [Eq. (9)] is twice the damping at q = 0 [Eq. (8)]. In fig. 10a the agreement with the data is only qualitative : in particular the pre-factor of Eq. (9) is not accounted for. Recently, the damping of the uniform (q = 0) L-magnon mode has been evaluated from Antiferromagnetic Resonance measurements (AFMR) [16]. The experimental values $\delta_o H$ are compared with the theoretical description -Eq. (8) with $E_s^o = \alpha H$- shown as the full line in fig. 10b. The agreement with the collision model is very good and one deduces $\alpha = 0.30 \pm 0.03$ K/kOe, a value close to previous determinations ($\alpha = 0.26 \pm 0.02$ K/kOe).

4.2 Soliton dampings : $CsCoCl_3$ and TMMC

For AF quantum spin chains, the non-interacting soliton gas model predicts the soliton modes to exhibit a "diverging" peak at the characteristic frequency $\Omega_q = 4\varepsilon J \sin(\pi q)$ [5] ($\varepsilon \simeq 0.12$ in $CsCoCl_3$).

Experimentally, the soliton modes are observed to be drastically overdamped in such materials. However, it seems possible to explain this damping in terms of collision effects [4]. The full lines in fig. 11 correspond to a phenomenological model which considers the collisions between the solitons. The collision rate is taken to vary essentially as

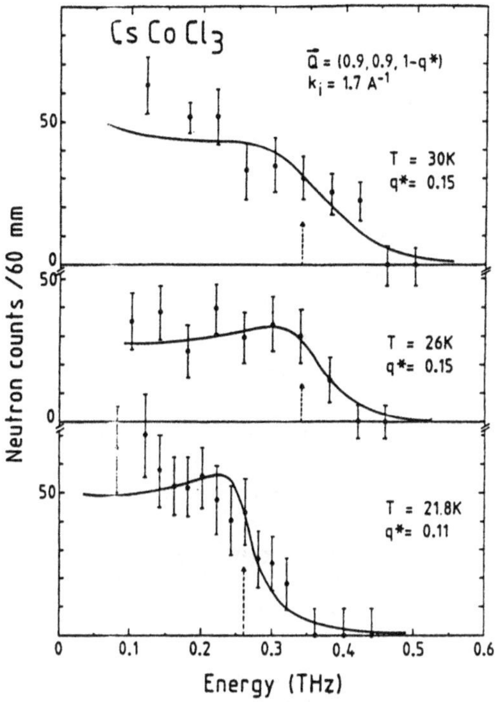

Fig. 11 – Soliton modes probed by INS from $CsCoCl_3$. The solid lines account for possible effects of collisions between solitons [4].

$$\Delta_{ss} \simeq 2\, n_s\, v_o$$

The agreement of the model with the neutron data for $CsCoCl_3$ is quite good in the full explored temperature range.

The soliton fluctuations in TMMC have recently been observed by INS experiment with polarisation analysis [17]. It has been established that as $q^* \to 0$ ($q^* = 1 - q$) there is a residual damping of the soliton modes. For ballistic solitons the frequency width is therefore better described by (compared with Eq. 5)

$$\Delta_\omega = \Delta_\omega^o + \sqrt{\pi^3}\; v_o\, q^* \tag{10}$$

The residual damping Δ_ω^o has been tentatively analyzed in terms of collision effects. Two collision rates have been defined for the collisions between solitons and between solitons and magnons, $\Delta_{ss} \simeq n_s v_o$ and $\Delta_{sm} \simeq n_m v_o$ (n_m is the magnon density), respectively.

In fig. (12) the dashed line corresponds to the non-interacting soliton gas model; the solid line takes into account the collision effects : the agreement with the experiments is very good. However, according to Sasaki and Maki (in this proceedings) no dynamical damping of the soliton modes should be expected from the collisions between solitons and magnons. The analysis of the data must now be considered in view of this new theory.

The concept of a dynamical damping which has been discussed in the present section appears to be an essential ingredient for any realistic des-

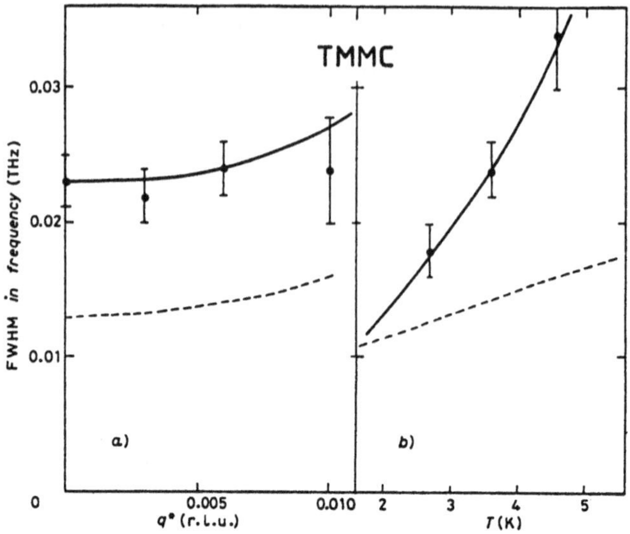

Fig. 12 - Dynamical damping of the soliton modes in TMMC. The solid lines account for collisions between solitons and between solitons and magnons [17].

cription of the fluctuations in such soliton gas systems. More investigation of this basic question is definitely required, theoretically and experimentally.

5. CONCLUSION

The main purpose of the present review was to show that magnetic chains offer very good examples of soliton excitations and that basic questions can be investigated. The dynamical behaviour of the solitons and the associated collision effects are two current problems which deserve to be better understood. More sensitive probes must also be found to go further in such investigations. To probe soliton modes, the electron spin resonance technique -which then can be referred to as Soliton Magnetic Resonance (SMR)- might appear to be a quite powerful tool. An interesting example is given by $CsCoCl_3$. As discussed above, in such quantum spin chains, the soliton regime is reached in zero field. The application of a magnetic field H does not affect the soliton picture. However it changes the soliton behaviour. For H perpendicular to the chains one expects a splitting of the soliton modes as shown in fig. 13. This splitting is associated with the fact that solitons can now move with two different velocities [18] (in a field parallel to the chains, a splitting of the soliton modes is also expected for some spin components).

In fig. 13 one observes that at q = 0 fluctuations occur at twice the Zeeman frequency : $\omega = 2\, g_\perp \mu_B H_\perp$. They can be detected by ESR. The first observation of such SMR signal has been made by Adachi, at $\omega \simeq 9$ GHz [19]. Fig. (14) shows a SMR signal observed at a higher frequency $\omega \simeq 35$ GHz [18]. The observed lineshape is characteristic of the quantum soliton model. This SMR signal can be compared with the INS data observed at about the same temperature T = 26 K (fig. 11). The experimental accuracy is so much better that any analysis of SMR line requires a very precise theoreti-

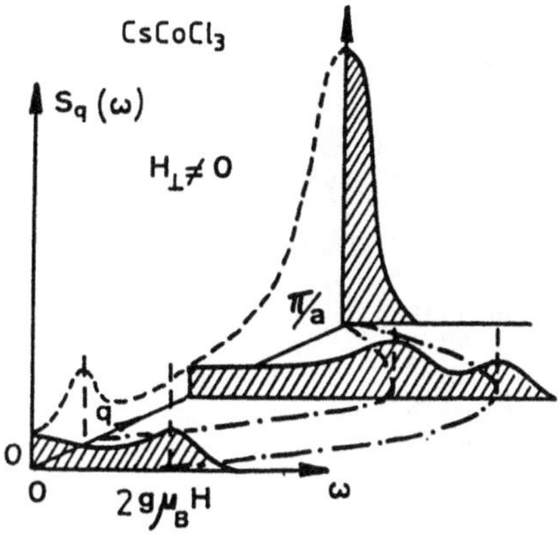

Fig. 13 - Splitting of the soliton modes $S_q(\omega)$ expected for Ising quantum spin chains in a transverse field $H_{\perp}$ [18].

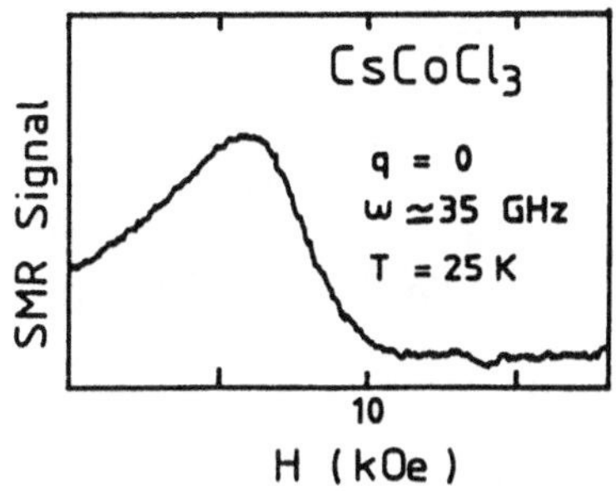

Fig. 14 - Soliton Magnetic Resonance in $CsCoCl_3$ [18].

cal description of the soliton fluctuations. In particular, the question of the soliton mode damping -is it due to collisions or not?- in such quantum spin chains must be considered with great attention.

This example with $CsCoCl_3$ gives an illustration of what improvement in the soliton studies can now be expected. Moreover, it shows that a direct coupling between soliton excitations and an external oscillating field can be achieved in magnetic chains. This suggests that it could become possible to induce chaos in strongly coherent spin systems, directly via non-linear excitations. Such a result would open new and promising experimental perspectives for the future.

REFERENCES

1. Recent examples can be found in "Magnetic Excitations and Fluctuations", edited by S. Lovesey, U. Balucani, F. Borsa and V. Tognetti (Springer-Verlag, Berlin, Heidelberg, New York, Tokyo, 1984).
2. M. Steiner, K. Kakurai and J.K. Kjems, Z. Phys. B, Condensed Matter 53, 117 (1983).
3. L.P. Regnault, J.P. Boucher, J. Rossat-Mignod, J.P. Renard, J. Bouillot and W.G. Stirling, J. Phys. C : Solid State Phys. 15, 1261 (1982).
4. J.P. Boucher, L.P. Regnault, J. Rossat-Mignod, Y. Henri, J. Bouillot and W.G. Stirling, Phys. Rev. B 31, 3015 (1985).
5. J. Villain, Physica (Utrecht) b79, 1 (1975).
6. J.P. Boucher, L.P. Regnault, J. Rossat-Mignod and Y. Henri in "Magnetic Excitations and Fluctuations" edited by S. Lovesey, U. Balucani, F. Borsa and V. Tognetti (Springer-Verlag, Berlin, Heidelberg, New York, Tokyo, 1984) pp. 6-11.
7. G.M. Wysin, A.R. Bishop and J. Oitmaa, J. Phys. C : Solid State Phys. 19, 221 (1986).
8. P. Kumar, Phys. Rev. B, 25, 483 (1982).
   E. Magyari and H. Thomas, Phys. Rev. B, 25, 531 (1982).
9. K. Kakurai, M. Steiner, R. Pynn, B. Dorner, J. Mag. and Mag. Mat., 54-57, 835 (1986).
10. J.P. Boucher, F. Mezei, L.P. Regnault, J.P. Renard, Phys. Rev. Lett., 55, 1778 (1985) and erratum Phys. Rev. Lett. 55, 2370 (1985).
11. L. Gunther and Y. Ymri, Phys. Rev. Lett., 44, 1225 (1980).

12. M. Buttiker and R. Landauer, Phys. Rev. Lett., 46, 75 (1981).
13. E. Allroth and H.J. Mikeska, Z. Phys. B, 43, 209 (1981).
14. H. Benner, H. Seitz, J. Wiezse and J.P. Boucher, J. Mag. and Mag. Mat., 45, 354 (1984).
15. L.P. Regnault, J.P. Boucher, J. Rossat-Mignod, J. Bouillot, R. Pynn,; J.Y. Henry and J.P. Renard, Physica, 136B, 329 (1986).
16. J. Wiese and H. Benner, unpublished.
17. J.P. Boucher, L.P. Regnault, R. Pynn, J. Bouillot, J.P. Renard, Europhys. Lett., 1 (8), 415 (1986).
18. G. Rius, F. Devreux and J.P. Boucher, unpublished.
19. K. Adachi, J. Phys. Soc. Japan, 50, 3904 (1981).

Soliton Dynamics in One-Dimensional Antiferromagnets

K. Sasaki and K. Maki

Department of Physics, University of Southern California,
Los Angeles, CA 90089, USA

1. Introduction

Recent neutron-scattering experiments by BOUCHER et al.[1] on soliton dynamics of the antiferromagnetic chain TMMC $[(CH_3)_4NMnCl_3]$ indicates the effect of the soliton-soliton collision. In this paper we study theoretically the effect of soliton-soliton and soliton-magnon collisions on correlation functions of antiferromagnetic chains based on a sine-Gordon model.

We find a unique flipping mechanism, which is present only in antiferromagnets and absent in ferromagnets, associated with the soliton-soliton scattering. This mechanism causes a damping of longitudinal (parallel to a magnetic field) spin component. Furthermore a soliton displaces its position randomly due to scatterings with other solitons and magnons, but does not change its momentum. This leads to anomalous diffusion: The diffusion is dominant only for short-time correlation, while the ballistic motion with constant (average) velocity dominates the long-time behavior. This is in sharp contrast to the ordinary diffusion, where the short-time behavior is ballistic while the long-time behavior is dominated by diffusion. The effect of the diffusive motion (Brownian motion) of solitons may be observed directly by the neutron spin echo technique [2]. The Brownian motion of a soliton due to collisions with magnons was studied theoretically by several authors [3-5]. We re-examined this problem by using the distribution function of the "breather," which is considered as an alternative but equivalent representation of the magnon in the sine-Gordon model, and obtained a different result from previous ones.

2. Damping and Diffusion

We shall first describe the flipping phenomena associated with the soliton-soliton collision and calculate the damping constant Γ_0 resulting from this effect. Then we study the Brownian motion of the soliton and obtain the diffusion constants D_S and D_M, which come from collisions with solitons and magnons, respectively.

In the sine-Gordon approximation of a planar antiferromagnetic chain in a symmetry-breaking magnetic field, the staggered magnetization $\vec{M}$ (i.e. the momentum measured from π/a) is expressed in terms of the sine-Gordon field ϕ as

$$\vec{M} = [-\sin(\phi/2),\ \cos(\phi/2), 0]. \tag{1}$$

We have chosen the xy plane parallel to the easy plane, and the x axis along the magnetic field. Out-of-plane component of $\vec{M}$ is absent in this approximation. In the ground state $\vec{M}$ lies perpendicular to the magnetic field (M_y = 1 or -1).

In the presence of a soliton (or an antisoliton) propagating with velocity v, the component of $\vec{M}$ parallel to the magnetic field is given by

$$M_x = \pm\text{sech}[m(z-vt-z_0)], \tag{2}$$

where m^{-1} is the width of the soliton, and z_0 is the position at t = 0. We have neglected the "Lorentz factor," since we consider only low-energy solitons ($v^2 \ll 1$). The value of ϕ jumps by 2π or -2π as a soliton or an antisoliton passes. Therefore, because of $\sin(\phi/2)$ in Eq. (1), the local magnetization carried by a soliton, Eq. (2), changes its sign when it collides with another soliton (see Fig. 1). As far as we know, this flipping mechanism has not been noticed before in the literature. On the contrary the local magnetization does not change the sign by the collision with a magnon, since the magnon is small fluctuation of ϕ around $\phi=0$ (mod 2π). We note that this flipping of the local magnetization is absent in the ferromagnetic case, where the local magnetization is proportional to $\sin\phi$ or $\cos\phi$.

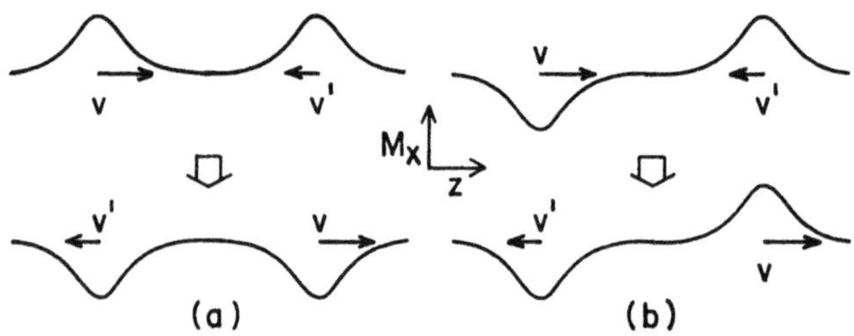

Fig. 1 Collision of two solitons with (a) the same and (b) the opposite signs of the staggered magnetization

The flipping of the magnetization carried by a soliton gives rise to Lorentzian-like broadening of the soliton peak in the dynamical structure factor (see Sec. 3). From the rate of collisions the damping constant Γ_0 is calculated as

$$\Gamma_0(v) = 2n\int dv'P(v')\,|v-v'|, \tag{3}$$

where the factor of 2 is due to the flipping mechanism. The total density of solitons and antisolitons, n, and the velocity distribution function, P(v), are given by

$$n \simeq 2m\left(\frac{2\beta M}{\pi}\right)^{\frac{1}{2}}e^{-\beta M}\left[1-4\left(\frac{2\beta M}{\pi}\right)^{\frac{1}{2}}e^{-\beta M}\ln(4\gamma\beta M)\right], \tag{4}$$

$$P(v) \simeq \left(\frac{\beta M}{2\pi}\right)^{\frac{1}{2}}e^{-\frac{1}{2}\beta M v^2}, \tag{5}$$

where $\beta = T^{-1}$, M is the soliton mass, and γ = 1.781... is the Euler constant. Equation (3) is well approximated by

$$\Gamma_0(v) \simeq 2nv_0[1+(v/v_0)^2]^{\frac{1}{2}},\quad v_0=\left(\frac{2}{\pi\beta M}\right)^{\frac{1}{2}}. \tag{6}$$

A similar $\Gamma_0(v)$ has been conjectured by BOUCHER et al. [1] in order to interpret their experimental result. However, they did not provide the physical origin of the damping. Furthermore they included incorrectly the soliton-magnon scattering term in Γ_0.

Now we study the effect of collisions on the motion of a soliton. The position of a soliton with velocity v is displaced by a collision with a soliton with velocity v' by an amount

$$\Delta_S(v,v') = 2m^{-1}\ell n(2/|v-v'|). \tag{7}$$

Although the velocity (or momentum) of each soliton is conserved during the collision, the average velocity $\overline{v}$ of a soliton is different from v since the number of head-on collisions is larger than that of head-to-tail collisions;

$$\overline{v}(v) = \lim_{t\to\infty} \frac{1}{t}\langle z(t)-z(0)\rangle = v+\sigma_S(v) \tag{8}$$

where z(t) is the soliton position at time t. The shift σ_S is calculated as

$$\begin{aligned}\sigma_S(v) &= n\int dv'P(v')(v-v')\Delta_S(v,v')\\ &\simeq m^{-1}nv[\ell n(8\gamma\beta M)-2].\end{aligned} \tag{9}$$

The randomness of collisions gives rise to the diffusive motion of a soliton. The diffusion constant D_S is given by

$$\begin{aligned}D_S(v) &= \lim_{t\to\infty} \frac{1}{2t} \langle(z(t)-z(0)-\overline{v}t)^2\rangle\\ &= \tfrac{1}{2}n\int dv'P(v')|v-v'|\ \Delta_S^2(v,v')\\ &\simeq \frac{n}{m^2}(2\pi\beta M)^{-\frac{1}{2}}\{[\ell n(2\gamma\beta M)]^2 + \frac{\pi^2}{6}\}\quad .\end{aligned} \tag{10}$$

So far we have considered the soliton-soliton collision. The soliton-magnon collision also causes the shift of the average velocity and the diffusive motion of a soliton. Therefore σ_S and D_S in Eqs. (8) and (10) should be replaced by $\sigma_S + \sigma_M$ and $D_S + D_M$, where σ_M and D_M are contributions from the soliton-magnon collision. As already noted, however, the soliton-magnon collision does not flip the magnetization carried by a soliton, and hence this collision does not contribute to the damping constant Γ_0.

Let us calculate σ_M and D_M. We use the position shift of a soliton with velocity v caused by a collision with a "j-th breather" with momentum q,

$$\Delta_M(v,q,j) = \frac{2mj}{M}\,\frac{1}{\omega_q-vq}\ ,\ \omega_q = (m^2+q^2)^{\frac{1}{2}} \tag{11}$$

and the density function of the breather $\rho_j(q)$, the number of j-th breathers in dq being given by $L\rho_j(q)dq$ where L is the system size. We obtain

$$\sigma_M(v) = \sum_j\int dq\rho_j(q)(v - \frac{d\omega_q}{dq})\Delta_M(v,q,j) \simeq \frac{v}{2\beta M}\ , \tag{12}$$

$$D_M(v) = \tfrac{1}{2}\sum_j\int dq\rho_j(q)|v - \frac{d\omega_q}{dq}|\ \Delta_M^2(v,q,j) \simeq \frac{2\pi}{9}\,\frac{1}{m(\beta M)^2}\ , \tag{13}$$

where we have used the identities [6]

$$2\pi \sum_{j=1}^{\infty} j\rho_j(q) = \frac{1}{\beta\omega_q} \quad , \quad 2\pi \sum_{j=1}^{\infty} j^2 \rho_j(q) = \frac{\pi^2}{3} \frac{1}{(\beta\omega_q)^2} \quad , \tag{14}$$

which are valid in the classical limit ($\beta\omega_q \to 0$).

We note that Eq. (12) agrees with the earlier result [3], but the numerical factor of Eq. (13) differs from the earlier ones [3-5]; our result for D_M is larger by the factor $\pi^2/3$ than that of [5]. In ref. [4] only two-magnon and three-magnon processes are considered, while in refs. [3,5] incorrect breather distribution functions are used.

3. Correlation Functions

The contribution from freely moving solitons to the longitudinal correlation function is given [7,8] by

$$\begin{aligned} S_{||}(q,\omega) &= \frac{1}{2\pi} \int dz dt \; e^{i(\omega t - qz)} \langle M_x(z,t) M_x(0,0) \rangle \\ &\simeq nF(q) \int dv P(v) \; \delta(\omega - vq) \quad , \end{aligned} \tag{15}$$

where F(q) is the form factor of the soliton, which is given by

$$F(q) = \left(\frac{\pi}{m}\right)^2 \operatorname{sech}^2\left(\frac{\pi q}{2m}\right) \quad . \tag{16}$$

In the presence of the damping (flipping) and the diffusion discussed in the preceding section, the correlation function is given by the following replacement in Eq. (15):

$$\delta(\omega - vq) \to \frac{1}{\pi} \frac{\Gamma}{(\omega - \bar{v}q)^2 + \Gamma^2} \quad , \tag{17}$$

where $\bar{v} = v + \sigma_S + \sigma_M$, and

$$\Gamma = \Gamma_0 + q^2 (D_S + D_M) \quad . \tag{18}$$

In a recent neutron scattering experiment BOUCHER et al. [1] found the broadening of the soliton peak, which is consistent with Eq. (17). They proposed a theoretical model to account for Γ, which is somewhat different from ours.

The soliton diffusion affects the transverse correlation also. We obtain the following result:

$$\langle M_y(z,t) M_y(0,0) \rangle \simeq e^{-2N(z,t)} \tag{19}$$

$$\begin{aligned} N(z,t) &= n\xi \left[e^{-z^2/\pi\xi^2} + \frac{z}{\xi} \operatorname{erf}\left(\frac{z}{\sqrt{\pi}\xi}\right) \right] \\ &\simeq n\sqrt{\xi^2 + z^2} \end{aligned} \tag{20}$$

where erf(z) is the error function, and

$$\xi = \left(\bar{v}_0 t^2 + \frac{4}{\pi} D|t|\right)^{\frac{1}{2}} \; , \quad D = D_S + D_M , \; \bar{v}_0 = \left(\frac{2}{\pi\beta M}\right)^{\frac{1}{2}} \; \frac{\bar{v}}{v} \quad . \tag{21}$$

Here N(z,t) is the average number of solitons and antisolitons which cross the line (0,0)-(z,t) in the zt plane. Our result (19) reduces to the earlier result, both in the free soliton limit ($\bar{v} \to v$, $D \to 0$) [7,8] and in the strong damping limit ($\bar{v}_0 \to 0$) [9].

The effect of the soliton diffusion due to impurities on the transverse correlation function of TMMC was observed by BOUCHER et al.[2] using the neutron spin echo technique. In principle, the soliton diffusion due to the soliton-soliton and the soliton-magnon collisions in a pure sample of TMMC can be detected by the same technique. However, the observation may be difficult in practice, since the diffusion constant predicted by the present theory is two orders of magnitude smaller than that observed in the experiment in impure samples [2].

4. Conclusion

We study the effects of soliton-soliton and soliton-magnon collisions in an antiferromagnetic chain within a sine-Gordon model. We find that the soliton-soliton collision leads to damping and diffusion of magnetization carried by solitons, while the soliton-magnon collision leads to magnetization diffusion. The present theory describes the Lorentzian-like broadening of the soliton peak observed in TMMC in a magnetic field.

We would like to thank Jean Paul Boucher for informing us of their experimental results [1] prior to publication and for useful discussions. This work was supported by the National Science Foundation under grant No. DMR-82-14525. One of us (K.S.) acknowledges financial support from the Japan Society for the Promotion of Science.

References

1. J.P. Boucher, L.P. Regnault, R. Pynn, J. Bouillot, and J.P. Renard: Europhys. Lett. 1, 415 (1986)
2. J.P. Boucher, F. Mezei, L. Regnault, and J.P. Renard: Phys. Rev. Lett. 55, 1778 (1985)
3. K. Fesser: Z. Phys. B39, 47 (1980). The diffusion constant obtained by Fesser is twice the result of Ref. [5].
4. Y. Wada and H. Ishiuchi: J. Phys. Soc. Japan 51, 1372 (1981). An improved calculation [M. Ogata and Y. Wada: J. Phys. Soc. Japan 55, Apr. (1986)] will modify the diffusion constant given by Wada and Ishiuchi by a factor 4 which brings their result in agreement with Ref. [5]. We would like to thank Professor Y. Wada for informing us of their new calculation.
5. N. Theodorakopoulos: In Ordering in Strongly Fluctuating Condensed Matter Systems, edited by T. Riste (Plenum, New York, 1980)
6. M. Fowler: Phys. Rev. B26, 2514 (1982)
7. H.J. Mikeska: J. Phys. C13, 2913 (1980)
8. K. Maki: J. Low Temp. Phys. 41, 327 (1980)
9. M. Buttiker and R. Landauer: J. Phys. C13, L325 (1980); and in Nonlinear Phenomena at Phase Transitions and Instabilities, edited by T. Risté (Plenum, New York, 1982)

One-Dimensional Easy-Plane Magnets: Classical Sine-Gordon Theory or a Quantum Model?

G.M. Wysin and A.R. Bishop

Los Alamos National Laboratory, Los Alamos, NM 87545, USA

A classical mechanics description of one-dimensional easy-plane ferro- and antiferromagnets predicts the existence of sine-Gordon kink excitations in these systems, for the case of a small applied field within the ("strong") easy plane. Here we consider:

i) the deviations from the sine-Gordon model, due to stronger fields or weaker anisotropy, which result in modified kink properties, including negative effective masses and kink-antikink annihilation and reflection during collisions, and,

ii) equilibrium thermodynamics, especially specific heat, for quantum spin $S = 1/2$, and $S = 1$ models, comparing with the predictions of classical sine-Gordon theory and with available experimental results.

The applicability of a sine-Gordon model versus a quantum model or the full classical magnetic Hamiltonian is discussed, for materials such as $CsNiF_3$, CHAB $[(C_6H_{11}NH_3)CuBr_3]$ and TMMC $[(CD_3)_4NMnCl_3]$.

I. Introduction

There is continued strong experimental and theoretical interest in nonlinear excitations supported by low-dimensional magnetic materials. This is because of their beguilingly direct connections to model theories, such as the sine-Gordon (SG) equation in 1-dimension [1] or the Kosterlitz-Thouless scenario of vortex unbinding in 2-dimensions [2]. As the comprehensive review of Boucher (these proceedings) shows, the experimental support for the basic predictions of rather simple soliton models in ferromagnetic and antiferromagnetic, easy-plane chain cases is indeed very strong, even to the extent that controlled studies of impurity effects on soliton diffusion are possible. Furthermore, a similarly satisfying picture is beginning to emerge for vortex gas phenomenologies applied to quasi-two-dimensional magnets [3]. Nevertheless, theoretical analysis persistently suggests that corrections to ideal (e.g. sine-Gordon) descriptions [4] are very strong in experimentally relevant circumstances. In the absence of a complete theoretical analysis of quantum statistical mechanics for finite spin systems, it is important that we gain an understanding of dominant individual corrections to sine-Gordon approximations.

Here, we briefly review the effects of: (i) nonlinear out-of-easy-plane spin motions on classical dynamics (for continuous or discrete spin fields); and (ii) quantum spin $S = 1/2$ or 1 on thermodynamic properties (e.g. specific heat, magnetic susceptibility). For definiteness, we restrict ourselves to one-dimensional Hamiltonians of the form [5]

$$H = \sum_n [J\, \vec{S}_n \cdot \vec{S}_{n+1} + A(S_n^z)^2 - g\mu_B B_x S_n^x] \tag{1}$$

and [6]

$$H = \sum_n [J_x S_n^x S_{n+1}^x + J_y S_n^y S_{n+1}^y + J_z S_n^z S_{n+1}^z - g\mu_B B_x S_n^x] \quad , \tag{2}$$

where the notation is conventional. Hamiltonians (1) and (2) are generally considered appropriate for materials such $CsNiF_3$ (S = 1 ferromagnet), TMMC ($(CD_3)_4NMnCl_3$; S = 5/2 antiferromagnet) and CHAB ($(C_6H_{11}NH_3)CuBr_3$; S = 1/2 ferromagnet). In all cases the exchange or local spin anisotropies are chosen such that there is an easy-plane and the magnetic field is applied in that plane. It has been customary to apply, as a first approximation, classical statistical mechanics with a field which is small compared with the anisotropy (e.g. $g\mu_B B \ll 2AS$). At low temperatures ($T \lesssim J$), the continuum limit is then approximately described by the sine-Gordon equation [5], which admits as elementary excitations small amplitude spin-waves, and nonlinear solitons (kinks or breathers).

In section II we describe the effects of out-of-plane spin motions (beyond the linear regime) on (1) for an antiferromagnet, i.e. J, A > 0. In particular we describe possible single kink-solitons and their collisions. Similar results for the ferromagnetic case are given in [7]. Section III summarizes recent results [8,9] for the thermodynamics of quantum S = 1/2 ferromagnets of the form (2) (with $|J_x| = |J_y| > |J_z|$). Some similar discussion of S = 1 ferromagnets is given in [9,10]. Section IV contains remarks on the outlook for nonlinear effects in spin chains.

II. Classical Dynamics in an Easy-Plane Antiferromagnetic Chain

Neutron scattering experiments [11] on TMMC have shown an interesting "cross-over" behavior of the in-plane and out-of-plane spin-wave dispersions as an in-plane magnetic field is increased. In model (1) this can be understood as a switching of the hard axis from the dipole anisotropy axis to the applied field axis [12].

This switching is also expected to affect the relative energies of the nonlinear soliton, or "kink" excitations. These kink excitations have been described in a continuum limit by an approximate mapping to the sine-Gordon (SG) equation, and are rotations of the spins through π around either the z-axis (XY kinks) or the field axis (YZ kinks) [13]. At the critical field the stationary XY and YZ kinks have equal energies, as do the in-plane and out-of-plane spin waves at the zone boundaries.

Previously XY and YZ kinks have been considered to be distinct solutions of the equations of motion, belonging to separate energy dispersion branches. Alternatively, here we review an Ansatz which includes the two as distinct limits of a single continuously connected dispersion curve [14,15]. Also we compare with results of a numerical integration of the discrete equation of motion, as well as presenting a linear stability analysis for YZ kinks.

Properties of continuum limit XY and YZ kinks have been given elsewhere [5,13]. Here we present a slightly different analysis for the YZ kinks.

In order to look for the YZ kink solution, it is best to use spherical polar coordinates where the field direction is the polar axis, in order to take the symmetry of the desired solution properly into account. So if the spins are parameterized in terms of x-polar spherical coordinates, then the equation of motion governing the YZ kinks has been shown to be [14,15]:

$$2(\Phi_{zz} - \tfrac{1}{4}\Phi_{tt}) = \alpha \sin 2\Phi; \quad \alpha \equiv 2A/J \quad . \tag{3}$$

This is a SG equation with c = 2, with the small angles θ, ϕ given in terms of the large angles Θ and Φ. See [14,15] for exact definitions of these angles. The YZ kink solution is

$$\Theta = 1/2\pi; \quad \Phi = \pm 2 \tan \exp \zeta \quad ; \tag{4a}$$

$$\zeta \equiv \gamma\sqrt{\alpha}\,(z-vt) \quad ; \quad \gamma \equiv (1-\tfrac{1}{4}v^2)^{-1/2} \quad . \tag{4b}$$

The YZ kink energy is approximately $E_{YZ} = 2\gamma\sqrt{\alpha} \times JS^2$. Note that the field determines the width and energy of XY kinks, whereas the anisotropy does so for YZ kinks. Both XY and YZ SG kink energies increase monotonically with velocity.

A linear stability analysis [14,15] of YZ SG kinks leads to the following decoupled eigenvalue equation for $\tilde{\Theta}$, where one has assumed small perturbations $\tilde{\Theta}$, $\tilde{\Phi}$, $\tilde{\theta}$ and $\tilde{\phi}$ from the SG profile, and $\beta \equiv g\mu_B B_x/JS$:

$$-\gamma^2\tilde{\Theta}_{\zeta\zeta} + (1-2\,\mathrm{sech}^2\zeta - \frac{\beta\gamma v}{2\sqrt{\alpha}}\,\mathrm{sech}\,\zeta)\tilde{\Theta} = \lambda\tilde{\Theta} \quad . \tag{5}$$

An instability is indicated by a bound state solution to this Schrödinger problem with an imaginary eigenfrequency, where the eigenfrequency ω is related to the eigenvalue by $\omega^2 = 4\alpha(\lambda-1) + \beta^2$. For v = 0, there is a sech ζ bound state, with zero eigenvalue. This corresponds to $\omega^2 = \beta^2-4\alpha$, however, and can have $\omega^2 < 0$ if $\beta < \beta_c \equiv 2\sqrt{\alpha}$, indicating a structural instability for fields less than the critical field.

For nonzero v, the potential for $\tilde{\Theta}$ is modified. The effect of nonzero v can be taken into account through first order perturbation theory. A short calculation shows that the stability criterion becomes $v/c > (4\alpha - \beta^2)/(\pi\beta\sqrt{\alpha})$.

The XY and YZ kinks can be connected through a variational Ansatz, made in terms of the xyz spin components in a manner similar to that done for the easy-plane ferromagnetic kinks [16]. Further motivation and details of this Ansatz calculation are given elsewhere [14,15]. Essentially, a kink profile can be represented by a trajectory of the spin vectors on the unit sphere, plus information about the distribution of the spin vectors along the trajectory. Specifically, there will be one trajectory for each sublattice, and we assume that the A sublattice spin trajectory lies in a plane at an angle θ_A to the easy plane, while the B sublattice spin trajectory lies in a plane at angle θ_B. The distribution of the spins along these trajectories is taken to be SG-like; the angular position within these tilted planes being $\phi = 4 \tan^{-1}\exp[(z-vt)/w]$, where w is a width-variational parameter. Thus the Ansatz involves three parameters: θ_A, θ_B and w, and for given α and β, these are determined by locating the extrema of the Lagrangian.

In Fig. 1 we show E(v) as obtained from this Ansatz using $\alpha = 0.04$ as appropriate for TMMC and for fields $\beta = 0.3$, 0.4 and 0.5 ($\beta_c = 0.4$), and for $0 < \theta_A < \pi$. We also show results from numerical integration of the equations of motion on a discrete lattice, using Ansatz profiles as initial conditions, and time averaging the kinks in their own reference frame to remove spin wave contributions to the energy (a result of having only an approximate traveling wave initial condition).

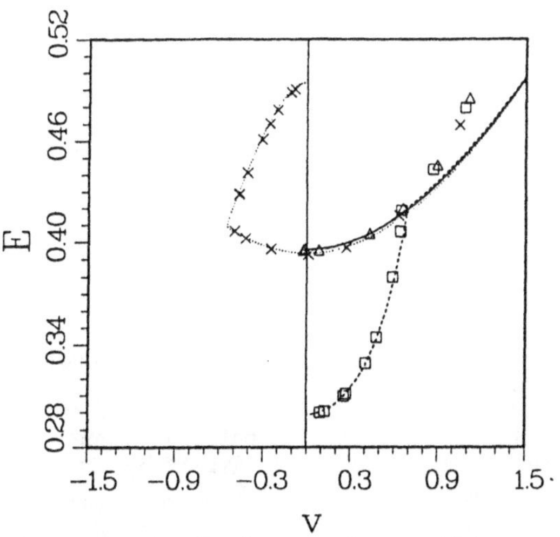

Fig.1. Some typical easy-plane antiferromagnet kink energy-velocity dispersion relationships, at three different fields. The data points were obtained from numerical integrations with $A/J = 0.02$, and $B/B_c = 0.75$ (□), 1.0 (Δ), and 1.25 (×). The curves are the corresponding results from a variational Ansatz calculation. The energy is measured in units of JS^2, and the velocity in units of $JS/\hbar$.

It is found from these numerical studies that the stable kink profile given by the Ansatz can correspond to either an XY or YZ kink, depending on θ_A; the XY and YZ branches are continuously connected. Also, for $\beta < \beta_c$, where static YZ kinks are unstable, there can be stable moving YZ kinks, consistent with the linear stability analysis results.

From the numerical integrations XY kinks are stable above and below the critical field, for $\alpha = 0.04$, $0.08 \leq \beta \leq 0.6$ and $|v/c| < 1$. Even for $\beta > \beta_c$, they show no tendency to decay to lower energy YZ kinks. Whether they can be viewed as slightly perturbed SG kinks is doubtful, especially for $\beta > \beta_c$. The situation is the same as in the ferromagnet above the critical field, where the kinks are dynamically stable but move in a direction opposite to that predicted by SG theory [7]. We conclude the XY kinks are not adequately described by SG theory, and that there is no structural instability at the critical field.

For small velocities $v \ll c$, a two-parameter Ansatz for YX kinks (putting $\theta_A + \theta_B = \pi$) reproduces the velocity dependence of the energy given by SG theory. SG theory adequately describes the YZ branch. Static YZ kinks are stable only if $\beta > \beta_c$. Dynamics YZ kinks require a minimum applied field to be stable, where this minimum field decreases with increasing velocity.

The limited extent of the XY branch could be interpreted to mean that SG-like XY kinks with larger (absolute value) velocities are dynamically unstable. Note that it is possible to estimate the stability limit for the XY kinks from the linear stability analysis for YZ kinks, since the two dispersion curves end where they intersect. First order perturbation theory for the moving YZ kink linear stability problem estimates the velocity v^*, at which the XY branch meetings the YZ branch, as

$$\nu^*/c_0 = \frac{2}{\pi}\frac{B_c}{B}\left[1 - \left(\frac{B}{B_c}\right)^2\right] \quad , \tag{6a}$$

where

$$c_0 = 2JS/\hbar \quad , \tag{6b}$$

and the critical field is

$$B_c = [8AJS^2]^{1/2}/(g\mu_B) \quad . \tag{7}$$

The lattice spacing is taken as the unit of length here. Note that for $B < B_c$, when $\nu^* > 0$, the XY effective mass is positive, while for $B > B_c$, when $\nu^* < 0$, the XY effective mass is negative. Equation (6a) is an approximate expression which is most accurate for B near B_c. Also note that one cannot determine ν^* by equating the predicted SG XY and YZ energies for a given field; the SG theory predicts that the branches do not cross except for B very near B_c. The non sine-Gordon behavior manifests itself by strongly changing the effective masses of the XY kinks.

The XY kinks are in many ways analogous to the kinks of the easy-plane ferromagnet. The ferromagnetic kink's effective mass changes sign at a corresponding critical field, the absolute values of the effective masses are much smaller than predicted by SG theory, and they are also dynamically stable even for fields greater than the critical field. Also, while the XY kinks obey dynamics very different from SG-like, the YZ kinks, on the contrary, can be described quite accurately using SG dynamics outside of the unstable regimes mentioned. The YZ kinks have no natural analogue in the ferromagnet.

Turning to kink-antikink ($K\bar{K}$) collisions, only numerical results are presently available. We have used chains with 101 to 501 lattice sites [17], where the kink width varies as $w \simeq 2JS/(g\mu_B B)$ for XY kinks and as $w \simeq \sqrt{J/(2A)}$ for YZ kinks. Using a fixed ratio $2A/J = 0.04$, the field ranged from $B/B_c = 0.10$ to $B/B_c = 1.50$. For TMMC, this corresponds to $9.0\text{ kG} \leq B \leq 140\text{ kG}$, with $B_c \simeq 90$ kG. The initial condition was an Ansatz profile for some specified parameter θ_A, where θ_A determines the tilt of the spins out of the easy plane on the A-sublattice [14,15]. The resulting profile could correspond to either an XY or YZ kink, depending on whether θ_A was near zero or $\pi/2$. At some intermediate value of θ_A, the Ansatz kink switched from the XY branch to the YZ branch. A given combination of θ_A and B/B_c then determined the initial velocity, energy, and width of the kink.

Neumann boundary conditions were applied to the xyz spin components with the spatial derivatives on each sublattice separately set to zero at the boundaries. Classification of the type of collision was based on viewing the time evolution of the spin profile and the spatial averages of in-plane and out-of-plane angles. The tilt of the two spins at the center of one kink, one on each sublattice, measured from the easy plane, provided an additional diagnostic.

Possible outcomes of collisions include SG-like transmission, annihilation, and reflection. Some typical cases of each of these are shown in Figs. 2-4, in terms of the in-plane (ϕ) and out-of-plane (θ) angles on one sublattice. Results from these simulations are summarized in the final state phase diagram of Fig. 5.

The results for XY kinks are similar to ferromagnetic kinks. Generally, for low fields $B < B_c$ there is SG-like transmission (Fig. 2). At higher

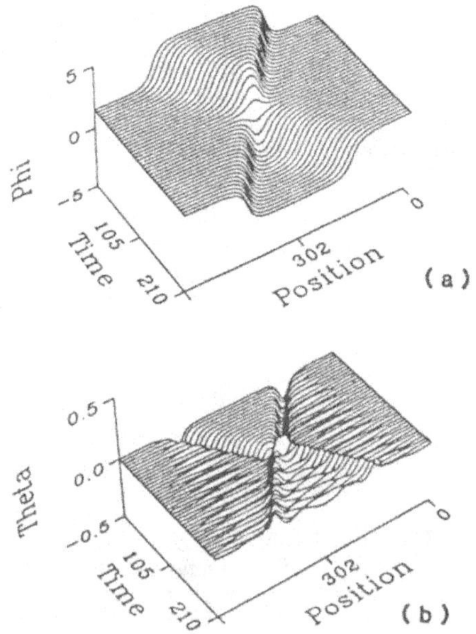

Fig. 2. K$\bar{K}$ collision of an XY-like pair, resulting in SG-like transmission; at field $B/B_c = 0.25$, with initial velocity $v_0 = 1.4$ (or $\theta_A = 0.180243$). The angle ϕ within the easy plane (on one sublattice) is shown in (a), and the associated out of easy plane angle θ is shown in (b).

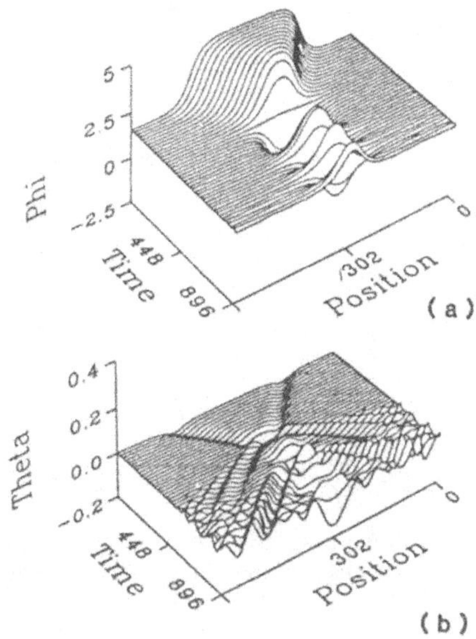

Fig. 3. K$\bar{K}$ collision of an XY-like pair resulting in annihilation; at field $B/B_c = 0.25$, with initial velocity $v_0 = 0.37$ (or $\theta_A = 0.024673$). Parts (a) and (b) are as described in Fig. 2. Note the oscillations, suggestive of formation of a breather-like bound state.

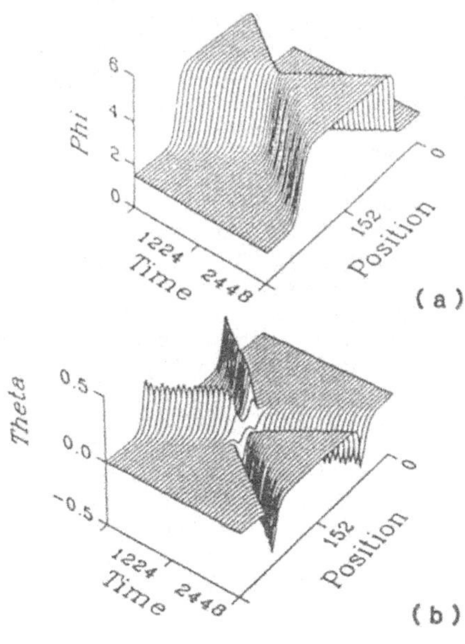

Fig. 4. K$\bar{K}$ collision of an XY-like pair also resulting in reflection, but for $B/B_c = 1.075$, with $v_0 = 0.06$ ($\theta_A = 0.364563$). Parts (a) and (b) are as described in Fig. 2.

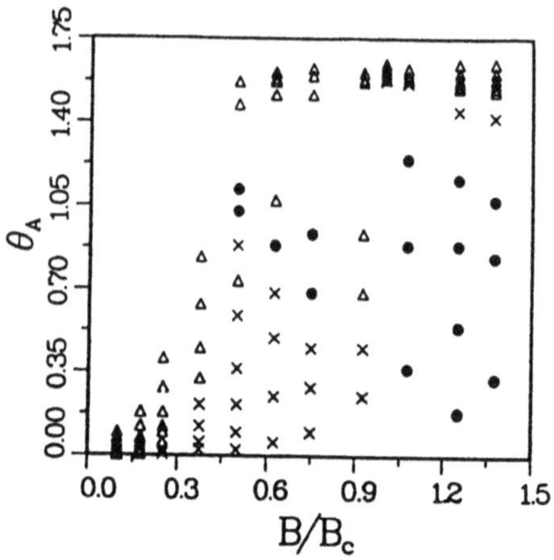

Fig. 5. An output state "phase diagram" in terms of the applied field B/B_c and the Ansatz parameter θ_A (approximately equal to the tilt of the spins out of the easy plane at the kink center). The symbols refer to transmission (Δ), annihilation (×) and reflection (o).

fields, but still with $B < B_c$, the low velocity pairs annihilate, or possibly form breathers, (Fig. 3) while the higher velocity XY kinks undergo SG-like transmission. Some cases with $B < B_c$ resulted in reflection of XY kinks. For $B > B_c$, the negative effective mass XY kinks reflect, as in the ferromagnet (Fig. 4). Most of the cases tested for YZ $K\bar{K}$ pairs resulted in transmission, consistent with their nearer-SG behavior. The exceptions included some cases at small velocity for $B > B_c$, where annihilation occurs. See [17] for further details of these results.

III. Quantum Thermodynamics for Easy-Plane Ferromagnetic Chains

When using a model such as (2) to describe materials such as $S = 1/2$ CHAB [[6]; $J_x = J_y = -110$ K, $J_z/J_x = 0.95$], it is natural to consider that quantum effects may be important. One approach is to reintroduce quantum mechanics by replacing the classical sine-Gordon equation with its quantized version, leading to a reduction of the effective SG soliton rest mass. Alternatively, Johnson and Wright [18] reviewed the Bethe Ansatz method [19] applied to solving the quantized SG equation relevant to easy-plane ferro- and antiferromagnets -- a similar rest mass reduction is found, but still theory and experiment for CHAB, $CsNiF_3$ and TMMC disagree (for specific heat, and therefore probably for other thermodynamic properties). These authors point out, in particular, that the corrected classical SG theory, including kink-kink interactions [20], would require a rest mass increase to bring the calculated specific heat into agreement with experiment for CHAB. This approach of quantizing a particular limit of the full classical Hamiltonian (the SG limit) seems questionable. By so doing, the out-of-plane degree of freedom is not treated properly; it is essentially transformed to a linear degree of freedom. In view of continuing concerns over the importance of out-of-plane classical motions (c.f. section II) versus the quantization of the SG model, it seems necessary to include both out-of-plane and quantum aspects simultaneously.

One way of achieving this is to use the recently developed Trotter-Suzuki transformation, whereby the thermodynamics of the original 1-D quantum system is mapped onto the thermodynamics of a 2-D classical system [21]. Numerical evaluation of the internal energy, specific heat, etc. is carried out by using either Monte Carlo or transfer matrix methods [22]. Although this will give no direct information about the excitations, (e.g., the question of existence of solitons) it can nevertheless give crucial indications of the importance of quantum effects and the validity of the assumed Hamiltonians. Attempts to implement Trotter-Suzuki formalisms for S = 1/2 and S = 1 ferromagnets are reported in [8,9,14,22]. Here we limit the discussion to an alternative "numerically exact" quantum transfer matrix (QTM) method [9,23] applied to S = 1/2 CHAB thermodynamics (model 2). We have used a technique to extrapolate from the finite size lattice to the infinite limit in both directions on the 2-D lattice, thereby making this preferred over the previous S = 1/2 quantum Monte Carlo (QMC) method [8]. We find that there is <u>no</u> value of exchange anisotropy from 4% to 10% for which the QTM results for specific   peaks will agree with experiment. We have, however, tested that the QTM calculation gives results consistent with the QMC calculation. The computing method used was given by Betsuyaku [23], who adapted that of Morgenstern and Binder [24] as originally applied to spin glass models, by allowing for the four-spin interactions. It is necessary to choose free boundary conditions in the spatial direction (N), while periodic boundary conditions are imposed in the Trotter direction (m) as a result of the trace. The method requires storing 2^{2m} Boltzmann factors -- for this calculation we have used $1 \leq m \leq 9$. (The integer m is the lattice size in the Trotter direction.) Computing time rises exponentially with m and linearly with N. Presently the practical limit is m = 9 for storage as well as CPU time using a CRAY-1 800 K word machine, while $N > 100$ is no practical problem. Extrapolations for N, $m \to \infty$ are discussed in [9,23].

First the method was tested for m = 8, N = 32, at 5% anisotropy (J_z/J_x = 0.95) to compare with previous spin-$\frac{1}{2}$ QMC data [8,14]. Results for internal energy, specific heat, magnetization and susceptibility all agreed to within about 5% over the temperature range 4 K to 20 K. Then we applied this method to model (2) with anisotropy ranging from 4% to 10%, in order to compare with the experimental specific heat data of Kopinga et al. [6]. The <u>difference</u> specific heat C(B) - C(0) = ΔC, is plotted versus field for a series of temperatures, and then the peak position and height are determined and plotted versus T^2 and T respectively. Some representative ΔC versus B curves are shown in Figure 6, for the case of 5% anisotropy. The data lie on smooth curves, making the determination of peak positions and heights possible. Interpolation, using a parabolic fit to the peaks, provided a simple accurate way to determine the heights and positions. In Figure 7 the resulting B_{peak} and ΔC_{max} are shown, for anisotropies 4%, 5%, 6%, 8%, and 10%, and compared with classical SG theory and experiment.

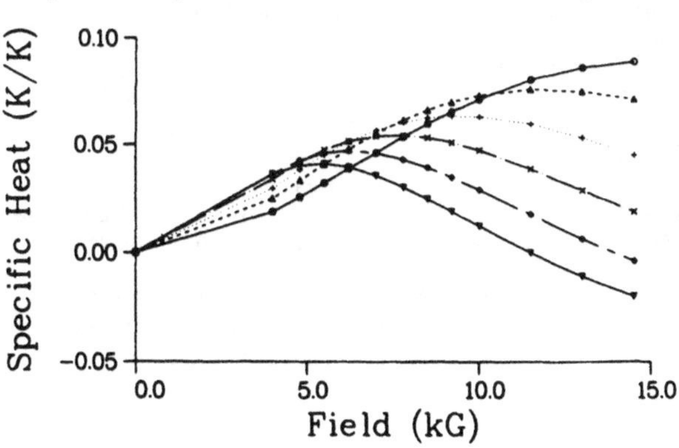

Fig. 6. Some typical results for ΔC vs. B as obtained with the spin-1/2 QTM calculation using CHAB parameters (5% anisotropy). The data correspond to temperatures T = 4.0 K (∇), 4.4 K (◇), 4.9 K (x), 5.5 K (+), 6.2 K (Δ) and 7.2 K (0), and have been extrapolated to $N \to \infty$, $m \to \infty$.

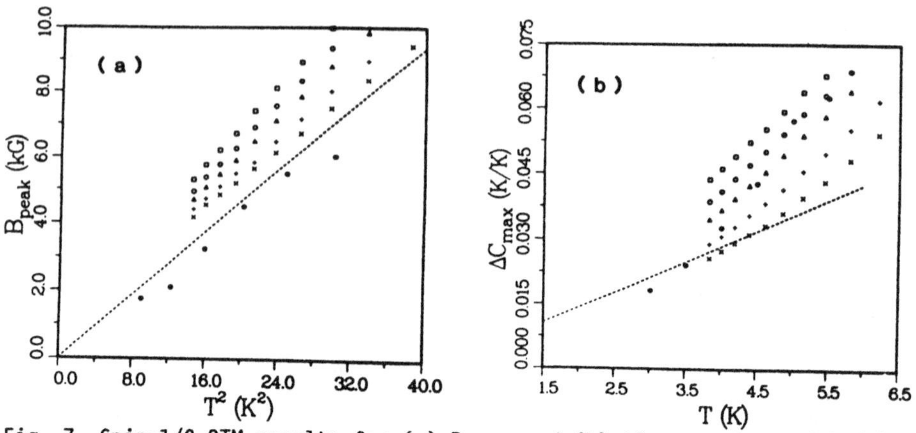

Fig. 7. Spin-1/2 QTM results for (a) B_{peak} and (b) ΔC_{max}, using model (2) with $J_x = J_y = -110$ K, for a series of values of anisotropy $J_z/J_x = 0.96$ (□), 0.95 (o), 0.94 (Δ), 0.92 (+) and 0.90 (x). These are all data from the extrapolation to $N \to \infty$, $m \to \infty$. The solid data points (•) are the experimental data on CHAB by Tinus et al. [25]. The dashed lines are the classical SG theory of Sasaki and Tsuzuki [20].

The dashed lines in Figure 7 are classical SG theory results using a soliton rest mass $E^{o}_{sG} = 8(JS^3 g\mu_B B)^{1/2}$, that is, with no adjusted parameters, as in [20]. The predictions of classical SG theory are independent of the anisotropy. Expressions given by Sasaki and Tsuzuki [20] include contributions from spin waves, solitons, and soliton-soliton interactions. Their calculations predict that the general result for a SG ferromagnetic is

$$B_{peak} = AT^2 \quad , \quad A = (64t_m^2 g\mu_B JS^3)^{-1} \quad , \tag{8}$$

where $t_m = T/E^{o}_{sG} = 0.190$ has determined the peak position. The corresponding peak height is found to be given by

$$\Delta C_{max} = A'T \quad , \quad A' = 0.196/JS^2 \quad . \tag{9}$$

We see that agreement between this classical theory and experiment is fair for B_{peak} but not as good for ΔC_{max}. None of the chosen values of anisotropy for the QTM fit well to the experimental CHAB data over this temperature range. If the SG soliton rest mass is ad-hoc renormalized such that the slopes of the SG theory B_{peak} curves agree with the experimental slope, then the implied changes in the SG theory ΔC_{max} are <u>not</u> adequate to cause them to simultaneously fit the experimental data. It has not been apparent how to resolve this problem with classical SG theory. The QTM data presented here obviously should require no such quantum renormalization, but nevertheless systematically disagree with experiment, casting some doubt on the adequancy of the customary model (2). Quantum Monte Carlo studies for S = 1 $CsNiF_3$ [9,14] raise similar questions. (QTM methods of the form used for S = 1/2 are presently impractical for S = 1 because of computer memory limitations.)

IV. Outlook

In view of the results summarized in sections II and III, we should be surprised how well classical sine-Gordon theory can explain experimental data on materials such as CHAB, TMMC and $CsNiF_3$ with easy-plane applied magnetic fields. For instance, we have seen that the quantized version of the ferromagnet Hamiltonian gives approximately the same low-T thermodynamics as the classical SG Hamiltonian, for the case of S = 1/2 CHAB. This can be compared with classical transfer matrix calculations [25] for the ferromagnet Hamiltonians, which give much larger low-temperature specific heat peaks. We tentatively conclude that the quantum mechanics plays a strong role in restricting the spins to the easy plane (including a zero-point out-of-plane component) thereby making the classical theory more appropriate than might at first be expected. It will be interesting to see whether such a restriction also controls dynamics quantum properties, e.g. as measured in inelastic neutron scattering. In the context of quantum Monte Carlo simulations, this will require solving the outstanding problem of Laplace inversion from a finite imaginary time interval.

We anticipate that low-dimensional magnetic materials will continue to develop as accessible contexts in which to investigate fundamental nonlinear and nonequilibrium processes, including: (i) effects of impurities and damping mechanisms on soliton transport (c.f. Boucher); space-time coherence and chaos (e.g. breather selection and synchronization by an external oscillatory field [26], or competing spin interactions leading to inhomogeneous structures and associated dynamics, or chaotic dynamics [26], or even "quantum chaos" [27]); and (iii) vortex-spin-wave dynamics in quasi-two-dimensional magnets [3], which is especially timely in view of the emergence of many well-characterized materials and improved resolution inelastic neutron scattering data.

References

1. M. Steiner, K. Kakuri and J. K. Kjems, Z. Phys. B 53, 117 (1983); J. P. Boucher, L. P. Regnault, J. Rossat-Mignod, J. P. Renard, J. Bouillot, W. G. Stirling, and F. Mezei, Physica 120B, 241 (1983).
2. J. M. Kosterlitz and D. J. Thouless, J. Phys. C 6, 1181 (1973); and Prog. Low Temp. Phys. (D. F. Brewer, ed.), Vol. VIII B, North-Holland, Amsterdam (1978).
3. C. Kawabata, M. Takenchi and A. R. Bishop, J. Magn. Magn. Mat. 54-57, 871 (1986); A. R. Bishop et al., in preparation.
4. See, for example, P. Kumar, Phys. Rev. B 25, 483 (1982); Physica 5D, 359 (1982); E. Magyari and H. Thomas, Phys. Rev. B 25, 531 (1982); J. Phys. C 16, L535 (1983).
5. H. J. Mikeska, J. Phys. C 11, L29 (1978); 13, 2913 (1980).
6. K. Kopinga, A. M. C. Tinus and W. J. M. de Jonge, Phys. Rev. B 29, 2868 (1984).
7. G. M. Wysin, A. R. Bishop and P. Kumar, J. Phys. C 17 , 5975 (1984); 15, L337 (1982).
8. I. Satija, G. M. Wysin and A. R. Bishop, Phys. Rev. B 31, 3205 (1985).
9. G. M. Wysin and A. R. Bishop, Phys. Rev. B 34, 3377 (1986).
10. G. Kamienarz and C. Vanderzande, preprint (1986).
11. I. U. Heilmann, R. J. Birgeneau, Y. Endoh, G. Reiter, G. Shirane and S. L. Holt, Solid State Commun. 31, 607 (1979).
12. I. Harada, K. Sasaki and H. Shiba, Solid State Commun. 40, 29 (1981).
13. N. Flüggen and H. J. Mikeska, Solid State Commun. 48, 293 (1983).
14. G. M. Wysin, PhD. Thesis, Cornell University (1985).
15. G. M. Wysin, A. R. Bishop and J. Oitmaa, J. Phys. C 19, 221 (1986); J. Magn. Magn. Mat. 54-57, 831 (1986).
16. R. Liebmann, M. Schöbinger and D. Hackenbracht, J. Phys. C 16, L633 (1983).
17. G. M. Wysin and A. R. Bishop, in preparation.

18. M. D. Johnson and N. F. Wright, Phys. Rev. B 32, 5798 (1985).
19. M. Fowler and X. Zotos, Phys. Rev. B 25, 2805 (1982).
20. K. Sasaki and T. Tsuzuki, J. Magn. Magn. Mat. 31-34, 1283 (1983).
21. M. Suzuki, Prog. Th. 1454 (1976); M. Barma and B. S. Shastry, Phys. Rev. B 18, 3351 (1978).
22. For examples, see J. J. Cullen and D. P. Landau, Phys. Rev. B 27, 297 (1983); H. DeRaedt, A. Lagendijk and J. Fivez, Phys. Rev. B 46, 261 (1982).
23. H. Betsuyaku, Prog. Th. Phys. 73, 319 (1985); Phys. Rev. Lett. 53, 629 (1984).
24. I. Morgenstern and K. Binder, Phys. Rev. B 22, 288 (1980).
25. M. G. Pini and A. Rettori, Phys. Rev. B 29, 5246 (1984); A. M. C. Tinus, W. J. M. de Jonge and K. Kopinga, preprint (1985); Phys. Rev. B 32, 3154 (1985).
26. A. R. Bishop and P. S. Lomdahl, Physica D 18, 54 (1986); G. M. Wysin and A. R. Bishop, J. Magn. Magn. Mat. 54-57, 1132 (1986).
27. K. Nakamura et al., Phys. Rev. B 33, 1963 (1986); Phys. Rev. Lett. 57, 5 (1986).

Towards a Quantitative Theory of Solitons in One-Dimensional Magnets: Quantum Effects, Out-of-Plane Fluctuations and the Specific Heat

H.J. Mikeska and H. Frahm

Institut für Theoretische Physik, Universität Hannover,
D-3000 Hannover, Fed. Rep. of Germany

In recent years the investigation of quasi-onedimensional (1d) magnets has contributed significantly to our knowledge about nonlinear excitations in solids. In particular a large amount of both theoretical and experimental work has been devoted to the study of soliton-like excitations in 1d magnets with an easy-plane (xy-like) symmetry in a magnetic field which breaks this xy-symmetry, as described by the Hamiltonian

$$H = -J\sum_n \vec{S}_n\vec{S}_{n+1} + A\sum_n (S_n^z)^2 - \mu B \sum_n S_n^x \tag{1}$$

This Hamiltonian is realised approximately by the compounds $CsNiF_3$ and $(C_6H_{11}NH_3)\ CuBr_3$ (CHAB) (both with ferromagnetic coupling along the chain) and by the antiferromagnet $(CH_3)_4NMnCl_3$ (TMMC). Following theoretical predictions /1,2/, solitonlike excitations have been observed in all these systems, using a variety of experimental techniques: neutron scattering /3,4/, NMR and ESR /5,6/, Mossbauer effect /7/, shift of T_N /8/ and magnetic specific heat /9-11/.

Qualitative agreement with the experimental results has been obtained on the basis of the simplest theoretical approach: A treatment of the spin chain in the classical continuum approximation, assuming an ideally strong xy-like anisotropy. In this limit the spin chain can be mapped to the well-known Sine-Gordon (SG) chain; using the phenomenological approach to the static and dynamic properties of this system in thermal equilibrium /12/, the soliton solutions of the SG chain are found to account qualitatively for experimentally observed features: the presence of a central peak in inelastic neutron scattering data, an additional linewidth in resonance experiments and the additional maximum in the magnetic specific heat as a function of temperature or magnetic field.

Following the era of qualitative confirmation of the soliton picture in 1d magnets, an important goal has been to establish quantitative agreement of experimental and theoretical results. This requires a realistic treatment of the spin chain, relaxing the simplifying assumptions which led to the SG approximation. In this respect the specific heat has become the most important experimental quantity, since theoretical results for this static quantity can be obtained much more completely than for dynamic quantities.

Before considering the specific heat it is instructive to review the status of quantitative description for the integrated intensity of the central peak observed in neutron scattering as it has developed in recent years. A quantitative comparison of theory and experiment has become possible by measuring the integrated intensity in units of the spin wave intensity at T=0, which is assumed to be given unambiguously from the linear theory. From a comparison of the experimental results /3/ to the various stages of the theory the following picture emerges:

1) In the simplest theoretical approach one considers single (i.e. nonoverlapping) solitons in the SG approximation at low temperatures. The reduced central peak intensity at q=o is then given by /13/

$$I/I_0 = 256(mJ^2/2\pi Ak_BT)^{1/2} e^{-8\beta m} \tag{2}$$

$$\beta = J/k_BT, \quad m = (\mu B/J)^{1/2} \tag{3}$$

For $CsNiF_3$ at B = 1T and T = 12K this gives $I/I_0 \cong 2.8$ whereas the experimental value is of the order of 1 /3/.

2) In order to cover finite temperatures, the existence of spin waves and the overlap of solitons have to be taken into account. This leads to a contribution to the central peak from two magnon processes /13,14/ and to interference effects, which decrease the central peak intensity. In the frame of the SG approximation this decrease is described by reduction factors /13,15/

$$\alpha_{sw} = 1 - T/T_0 + O(T/T_0)^2, \; T_0 = 2mJ \tag{4}$$

$$\alpha_{sol-sol} = 1 - 16m(\beta m/\pi)^{1/2} e^{-8\beta m} \ln \frac{\gamma T_0}{T}, \; \gamma \cong 28.5 \; ... \tag{5}$$

$\alpha_{sol-sol}$ is given here for the spin correlations perpendicular to the magnetic field, an additional reduction occurs for the parallel correlations.

3) For a realistic treatment of the spin chain one has to go beyond the SG approximation and to include fluctuations of the spin component perpendicular to the easy plane (out-of-plane fluctuations). This leads to a decrease in soliton energy and a corresponding increase in soliton density which to first order in $k_BT/(AJ)^{1/2}$ is given by /16,17/

$$I/I_0 \rightarrow I/I_0 \exp(3/\lambda^{1/2}), \quad \lambda = 2AS/\mu B \tag{6}$$

Changes in the soliton profile with decreasing anisotropy lead to further corrections /18/, the main effect, however, is an increase of the central peak intensity by a factor of the order of 3, as given by eq. (6).

A quantitative comparison between this more realistic theoretical approach and neutron scattering data suffers from the fact that these data are available only at comparatively high temperatures, e.g. $T \cong T_0$. This implies that the reduction factor α cannot be calculated reliably in the order which is available theoretically. The practice therefore has been to interpret the experimental results as measuring the total reduction factor α, as obtained from α_{sw}, $\alpha_{sol-sol}$ and the correction from out-of-plane fluctuations /19/. The values obtained ($\alpha_{yy} \cong 0.36$) are consistent with the theoretical results. Owing to the uncertainty from the higher order contributions to α, however, it is not possible to draw conclusions about the importance of further corrections to the theoretical picture.

Apart from the uncertainty in the treatment of out-of-plane fluctuations the most important remaining approximations are the neglect of the discrete lattice and of quantum effects. Whereas recently it has become clear theoretically that the effect of the discrete lattice is small at the magnetic fields applied in experiments so far /18,20/, very little is known quantitatively on quantum effects. From a semiclassical calculation of the soliton energy /21/ it has become clear that quantum corrections are more

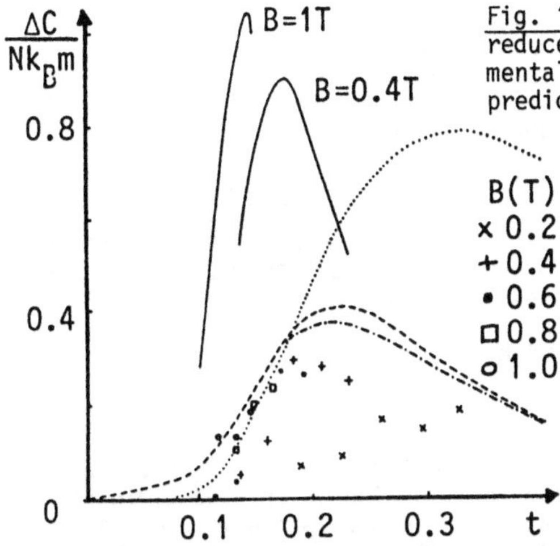

Fig. 1 Reduced magnetic specific heat vs reduced temperature $t = k_B T/8mJ$. Experimental points /9/ and various theoretical predictions (see text).

important for the realistic magnetic chain (including out-of-plane fluctuations) than for the planar (SG) case, but no calculations of this type exist for correlation functions.

A quantitative comparison between experiment and theory is much more promising for the soliton contribution to the magnetic specific heat, since the specific heat as a static quantity allows a considerably more complete theoretical treatment. The various stages of the theoretical approach are shown in Fig. 1 together with experimental points for $CsNiF_3$ /9/. The discussion /11,22/ parallels closely the preceding discussion for the central peak intensity:

1) The single soliton theory in the SG approximation at low temperatures leads to a specific heat maximum which is above the experimental values by a factor of the order of 3 (dotted curve in Fig. 1).

2) Finite temperature effects can be discussed sufficiently well by expansions to high order in T/J /23/ or numerically using the transfer matrix approach /24/ (dashed curve in Fig. 1). Thus the equivalent of the reduction factor α for the central peak intensity is known quantitatively.

3) The out-of-plane fluctuations characteristic for the realistic magnetic Hamiltonian can also be taken into account using the transfer matrix approach /25/. The result, which is confirmed by a Monte-Carlo calculation /26/, is an increase of the specific heat maximum by a factor of the order of 4 (full lines in Fig. 1), in qualitative agreement with the corresponding result for the central peak intensity. For the specific heat, however, the uncertainty in the finite temperature effects and from inadequacies in the treatment of out-of-plane fluctuations is absent and from the quantitative result one reaches the conclusion that the remaining approximation, i.e. in particular the neglect of quantum effects, must be responsible for the large discrepancy between theory and experiment.

On the other hand it is known from several independent approaches /27-29/ that quantum corrections to the specific heat are practically negligible in the frame of the SG approximation (dashed-dotted curve in Fig. 1). Thus the properties of the magnetic chain beyond the SG approximation, i.e. the presence of out-of-plane fluctuations, in combination with quantum effects must be responsible for the observed strong reduction of the specific heat. This conclusion is supported by the results of Quantum Monte Carlo computations for the Hamiltonian (1) with S=1, which give the magnitude of the magnetic specific heat in reasonable agreement with experiment /30/.

In an effort to understand the reduction of the specific heat in terms of the soliton picture we have performed a semiclassical calculation for the Hamiltonian (1) /22/. From previous work /21/ it is known that out-of-plane fluctuations have a strong influence on quantum corrections to the soliton energy and our calculation is an extension of that approach to finite temperatures. In the spirit of the semiclassical theory of the SG chain /31/ we expand in the coupling constant g^2, which in our case is given by

$$g^2 = (2A/JS(S+1))^{1/2} \tag{7}$$

and in the strength of the out-of-plane fluctuations, $1/\lambda = \mu B/2AS$. The free energy is calculated including the contribution of spin waves in the presence of a soliton; the important quantity is the spin wave phase shift, which so far can be calculated only for a static soliton and to first order in $1/\lambda$.

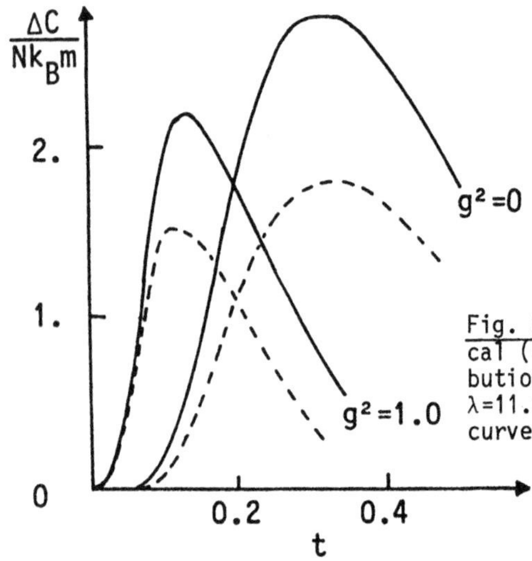

Fig. 2 Semiclassical (g^2=1.0) and classical (g^2=0) results for the soliton contribution to the reduced specific heat for λ=11.2 (broken curves) and λ=5.6 (solid curves)

For the details of the calculation we refer to the original work /22/; here we only discuss the results: For g^2 = o our approach reproduces the increase in the classical specific heat owing to out-of-plane fluctuations. For $g^2 \neq$ o a reduction of the specific heat from the classical result is obtained, which strongly depends on the strength of out-of-plane fluctuations $1/\lambda$, as shown in Fig. 2. On the basis of these results we expect that a complete treatment of the soliton degrees of freedom of the full magnetic Hamiltonian including its quantum nature will be able to describe the measured magnetic specific heat quantitatively in terms of the soliton picture; this quantitative agreement, however, cannot be expected from our semiclassical approach for the following reasons:

i) The expansion in $1/\lambda$ to first order is not sufficient for realistic values of λ /17/.

ii) The quantitative result suffers from the neglect of soliton motion and of the difference of the soliton spectrum in the SG and the magnetic chain.

iii) The renormalisation of the soliton energy from quantum effects becomes very strong at coupling constants $g^2 \cong O(1)$. This renormalisation leads to the striking shift of the peak position to lower temperatures as shown in Fig. 2. This shift will be compensated by a corresponding renormalisation of the parameters of the Hamiltonian (1). A quantitative treatment, however, which would also have to take into account terms of higher order in g^2, has not been possible so far.

We would also like to draw attention to the work of FOGEDBY et al. /32/ on quantum effects in the specific heat, which in principle parallels our approach. In this work, however, an expression for the soliton energy is used, which does not take into account the strong influence of out-of-plane fluctuations on this quantity as found in /21/.

Although evidently a quantitative theory is not yet available, we conclude from the present results that the soliton picture is consistent with experimental observations and that a quantitative treatment appears to depend on the solution of difficult technical problems. In particular we conclude:

(i) The combined effect of quantum corrections and out-of-plane fluctuations has to be treated in more detail for a complete theory of the magnetic specific heat.

(ii) The interpretation of neutron scattering results as measuring the reduction coefficient α may be inadequate, since the closely parallel discussions of specific heat and central peak intensity suggest an important influence of quantum effects also on correlation functions.
More work is clearly required before final conclusions can be drawn.

References

1. H.J. Mikeska: J. Phys. C11, L29 (1978)
2. H.J. Mikeska: J. Phys. C13, 2913 (1980); K.M. Leung, D.W. Hone, D.L. Mills, P.S. Riseborough and S.E. Trullinger: Phys. Rev. B21, 4017 (1980)
3. J.K. Kjems and M. Steiner: Phys. Rev. Letters 41, 1137 (1978); M. Steiner, K. Kakurai and J.K. Kjems: Z. Phys. B53, 117 (1983)
4. L.P. Regnault, J.P. Boucher, J. Rossat-Mignod, J.P. Renard, J. Bouillot and W.G. Stirl ng: J. Phys. C15, 1261 (1982)
5. J.P. Boucher and J.P. Renard: Phys. Rev. Letters 45, 486 (1980)
6. H. Benner, H. Seitz, J. Wiese and J.P. Boucher: J. Magn. Magn. Mat. 45, 354 (1984)
7. L.J. de Jongh: J. Appl. Phys. 53, 8018 (1982)
8. H.J.M. de Groot and L.J. de Jongh: Physica 141B, 1 (1986)
9. A.P. Ramirez and W.P. Wolf: Phys. Rev. Letters 49, 227 (1982); Phys. Rev. B32, 1639 (1985)
10. F. Borsa: Phys. Rev. B25, 3430 (1982)
11. K. Kopinga, A.M.C. Tinus and W.J.M. de Jonge: Phys. Rev. B29, 1924 (1984); Phys. Rev. B32, 3154 (1985)
12. J.A. Krumhansl and J.R. Schrieffer: Phys. Rev. B11, 3535 (1975)
13. E. Allroth and H.J. Mikeska: Z. Phys. B43, 209 (1981)
14. G. Reiter: Phys. Rev. Letters 46, 202 (1981); Erratum 46, 518 (1981)
15. K. Sasaki: Dynamical Problems in Soliton Systems, ed. S. Takeno (Springer, Berlin, Heidelberg 1985)

16. H.J. Mikeska: J. Appl. Phys. 52, 1950 (1981)
17. K. Osano: J. Phys. C17, L511 (1984)
18. C. Etrich: Ph.D. thesis, Universität Hannover (1986)
19. K. Kakurai and M. Steiner: In Proc. of the Int. Conf. on Neutron Scattering, Santa Fé (1985)
20. C. Etrich, H.J. Mikeska, E. Magyari, H. Thomas and R. Weber: Z. Phys. B62, 97 (1985)
21. H.J. Mikeska: Phys. Rev. B26, 5213 (1982)
22. H.J. Mikeska and H. Frahm: J. Phys. C19 (1986) in print
23. K. Sasaki and T. Tsuzuki: Solid State Commun. 41, 521 (1982)
24. T. Schneider and E. Stoll: Phys. Rev. B22, 5317 (1980)
25. M.G. Pini and A. Rettori: Phys. Rev. B29, 5246 (1984)
26. O.G. Mouritsen, H. Jensen and H.C. Fogedby: Phys. Rev. B30, 498 (1984)
27. M. Fowler and X. Zotos: Phys. Rev. B25, 5806 (1982); M.D. Johnson and N.F. Wright: Phys. Rev. B32, 3169 (1985)
28. T. Tsuzuki: Progr. Theor. Phys. Suppl. 80, 146 (1985)
29. R. Giachetti and V. Tognetti: Phys. Rev. Letters 55, 912 (1985)
30. G.M. Wysin: Ph.D. thesis, Cornell University (1985)
31. R.F. Dashen, B. Hasslacher and A. Neveu: Phys. Rev. D11, 3424 (1975)
32. H.C. Fogedby et al.: Preprint, Univ. of Aarhus (1985)

Quantum Corrections to the Specific Heat of the Easy-Plane Ferromagnetic Chain [1]

H.C. Fogedby, K. Osano [2], and H.J. Jensen [3]

Institute of Physics, University of Aarhus, DK-8000 Aarhus C, Denmark

1. Introduction

One-dimensional non-linear soliton-bearing condensed matter systems have received much attention in recent years [1]. In this context the properties of the magnetic salts $CsNiF_3$ and $(C_6H_{11}NH_3)CuBr_3$ (CHAB) are of particular interest. In the long wavelength limit, $k \to 0$, both these salts are described by the one-dimensional easy-plane continuum ferromagnetic spin model [2,3]

$$H = \int dx \left[\frac{1}{2} j \left(\frac{d\bar{S}}{dx}\right)^2 + a(S^z)^2 - hS^x \right] . \tag{1.1}$$

To leading order in the lattice constant this model is obtained from the corresponding discrete spin-1 and spin-1/2 Hamiltonians for $CsNiF_3$ and CHAB. Choosing the lattice constants as length units, the conventional model parameters for $CsNiF_3$ and CHAB are [2,4]

$$j/k = 23.6\text{K}, \quad a/k = 4.5\text{K}, \quad h/k = .16H\ \text{K}\ (H \text{ in kG}) \tag{1.2}$$

and [3,5]

$$j/k = 110.\text{K}, \quad a/k = 5.5\text{K}, \quad h/k = .13H\ \text{K}\ (H \text{ in kG}) \tag{1.3}$$

For a purely <u>classical</u> spin field, i.e., the limit $S \to \infty$ keeping jS^2, aS^2, and hS fixed, the non-linear model in eq. (1.1) is <u>soliton bearing</u> [2] and has been the subject of numerous investigations, see e.g. [6]. Most recently, specific heat data obtained for $CsNiF_3$ [7] and CHAB [3] have, in the absence of specific heat calculations for the discrete models, been compared with the theoretical predictions for the

---

1. Supported in part by grants from the Danish Natural Science Research Council.

2. Present address: Department of Physics, Tohoku University, Sendai 980, Japan.

3. Present address: Institute for Materials Research, McMaster University, 1280 Main Street West, Hamilton, ONT L8S 4M1, Canada.

classical Sine-Gordon model [8], which is taken as an approximation for eq. (1.1) in the limit of large anisotropy [2], and the observed peak features have been attributed to the non-linear soliton excitations of that model [2,8].

The anisotropy parameters for $CsNiF_3$ and CHAB are, however, not large enough to justify the use of the Sine-Gordon model [5,9,10,11] and a variety of investigations have examined the importance of out-of-plane fluctuations [12-16]. The discrepancy between the experimental data and the classical predictions, however, remains large and it has become clear [5,12-16] that quantum effects must be included in order to approach quantitative agreement.

Here we report the results of a semi-classical evaluation of the quantum corrections to the specific heat of $CsNiF_3$ and CHAB. The calculation is carried out within the ideal gas soliton picture [8] and is based on the continuum model in eq. (1.1).

2. Classical Calculation of the Specific Heat

We take as our starting point the steepest descent analysis [6] of the classical partition function $Z = \int \Pi d^3S \, \exp(-H/kT)$. In the low-temperature limit the leading contributions to Z arise from static linear spin wave configurations and thermally excited static non-linear soliton configurations, corresponding to the local minima and their neighbourhoods of the Hamiltonian H in phase space $\{\bar{S}(x)\}$. Introducing a local spin frame [6] associated with the soliton, determined by the static Euler-Lagrange equation $\delta H/\delta\bar{S}=0$, and considering small in-plane and out-of-plane fluctuations about the soliton, we obtain the quadratic form.

$$H = E_0 + \frac{1}{2} \int dx \left[\Delta\phi M_{in} \Delta\phi + \Delta S^z M_{out} \Delta S^z\right] . \quad (2.1)$$

Here $E_0=8S(jhS)^{1/2}$ is the soliton energy [2,8], that is the soliton gap. The azimuthal angle $\Delta\phi$ and the longitudinal spin component ΔS^z, characterizing the in-plane and out-of-plane fluctuations, form action-angle variables satisfying the Poisson bracket $\{\Delta S^z(x),\Delta\phi(y)\} = (1/\hbar)\delta(x-y)$. For later purposes and in accordance with the correspondance principle [17] we have introduced $\hbar$ as unit of action. The stability matrices M_{in} and M_{out} for in-plane and out-of-plane fluctuations have the form [6]

$$M_{in} = jS^2\left[- \frac{d^2}{dx^2} + \kappa^2(1-2/\cosh^2(\kappa x))\right] \quad (2.2a)$$

$$M_{out} = j[-\frac{d^2}{dx^2} + 2\alpha + \kappa^2(1-6/\cosh^2(\kappa x))] \quad , \tag{2.2b}$$

where $\kappa=(h/jS)^{1/2}$ is the inverse correlation length and $\alpha=a/j$ the reduced anisotropy parameter. We also define the derived strength parameter $\beta=2aS/h$. The evaluation of Z is carried out by inserting eq. (4) in Z, diagonalizing the stability matrices M_{in} and M_{out} independently, and performing the Gaussian integrations over the static in-plane and out-of-plane fluctuations. We find to leading order the following expressions for the classical specific heat per spin, $c=-Td^2(F/N)/dT^2$, $F=-kT\log Z$ (N is the number of spins):

$$c = k + k(8\pi)^{-1/2}g(\beta)(E_0/jS^2)(E_0/kT)^{5/2}\exp(-E_0/kT) \tag{2.3a}$$

$$g(\beta) = [1 + (1+\beta)^{1/2}][2 + (1+\beta)^{1/2}]/[\beta-3]^{1/2}\beta^{1/2} \quad . \tag{2.3b}$$

For large anisotropy $\beta>>1$, $g(\beta)=1+3\beta^{-1/2}$, the out-of-plane fluctuations are suppressed and we obtain the <u>classical</u> Sine-Gordon result for c, <u>independent</u> of a Ref. [8].

3. <u>Quantum Calculation of the Specific Heat</u>

In the quantum mechanical case the aim is to evaluate the <u>quantum</u> partition function $Z=\mathrm{Tr}[\exp(-H/kT)]$, ideally by diagonalizing H and performing the trace. This is, however, a much more difficult problem and progress beyond linear spin wave theory has been slowly and most restricted to the Sine-Gordon model [18-25]. Here we limit ourselves to a <u>semi-classical</u> evaluation of Z valid for <u>large</u> S. In the ideal gas approximation [6,8] which holds for $E_0<<kT$ it is justified to treat the soliton as a <u>classical</u> excitation and only quantize the linear vibration spectrum about the soliton, characterized by the quadratic form in eq. (2.1). To leading order in $\hbar$ the action-angle operators ΔS^z and $\Delta\phi$ satisfy the commutator $[\Delta S^z(x),\Delta\phi(y)]= (-i)\delta(x-y)$. Diagonalizing the form in eq. (2.1) by means of a unitary transformation, the Hamiltonian H is represented by a set of quantum oscillators

$$H = E_0 + \frac{1}{2}\sum_n [|p_n|^2 + \omega_n^2|q_n|^2] \tag{3.1}$$

where $[p_n,q_m]=(-i)\hbar\delta_{nm}$. The density of oscillator states, the eigenfunctions Ψ_n, and eigenfrequencies ω_n are given by the Hermitian eigenvalue problem

$$M_{out}^{1/2}M_{in}M_{out}^{1/2}\Psi_n = (\hbar\omega_n)^2\Psi_n \tag{3.2}$$

which, of course, also is obtained [26] by assuming a harmonic time dependence in the equations of motion following from eq. (2.1).

The spectrum has the form [26] $\{0,\omega,\omega_k\}$. The zero frequency mode with coordinate and momentum q_0 and p_0 corresponds to the classical translational motion of the soliton [6,8]. Noting that the eigenstate $M_{out}^{1/2}\Psi_0$ is proportional to $d\phi/dx$ where $\phi=4\tan^{-1}(\exp(\kappa x))$ is the static soliton (kink) solution, it is straightforward to determine the soliton mass [27,28] (Ψ' is the trigamma function)

$$M = \frac{64(S\hbar)^2}{E_0(\beta-3)}\left[\frac{3}{4}+\frac{1}{8}\frac{\beta}{(1+\beta)^{1/2}}\Psi'\left(\frac{1}{2}+\frac{1}{2}(1+\beta)^{1/2}\right)\right] \qquad (3.3)$$

The discrete finite frequency solution is a localized vibration or precession mode of the soliton, essentially the dynamical counterpart of the static rotational and meridional modes in the steepest descent analysis [6]. The energy of the mode lies below the spin wave band

$$\hbar\omega_k = jS[(k^2+\kappa^2+2\alpha)(k^2+\kappa^2)]^{1/2} \qquad (3.4)$$

and is to leading order in β^{-1} given by [26]

$$\hbar\omega = \hbar\Delta\left(1-\frac{8}{9}\beta^{-2}\right), \qquad (3.5)$$

where $\hbar\Delta=\hbar\omega_{k=0}$ is the spin wave gap. The spin waves are phase shifted by the soliton in the usual manner [6,8] giving rise to a reduction of the density of free spin wave states, $\varrho_k=N/2\pi+\Delta\varrho_k$. By means of a distorted wave approximation equivalent to the method in [26] about the Sine-Gordon limit $\beta=\infty$ we have within the Born approximation [17] determined the symmetrized part $\Delta\varrho_k$ to leading order in β^{-1} [29]

$$\Delta\varrho_k = \frac{1}{2\pi}\frac{d}{dk}(\delta_0(k)+\delta_1(k)) \qquad (3.6a)$$

$$\delta_0(k) = 2\tan^{-1}(\kappa/k) \qquad (3.6b)$$

$$\delta_1(k) = 2\,\frac{k^2+\frac{1}{3}\kappa^2}{k^2+\alpha+\kappa^2}\,\frac{\kappa}{k}\,. \qquad (3.6c)$$

Here the Sine-Gordon phase shift δ_0 holds for all k and approaches $\pm\pi$ for $k\to 0^{\pm}$, corresponding to the removal of the translation mode from the spin wave band [6,8], whereas δ_1, evaluated in the Born approximation, only is valid for large k.

However, invoking Levinson's theorem [8] δ_1 must also approach $\pm\pi$ for $k \to 0^{\pm}$, corresponding to the localized mode emerging below the band, and we obtain $\int dk\Delta\varrho_k = -2$.

The physical momentum of the soliton is $p = M^{1/2} p_0$, introducing furthermore the Bose occupation number operators n and n_k for the localized mode and the spin waves, and the density of states ϱ_k we can express H in eq. (3.1) in the form

$$H = E_0 + p^2/2M + \hbar\omega(n+\gamma) + \int dk\varrho_k \hbar\omega_k(n_k+\gamma_k) \quad . \tag{3.7}$$

The localized mode and spin wave zero point energies $\hbar\omega\gamma$ and $\hbar\omega_k\gamma_k$ are <u>undetermined</u> parameters within the semi-classical quantization scheme [17]. Inserting the density of states ϱ_k and rearranging eq. (3.7), we finally obtain the effective Hamiltonian describing the interaction between the localized mode, the spin wave gas and a single soliton

$$H = \Delta_{sw} + E + p^2/2M + \hbar\omega n + N \int \frac{dk}{2\pi} \hbar\omega_k n_k + \int dk\Delta\varrho_k \hbar\omega_k n_k \quad . \tag{3.8}$$

Here $\Delta_{sw} = N\int(dk/2\pi)\hbar\omega_k\gamma_k$ is the spin wave contribution to the ground state energy. This term diverges in the thermodynamic limit $N \to \infty$ and is usually removed in field theoretical treatments by means of an appropriate normal ordering [26]. In the present context Δ_{sw} does not contribute to the specific heat c and we set $\Delta_{sw}=0$. More importantly, the localized mode and the phase-shifted spin waves give rise to a correction to the soliton gap,

$$E = E_0 + \hbar\omega\gamma + \int dk\Delta\varrho_k \hbar\omega_k\gamma_k \quad . \tag{3.9}$$

Since the occupation number operators n and n_k are constants of motion, the Hamiltonian H in eq. (3.8) is in diagonal form and it is straightforward to evaluate the quantum function Z. In accordance with the ideal gas phenomenology [2,8], we treat the the solitons as a Boltzmann gas of particles with gap energy E and mass M. On the other hand, the local vibration mode associated with the soliton and the spin wave gas are treated according to Bose statistics. Noting also that the soliton has two parities, corresponding to the kink and antikink solutions [2,8] in terms of the phase variable ϕ, accounting for the indistinguishability by means of a Gibbs factor 1/n!, integrating over soliton positions, and including the classical phase space volume $2\pi\hbar$ per soliton, we have

$$\begin{aligned} Z = & \exp([-N \int \frac{dp}{2\pi} \log(1-\exp(-\hbar\omega_p/kT))] \\ & \sum_{n=0}^{\infty} (1/n!) 2^n N^n (2\pi MkT)^{n/2} (2\pi\hbar)^{-n} \exp(-nE/kT) \\ & [1-\exp(-\hbar\omega/kT)]^{-n} \exp[-n\int dp\Delta\varrho_p \log(1-\exp(-\hbar\omega_p/kT))] \end{aligned} \tag{3.10}$$

in complete analogy with the classical expression derived in [12, 13]. For the *quantum* specific heat per spin we finally obtain $c=c_{sw}+c_{sol}$, where the spin wave part

$$c_{sw} = k\int \frac{dp}{2\pi}\left[\frac{\hbar\omega_p/2kT}{\sinh(\hbar\omega_p/2kT)}\right]^2 \qquad (3.11)$$

and to leading order [6,12] the soliton part

$$c_{sol} = k\,\frac{2}{2\pi\hbar}\,(2\pi MkT)^{1/2}[1-\exp(-\hbar\omega/kT)]^{-1}(E/kT)^2$$
$$\exp(-E/kT)\exp[-\int dp\Delta\varrho_p \log(1-\exp(-\hbar\omega_p/kT))] \quad . \qquad (3.12)$$

The above expressions for c_{sw} and c_{sol} together with eqs. (3.3)-(3.6a-c) for M, ω_p, ω and $\Delta\varrho_p$ are expected to hold in the semi-classical - ideal gas - long wavelength - large anisotropy limit, that is for $S\gg1$, $T\ll E/k$, $\kappa\ll1$, and $\beta\gg1$.

4. Discussion

The *classical limit* is attained by letting $S\to\infty$ keeping jS^2, κ, and α fixed. Consequently, $\hbar\omega\to0$, $\hbar\omega_p\to0$, and $E\to E_0$, and c takes the form

$$c=k+k\,\frac{2}{2\pi\hbar}\,(2\pi MkT)^{1/2}\,\frac{kT}{\hbar\omega}\,\left(\frac{E_0}{kT}\right)^2\exp\left[-\int dp\Delta\varrho_p\log\left(\frac{\hbar\omega_p}{kT}\right)\right]\exp\left(-\frac{E_0}{kT}\right) \qquad (4.1)$$

in accordance with the classical ideal gas calculation [14] of c. Closer inspection also shows that c given by eq. (4.1) has the same form as the steepest descent result in eqs. (2.3a-b), thus demonstrating the complete equivalence between the static *steepest descent method* [12,13] and the *dynamical ideal gas approach* [8,14]. In the *Sine-Gordon* limit $\beta\to\infty$, $\hbar\omega\to\hbar\Delta$, $\hbar\omega_p\to jS(2\alpha)^{1/2}(p^2+\kappa^2)^{1/2}$, $M\to4\hbar^2\kappa/a$, $\Delta\varrho_k\to-1/\pi)\kappa/(k^2+\kappa^2)-\delta(k)$ and we readily obtain the Sine-Gordon result [8] for $g(\beta)\to1$. Inserting eq. (3.4) for $\hbar\omega_p$ and eqs. (3.6a-c) for $\Delta\varrho_k$ we have also identified the leading correction to $g(\beta)$, i.e. $g(\beta)=1+3\beta^{-1/2}$.

In the *large anisotropy* limit $\alpha\to\infty$ the spin wave band $\hbar\omega_p$ diverges and the spin wave contribution to the *quantum* specific heat given by eqs. (3.11) and (3.12) vanishes exponentially. The soliton mass M goes to zero as α^{-1} and the soliton contribution to c behaves like $T^{-3/2}\alpha^{-1/2}\exp(-E/kT)$. Since the shifts γ and γ_k are undetermined within the presence scheme it is difficult to assess the behaviour of E except for the thermodynamic stability requirement that E must be positive.

However, choosing $\gamma=\gamma_k=1/2$ (harmonic oscillator values) we obtain for the <u>renormalized</u> soliton gap

$$E = E_0[1-(2\alpha)^{1/2}\log(2\pi/\kappa)/8S\pi] \ . \tag{4.2}$$

To the extent that we can compare, the expression in eq. (4.2) is to leading order in partial agreement with the <u>quantum Sine-Gordon</u> treatments in Refs. [18-20], and [26]. We believe the discrepancy is due to the difficulty of mapping the easy-plane magnet onto the quantum Sine-Gordon model [30].

In a recent paper Mikeska and Frahm [31] have also discussed the specific heat of the easy-plane ferromagnet within a semi-classical quantization scheme. Drawing on the results in [26], their analysis is parallel to the present one. However, since they do not provide explicit expressions for the specific heat, we have been unable to make a detailed comparison. Comparing our data with their specific heat plots, we find some disagreement which we believe is due to a different treatment of phase shift effects.

Choosing $\gamma_k=\gamma=1/2$ for the spin waves and the localized mode, the small-k parametrization $\delta_1(k)\to 2\tan^{-1}[\delta_1(k)/2]$ satisfying Levinson's theorem, and parameters appropriate for $CsNiF_3$ given in eq. (1.2) we have in Fig. 1 depicted the soliton contribution to the classical and quantum expressions for the specific heat. The figure clearly shows the considerable reduction in c_{sol} caused by the inclusion of quantum effect albeit only in the semi-classical approximation.

In Fig. 2 we have shown $\Delta c=c_T(H)-c_T(0)$ using parameters for $CsNiF_3$ at two temperatures (in order to account for the uncertainty in c(H) in the gapless low-field region we have calibrated the theoretical expressions to the experimental data in

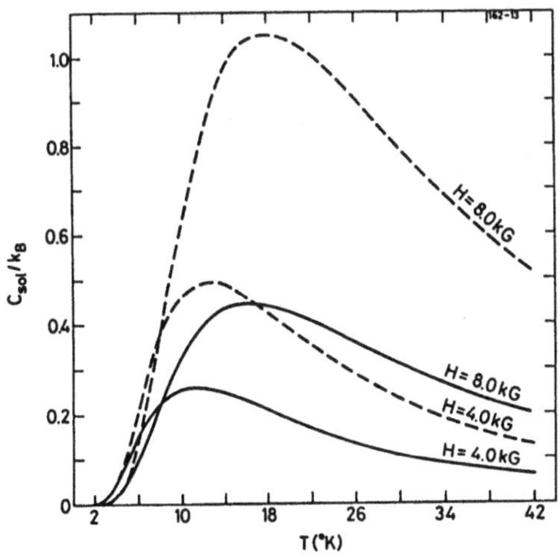

<u>Fig. 1.</u> Soliton contribution c_{sol} to the specific heat as a function of temperature for two different fields. The dashed curves show the <u>classical</u> results; the full-drawn curves the <u>quantum</u> results.

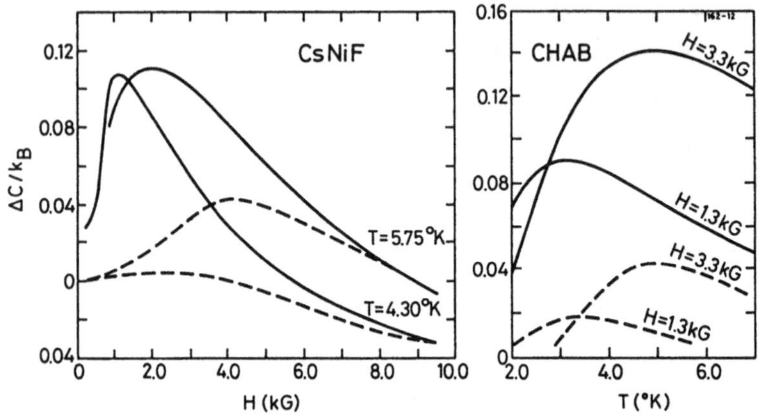

Fig. 2. Specific heat differences $\Delta c=c_T(H)-c_T(O)$ as a function of magnetic field for two different temperatures. The dashed curves are the experimental results; the full-drawn curves the theoretical quantum results. Specific heat differences $\Delta c=c_H(T)-c_{H=O}(T)$ as a function of temperature for two different fields. The dashed curves are the experimental results; the full-drawn curves the theoretical quantum results.

the high-field limit) and $\Delta c=c_H(T)-c_{H=O}(T)$ in the case of CHAB for two fields.

5. Conclusion

We have studied the low temperature thermodynamics of the easy-plane ferromagnet in the semi-classical approximation, including non-linear soliton excitations within the ideal gas scheme. The quantum specific heat is reduced compared to the corresponding classical result and leads to a better qualitative fit to experiments. The remaining substantial quantitative discrepancy is, of course, to be expected given the approximations in the present scheme. In that context we note that the quantum Sine-Gordon treatment in [20] yields a better fit to experiments despite the neglect of out-of-plane fluctuations. Thermodynamics sample fluctuation at all wavelengths and it might be a severe limitation to employ a continuum description. Also, the quantum spin values for $CsNiF_3$ and CHAB are S=1 and S=1/2 and the semi-classical approximation valid for large S probably underestimates quantum effects. Finally, we are only including solitons plus linear spin waves; spin wave - spin wave interaction effects are entirely ignored. We nevertheless believe that the present calculation has a certain methodological interest and shows a significant trend as regards comparison with experiments. We also think that analytical attempts to proceed beyond a semi-classical scheme, including the non-linear effects necessary to describe the specific heat anomalies, will prove very difficult. In that context it is of particular interest that quantum Monte Carlo calculations for the easy-plane discrete spin-1 model are now emerging [32], providing a theoretical calibration.

References

1) Physics in One Dimension. Ed. by J. Bernasconi and T. Schneider (Springer Verlag, Heidelberg, 1981)
2) H.J. Mikeska, J.Phys. C 11, L29 (1978).
3) K. Kopinga, A.M.C. Tinus, and W.J.M. de Jonge, Phys.Rev. B 29, 2868 (1984).
4) J.K. Kjems and M. Steiner, Phys.Rev.Lett. 41, 1137 (1978).
5) A.M.C. Tinus, W.J.M. de Jonge, and K. Kopinga, Phys.Rev. B 32, 3154 (1985).
6) H.C. Fogedby, P. Hedegård, and A. Svane, Physica 132 B, 17 (1985).
7) A.P. Ramirez and W.P. Wolf, Phys.Rev.Lett. 49, 227 (1982) and Phys.Rev. B 32, 1639 (1985).
8) J.F. Currie, J.A. Krumhansl, A.R. Bishop, and S.E. Trullinger, Phys.Rev. B 22, 477 (1980).
9) J.M. Loveluck, T. Schneider, E. Stoll, and H.R. Jauslin, Phys.Rev.Lett. 45, 1505 (1980) and J.Phys. C 15, 1721 (1982).
10) R. Tammetta and J. Oitmaa, J.Phys. C 16, L809 (1983).
11) T. Schneider and E. Stoll, J.Appl.Phys. 53, 8024 (1982).
12) H.C. Fogedby, P. Hedegård, and A. Svane, Phys.Rev. B 28, 2893 (1983); ibid. Phys.Rev. B 29, 2861 (1984).
13) K.M. Leung and A.R. Bishop, J.Phys. C 16, 5893 (1983).
14) K. Osano, J.Phys. C 17, L511 (1984).
15) O.G. Mouritsen, H.J. Jensen, and H.C. Fogedby, Phys.Rev. B 30, 498 (1984).
16) M.G. Pini and A. Rettori, Phys.Rev. B 29, 5246 (1984).
17) L.D. Landau and E.M. Lifshitz: Quantum Mechanics (Pergamon Press, London 1959).
18) K. Maki, Phys.Rev. B24, 3991 (1981).
19) M. Fowler and X. Zotos, Phys.Rev. B 25, 5806 (1982).
20) M.D. Johnson and N.F. Wright, Phys.Rev. B 32, 5798 (1985).
21) T. Tsuzuki, Prog.Theor.Phys. 70, 975 (1983); ibid. Suppl. 80, 146 (1984).
22) R. Giachetti and V. Tognetti, Phys.Rev.Lett. 55, 912 (1985).
23) M. Moraldi, A. Rettori, and M.G. Pini, Phys.Rev. A 31, 1971 (1985).
24) M. Imada, K. Hida, and M. Ishikawa, J.Phys. C 16, 35 (1983).
25) N. Flüggen, Diplom Thesis, Hannover (1984).

26) H.J. Mikeska, Phys.Rev. B 26, 5213 (1982).
27) H.J. Mikeska, Physica 120 B, 235 (1982).
28) H.J. Mikeska and K.Osano, Z.Physik B 52, 111 (1983).
29) Only the symmetrized part of $\Delta\varrho_k$ enters in the evaluation of Z. The mode counting in the present case of a reflecting potential is a little subtle and will be discussed in another context. See also S.E. Trullinger and R.M. DeLeonardis, Phys.Rev. A 20, 2225 (1979) and [26].
30) J. Timonen and A. Luther, J.Phys. C 18, 1439 (1985).
31) H.J. Mikeska and H. Frahm, preprint (1985).
32) G.M. Wysin, Ph.D. Thesis, Cornell University (1985).

Monte Carlo Calculations of Quantum Systems

H. De Raedt[1], *A. Lagendijk*[2], *and P. de Vries*[2]

[1]Physics Department, University of Antwerp (U.I.A.), Universiteitsplein 1, B-2610 Wilrijk, Belgium

[2]Natuurkundig Laboratorium, University of Amsterdam, Valckenierstraat 65, NL-1018 XE Amsterdam, The Netherlands

1. INTRODUCTION

The thermodynamic behavior of classical systems can be studied by means of analytical methods and computer simulations, e.g. high-temperature expansions and renormalization techniques. In the last twenty years much progress has been achieved. Classical systems under study include Ising models, hard sphere gases, fluids etc. The situation for quantum systems is different. The only method which is generally applicable is the high-temperature expansion, but this is in many cases not sufficient and certainly not sufficient to describe for example phase transitions. Today, because of modern developments, computer simulations of the thermodynamics of quantum systems can be performed in a number of cases.

In this paper we discuss a $T\neq 0$ Monte Carlo technique for quantum statistical models. (This technique is complementary to the $T=0$ Monte Carlo methods, see e.g. CEPERLY and KALOS [1]. In the following section we outline the basic concepts and mathematical ingredients underlying the method. We illustrate the general strategy by considering explicit examples in the subsequent sections. Topics treated include the two-dimensional spin-½ XY model, a two-level system with non-linear coupling and bipolaron formation.

2. STATISTICAL MECHANICS

2.1 Classical systems

In statistical mechanics one wants to calculate thermodynamic properties of many-body systems starting from a microscopic point of view. The thermodynamical equilibrium values of physical quantities (like the energy) are given by:

$$\langle A\rangle = \frac{\int d^N p \int d^N q \; A(\{p\},\{q\}) e^{-\beta H(\{p\},\{q\})}}{\int d^N p \int d^N q \; e^{-\beta H(\{p\},\{q\})}} , \tag{1}$$

where H is the Hamiltonian of the system and β is inversely proportional to the temperature T. In (1) A is a quantity which depends on the positions and momenta of the particles. Formula (1) gives the mean value of A in phase space, where each phase point $\Gamma=(\{p\},\{q\})$ has a weight $\exp(-\beta H(\Gamma))$.

The idea behind computer simulations is in order to give a description of the properties of a system. It usually is sufficient to take a few hundred particles instead of a number of the order 10^{23}. Further, one can make use of statistical properties. Namely, that a large part of the phase space does not contribute (significantly) to result (1). A computer simulation then proceeds as follows: one takes a limited number of particles (or more generally degrees of freedom) and one generates a subset of all possible configurations (realizations of microstates). In practice one uses importance sampling techniques (BINDER [2]), where a configuration is generated with a probability determined by $\exp(-\beta H)$ (see (1)). Configurations that hardly contribute to $\langle A\rangle$ are consequently (almost) not generated at all. This tremendous simplification (the Metropolis-method) is in principle always possible because the weight factor $\exp(-\beta H)$ is always positive.

2.2 Quantum systems

If we want to apply these concepts to quantum systems several fundamental problems arise. The quantum mechanical version of (1) is given by:

$$\langle A\rangle = \frac{\mathrm{Tr}(Ae^{-\beta H})}{\mathrm{Tr}(e^{-\beta H})} . \tag{2}$$

In (2) A and H are not real numbers, but linear operators (think of matrices). Tr implies the calculation of the trace, that is the summation of all diagonal matrix elements. The evaluation of matrix elements of the exponent of an operator (defined by its Taylor expansion) generally requires the diagonalization of the operator in question. For the Hamiltonian H a diagonalization is equivalent to requiring the exact solution. This implies that e.g. for quantum spin systems containing more than about ten spins the direct application of computer simulation methods is impossible.

A possible solution to this problem is the use of path-integral methods. The starting point of the method is the Trotter product formula:

$$e^{A+B} = \lim_{m\to\infty} (e^{A/m}e^{B/m})^m . \tag{3}$$

This equation also holds if the operators A and B do not commute. Equation (3) can easily be generalized to the case of more than two non-commuting operators. The key point being that in many cases A and B can be diagonalized separately. There is a freedom of choice in the way one breaks up the Hamiltonian. If the Hamiltonian is written as a sum of kinetic and potential energy one eventually gets the well-known Feynman path integral [3]. In the case of spin-Hamiltonians such a decomposition is not applicable. One then can use some kind of real-space break-up, e.g. the Hamiltonian is written as a sum of pair interactions in which single-site interactions are partitioned equally over the pair interactions [4]. The resulting path-integral representations are obtained by introducing the appropriate resolutions of the identity operator and working out all matrix elements, e.g. for the partition function:

$$\begin{aligned} Z &= \mathrm{Tr}(e^{-\beta H}) \\ &= \mathrm{Tr}(\text{product of operators}) \\ &= \lim_{m\to\infty} \sum_{\{i_1\}} \cdots\cdots \sum_{\{i_m\}} (\text{product of matrix elements}) \;, \end{aligned} \tag{4}$$

where the summations are over complete sets of states.

In practice one can never take the limit $m\to\infty$, so taking a finite value for m constitutes the only approximation being made. The implementation of representations like (4) on a computer, however, can still give rise to complications. Owing to a high symmetry in the system many numbers in a product (the matrix elements of the density operator) may be zero, thus leading to an extremely inefficient use of computer time. An even more serious problem is the possibility that these numbers may become negative. This would impede the application of importance sampling techniques, because the statistical weights would not all be positive. Sometimes a canonical transformation removes these non-positive weights. Negative weights are a (still largely unresolved) problem in fermion systems with dimension two or higher.

How does a Monte Carlo calculation proceed given a model Hamiltonian ?

1) Choose the appropriate "trottering" of the Hamiltonian. (In principle any break-up will converge).
2) Determine the matrix-elements by means of resolutions of the identity operator. If there are many zero's or negative signs try to find a transformation to remove them.
3) Perform exact finite-size calculations (diagonalize the Hamiltonian).
4) Calculate quantities like the energy etc. in the weak and/or strong coupling limit.
5) Perform rigorous finite-m summations. Study the convergence as function of m with the help of 3 and 4. (In practice the convergence depends weakly on the system size).
6) Check the Monte Carlo algorithm on the points 3, 4 and 5. The ergodicity of the algorithm is crucial for obtaining correct results.
7) Start of Monte Carlo calculations.

3. Two-dimensional quantum spin-1/2 XY model

In this section we consider a two-dimensional (2D) spin system with nearest-neighbor interactions. The Hamiltonian is given by:

$$H = \sum_{<i,j>} H_{i,j} = -J \sum_{<i,j>} (\sigma_i^x \sigma_j^x + \sigma_i^y \sigma_j^y) , \tag{5}$$

where σ_i is the Pauli spin-operator at site i. It is known that in this 2D spin-1/2 XY model there cannot exist long-range order at non-zero temperature, although high-temperature series expansions suggest the existence of critical behavior. For classical versions of the model (planar rotator models) it is well established that although there is no spontaneous magnetization there is a phase transition at non-zero temperature [5]. This phase transition is attributed to topological excitations in the system. There are certain configurations (spin vortices) that are important. At low temperature there are few, so called, vortex anti-vortex pairs, at higher temperatures more of such pairs are created. The unbinding of pairs of vortices is responsible for the occurrence of a phase transition.

Little is known about the quantum model mainly because most techniques which have been succesfully applied to the classical model are very difficult to apply to the quantum case. The thermodynamical properties of this particular spin model have been calculated using a generalized Trotter formula in which the Hamiltonian is written in its real-space decomposition, see DE RAEDT et al [6]. The m-th approximation to the partition function is

$$Z_m = \mathrm{Tr}[\prod_{<i,j>} \exp(-\beta H_{i,j}/m)]^m . \tag{6}$$

The basic idea of this approach is to calculate the thermodynamical properties of approximation (6) and to study convergence of the results as a function of m. Two of us have given a rigorous solution for the m=1 approximation [7]. Central in this calculation is the mapping of this approximant onto a staggered 8-vertex model which can be solved analytically. The result is an Ising-like phase transition (i.e. no spontaneous magnetization but a logarithmic divergence of the specific heat).

Extending the staggered 8-vertex formulation to m>1 is not useful if Monte Carlo simulation methods have to be used, because of the occurrence of zeroes in the products of matrix elements in (6). By transforming the XY model into the XZ model a representation is found in which all relevant zeroes are removed. The m-th approximant to the partition function of the two-dimensional spin-1/2 model can now be rewritten as a three-dimensional Ising model with complicated many-spin interactions. The couplings depend

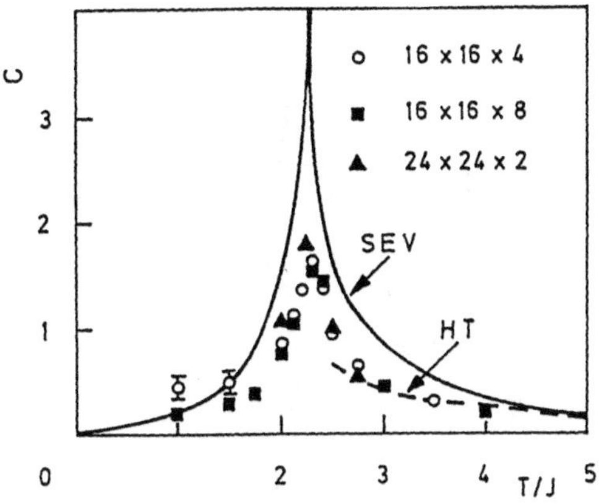

Fig. 1. Specific heat per site. Solid line: rigorous staggered 8-vertex solution; broken line: high-temperature expansion

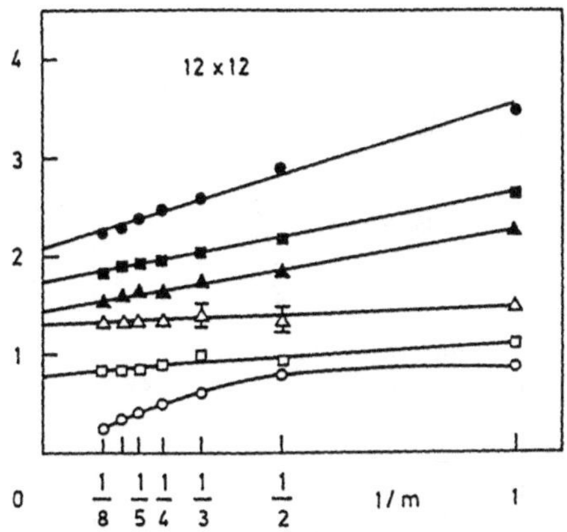

Fig. 2. Minus the energy per site (full symbols), the specific heat per site (open symbols). Circles: T/J=1; squares: T/J=2 and triangles: T/J=2.25

on the lattice size m in the new "Trotter" direction. There still are constraints on the possible configurations, but by introducing string variables (products of rows or columns of spins) an efficient and ergodic Monte Carlo algorithm can be constructed.

Simulations for m values up to eight have been performed. The data for the specific heat is shown in Fig.1. At $T/J\approx2.3$ the specific heat exhibits a maximum in concert with the exact solution for m=1. At high temperatures the simulation results are in very good agreement with those obtained from high-T series. For $T/J>2$ the specific heat, which from the point of view of convergence is the most difficult quantity to calculate, depends weakly on the particular value of m. More convincing evidence that the convergence of the energy and specific heat is very good is given in Fig.2.

To study the role of vortex-like excitations a quantum version of Swendsen's vortex detector D [8] has been measured. Simulation data for the m-th approximant D_m is shown in Fig.3. The rapid change of the disorder-parameter D as the temperature approaches T_c also suggests that there is phase transition. From Fig. 4 we learn that the in-plane structure factor $S_{\parallel}$ (sum over all two-spin correlations) shows a clear dependence on the system size for low temperatures. This implies the existence of long-range correlations, which we attribute to the spin vortices.

4. TWO-LEVEL SYSTEM WITH NON-LINEAR COUPLING

In this section we consider a model for a special kind of polaron. In solids and liquids the polaron is a fundamental concept. The polaron is one electron which is coupled to some kind of harmonic lattice degree of

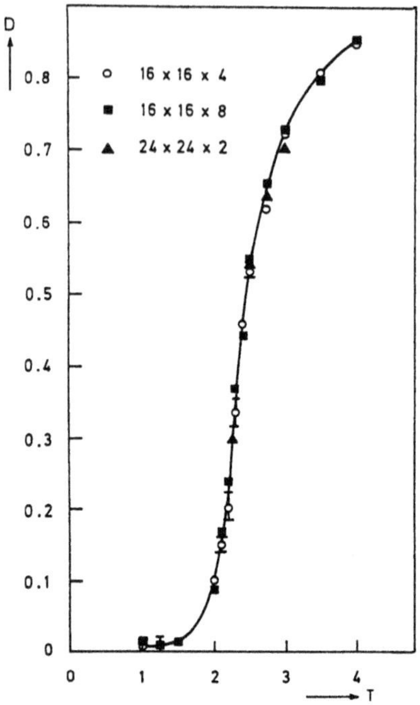

Fig. 3. The vortex detector D for lattices of different size and different m

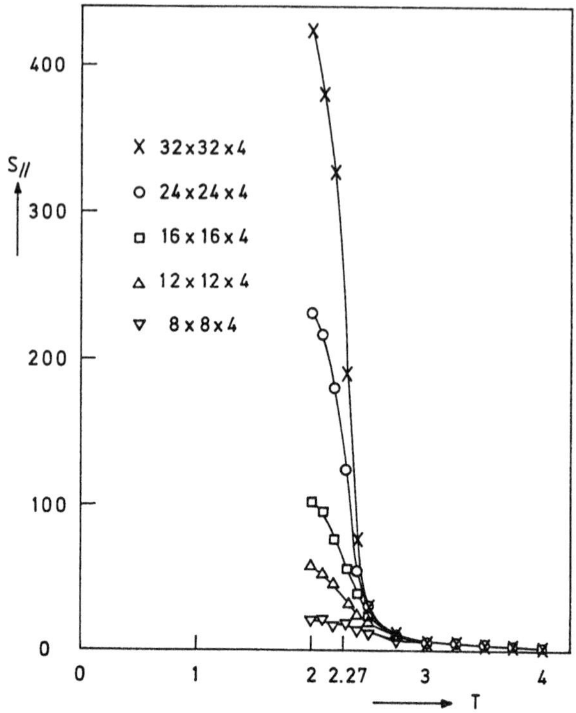

Fig. 4. The in-plane structure factor $s_{||}$ for lattices of different size

freedom. There are many classes of polaron models; one model which has been studied extensively is the Molecular Crystal Model (MCM) [9]. In this model the electron is described by a tight-binding Hamiltonian interacting (linearly) with harmonic lattice vibrations. A key issue in the polaron model is the notion of self-trapping. For the two-site MCM the presence of a ground state transition between an extended state and a self-trapped state can be proven rigorously in the adiabatic limit [10]. It has been shown that the many-site MCM in all lattice dimensionalities have the same qualitative behavior [11].

We now consider a two-level model in which the electron is coupled quadratically to a harmonic oscillator. The Hamiltonian we study is given by

$$H = -h\sigma^x + p^2/2M + \tfrac{1}{2}M\Omega^2x^2 + \lambda x^2\sigma^z \quad , \tag{7}$$

where the electron is described by means of pseudo-spin operators and h is the transfer integral for the hopping motion of the electron. The coupling constant λ ($\lambda>0$) has to be smaller than $\tfrac{1}{2}M\Omega^2$, so that the Hamiltonian is bounded from below. A nice feature of the Hamiltonian (7) is that it is quadratic in p and x, so that in a path-integral representation of the partition function the oscillator degrees of freedom can be eliminated analytically.

We will study (7) in the adiabatic limit ($M\to\infty$, $M\Omega^2$ finite). Qualitatively the adiabatic limit shows the same behavior as the non-adiabatic limit. In practice working in the adiabatic limit has the advantage of simpler expressions for the approximants of quantities one wants to calculate. More importantly, however, in this special case working in the adiabatic limit not only permits us to calculate the ground state exactly but also allows us to compute numerically all thermodynamical properties for every temperature T. So we will be able to compare the results obtained from a path-integral formulation with the exact results.

First we discuss the T=0 case. The exact ground state properties are:

$$E_0 = -h \ , \tag{8.a}$$

$$|0\rangle = |s_x=+\tfrac{1}{2},\ x=0\rangle \ , \tag{8.b}$$

$$\langle\sigma^z\rangle = 0 \ , \tag{8.c}$$

independent of the coupling parameter λ. We conclude that this quadratic model does not exhibit a ground state transition as a function of λ. There is no self-trapping for T=0.

To study the T≠0 case we perform exact numerical calculations and computations based on a path-integral representation of (7). The m-th approximant of the partition function is given by (the oscillator part is factored out)

$$Z_m = \sum_{\{s_i\}} \left(1 + \lambda \sum_i s_i/m\right)^{\frac{1}{2}} \exp\left(J \sum_i s_i s_{i+1}\right) \ , \tag{9}$$

where $J=\frac{1}{2}\ln\coth(\beta h/m)$ and $C=2\lambda/M\Omega^2$ (range of C: $0\le C\le 1$). Equation (9) resembles the partition function of an Ising system, the pre-exponential factor originates from the elimination of the oscillator degree of freedom. The sum over the spin states can easily be performed exactly for m values up to 20. Thus results presented here for $m\le 20$ are numerically exact. For larger m values we have performed Monte Carlo calculations. In Fig. 5 the results are presented for the derivative of the free energy with respect to the normalized coupling constant C as a function of temperature. The drawn line is the numerically exact result. The first derivative of the free energy is related to the expectation value of the coupling term of the Hamiltonian. A discontinuity in $\partial F_m/\partial C$ as a function of C would mean that the free energy is not an analytic function of the coupling C and in analogy with the theory of phase transitions this indicates that the system undergoes a transition.

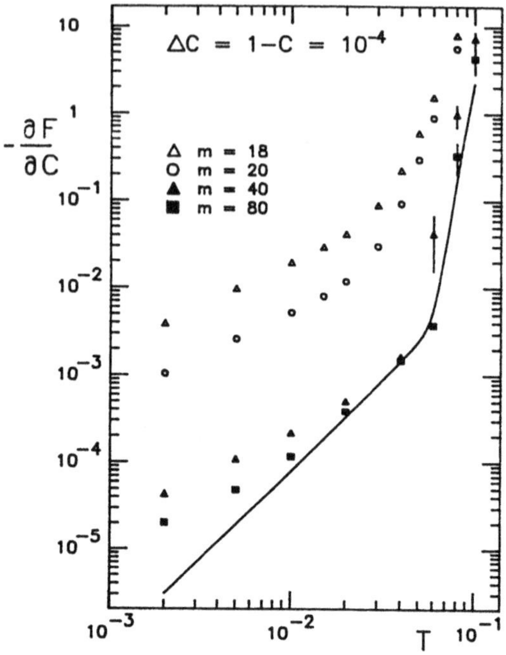

Fig. 5. The derivative of the free energy with respect to the normalized coupling constant C for C = 0.9999. Open symbols: exact summations. Closed symbols: Monte Carlo results (error bars not shown have the size of symbols). Drawn line: exact result

For brevity we only present the results for one value of C which we take close to 1 (the energy scale is set by h=1). As C approaches 1 the system becomes polarized ($\langle\sigma^z\rangle \to -1$) and the oscillator mode becomes soft. For all C there is a weak-coupling regime at sufficiently low temperatures. Note that for small values of m the deviation from the exact results is enormous. This is an intrinsic property of the approximation. By construction letting the temperature go to zero implies that we must take the limit of m to infinity first. Also the qualitative behavior deviates from exact results at low temperatures. As the temperture is increased the convergence as a function of m increases. At a certain temperature the derivative of the free energy rapidly increases (if C is very close to one). Clearly there is a divergence at this temperature which becomes lower if C is closer to 1. So C=1 is a critical value for T>0. At T=0 C is not a critical value.

5. BIPOLARON FORMATION

The ultimate goal of the study of polarons is undoubtly the complete understanding of the many-polaron system. A first and important step towards this ambitious goal is the investigation of bipolaron formation. A bipolaron is a self-trapped entity consisting of two polarons. In a two-polaron problem there are in addition states possible consisting of two separate self-trapped polarons and also of two (quasi) free electrons. Bipolaron formation in continuum space has been investigated recently

[12] in order to explain experimental results. Lattice polaron models may behave different from their continuum counterparts (with or without a lattice cutoff) [11,13].

In this section we consider bipolaron formation on a lattice. The generic Hamiltonian of the lattice model reads

$$H = \sum_{k,s} \epsilon(k) c^+_{k,s} c_{k,s} + \sum_{k} V(k) \rho_k \rho_{-k} + \sum_{k} Q(k)(a^+_{-k} + a_k)\rho_k + \sum_{k} \omega(k) a^+_k a_k \ . \tag{8}$$

The operators a^+_k (a_k) create (annihilate) a phonon excitation of wave vector k, $c^+_{k,s}$ $(c_{k,s})$ create (annihilate) an electron with spin s and ρ_k is the Fourier transformed electron density. For simplicity we assume a simple cubic lattice. We will confine ourselves to the case where one electron has spin up and the other has spin down. Then the electrons may be considered indistinguishable.

In writing down an expression for the approximant Z_m to the polaron partition function Z, using a generalization of the Trotter formula (3), we can exploit the fact that (8) is a quadratic form in the phonon operators. As in the previous section this fact allows us to integrate out all phonon coordinates analytically. This feature makes the use of path-integral concepts ideally suited to study electron-phonon models. The combination of analytical and numerical techniques makes the Monte Carlo simulation very efficient. The number of variables is greatly reduced and, by factoring out the free boson part of the partition function, the properties of the electrons can be calculated directly. In other methods the polaron contribution would be hidden in the statistical noise of the bosons. The elimination of the phonons leads to an effective electron-electron interaction which in the adiabatic limit can be simplified to an expression that depends on $Q^2(k)/\omega(k)$. See DE RAEDT et al [14] for details. This is an important result, since models with different $Q(k)$ and $\omega(k)$ can have the same effective interaction in the adiabatic limit. We have studied (8) in the adiabatic limit and have confined ourselves to the case where $Q^2(k)/\omega(k)$ does not depend on k. Then the strength of the electron-phonon coupling can be parametrized by one constant C. In addition we take an on-site repulsive interaction between the two electrons:

$$V(\mathbf{x}) = U\delta_{\mathbf{x},0} \ . \tag{10}$$

Calculations using a long-range interaction have been performed and show qualitatively the same picture.

The behavior one expects is the following: when there is no direct repulsion between the electrons an increase of the electron-phonon coupling constant will result in a trapping of both electrons in a combined state. If the repulsion is sufficiently large one can expect an intermediate

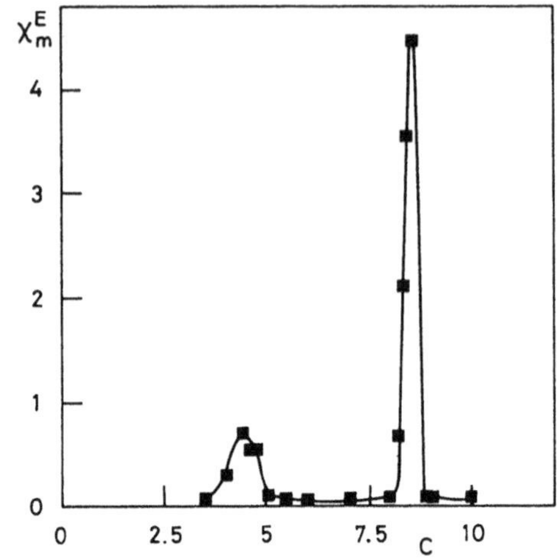

Fig. 6. Simulation results for the coupling-energy susceptibility as a function of C for an on-site Coulomb repulsion U=15

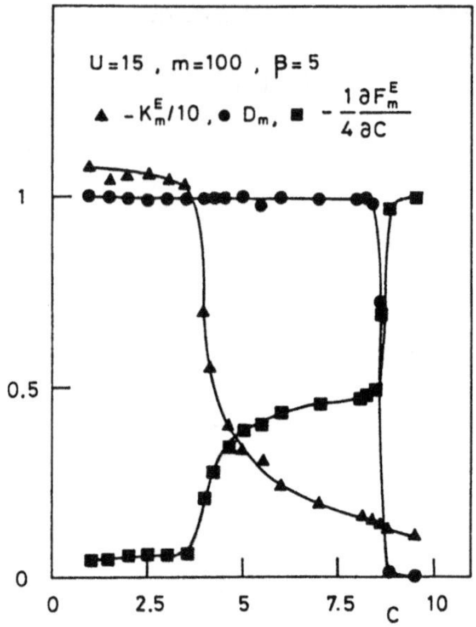

Fig. 7. Simulation results for the kinetic energy (triangles), the distance between the electrons (dots), and the coupling energy (squares)

state, namely the appearance of two independant self-trapped polarons. A strong coupling argument shows that the bipolaron state has the lowest energy when U<2C.

In Fig. 6 the results are presented for the coupling-energy susceptibility $\chi_m^E = -\partial^2 F_m^E/\partial C^2$ as a function of the coupling constant. The simulation was performed for a 32x32x32 model with U=15. The occurrence of two pronounced maxima is very clear. In Fig. 7 we have calculated the m-th approximant D_m for the RMS distance between the electrons (normalized to the case of two free particles). The sharp drop in distance is connected with the second transition in Fig. 6. We conclude that the second transition is the condensing into a bipolaron state and the first is the self-trapping of the two individual polarons. In the same figure we have presented data on the kinetic energy and the coupling energy $-C\partial F_m^E/\partial C$. It follows that the kinetic energy drops quickly when the two independant polarons are formed but changes smoothly when the transition to the bipolaron state takes place. The relation between the coupling energy and the susceptibility is of course that the first is the derivative of the other. Both quantities are measured independently. Figure 8 shows the phase diagram of the two-polaron problem.

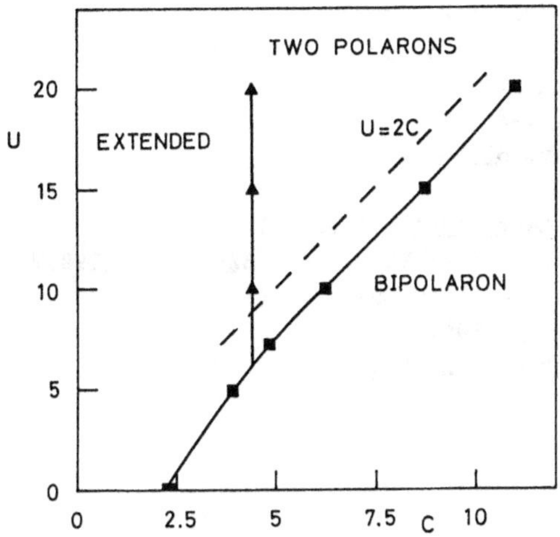

Fig. 8. U-C phase diagram for an on-site Coulomb repulsion. Dashed line is the prediction of simple strong-coupling theory for the two-polarons - bipolaron transition. Solid lines are guide to the eyes

6. CONCLUDING REMARKS

We have discussed the application of the Monte Carlo method to quantum systems. The implementation of a Monte Carlo algorithm for path-integral representations of quantum models is not straightforward. There are still a number of fundamental problems to be solved. The simulation of spin models requires rather sophisticated and dedicated algorithms in order to achieve ergodicity and an efficient use of computer time. Coupled fermion-boson systems form a class of models which can be treated especially with a path-integral formulation. The combination of the Monte Carlo method and the Trotter-formula approach to statistical mechanics can be very fruitful. At present Monte Carlo calculations on quantum systems cannot (yet) be treated as a black box computer program with results.

This work is supported by the Belgian National Science Foundation (NFWO) and the Dutch Stichting voor Fundamenteel Onderzoek der Materie (FOM).

REFERENCES

1. D.M. Ceperley, M.H. Kalos: In Monte Carlo Methods in Statistical Physics, ed. by K. Binder, Topics in Current Physics, Vol. 7 (Springer, New York 1979) p.145
2. K. Binder: In Monte Carlo Methods in Statistical Physics, ed. by K. Binder, Topics in Current Physics, Vol. 7 (Springer, New York 1979)
3. R.P. Feynman, A.R. Hibbs: Quantum Mechanics and Path Integrals (McGraw-Hill, New York 1965)
4. M. Suzuki, S. Miyashita, A. Kuroda: Prog. Theor. Phys. 58, 1377 (1977)
5. J.M. Kosterlitz, D.J. Thouless: J. Phys. C6, 1181 (1973)

6. H. De Raedt, B. De Raedt, A. Lagendijk: Z. Phys. B57, 209 (1984)
7. A. Lagendijk, H. De Raedt: Phys. Rev. Lett. 49, 602 (1982)
8. R.H. Swendsen: Phys. Rev. Lett. 49, 1302 (1982)
9. T. Holstein: Ann. Phys. (N.Y.) 8, 343 (1959)
10. H.H. Shore, L.M. Sander: Phys. Rev. B7, 4536 (1973);
P. Prelovsek: J.Phys. C12, 1855 (1979)
11. A. Lagendijk, H. De Raedt: Phys. Lett. 108A, 91 (1985)
12. M.H. Cohen, E.N. Economou, C.M. Soukoulis: Phys. Rev. B29, 4496 (1985);
H. Hiramoto, Y. Toyozawa: J. Phys. Soc. Jpn. 54, 245 (1985)
13. S. Das Sarma: Solid State Commun. 54, 1067 (1985)
14. H. De Raedt, A. Lagendijk: submitted to Z. Phys. B

Part III

Conducting Polymers

Photogeneration of Nonlinear Excitations in Semiconducting Polymers: Fast Response Nonlinear Optical Phenomena

A.J. Heeger, D. Moses, and M. Sinclair

Department of Physics and Institute for Polymers and Organic Solids, University of California, Santa Barbara, CA 93106, USA

I. Introduction: The Origin of Nonlinear Optical Properties of Semiconducting Polymers

In traditional three-dimensional semiconductors, the four-fold (or six-fold, etc.) coordination of each atom to its neighbor through covalent bonds leads to a rigid structure. In such systems, therefore, the electronic excitations can usually be considered in the context of this rigid structure leading to the conventional concepts of electrons and holes as the dominant excitations. The situation in semiconductor polymers is quite different; the two-fold coordination makes these systems generally more susceptible to structural distortion. As a result, the dominant "electronic" excitations are inherently coupled to chain distortions. Thus, solitons, polarons and bipolarons are the excitations of major importance in this novel class of one-dimensional polymer semiconductors. These semiconducting polymers are, therefore, fundamentally and inherently nonlinear. Moreover, the major structural rearrangements (to form the solitons, bipolarons, etc.) which lead to the nonlinear optical properties are known to occur on time scales of order 10^{-13} seconds [1].

We illustrate these concepts through a brief summary of relevant experimental results on two systems: 1. Polyacetylene, $(CH)_x$: the degenerate ground state leads to solitons as the important excitations and the dominant charge storage species. 2. Poly(thiophene): the ground state degeneracy is lifted so that polarons and bipolarons are the important excitations with charge storage in bipolarons.

As shown in Fig. 1, trans-$(CH)_x$ is a two-fold degenerate Peierls insulator which allows for the possibility of nonlinear excitations in the form of soliton-like bond-alternation domain walls, each with an associated electronic state at the center of the energy gap [2-6]. For polythiophene, on the other hand, the two structures sketched in Fig. 2 are not energetically equivalent. Poly(thiophene) (PT) can be viewed as an sp^2p_z carbon chain in a

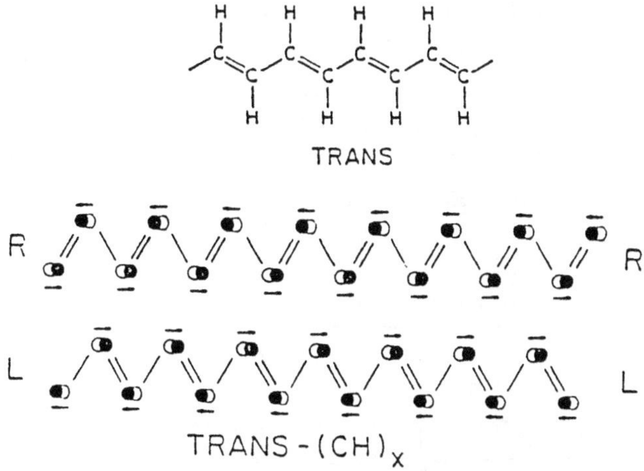

Fig. 1. a) *trans*-$(CH)_x$; b) The two degenerate ground states of *trans*-$(CH)_x$

structure somewhat analogous to that of cis-$(CH)_x$ but stabilized in that structure by the sulfur which covalently bonds to neighboring carbons to form the heterocycle.

The results of a series of experiments have demonstrated that solitons are important excitations in *trans*-$(CH)_x$ and that the properties of these nonlinear excitations can be directly studied through measurements on *trans*-$(CH)_x$ samples either *during photo-excitation* or after doping. The relevant concepts have been generalized to confined soliton pairs (bipolarons) and applied to a wide variety of conjugated polymers in which the ground state degeneracy is not present. Polythiophene (Fig. 2) is a prototype example of such systems.

The soliton in $(CH)_x$ is a topological kink in the electron-lattice system, a bond alternation domain wall connecting the two phases with opposite bond alternation. Since there is translational symmetry (the kink can be anywhere)

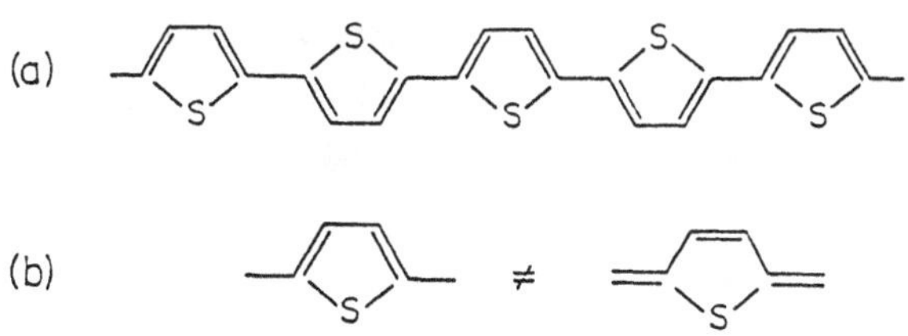

Fig. 2. a) Chemical structure of polythiophene; b) The two valence bond configurations are not equivalent

and since the mass is small, the soliton should be mobile. The competition of elastic and condensation energies spreads the domain wall over a region of about 12 - 14 a, where a is the C-C distance along the $(CH)_x$ chain. Although single soliton defects can exist on imperfect chains (and have been studied extensively) [8a,b], intrinsic exciations, either photo-produced or doping induced, must occur in the form of soliton-antisoliton (S-S) pairs.

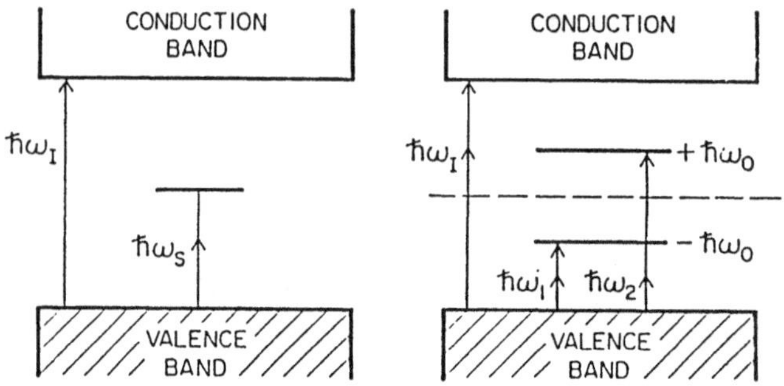

Fig. 3. a) Solid band diagram showing interband transition ($\hbar\omega_I$) and mid-gap transition ($\hbar\omega_S$); b) Bipolaron band diagram showing interband transition ($\hbar\omega_I$) and sub-gap transitions ($\hbar\omega_1$ and $\hbar\omega_2$)

Associated with the structural kink is a localized electronic state with energy at mid-gap (see Fig. 3a). This electronic state is a solution of the Schrodinger equation in the presence of the structural domain wall and can therefore accommodate 0, 1 or 2 electrons [1,2]. The neutral soliton has one electron in the mid-gap state; the positive and negative charged solitons have zero or two electrons, respectively, in the mid-gap state. Consequently, the spin-charge relations of solitons are reversed; charged solitons are nonmagnetic, whereas a neutral soliton has spin 1/2.

A charged soliton represents the lowest energy configuration for an excess charge on a trans-$(CH)_x$ chain,[1,2] i.e., $E_S < \Delta$ where E_S is the energy for creation of a soliton and Δ is the energy for creation of an electron or hole, $\Delta \doteq E_g/2$. Both numerical calculations using a discrete lattice model [3] and analytical results for a continuum model indicate the $2E_S = (4/\pi)/\Delta$. Charged solitons are therefore formed by electron transfer onto or off the polymer chains when doping with electron donors or acceptors,or by injection of an electron-hole pair through photoexcitation.

An analogy can be constructed between PT and $(CH)_x$. If one purposely leaves out the sulfur heteroatom, the resulting backbone structure is that of an sp^2p_z polyene chain consisting of four carbon all-trans segments linked through a cis-like unit. In such a structure, the ground state is not degenerate (as sketched in Fig. 2). However, the energy difference per bond, $(\Delta E/\ell)$, might be expected to be small, i.e., greater than zero (as in trans-$(CH)_x$) but perhaps comparable to that of cis-$(CH)_x$. An obvious consequence of the lack of degeneracy is that the schematic PT structure of Fig. 2 cannot support stable soliton excitations [1,2], since creating a soliton pair separated by a distance d would cost energy ~ $d(\Delta E/\ell)$. This linear "confinement" energy leads to bipolarons as the lowest energy charge transfer configurations in such a chain with creation energy somewhat greater than $(4/\pi)\Delta$ (the creation energy for a bipolaron goes to Eq. 1 in the limit of zero confinement). The corresponding energy level diagram (for a positive bipolaron) is sketched in Fig. 3b. The two gap states are empty for a positive bipolaron (charge 2 e) and filled for a negative bipolaron (charge -2 e). Polaron excitations in such systems have an energy level diagram similar to that sketched in Fig. 3b. In the polaron case, however, the gap states are partially filled (e.g., one electron in the lower level for a hole polaron) leading to additional sub-gap absorption subsequent to polaron formation.

The energy level diagrams sketched in Fig. 3 are of principal importance to the nonlinear optical properties of this class of polymers. Prior to charge injection by photo-excitation of an electron-hole pair, all of the oscillator strength is in the interband absorption, $\hbar\omega_I$. After photo-excitation, there is a major redistribution in oscillator strength to $\hbar\omega_s$ (or $\hbar\omega_1$, $\hbar\omega_2$, etc.) as shown in Fig. 4. These shifts in oscillator strength occur on a time scale ~ 10^{-13} seconds. Thus, these conjugated semiconductor polymers are intrinsically fast nonlinear optical materials.

II. Photoexcitation: Photo-induced Absorption, Photo-induced Bleaching, Photo-luminescence, Photoconductivity

Photo-excitation studies of conjugated semiconductor polymers were stimulated by the calculations of Su and Schrieffer [3] which demonstrated

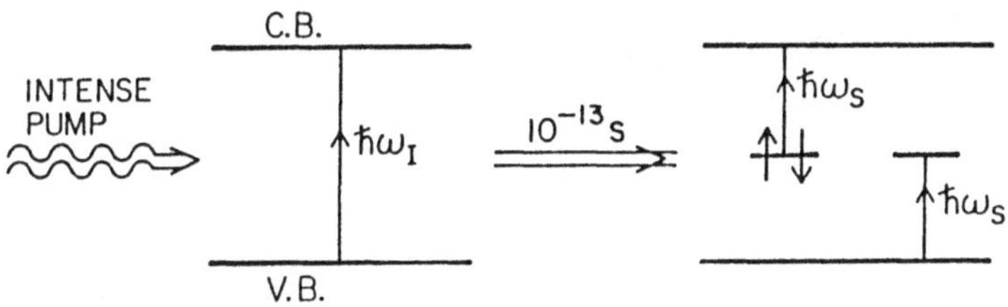

Fig. 4. A photo-pump makes e-h pairs which evolve in 10^{-13} seconds to soliton pairs with states at mid-gap. The oscillator strength shifts accordingly

that in trans-$(CH)_x$ an e-h pair should evolve into a pair of solitons within an optical photon period or about 10^{-13} sec. Thus, the absorption spectrum was predicted to shift from $\hbar\omega_I$ to $\hbar\omega_S$ (see Fig. 4) on a time scale of 10^{-13} seconds after photo-excitation.

The photo-generation of soliton-antisoliton pairs implies formation of states at mid-gap. Time-resolved spectroscopy [4] has been used to observe the predicted absorption due to photo-generated intrinsic gap states in trans-$(CH)_x$. Moreover, the time scale for photo-generation of these gap states has been investigated [5,6]. Using sub-picosecond resolution, these studies demonstrated that the gap states and the associated interband bleaching are produced in less than 10^{-13} seconds, consistent with the theoretical predictions. The data showing sub-picosecond photo-induced bleaching of the interband transition are shown in Fig. 5. Vardeny, et al. [7] and Blanchet, et al. [8] have observed the photo-induced absorption arising from both the mid-gap electronic transition (sketched in Fig. 3) and the associated infrared active modes introduced by the local lattice distortion.

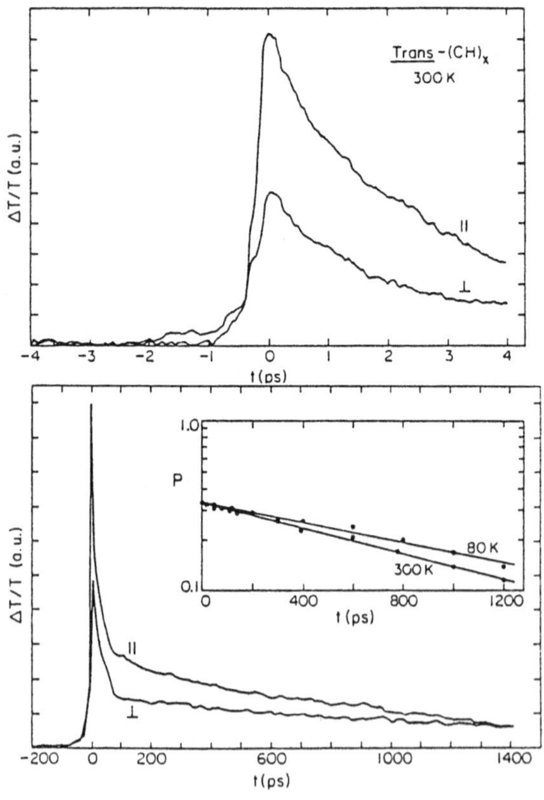

Fig. 5. Transient photoinduced bleaching of the interband transition in polyacetylene (Ref. 6)

Infrared spectroscopy [9-12] of lightly doped trans-$(CH)_x$ has demonstrated that the same spectroscopic features arise upon doping. Moreover, these doping-induced absorptions are independent of the dopant and are therefore identified as intrinsic features of the doped trans-$(CH)_x$ chain. These important results demonstrate that both the photo-induced spectroscopic features and those induced by doping are associated with the same charged state. Moreover, the observed frequencies and line shapes are consistent with those expected for charged soliton excitations [11,12]. It has been argued [13] that these changes in infrared absorption are general features of charged localized states on the trans-$(CH)_x$ chain (solitons or polarons) and thus not definitive proof of the photo-generation of solitons. However, since these excitations have the reversed spin-charge relation of solitons, this ambiquity is settled [14]. Thus, for example, the PI modes can be used as signatures of soliton formation.

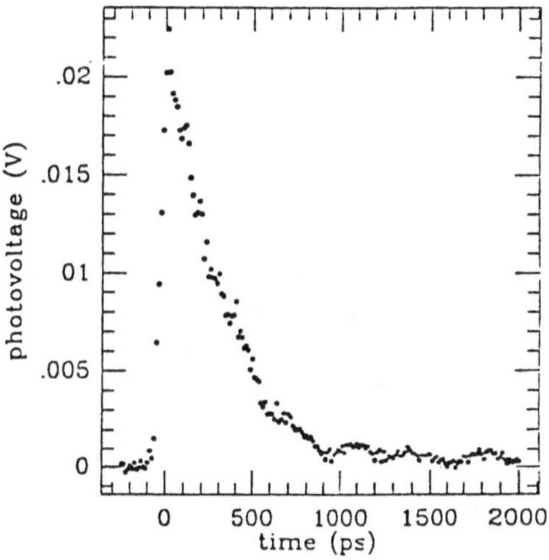

Fig. 6. Transient photovoltage across 50 Ω of trans-$(CH)_x$ in the Auston switch configuration

Figure 6 shows the transient photoconductivity [15] due to a 1 μJ pulse at 2.1 eV with a bias voltage of 300 V. The rise time of the signal is ~ 50 ps. The observed rise time represents a convolution of the sampling head response time (25 ps), the laser pulse rise time (15 ps), the trigger jitter (15 ps), the rise time of the transmission line and the rise time of the photocurrent. Thus, the charge carriers are produced within picoseconds of optical excitation. The fast rise is followed by (approximately exponential) decay with a time constant of ~ 300 ps; the ratio of the photocurrent at 1500 ps to that at 50 ps is 0.03. The experimental demonstration that the photoconductivity and the photoinduced change in optical absorption [5,6], $\delta\alpha(t)$, decay on the same time scale implies that the photogenerated charge carriers involve a major shift in oscillator strength, consistent with the proposed photogeneration of charged solitons.

The magnitude of σ_{ph} is large. For an absorbed photon flux of 10^{15} cm^{-2} per pulse, the photocurrent (at ~ 50 ps) is 4×10^{-4} A (i.e., current density ~ 10^3 A/cm^2) with good reproducibility from sample to sample. This corresponds to a conductivity of about 0.3 S/cm with a possible error of a factor of two due to uncertainty in αd in the top-illumination measurement geometry. The increase in σ_{ph} over the dark room temperature value is five to six orders of magnitude!

The relationship between the photocurrent and the incident photon flux is given by

$$\sigma_{ph} = (E/\hbar\omega)e\eta\varphi\mu \tag{1}$$

where $(E/\hbar\omega)$ is the number of absorbed photons per unit volume, η is the quantum efficiency, φ is the probability to escape geminate (or early time) recombination and μ is the mobility. Using σ_{ph}(50 ps) ~ 0.3 S/cm and assuming both that $\eta = 1$ and that all the initial carriers live to at least 50 ps ($\varphi = 1$), the calculated mobility would be $\mu \sim 10^{-2}$ cm^2/v-s. However, the picosecond [5,6] and sub-picosecond [16] $\delta\alpha(t)$ data indicate that the number of photoexcitations has already decayed to less than 10^{-2} of the initial value at 50 ps. Thus, the carrier yield at 50 ps is $\varphi \sim 0.01$. Using this value for φ results in a mobility of approximately 1 cm^2/V-s. With this value, the net distance drifted in the measured decay time (300 ps) is about 400 Å, in good agreement with that inferred from the picosecond decay of photoinduced dichroism [1].

Photo-excitation spectroscopy has also been carried out for polythiophene [17,18]. From resonant Raman scattering, photo-induced absorption and photo-luminescence, a fully consistent picture of the ground state and photo-excitations of polythiophene was obtained [18]. The non-degenerate ground state causes relatively strong confinement (indicated by the Raman data) and fast radiative recombination (observed via luminescence with maximum at 1.9 eV). As a result of the confinement (due to the nondegenerate ground state), the time scale for photo-luminescence decay is in the picosecond regime. Thus, all aspects of the photoprocesses occur on this short time scale. The photoinduced absorption results obtained from polythiophene imply a shift in oscillator strength from $\hbar\omega_I$ into the gap state transitions ($\hbar\omega_1$ and $\hbar\omega_2$, see Fig. 3b). Since polythiophene is the prototype example of the non-degenerate ground state semiconductor polymer, we expect these features to be universal.

III. Relationship of the observed Nonlinear Photo-excitations Processes in Semiconducting Polymers to the Phenomenology of Nonlinear Optics

In bulk media, there is a nonlinearity in the constitutive relationship between the induced polarization (P_i) proportional to the amplitude of the electric field of the incident light; the induced polarization may be expanded in a power series of the electric field components:

$$P_i = \chi_{ij}{}^{(1)}E_j + \chi_{ijk}{}^{(2)}E_jE_k + \chi_{ijk\ell}{}^{(3)}E_jE_kE_\ell \qquad (2)$$

In this expression, $\chi^{(1)}$ represents the linear optical properties; $\chi^{(2)}$ and $^{(3)}$ are. respectively, the second and third order nonlinear susceptibilities. In materials which possess inversion symmetry, the lowest order nonlinear susceptibility is $\chi^{(3)}$.

We focus here on $\chi^{(3)}$ which will be large whenever there is a major shift of oscillator strength in response to a pump photo-excitation. The data obtained from polyacetylene and polythiophene demonstrate that such phenomena are observed in the class of semiconducting polymers. Moreover, the fundamental mechanisms sketched in Figs. 3 and 4 imply that these features are general and can be expected to occur in a wide (and growing!) class of such materials.

As noted above, ultrafast ($< 10^{-13}$ s) bleaching of the interband transition has been demonstrated [5,6,16] with a shift of oscillator strength into the infrared. Thus, we can write the intensity-dependent (complex) index of refraction as

$$n(\omega_2) = n_0(\omega_2) + \delta n(\omega_2,\omega_1)I(\omega_1) \qquad (3)$$

where the second term describes the nonlinear response at frequency ω_2 due to an intense pump at frequency ω_1. From Eqs. (2) and (3)

$$\delta n = 4\pi^2/c\varepsilon_2\chi^{(3)} \qquad (4)$$

where ε_2 is the dielectric constant at the probe frequency. Using the photoinduced absorption data and Eq. 4, we obtain Im $\chi^{(3)}(-\omega_2 = \omega_1 - \omega_1 - \omega_2) = 5 \times 10^{-8}$ esu for $\hbar\omega_1$ = 2.0 eV and $\hbar\omega_2$ = 1 4 eV, a truly impressive value even under resonant conditions. From a standard Kramers-Kronig analysis of these photoinduced absorption data, one concludes that the real parts of δn and $\chi^{(3)}$ are correspondingly large. Based upon these observations, we initiated an experimental study of third harmonic generation [20].

IV. Measurement of the Third Order Susceptibility of Trans-$(CH)_x$ by Third Harmonic Generation

The demonstration of third harmonic generation using thin films of polyacetylene, $(CH)_x$, as the nonlinear optical medium was recently reported [19-21]. With 100 ps pulses at 1.06 μ (1-100 MW/cm^2 peak power) incident on a 1000 Å nonoriented film, the conversion into 3ω was proportional to the cube of the incident power with proportionality constant 4..3 x 10^{-6} (W/MW^3). The implied value for the third order susceptibility is $\chi^{(3)}(3\omega) = 4 \times 10^{-10}$ esu. Since the trans-$(CH)_x$ films used in this experiment were non-oriented, the value for $\chi^{(3)}$ parallel to the polyene chain ($\chi_{\parallel}^{(3)}$) is an order of magnitude larger than the corresponding value for polydiacetylene.

Figure 7 is a log-log plot (see Ref. 20) of the third harmonic intensity (in W/cm^2) as a function of fundamental intensity (in MW/cm^2). The solid line is

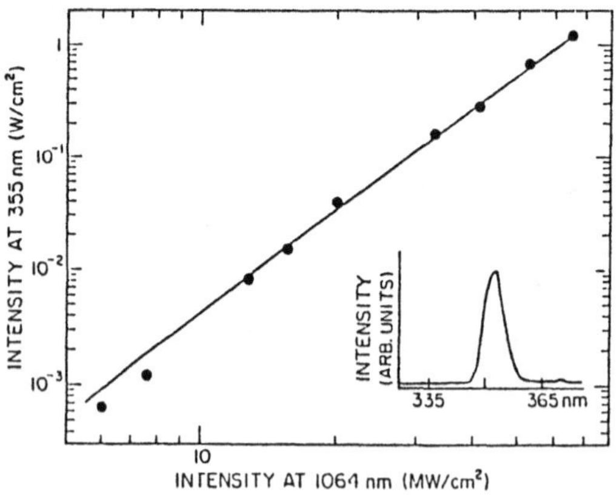

Fig. 7. Log-log plot of the third harmonic intensity (W/cm^2) as a function of the fundamental intensity (MW/cm^2). The solid line indicates a cubic dependence of the third harmonic intensity on the fundamental intensity. The inset shows that the spectrum of the third harmonic is resolution limited at 355 nm

a best fit to the data assuming a cubic dependence of the third harmonic intensity on the fundamental intensity. The constant of proportionality is 4.3 x 10^{-6} (W/MW^3). The inset shows the spectral content of the third harmonic intensity showing a resolution-limited peak at 355 nm. Using Q-switched, modelocked pulses, we have extended the measurement to peak pump powers in excess of 10 GW/cm^2 without damage to the sample.

In the presence of strong absorption at 3ω, the intensity of the third harmonic can be written in terms of the third order polarization as [21]:

$$I(3\omega) = 2\pi(3\omega)^2/c\sqrt{\varepsilon_3}\,|P^{(3)}(3\omega)|^2/[(\Delta k)^2 + \beta^2] \tag{5}$$

where $P^{(3)}(3\omega) = \chi^{(3)}(3\omega = \omega+\omega+\omega)E(\omega)E(\omega)E(\omega)$, β is the absorption coefficient at 3ω, Δk is the phase mismatch and ε_3 is the dielectric constant at 3ω. This equation applies in the case of negligible pump depletion which is valid for thin films (for thicker films, the effects of absorption and scattering of the fundamental beam must be included). Since

$$|P^{(3)}(3\omega)|^2 = (2\pi/c\sqrt{\varepsilon_1})^3\,(\chi^{(3)}(3\omega))^2 I^3(w), \tag{6}$$

one can relate the experimentally measured intensity at 3ω to the third order susceptibility:

$$\chi^{(3)}(3\omega) = (c/2\pi)^2 \left[\sqrt{\varepsilon_1{}^3\varepsilon_3}\;[(\Delta k)^2 + \beta^2]/(3\omega)\, I(3\omega)/I^3(\omega)\right]^{1/2} \tag{7}$$

Using $I(3\omega)/I^3(\omega) = 4.3 \times 10^{-38}$ (in esu), $\beta = 1 \times 10^5$ cm^{-1}, $\Delta k = 1.5 \times 10^5$ cm^{-1}, $\varepsilon_1 = 7$ and $\varepsilon_3 = 3$, we obtain $\chi^{(3)} = 4 \times 10^{-10}$ esu. This value of $\chi^{(3)}$ compares favorably to the $\chi^{(3)}$ values measured in the polydiacetylenes ($\chi^{(3)} \sim 10^{-10}$ esu in PTS) [22] and in inorganic semiconductors ($\chi^{(3)} \sim 10^{-10}$ for Ge) [22].

Since the optical properties of polyacetylene are highly anisotropic, the measured $\chi^{(3)}$ is related to $\chi_{\parallel}{}^{(3)}$ through the relation [23]:

$$\chi^{(3)} = \chi_{\parallel}{}^{(3)} \left[\langle\cos^6\theta\rangle\right]^{1/2} \tag{8}$$

where the angle θ is the angle between a given $(CH)_x$ chain and the E field and the brackets denote an angular average. Assuming the polymer chains are oriented randomly in the plane of the thin film, $\langle\cos^6\theta\rangle = 5/16$. The implied value of $\chi_{\parallel}^{(3)}$ is therefore about an order of magnitude greater than the corresponding value for polydiacetylene.

IV. Conclusion

In summary, semiconductor polymers such as polyacetylene and polythiophene have experimentally demonstrated nonlinear optical processes (photo-induced absorption, photo-induced bleaching and photo-luminescence) with characteristic time scales in the picosecond range or faster. These phenomena are intrinsic and originate from the instability of these conjugated polymers toward structural distortion. The major shifts in oscillator strength due to photoexcitation of solitons, polarons and bipolarons lead to relatively large third-order nonlinear optical processes ($\chi^{(3)}$) on time scales of order 10^{-13} seconds. Largely overlooked in earlier analyses, we believe these novel photoexcitations are key to understanding the nonlinear optical properties of this growing class of semiconducting (conjugated) polymers.

Acknowledgments

The photoinduced absorption and photo conductivity studies were supported by the Office of Naval Research. The third harmonic generation measurements were carried out in collaboration with K. Vilhelmsson, B. Valk and M. Salour of TACAN Aerospace Corporation, Carlsbad, CA.

References

1. A. J. Heeger, Polymer Journal, 17, 201 (1985) and references therein.
2. A. J. Heeger, Comments on Solid State Physics, 10, 53 (1981).
3. W. P. Su and J. R. Schrieffer, Proc. Nat. Acad. Sci. U.S.A. 77, 5626 (1980).
4. J. Orenstein and G. Baker, Phys. Rev. Lett. 49, 1043 (1982).
5. C. V. Shank, R. Yen, R. L. Fork, J. Orenstein and G. L. Baker, Phys. Rev. Lett. 49, 1660 (1982).
6. Z. Vardeny, J. Strait, D. Moses, T.-C. Chung and A. J. Heeger, Phys. Rev. Lett. 49, 1657 (1982).
7. Z. Vardeny, J. Orenstein and G. L. Baker, J. Phys. Colloq. 44, C3-325 (1983): Phys. Rev. Lett. 50, 2032 (1983).
8. G. B. Blanchet, C. R. Fincher, T.-C. Chung and A. J. Heeger, Phys. Rev. Lett. 50, 1938 (1983).
9. C. R. Fincher, Jr., M. Ozaki, M. Tanaka, D. Peebles, L. Lauchlan, A. J. Heeger and A. G. MacDiarmid, Phys. Rev. B 20, 1589 (1979) and references therein.

10. J. F. Rabolt, T. C. Clarke and G. B. Street, J. Chem. Phys. 71, 4614 (1979).
11. S. Etemad, A. Pron, A. J. Heeger, A. G. MacDiarmid, E. J. Mole and M. J. Rice, Phys. Rev. B 23, 5137 (1981).
12. A. Etemad, A. J. Heeger and A. G. MacDiarmid, Ann. Rev. Chem. Phys. 33, 433 (1982).
13. B. Horovitz, Solid State Commun. 41, 729 (1982).
14. a) J. D. Flood and A. J. Heeger, Phys. Rev. B 28, 2356 (1983).
    b) F. Moraes, Y. W. Park and A. J. Heeger, Syn. Mtls. 13, 113 (1986).
15. M. Sinclair, D. Moses and A. J. Heeger, Phys. Rev. Lett. (in press).
16. a) C. V. Shank, Science, 219, 1027 (March 4, 1984).
    b) C. V. Shank, R. Yen, J. Orenstein and G. L. Baker, Phys. Rev. B 28, 6095 (1983).
17. F. Moraes, M. Schaffer, M. Kobayashi, A. J. Heeger and F. Wudl, Phys. Rev. B 30, 2948 (1984).
18. a). Z. Vardeny, E. Ehrenfreund, O. Brafman, M. Nowak, H. Schaffer, A. J. Heeger and F. Wudl, Phys. Rev. Lett. 56, 671 (1986).
    b). K. Kaneto, Y. Kohno and K. Yoshimo, Solid State Commun. 51, 267 (1984).
19. D. M. Gookin and J. C. Hicks, SPIE Advances in Materials for Active Optics, 567, 41 (1985).
20. M. Sinclair, D. Moses, A. J. Heeger, K. Vilhelmsson, B. Valk and M. Salour, Appl. Phys. Lett. (in press).
21. S. Etemad, reported at the March Meeting of the American Physical Society, Las Vegas, April 1, 1986.
22. Y. R. Shen, The Principles of Nonlinear Optics (Wiley-Interscience, NY, 1984).
23. C. Flytzanis, in Nonlinear Optical Properties of Organic and Polymeric Materials, Ed. D. J. Williams, American Chemical Society Symposium Series 233, 1983, Ch. 8.
24. C. Sauteret, J. P. Hermann, R. Frey, F. Pradere, J. Ducuing, R. H. Baughman and R. R. Chance, Phys. Rev. Lett. 36, 956 (1976).

When Kinks are not Elementary Excitations

S. Mazumdar

GTE Laboratories Incorporated, 40 Sylvan Road, Waltham, MA 02254, USA

The single most important ingredient missing in the single-particle soliton theories of conjugated polymers is the Coulomb interaction between electrons. The ground state bond alternation, and as a result the bond-alternation domain walls and the spin-charge relationship, survive even when electron-electron interactions are included. The optical properties of the pristine and doped systems are, however, remarkably different for moderately strong electron-electron interactions. We believe that experiments indicate conjugated polymers to be Mott-Hubbard semiconductors, as opposed to Peierls semiconductors.

I. Introduction

Even though a fairly large number of theoretical /1-7/ and experimental /8-12/ works in the past three to four years have sought to indicate the role of electron-electron interactions in π-conjugated polymers, these systems continue to be described within the SU-SCHRIEFFER-HEEGER (SSH) model /13-14/ or the continuum version /15-16/ of it. Aside from the fact that it is always difficult to replace an "established" concept with a new one, there are several other important reasons for the above. First and foremost, of course, is the fact that one-electron theory is simple and familiar, conceptually as well as computationally. Within band theory, all experimental behavior can be linked to a single phenomenon, -- bond-alternation and the consequent solitons or polarons in such a Peierls semiconductor. Secondly, because of the unique nature of the half-filled band, a large number of experiments (especially those that probe the ground state) can be seemingly explained successfully within the Peierls model, with electron-electron interactions playing at the most a perturbative role (the best example of this is the fitting of the ground state spin densities of trans-polyacetylene /17/). These "successes" have led to various modified Peierls semiconductor models with weak electron-electron interactions /17-18/. A third very important reason is that even investigators who have emphasized the role of electron-electron interactions have either probed completely general problems related to one-dimensional broken symmetries, or have tried to fit experimental bond alternation or optical and magnetic properties with specific correlation parameters. Thus, these investigations have (at least, according to the present author) left unanswered a very important question in the present context, viz., are π-conjugated polymers Peierls semiconductors with weak electron-electron interactions merely shifting the single-particle levels, -- or are they Mott-Hubbard semiconductors, as originally proposed by OVCHINNIKOV /19/, but with Peierls-type broken symmetry?

In the present talk, I will argue that conjugated polymers are Mott-Hubbard semiconductors. I will briefly discuss the ground state broken symmetry in a correlated system. We will see that bond alternation persists for any realistic value of the correlation parameters, and, therefore, experiments that probe only the formation of domain walls in the bond-alternated ground state are not proofs of the single-particle model. I will briefly only mention a few of these experiments, and will then proceed to discuss those experiments that I believe more clearly distinguish between the two models. Necessarily this latter set probes optical properties. A central theme of the talk is then as follows: while the ground state can have various kinds of domain walls, optical absorption from this or the doped and photoexcited states cannot be just loosely classified as absorption from or to soliton or polaron levels. If indeed π-conjugated polymers are Mott-Hubbard semiconductors, complete reinterpretation of all optical data will be necessary. Since the effects of electron correlation are difficult to visualize, I will also present a simplified (not overly, one hopes) real space picture that has already been successfully used /1,7/ to solve the ground state broken symmetry problem. It is hoped that the real space picture can substitute the single-particle momentum space picture that is valid only in the band limit. The present work is valid for conjugated polymers in general, but only a few examples have been chosen here. These are illustrated in Fig. 1.

(a) (b) (c)

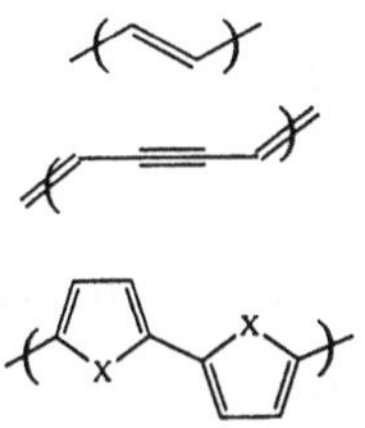

Fig. 1. Typical conjugated polymers: (a) trans-polyacetylene, (b) polydiacetylene, (c) X=NH: polypyrrole, X=S: polythiophene. The sp^2-carbons in (b) are attached to very large side groups

II. Theoretical Model

The model π- electron Hamiltonian we consider for polyacetylene (see later sections for descriptions of the other systems) is,

$$H = H_{1e} + H_{ee} \tag{1}$$

$$H_{1-e} = \sum_{i,\sigma} [t_0 - \alpha(y_i - y_{i+1})][c^{+}_{i\sigma} c_{i+1,\sigma} + c^{+}_{i+1,\sigma} c_{i\sigma}]$$

$$+ \frac{1}{2} K \sum_{n} (y_i - y_{i+1})^2 \tag{2a}$$

$$H_{e-e} = U \sum_{i} n_{i\uparrow} n_{i\downarrow} + \sum_{i,j} V_j n_i n_{i+j} \tag{2b}$$

Here H_{1e} is the standard SSH Hamiltonian /13/ in the adiabatic limit, $c^{+}_{i\sigma}$ creates an electron with spin σ (↑,↓) on the ith carbon $2p_z$ orbital, y_i the displacement of the ith CH unit, t_0 the nearest neighbor hopping integral and α and K are the electron-phonon coupling constant and the stretching constant respectively. H_{ee} is the many-electron term with $n_{i\sigma} = c^{+}_{i\sigma}c_{i\sigma}$, $n_i = \sum_{\sigma} n_{i\sigma}$, and U and V_j are the onsite and intersite Coulomb repulsions. In the absence of H_{ee} the band structure corresponding to H_{1e} consists of a gap at the Fermi level ($\pm\ \pi/2a$ for the half-filled band, where a is the lattice spacing) and the ground state has finite bond alternation ($\langle y_i \rangle = (-1)^i y_o$), for any nonzero α. We will consider here the possibility that the gap is dominated by H_{ee}/19/, in which case the band structure has broken down completely. Even when Coulomb interactions are included, only short-range interactions are investigated usually. We have included long-range interactions in accordance with the chemists' Pariser-Parr-Pople (PPP) model.

III. The Ground State Broken Symmetry

Probably the most important reason why the Peierls model was accepted for such a long time was the Hartree Fock (HF) result that bond alternation within (1) vanishes for strong H_{ee}, when the ground state has a spin density wave (SDW) with uniform bond lengths. Since finite bond alternation is found experimentally, H_{ee} had to be small within mean field and perturbative theories. Thus, even in quantum chemical calculations /14/, HF type approaches are justified on the basis of the above theoretical result. In a series of papers, the present author and his collaborators have shown /1,7/ that the above theoretical result, viz., bond alternation persists for only small H_{ee}, is completely erroneous and an artifact of approximations. This conclusion, by itself, does not settle the argument about conjugated polymers. However, I believe that the physics behind the broken symmetry problem is very instructive, particularly if one considers that essentially exact results have been obtained even when the <u>method</u> is not exact, and even when arbitrarily long-range interactions in H_{ee} are included. These results are given below.

To start with, there are three possible broken symmetries in a 1-d half-filled band: (i) the bond-order wave (BOW), which drives the bond-alternation for $\alpha \neq 0$, (ii) the charge density wave (CDW), which favors modulation of site charge densities, and (iii) the SDW. The possibilities depend on how many terms in H_{ee} are included:

(i) For $V_j = 0$ for all j (Peierls-Hubbard model) the competition is between a BOW and a SDW. The BOW is initially enhanced by U until a maximum in the bond-alternation is reached at $U \sim 4t_o$, beyond which the bond-alternation decreases smoothly until $U = \infty$ /1/. (Several authors have recently pointed out that the peak in the bond-alternation shifts to smaller U for large α, but this point was already discussed in /1/: we are interested only in $\alpha \to 0$ unconditional broken symmetry, since for large enough α bond alternation cannot disappear at any U, even within HF theory). Thus, the BOW and the SDW always coexist in the half-filled band.

(ii) The possibility of the vanishing of the BOW arises only when we have intersite interactions that may favor a CDW. With $V_1 > 0$, but $V_j = 0$ for $j \geq 2$ (extended Peierls-Hubbard model), the ground state is a BOW for $V_1 < U/2$

/1/). Nonzero V_1 can enhance the BOW very strongly, even for rather large U /1/.

(iii) For arbitrarily long-range V_j the solutions are the same as (ii) if the following inequality holds,

$$V_{j+1} + V_{j-1} \geq 2V_j \tag{3}$$

If (3) is true then the ground state is either a $2k_F$ BOW or a $2k_F$ CDW, the phase-boundary being given by,

$$\sum_j V_{2j+1} \lessgtr \frac{1}{2}U + \sum_j V_{2j} \tag{4}$$

where a smaller left-hand side favors the BOW, and a smaller right-hand side favors the CDW /7/. Irrespective of whether the chemists' PPP model is applicable to conjugated polymers, we find that substitution of the PPP-Ohno or Mataga-Nishimoto parameters in (3) and (4) give enhanced BOW for both parameterizations /7/.

To summarize this section, only a CDW destroys the BOW, which coexists with the SDW for any H_{ee}. Furthermore, the $2k_F$ BOW is the dominating broken symmetry for any realistic H_{ee}.

IV. Solitons in a Correlated System

We can expect domain walls between the so-called A and B-phases of bond alternation to persist as low-energy excitations as long as bond alternation itself persists. A convenient way of checking this is to calculate the energy of an odd chain, which has an even number of double bonds and, therefore, can exist in two forms, dimer (D) and soliton (S), as shown in Fig. 2. For the soliton to exist, a necessary condition is that the ground state of an odd chain resembles S. We consider here only the sharp or single-site soliton, which is higher in energy than the true optimally relaxed soliton. Therefore, if $E_S < E_D$, where E_S and E_D are the energies per site of the two configurations in Fig. 2, the true soliton is necessarily stable.

Fig. 2. The two possible configurations for a N=5 odd chain: dimer (D) and soliton (S)

For $H_{ee} = 0$ in (2), $E_S < E_D$, irrespective of the charge on the chain /20/. Furthermore, within the orbital approximation there is always a nonbonding orbital at exactly the Fermi energy for an odd chain, and therefore $E_D{}^o = E_D{}^{\pm}$, $E_S{}^o = E_S{}^{\pm}$, where the superscripts 0 and ± refer to the charge on the chain. For $H_{ee} > 0$, CAMPBELL, DEGRAND and MAZUMDAR, using a quantum Monte Carlo approach, have calculated the above electronic energies corresponding to a fixed bond alternation for a chain of length 21 atoms/5/. Again, $E_S < E_D$, but now both the absolute energies <u>and</u> the stabilization

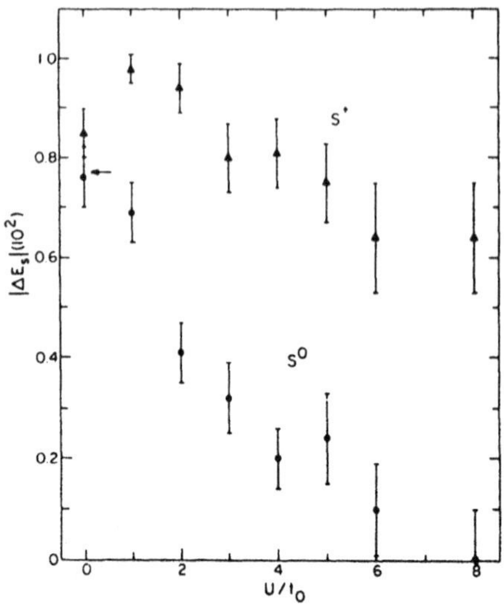

Fig. 3. The magnitude of the energy difference per site, $|\Delta E_s|$, between the soliton and the pure dimer configuration on an open N = 21 chain vs. U. The solitons are always lower in energy. The curve labeled S^+ is for the positively charged soliton, that labeled S^0 is for the neutral soliton. Both are in units of t_0. The arrow marks the exact $|\Delta E_s|$ for U = 0

energy of the soliton, defined as, $\Delta E_S = E_D - E_S$, depend on the charge on the chain. We show here only the result for the Peierls-Hubbard model (V_j = 0); the result for $V_1 > 0$ may be found elsewhere /5/. In Fig. 3, we have plotted $|\Delta E_S|$ against U/t_0 for $\alpha y_0 = 0.1t_0$, which is close to the experimental bond alternation. The stabilization energy ΔE_{S^0} for the neutral soliton decreases rapidly with U, but the stabilization ΔE_{S^+} for the positive soliton remains quite large even at $U = 8t_0$. (We emphasize that $E_{S^0} < E_{D^0}$ even at $U = \infty$, but $|\Delta E_{S^0}|$ is too small to be calculated within the Monte Carlo method beyond $U = 8t_0$) We show the result for S^+ only, since $\Delta E_{S^+} = \Delta E_{S^-}$.

We note that our calculations are only for the degenerate polymer trans-polyacetylene, but the results can be easily carried over to the case of nondegenerate polymers. More importantly, the rather large ΔE_{S^+} points to the strong stabilization of dopant-induced solitons even when H_{ee} is large, and, therefore, epr measurements that prove formation of spinless domain walls do not say anything about the magnitude of H_{ee}. The spin-charge relationship in commensurability 2 systems is more general than it is thought.

V. Excited States

As has been emphasized by SOOS and coworkers /4,21-23/, the effects of electron correlations will be most clearly visible on excited states. Among such effects, the occurrence of the two-photon 2^1A_g state below the optically allowed 1^1B_u state in polyenes is well understood /9,21/. I will, therefore, not spend time on this. I will rather discuss here three more recent developments, which involve (a) neutral soliton absorption in trans-polyacetylene /11,22/, (b) recent theoretical /23/ and experimental /24-26/ work on the polydiacetylenes and finite fragments of the polymer, and, in less detail for the time being, (c) the so-called polaron and bipolaron absorptions in doped and photoexcited nondegenerate polymers /27,28/. As in our original work on the ground state broken symmetries /1,7/, the emphasis will be on a general understanding, and not on fitting the data. We will use here the real space approach previously formulated. The chief motivation will be to demonstrate that electron correlations in conjugated polymers are large enough that they should be classified as Mott-Hubbard semiconductors.

A. Neutral Soliton Absorption in Trans-Polyacetylene

Within one-electron theory, an odd open chain (soliton) has a nonbonding orbital exactly at the Fermi level. Absorptions to or from the nonbonding level are expected in S^+ or S^-, while both can occur in S^o. Thus, irrespective of the charge on the chain, midgap absorptions are expected due to solitons in a degenerate polymer. Shifting of the midgap level occurs at the HF level of calculation, and even though $S^\pm$ and S^o absorptions are no longer degenerate, subgap absorptions are still expected in a Peierls semiconductor.

Calculations based on the PPP-Ohno model by SOOS et al, however, predicted that the neutral odd chain absorbs at nearly the same frequency as the homogeneous even chain /2/. This was confirmed experimentally by WEINBERGER et al. /11/. Subgap transitions that occur in the doped polymer correspond to $S^\pm$ absorptions. As discussed below, the pristine polymer must be a Mott semiconductor for this to happen.

For small H_{ee}, where Fermi liquid theory is valid, one visualizes electrons being replaced by quasi-particles. In the present case, however, we need to start from a real space view, where the ground state is thought to be covalent, i.e., each site is singly occupied by electrons. Of course, this is the solution to (1) only when $H_{1e}=0$. For $H_{1e} \neq 0$ (but $H_{ee} > H_{1e}$) one can conceive of H_{1e} being "switched on" in a "reverse Fermi liquid" sense, and the ground state can be thought of as a "dressed" covalent valence bond diagram. Optical absorption is best understood in terms of the conductivity operator,

$$\tau \sim \sum_{i,\sigma} (c^+_{i\sigma}c_{i+1,\sigma} - c^+_{i+1,\sigma}c_{i\sigma}) \tag{5}$$

where we have neglected the constants of proportionality. From (5), optical absorption in a Mott-Hubbard semiconductor is a transition from "dressed" covalent to a "dressed" singly ionic valence bond diagram, where the latter contains one doubly occupied and one empty site. For moderate to large H_{ee}, this energy difference ($\sim U-V_1$) is the same for both odd and even neutral chains. It should be obvious that the ground state in a charged chain can similarly be thought of as a "dressed" many-electron configuration with the least number of double occupancies, but now nearest neighbor transfers that do not create additional double occupancies are possible. These, and additional lower energy transfers that lead to subgap transitions, will be discussed later, in connection with "bipolaron absorptions".

B. Excited States in Polydiacetylenes

Polydiacetylenes, unlike other conjugated polymers, can be made as single crystals with virtually infinite chain lengths, very bulky side groups and large interchain distances. These, therefore, provide the means to identify properties that can be associated with a single chain. Within one-electron theory, the band structure now consists of four subbands /29/, of which the lower two are completely filled and comprise the valence band, while the upper two are empty and comprise the conduction band. Several problems arise immediately with such a theoretical description, as discussed below.

The most severe problem is that unlike what is seen in other conjugated polymers, optical absorption in the polydiacetylenes /24/ consists of a rather sharp transition, unexpected in a band-to-band transition. More importantly, the absorption edge is at 2.0 eV, while photoconductivity begins at 2.5 eV only, where optical absorption shows almost vanishing oscillator strength /30/. Clearly then, optical transition here is to an excitonic state, in contrast to band theory predictions.

Aside from the above, band theory predicts a single π-π^* transition, while experimental spectra indicate that beyond the sharp transition at 2 eV a broad feature peaks at 2.2-2.3eV. The second feature, until now, has been explained as a vibrational sideband (probably because there is no other choice within one-electron theory). Two recent experiments cast serious doubt on this. Firstly, ECKHARDT /25/ has determined that high resolution low-temperature piezomodulated reflection spectra for single crystals indicate two nearby 0-0 transitions. Secondly, finite chain analogs of polydiacetylene containing 6-38 carbon atoms have been recently synthesized by WUDL /26/. These clearly indicate two close-lying electronic absorptions, separated by 0.25 eV.

The spectra of polydiacetylene fragments show another unexpected feature. Within one electron theory an optical gap is supposed to increase with bond alternation in both finite as well as infinite chains. The lowest absorption in the smallest polydiacetylene fragments occurs at a frequency that is lower than the absorption frequency of the corresponding polyene containing the same number of carbons /26/. Clearly, bond alternation does not dominate the optical gap since, within the bond-alternation model, polydiacetylene oligomers are expected to have a larger absorption threshold. The situation reverses in long chains, where polydiacetylene fragments absorb at a higher frequency than the corresponding polyenes, and the correct theoretical model must also be able to explain this reversal. Our recent theoretical work, which explains all the above anomalies, at least qualitatively, is discussed below.

The occurrence of the exciton within (2) is expected, at least as long as the V_j are rapidly decaying. This is physically easiest to visualize when there is a substantial $V_1 < U/2$ but all other $V_j = 0$ from V_2 onwards. In the delocalized band limit of (2) the ground state of a N-carbon system is dominated by configurations with N/4 double occupancies (maximum number of double occupancies = N/2), and nearest neighbor charge transfers can both create and destroy additional double occupancies in real space. In the correlated case, the ground state consists of "covalent" configurations, and charge transfer leads to configurations with neighboring C^+ and C^- ions, which are at an energy $\sim U-V_1$ relative to the ground state. For conductivity, it is necessary to separate the C^+C^- pair, which now requires an additional energy V_1 that gives rise to the exciton binding. We emphasize that the above estimates of energy gaps etc. are strictly valid only in the limit $H_{1e} = 0$, but the qualitative picture remains valid for finite H_{1e} as long as the optical gap is a Mott-Hubbard gap and we think of the ground and excited states in the "reverse Fermi liquid" sense described above. One more important point to be noted here is as follows: the magnitudes of U and the V_j in polyacetylene and the polydiacetylenes are expected to be the same, while the bond-alternation contribution to the optical gap in the latter is larger. Therefore, any difference in the spectra (i.e., exciton vs. broad band) must originate in the crystalline nature of polydiacetylenes and the large separation between the strands.

The occurrence of two low-lying electronic absorptions in the polydiacetylenes, as well as a lower absorption threshold in the smaller polydiacetylene oligomers than the corresponding polyenes, however, requires additional Coulomb interactions, viz., the interaction between the electrons occupying the orthogonal π and π orbitals of the triple bond,

$$\sum_{i,j} V'_j n_i^{\pi} n_{i+j}^{\pi'} \qquad (6)$$

where the π and π' refer to the conjugated and localized π-bond, and the term corresponding to j=0 is the on-site repulsion between electrons occupying orthogonal p-orbitals on the same atom. The model may be considered a modified degenerate PPP-model, the terminology being borrowed from the degenerate Hubbard Hamiltonian of solid state physics. In our calculations, we have presently assumed that $V_j = V'_j$, but this need not be so, and further work on reasonable estimates of V'_j is in progress.

The chief effect of H_{ee} is to mix in "pseudocovalent" configurations in the ground state /23/, defined in Fig. 4.

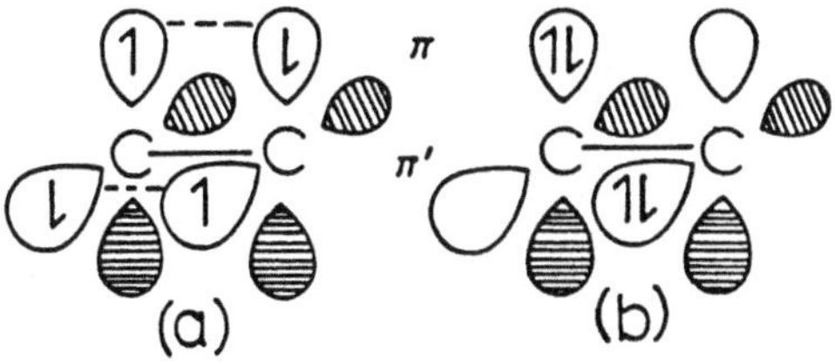

Fig. 4. The two possible low-energy configurations on the triple bond segment of polydiacetylene: (a) covalent diagram with π and π' bonds; (b) pseudocovalent diagram, with π and π' lone pairs on adjacent sites

Because the ground state now has both covalent and pseudocovalent configurations we have two different kinds of charge transfer leading to two low-lying absorptions. We have confirmed this in detailed calculations on finite ene-yne modules /23/. Furthermore, the ionic configuration reached from the covalent configuration now interacts with the low-energy pseudocovalent configuration and this lowers the optical threshold. Calculated as well as experimental optical thresholds in polydiacetylene oligomers are therefore smaller than those in the corresponding polyenes. The π' orbitals are localized while the π orbitals are delocalized, and as a result the $\pi \to \pi^*$ absorption energy decreases rapidly with chain length while the $\pi' \to \pi'^*$ absorption remains the same. Due to this, we expect the π-π' interaction to decrease with the chain length, and beyond a critical length the larger bond alternation in the polydiacetylene takes over so that at $N \to \infty$ the polydiacetylenes have a larger absorption energy. Indeed, the smaller absorption frequency in ene-yne molecules can also be explained within other mechanisms (for e.g., by including different site energies for the sp^2 and sp-carbons, but we believe that the crossing of absorption thresholds as a function of chain length necessarily indicates a chain-length-dependent interaction, and H_{ee} seems to be the only such realistic interaction. An alternative possibility involving transition from butatrienic to acetylenic structure in diradicals has been discussed by SIXL /26/, but this mechanism cannot apply to the stable oligomers discussed here.)

C. Subgap Absorptions in Nondegenerate Polymers

The one-electron description of nondegenerate polymers is as follows. Defects give rise to nondegenerate localized levels in the bandgap that are symmetrically shifted from the midgap. Single charge injection produces polarons, with either the lower localized level singly occupied (P^+), or the lower level completely filled with two electrons and the upper level singly occupied (P^-). When a second like charge is introduced, the two polarons are bound into a bipolaron (B^{2+} or B^{2-}) in which the defect levels are either completely empty or completely filled. Multiple subgap absorptions are, therefore, expected in doped nondegenerate polymers, as shown in Fig. 5. Furthermore, bipolarons are spinless, and therefore "bipolaron-doping" must be observable in magnetic experiments.

A large volume of recent work has claimed to have observed bipolarons in doped polypyrrole /27a/, poly-β-β'-dimethyl-pyrrole /27b/ and polythiophene /28/. Very recently, photogeneration of bipolarons in polythiophene has been claimed from the photoinduced spectrum of polythiophene /28/. Further-

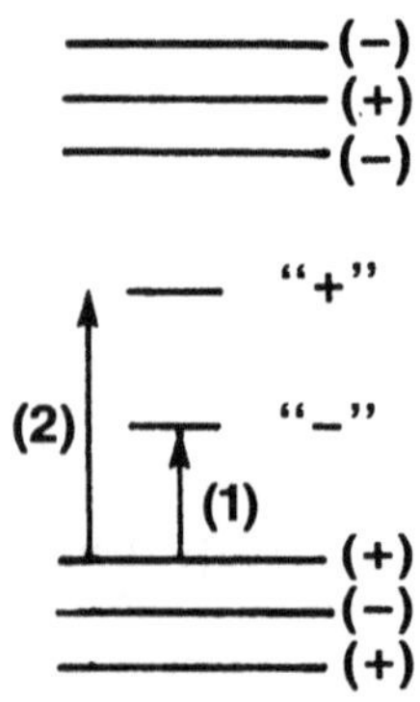

Fig. 5. Bipolaron transitions in one-electron theory. The (+) and (-) correspond to symmetry characters (see text); the subgap levels acquire this character the more they move away from the mid gap. The second absorption is "forbidden" as a consequence of the mirror plane symmetry. Additional transition between the subgap levels is expected in a polaron

more, it is generally assumed that a similar process is occurring in other conjugated polymers. Let us review here the experimental evidence and simultaneously present the problems with the above descriptions as well as a many-electron description of these systems.

First of all, as already emphasized in section IV, occurrence of domain walls and spinless charge carriers does not prove the validity of either the Peierls or the Mott-Hubbard model. Thus, epr measurements prove bipolaron-doping, but <u>not</u> the band model. On the other hand, the experimental optical spectra of doped polymers is absolutely different from theoretical predictions, -- it is not enough to just correlate the <u>number</u> of predicted transitions with experiments. The theoretical intensity ratio I_1/I_2, where I_1 and I_2 correspond to absorptions 1 and 2, increases progressively as the defect levels are shifted more and more from midgap. From the original calculations of FESSER et al /16/, and the experimental ω_1 and ω_2 in the polymers studied, one would predict $I_1/I_2 \geq 6$, -- while in reality $I_1/I_2 \leq 2$ /27,28/.

At this point, it may be conjectured that the anomalously large intensity of absorption 2 is due to the effect of the heteroatom, nitrogen or sulfur in Fig. 1(c). As originally implied by FESSER et al /16/, however, a large I_1/I_2 is a consequence of the <u>symmetry</u> in the problem, and as we qualitatively discuss here, this involves the <u>mirror-plane symmetry</u> that passes through the center of an even chain. All eigenfunctions can be classified as plus and minus with respect to this symmetry plane, as shown in Fig. 5, and the subgap levels have more of valence band-conduction band character the further they are from the midgap. Due to the alternation of the symmetry characters of the levels in Fig. 5, absorption 1 is strongly allowed while 2 is allowed only weakly. Now the effect of the heteroatom is to create inequivalent carbons, i.e., carbons which are directly bonded to the N or S (α-carbons) have electronegativities (site energies) that are different from the β-carbons, and the polymer can be classified as an A_2B_2 type. This destroys the <u>electron-hole symmetry</u>, but not the mirror plane symmetry, and, therefore, would still predict a large I_1/I_2.

How do we then explain comparable I_1 and I_2 ? In Fig. 6, we show a positively charged chain, for the case where H_{ee} dominates the optical gap. All sites but one are singly occupied. Aside from the regular charge transfers, there are two additional low-energy transfers here, indicated by the curved arrows 1 and 2 in Fig. 6. In a negatively charge chain, there are the same two charge transfers, only now in the opposite direction. We again approximate the effects of long-range interactions by a single V_1. Absorption 1 occurs at zero frequency in the limit when all bond lengths are equal, but occurs at finite frequency when bond alternation is finite. Absorption 2 occurs at $U-2V_1$ (as before, this is the Coulomb contribution only, and it gets modified by finite H_{1e}). The important point is that neither of the processes 1 and 2 are symmetry forbidden. Their intensities are directly proportional to the probabilities of the particular charge transfers (which are the same for both 1 and 2, -- 2/N) and inversely proportional to the energy gaps ω_1 and ω_2 (with again modifications due to finite H_{1-e} as well as the difference in the two transfer integrals). Given the ω_1 and ω_2 then, it is not at all surprising that $I_1/I_2 \lesssim 2$.

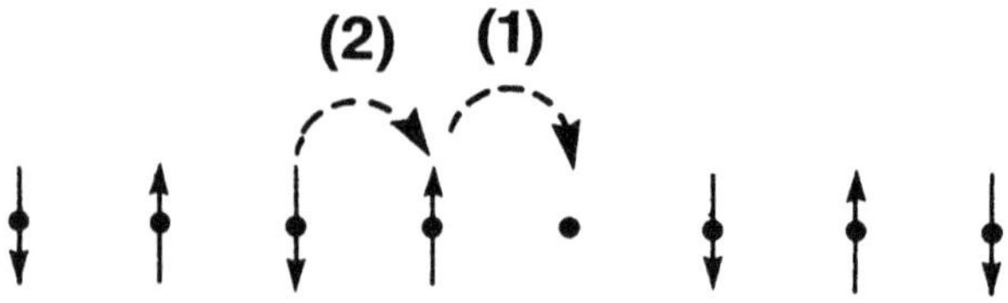

Fig. 6. Positively charged even chain; vertical solid arrows: electrons occupying sites (dots); curved dashed arrows: low-energy nearest neighbor electron transfers leading to subgap transitions (see text)

Conclusions

Only a brief summary could be presented here, due to space limitations. Also, the presentation necessarily dealt with the theoretical and experimental work I am most familiar with. For more detailed discussion of correlation effects, the reader is advised to consult the original references. The chief conclusion of the present work is that magnetic data which prove the existence of bond-alternation domain walls are not enough to prove the validity of the one-electron model. On the other hand, optical data consistently indicate that these systems are Mott-Hubbard semiconductors with bond alternation. Single isolated strands of any conjugated polymer are expected to behave like polydiacetylene. No attempt to explain the photoinduced absorptions in polyacetylene was made here, because theoretical analysis is far from complete. On the other hand, occurrence of the triplet state below the exciton in polydiacetylene /12/ clearly confirms the strength of Coulomb correlations.

Acknowledgements

Various parts of the ground state broken symmetry problem were done with S.N. Dixit, D.K. Campbell and T.A. DeGrand. The work on the neutral soliton absorption was done by Z.G. Soos and coworkers, and I have merely quoted their result. Investigation of the excited states in polydiacetylene is being pursued in collaboration with Z.G. Soos and S. Kuwajima. Innumerable fruitful discussions with D. Baeriswyl are gratefully acknowledged.

References

1. S. Mazumdar and S.N. Dixit, Phys Rev. Lett., 51, 292 (1983); S.N. Dixit and S. Mazumdar, Phys. Rev. B 29, 1824 (1984)
2. J.E. Hirsch, Phys. Rev. Lett., 51, 296 (1983); J.E. Hirsch and M. Grabowski, Phys. Rev. Lett., 52, 1713 (1984).
3. A.A. Ovchinnikov and I.I. Ukrainskii, Sov. Sci, Rev. 1984 and references therein.
4. Z.G. Soos and S. Ramasesha, Phys. Rev. B 29, 5410 (1984); S. Ramasesha and Z.G. Soos, J. Chem. Phys. 80, 3278 (1984).
5. D.K. Campbell, T.A. DeGrand and S. Mazumdar, Phys Rev. Lett., 52, 1717 (1984). See Mol. Cryst. Liq. Cryst. 118, 41 (1985) for $V_1>0$ results.
6. D. Baeriswyl and K. Maki, Phys. Rev. B. 31, 6633 (1985). See also D. Baeriswyl in present proceedings
7. S. Mazumdar and D.K. Campbell, Synthetic Metals, 13, 163 (1986), and Phys. Rev. Lett. 55, 2067 (1985).
8. H. Thomann, et al Phys. Rev. Lett., 50, 553 (1983).
9. B.S. Hudson and B.E. Kohler, Synth. Met., 9, 241 (1984).
10 J. Orenstein and G.L. Baker, Phys. Rev. Lett. 49, 1043 (1982).

11. B.R. Weinberger, et al, Phys. Rev. Lett. 53, 86 (1984).
12. L. Robins, J. Orenstein and R. Superfine, Phys. Rev. Lett., 56, 1850 (1986).
13. W.P. Su, J.R. Schrieffer, and A.J. Heeger, Phys. Rev. B 22, 2099 (1980).
14. J.L. Brédas and G.B. Street, Acc. of Chem. Res. 18, 309 (1985) and references therein.
15. H. Takayama, Y.R. Lin-Liu and K. Maki, Phys. Rev. B. 21, 2388 (1980).
16. K. Fesser, A.R. Bishop and D.K. Campbell, Phys. Rev. B 27, 4804 (1983).
17. M. Grabowski, D. Hone and J.R. Schrieffer, Phys. Rev. B 31, 7850 (1985).
18. S. Kivelson and W.K. Wu, Mol. Cryst. Liq. Cryst. 118, 9 (1985).
19. A.A. Ovchinnikov, I.I. Ukrainskii and G.V. Kventsel, Usp. Fiz. Nauk 108, 51 (1972) (Sov. Phys. - Usp. 15, 575 (1973)).
20. W.P. Su, Sol. St. Commun., 35, 899 (1980).
21. S. Ramasesha and Z.G. Soos, Synth. Met. 9, 283 (1984).
22. Z.G. Soos and L.R. Ducasse, J. Chem. Phys. 78, 4092 (1983).
23. Z.G. Soos, S. Mazumdar and S. Kuwajima, in Solid State Polymerization (ed. D.J. Sandman, ACS Symposium Series, 1986), in press.
24. Polydiacetylenes, NATO ASI series E102, edited by D. Bloor and R.R. Chance, Martinus Nijhoff, Dordrecht, The Netherlands, 1985.
25. C.J. Eckhardt in ref. 23.
26. F. Wudl in ref. 23. See also, Hans Sixl, Mol. Cryst. Liq. Cryst., 134, 65 (1986).
27(a) J.L. Brédas, et al, Phys. Rev. B 30, 1023 (1984);
(b) K. Yakushi, et al, J. Chem. Phys. 81, 4133 (1984).
28. T.C. Chung, et al, Phys. Rev. B. 30, 702 (1984); Z. Vardeny et al., Phys. Rev. Lett., 56, 671 (1986).
29. E.G. Wilson, J. Phys. C 8, 727 (1975).
30. K. Lochner, et al, Phys. Stat. Sol. (b) 88, 653 (1978).

Phonons and Soliton Diffusion in Polyacetylene

Y. Wada

Department of Physics, University of Tokyo, Bunkyo-ku, Tokyo 113, Japan
Institute for Theoretical Physics, University of California,
Santa Barbara, CA 93106, USA

1. Introduction

Dynamical properties of solitons have been interesting subjects in trans-polyacetylene. We have tried to study them with the help of soliton-phonon interactions. It has been necessary to find structure of the phonons, when there is a soliton. We have studied the continuum model by Takayama, Lin-Liu, and Maki (TLM) and the discrete model by Su, Schrieffer, and Heeger (SSH). It is found that a soliton generates a new localized phonon mode, which is infrared active, and a polaron does so as well. Using the amplitude mode formalism, the infrared activities of these modes are investigated to identify the two small peaks, observed by a recent photo-induced absorption experiment. Phase shift analyses for extended phonon modes show there are two mechanisms in the second order processes of the soliton-phonon collision. One is a shift in the soliton location and the other a momentum transfer to the soliton. The interplay of the two mechanisms is clarified by studying the dynamical diffusion coefficient of the soliton for classical phonon distribution.

2. Localized Phonon Modes

In TLM, the time-dependent order parameter is written

$$\Delta(x,t) = \Delta_s(x) + \Delta_1(x,t), \qquad (1)$$

where Δ_s is the static solution. Rewriting the electronic wave functions as well, we linearize the equations of motion with respect to the deviations Δ_1 and so on. We obtain an eigenvalue problem for the phonon modes, with the help of an adiabatic approximation for electronic motion, in a form of an integral equation for Δ_1 [1,2]. The kernel is given as a double sum over the occupied and unoccupied states of electrons. It is first estimated numerically as a double integral over two wave numbers k and k'. A cutoff Λ is introduced. It depends on the coupling strength. In the weak coupling limit, where TLM is valid, it becomes infinite. Solutions are obtained with the system size $L = 20\xi$ where ξ is the soliton width. The integral equation is replaced by linear coupled equations, using a mesh $dx = 0.1\xi$. The cutoff is assumed to be $\Lambda = \pi/dx$.

Numerical solution gives three localized phonon modes for the soliton. The third mode is a new discovery. It has a frequency close to the minimum frequency of extended optical phonons. It is infrared active, together with the Goldstone mode which has a small but non-vanishing frequency. It turns out that all the three eigenfrequencies reduce as Λ increases [3]. In particular, the lowest frequency vanishes in the large Λ limit, as a Goldstone mode should. For the polaron, we find five localized modes [4]. The Goldstone mode and the fourth mode are infrared active.

It is suggested in [5] that, when there is a soliton, the electronic wave functions have to satisfy a boundary condition different from the ordinary periodic condition [6] and it would decrease the numbers of the localized modes. The correct boundary condition modifies the kernel since the electronic wave number k has a small shift from the ordinary $2\pi n/L$. We find that the modified kernel still reproduces the three localized modes for the soliton [2]. The third mode has a very long tail. It is shown that a large system size $L > 50\xi$ is necessary to determine whether the third mode is localized or not, if use is made of the coupling strength appropriate for the trans-polyacetylene.

To study the phonons in SSH, TERAI and ONO [7] have used an iteration method to determine the static solutions for the soliton and the polaron. They have taken 201 lattice sites and 200 electrons for a charged soliton. The static order parameters are shown to involve large acoustic components. The eigenvalue problem for the phonons becomes a diagonalization of a matrix. The results are shown in Figs. 1 and 2 for the soliton and the polaron, respectively. In addition to the two branches of the extended modes, there are four and seven localized modes indicated by arrows. The soliton has one additional staggered localized mode at $k = \pi/2a$, and the polaron two, a being the lattice constant. They are infrared inactive.

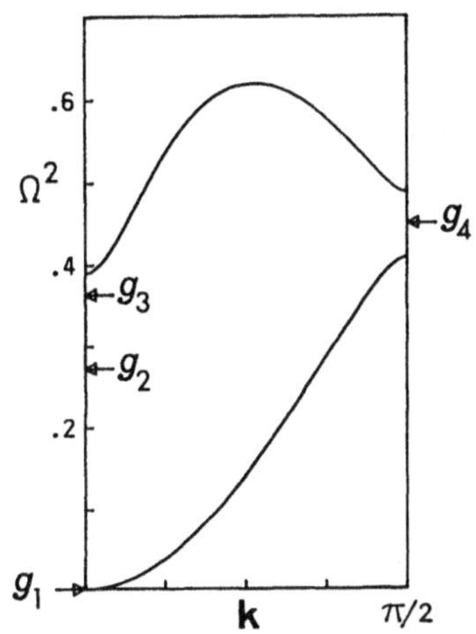

Fig. 1. The phonon dispersion relation for the soliton, g_i's being localized modes, from [7]: The ordinate is scaled by the square of bare phonon frequency and the abscissa by 1/a. TLM gives g_1, g_2, and g_3, while the staggered mode g_4 is characteristic of SSH.

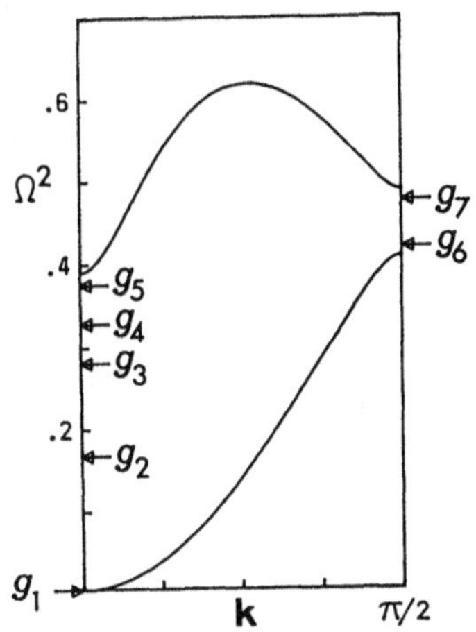

Fig. 2. The phonon dispersion relation for the polaron, g_i's being localized modes, from [7]: Scaled as in Fig. 1. TLM gives $g_1 \sim g_5$. The staggered modes g_6 and g_7 are characteristic of SSH.

There are three other groups who have been working in SSH. SUN, WU, and SHEN [8] have studied the soliton to obtain four localized modes at k=0 and one staggered mode at $k = \pi/2a$, using stronger coupling constants. CHAO and WANG [9] have obtained the results consistent with [7]. With the help of a semi-classical approximation, GAMMEL [10] has found the same result for the soliton, and even a possible sixth mode at k=0 for the polaron. The theoretical results are, thus, in good agreement with one another, as far as the third mode for the soliton and the five modes for the polaron are concerned.

3. Infrared Absorption

It has been shown in the amplitude mode formalism [11] that there are three physical phonon modes, corresponding to each optical mode in TLM. The third mode of the soliton thus generates three B modes. We have calculated the peak positions as well as their relative intensities of absorption for $(CH)_x$ and $(CD)_x$ [12]. HICKS and MELE [13] have calculated them, taking account of the fourth mode of the polaron. It turns out there are practically no significant differences between the two.

Recently, two experiments have been independently performed for photo-induced infrared absorption. EHRENFREUND and coworkers [14] have discovered two small peaks in partially isomerized trans samples, in addition to the large T mode peaks. It has, however, turned out that they are no longer observable in fully isomerized $(CH)_x$ sample. On the other hand, SCHAFFER and HEEGER [15] have observed the two peaks. Their peak positions and relative intensities are in rather good agreement with the theoretical results.

Although polaron's fourth mode gives an equivalent description, we believe the identification with the third mode of soliton is more plausible. The experiment has been carried out at the steady state condition, while photogenerated polarons are supposed to be transient.

4. Extended Phonon Modes

In the region far from the soliton, extended phonon functions with even and odd parities can be written

$$g_{eq}(x) = \cos(qx + \delta_e(q)/2), \text{ even}, \quad x \gg \xi, \tag{2}$$

$$g_{oq}(x) = \sin(qx + \delta_o(q)/2), \text{ odd}, \quad x \gg \xi, \tag{3}$$

respectively. Here δ_e and δ_o are even and odd phase shifts which are shown in Fig. 3. It is remarkable that they agree with each other at $q\xi > 1$. It means that the soliton gives these phonons an effective potential which is reflectionless. The phonons with $q\xi < 1$ suffer reflections.

5. Soliton-Phonon Interactions

The Goldstone mode of the soliton does not interact with the other phonon modes in the first order, since they are both normal modes. In the higher order processes, they interact with each other. We study the soliton-phonon collision [16], expanding the order parameter

$$\Delta(x,t) = \Delta_s(x) + \Delta_1(x,t) + \Delta_2(x,t) + \cdots, \tag{4}$$

where Δ_1 is the incoming phonon which includes the effect of the phase

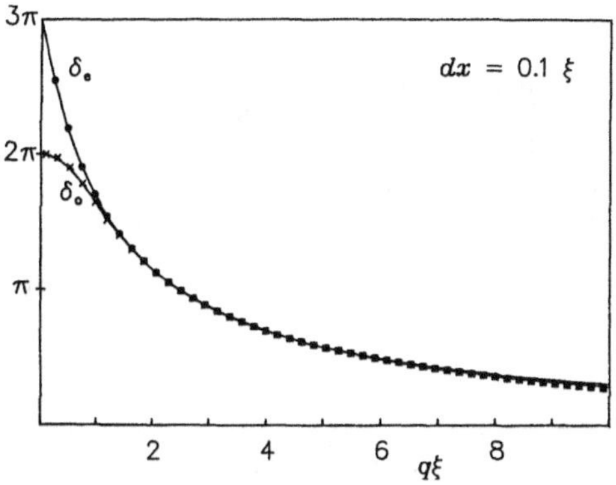

Fig. 3. Phase shifts in the phonon eigenfunctions: δ_e for the even function and δ_0 for the odd function, dx being the mesh, from [2]

shifts. The third term Δ_2 is quadratic with respect to the incoming phonon amplitude. It includes the Goldstone component. The coefficient of the component gives the shift in the soliton location in a form of a Fourier integral over phonon frequencies. For the phonons in the reflectionless regime, the integrand has a simple pole, giving rise to a finite shift. It turns out to be from four to six times as large as that of a ϕ^4-soliton. The phonons with reflection modify the integrand to have a pole of order two. The shift becomes proportional to t. It means the phonon transfers momentum to the soliton which begins to move with a velocity. The soliton would, then, be affected by a friction due to the phonon medium and would stop ultimately. In trans-polyacetylene, the optical phonons are not thermally excited, since their frequency ω_0 is as high as 0.18ev. It would, however, be interesting to consider the case with a small ω_0 to see how the two mechanisms influence the soliton motion. Using collective coordinate method, we can introduce soliton coordinate Q_0 [17,18]. Equation of motion for the coordinate is written in a form of a generalized Langevin equation. The dynamical diffusion coefficient $D(\omega)$ is given by

$$D(\omega) = \langle \dot{Q}_0 . \dot{Q}_0 \rangle / (i\omega + \Gamma(\omega)), \tag{5}$$

where $\Gamma(\omega)$ is estimated as a Fourier-Laplace transform of a total force correlation function. It is calculated, in low-frequency and low-temperature region, to be

$$\mathrm{Re}\Gamma(\omega) = (0.001 + 50(\omega/\omega_0)^2) k_B T\ \omega_0/\Delta_0 + O(T^2), \tag{6}$$

$2\Delta_0$ being the energy gap. The first term in the parentheses comes from the momentum transfer processes and the second from the finite shift. In the static limit $\omega = 0$, Eq.(5) gives

$$D(0) = (4 \times 10^2)\pi\xi^2\omega_0. \tag{7}$$

When $\omega > \Gamma(\omega)$ is satisfied, we get

$$\mathrm{Re}D(\omega) = (0.001(\omega_0/\omega)^2 + 50)(3\pi/8)(k_B T/\Delta_0)^2 \xi^2 \omega_0. \tag{8}$$

6. Summary

The structures of phonons around a soliton and a polaron in trans-polyacetylene are understood. Particularly, the third mode of the soliton is almost established experimentally as well as theoretically. On the other hand, it is evident that the optical phonons are irrelevant as far as the soliton diffusion is concerned, because of their high frequency. Presumably, acoustic phonons are playing important roles, however weak their coupling may be. We still have a long way to go in order to understand the soliton diffusion in trans-polyacetylene.

References

1. H. Ito, A. Terai, Y. Ono, Y. Wada: J. Phys. Soc. Japan 53, 3520 (1984)
2. Y. Ono, A. Terai, Y. Wada: J. Phys. Soc. Japan 55, 1656 (1986)
3. A. Terai, H. Ito, Y. Ono, Y. Wada: J. Phys. Soc. Japan 54, 4468 (1985)
4. A. Terai, H. Ito, Y. Ono, Y. Wada: J. Phys. Soc. Japan 54, 196 (1985)
5. J. C. Hicks, G. A. Blaisdell: Phys. Rev. B31, 919 (1985)
6. S. Kivelson, T.-K. Lee, Y. R. Lin-Liu, I. Peschel, L. Yu: Phys. Rev. B25, 4173 (1982)
7. A. Terai, Y. Ono: J. Phys. Soc. Japan 55, 213 (1986)
8. X. Sun, C. Wu, X. Shen: Solid State Commun. 56, 1039 (1985)
9. K. A. Chao, Y. Wang: J. Phys. C: Solid State Phys. 18, L1127 (1985)
10. J. T. Gammel: Proc. Intn'l Conf. on Science and Tech. of Synthetic Metals, Kyoto (1986)
11. B. Horovitz: Solid State Commun. 41, 729 (1982)
    E. J. Mele, J. C. Hicks: Phys. Rev. B32, 2703 (1985)
12. A. Terai, Y. Ono, Y. Wada: J. Phys. Soc. Japan 55, No. 8 (1986)
13. J. C. Hicks, E. J. Mele: to be published in Phys. Rev. B
14. E. Ehrenfreund, Z. Vardeny, O. Brafman, H. Fujimoto, J. Tanaka, M. Tanaka: Proc. Intn'l Conf. on Science and Tech. of Synthetic Metals, Kyoto (1986)
15. H. Schaffer, A. J. Heeger: private communication
16. A. Terai, M. Ogata, Y. Wada: J. Phys. Soc. Japan 55, No. 7 (1986)
17. M. Ogata, A. Terai, Y. Wada: J. Phys. Soc. Japan 55, No. 7 (1986)
18. M. Ogata, A. Terai, Y. Wada: Proc. Intn'l Conf. on Science and Tech. of Synthetic Metals, Kyoto (1986)

Ground State Properties of Polyacetylene with Off-Diagonal Random Disorder

Bing-Chang Xu and S.E. Trullinger

Department of Physics, University of Southern California, Los Angeles, CA 90089, USA

Abstract We employ a supersymmetric functional integral formalism to obtain exact results for the influence of "off-diagonal" disorder on the nature of the zero-temperature ground state of the continuum model of polyacetylene.

The problem of random (impurity) disorder influencing the ground state properties of the soliton-bearing[1-3] polymer polyacetylene has been investigated by several authors[4-10]. In general, the disorder will weaken the Peierls instability smearing out the transition between the normal and distorted phases. In this paper we report exact, zero-temperature results for the dependence of the dimerization amplitude,electronic density of states and localization length on the strength (or concentration) of "off-diagonal" random disorder (impurities). By off-diagonal or off-site we mean that the random disorder is introduced in the hopping matrix element connecting the tight-binding electronic orbitals on neighboring (CH) groups.

Our results are obtained with an adaptation of the supersymmetry method[11-14] used by Bohr and Efetov[12] to treat the case of random disorder in a one-dimensional free-electron model. In general, by taking the advantage of both Grassman and conventional integrals (supersymmetry integrals) to overcome technical obstacles, one is able to average over the disorder potential at an early stage in calculations and arrive at an effective quantum field theory with a supersymmetric Lagrangian.

Our starting point is the continuum model of the electronic properties of a polyacetylene chain introduced by Takayama, Lin-Liu and Maki[15]. The Hamiltonian with the presence of the random potential is given by

$$H = (\omega_Q^2/2g^2) \int dx\, \Delta^2(x) + H_{el} \quad , \tag{1}$$

where ω_Q is the maximum phonon frequency in the discrete lattice model of Su, Schrieffer and Heeger[1](SSH), g is a scaled electron-phonon coupling constant and $\Delta(x)$ is proportional to the amplitude of the dimerization which is regarded as a static, classical field in the TLM model. The electronic part of the Hamiltonian, H_{el}, is given by (setting v_F =1 in the TLM model)

$$H_{el} = \int dx\, \Phi^+(x) \{ -i\sigma_3 \partial/\partial x + [U(x) + \Delta(x)]\, \sigma_1 \}\, \Phi(x) \quad , \tag{2}$$

where the $\{ \sigma_i \}$ are Pauli matrices, and $\Phi(x)$ is a two-component pseudospinor field composed of right- and left-moving electron fields, u(x) and v(x). The uniform, dimerized Peierls state is characterized by a constant value of Δ and we shall regard it as a constant in what follows. The random disorder potential, U(x), is assumed to be Gaussian randomly distributed with zero average, ie., $<U(x)> = 0$, and $<U(x)U(x')>= \gamma\, \delta(x-x')$, where γ measures the strength of the disorder.

The component functions u(x) and v(x) satisfy Bogoliubov-de Gennes (BdG) equations, obtained by varying Eq.(2), which can be decoupled by a linear transformation ($f_{\pm} = u \pm iv$). However, due to the presence of the disorder potential U(x), the engergy spectrum cannot be found in general. Fortunately, one is usually only interested in quantities which are averaged over the disorder. For example, the calculation of thermodynamic quantities requires only knowledge of the averaged density of states which can be expressed in terms of the imaginary part of the retarded Green function. By representing[12] the retarded Green function in supersymmetric form, one is able to average over the disorder easily and the averaged Green function is given by[16]

$$< G^R(x_1,x_2;E) > = i \int D\Psi^* D\Psi \; \Psi_\alpha(x_1)\, \Psi^*_\alpha(x_2) \exp\left[- \int L\, dx\right] \quad , \tag{3}$$

where the supersymmetric Lagrangian is defined by

$$L = \mu\Psi^+\{ E^2 - \Delta^2 + (\partial/\partial x) - i\delta \}\Psi + (\{4\Delta^2\gamma + \gamma^2\}/2)\, (\Psi^*\Psi)^2 \tag{4}$$

and the constant μ will be set equal to i later on. The subscript α takes either of two values which indicate the fermonic (χ) or bosonic (S) components of Ψ_k [each of the components χ and S is itself a two-component vector containing "right- and left-moving" fields so that the superfield Ψ_k is a four-component object]. The functional integrations over the superfields in Eq.(3) may be performed with the aid of a transfer integral operator technique[17], which leads to an effective Schrödinger problem involving the supersymmetric effective Hamiltonian

$$H_{eff} = -\mu^{-1}\{ \partial^2/\partial S^*\partial S + \partial^2/\partial\chi^*\partial\chi \} + \mu(E^2 - \Delta^2)\Psi^*\Psi + (2\Delta^2\gamma + \gamma^2/2)(\Psi^*\Psi)^2 \quad , \tag{5}$$

where we note that $\Psi^*\Psi = (|S|^2 + |\chi|^2)$.

The averaged density of electronic states is obtained from the imaginary part of the Green function. This in turn can be expressed in terms of the ground-state eigenfunction of H_{eff} to yield the exact result,

$$< \rho(E) > = (4|E|/\pi^2)(2/\gamma')^{-1/3}\, \frac{Ai'(z)Ai(z) + Bi'(z)Bi(z)}{[\, Ai^2(z) + Bi^2(z)\,]^2} \quad , \tag{6}$$

where $\gamma' = 4\Delta^2\gamma + \gamma^2$, $z = (\Delta^2 - E^2)(2/\gamma')^{2/3}$, Ai(z) and Bi(z) are Airy functions[18] of the first and second kinds, respectively. The density of states given by Eq.(6) depends on the dimerization amplitude (order parameter), Δ,which remains to be determined. This is accomplished by minimizing the total energy with respect to Δ with due regard given to the fact that the cutoff required for the electronic energy spectrum in the TLM model depends on Δ and the disorder strength γ because the number of electrons to be accomodated is the same as in the unperturbed TLM model. We find that Δ satisfies the self-consistency condition

$$\int_0^{E_c} dE \left\{ \frac{<\rho(E)>}{E}\left[1 - \frac{8\gamma'}{3\gamma}(\Delta^2 - E^2)\right] - \frac{8\gamma'}{3\gamma} n(E) \right\} = \frac{1}{2\pi\lambda} \quad , \tag{7}$$

where λ is the dimensionless electron-phonon coupling constant [15] , $n(E) = (\pi^{-2})(\gamma'/2)^{1/3}\{Ai^2(z)+Bi^2(z)\}^{-1}$,and E_c is the cutoff energy which is determined from the condition,

$$n(E_c) = \pi^{-2} (\gamma'/2)^{1/3} \{ Ai^2(z_c) + Bi^2(z_c) \}^{-1} = \pi^{-1} \quad , \tag{8}$$

where $n(E_c)$ is the density of electrons to be accommodated by the cutoff and $z_c=(2/\gamma')^{2/3}(\Delta^2 - E_c^2)$. It is straightforward to conduct a numerical search for the value of Δ which satisfies both Eqs.(7) and (8) and in Fig. 1 we present the results for Δ as a function of the dimensionless disorder strength parameter, γ, for a typical value of the electron-phonon coupling λ. We note that Δ decreases along an almost BCS-like curve until a critical value γ_c is reached above which the dimerization amplitude is identically zero , implying the destruction of the Peierls state.

In Fig. 2 we plot the dimensionless density of states obtained from Eqs.(6-8) for several values of γ. Note that for small values of γ there exists a "psuedogap" in the energy region where a gap would exist in the pristine Peierls state. The exact zero appearing in the density of states at E=0 for all values of γ is a consequence of the

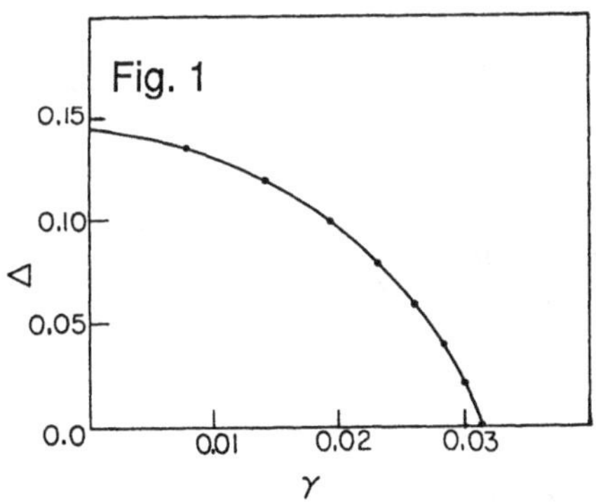

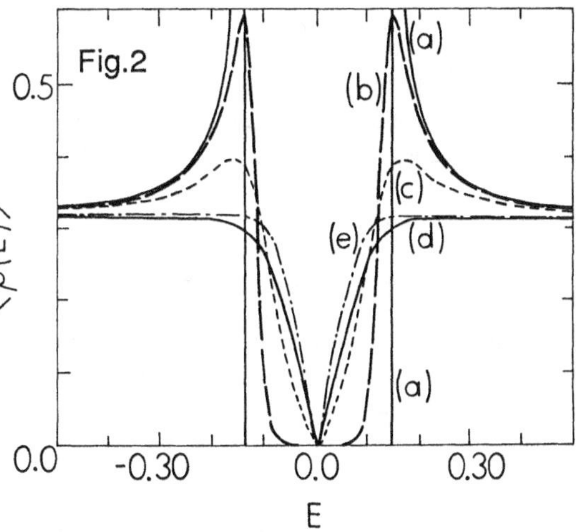

Fig.1 Dimensionless dimerization amplitude, Δ, as a function of dimensionless disorder strength γ for a typical value of the electron-phonon coupling(λ=0.19). Above the critical value $\gamma_c \approx 0.0311$ the dimerization amplitude vanishes.

Fig.2 Dimensionless density of electronic states per unit length $<\bar{\rho}(E)>=<\rho(E)>/L$ for several disorder strengths with λ=0.19: (a) γ=0,Δ=0.143; (b) γ=0.005,Δ=0.136; (c) γ=0.188, Δ=0.101; (d) γ=0.031, Δ=0.0; (e) γ=0.05, Δ=0.0.

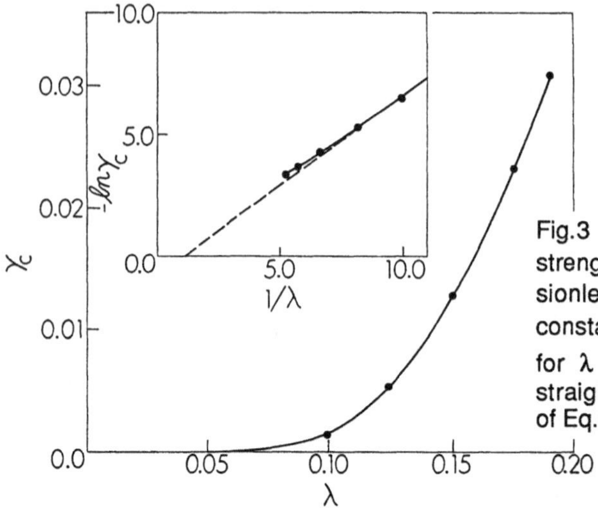

Fig.3 Dimensionless critical disorder strength γ_c as a function of the dimensionless electron-phonon coupling constant λ. Inset: $-\ln\gamma_c$ vs. $1/\lambda$ deviates for λ larger than about 0.1 from the straight-line (dashed) representation of Eq.(9).

facts that the disorder function, U(x), is symmetrically distributed about zero and that its "off-diagonal" presence does not destroy the electron-hole symmetry[3] possessed by the TLM Hamiltonian[19]. We also note that the local maxima appearing as remnants of the square-root divergences in the pure system are gradually suppressed as γ is increased and they disappear altogether for $\gamma > \gamma_c$. In Figure 3 we plot the dependence of γ_c on the dimensionless electron-phonon coupling constant, λ. These numerical results are very well fit by the expression

$$\gamma_c = (3.78 - 10.66\,\lambda)\exp[-3/(4\lambda)] \tag{9}$$

for small λ. In fact, the exponential dependence agrees precisely with the behavior obtained from a small-λ (small-γ) analysis based on the asymptotic properties [18] of the Airy functions appearing in the above formulae. For λ values larger than about 0.1 there are significant corrections to (9) as can be seen in the inset of Fig.3 where we plot $-\ln(\gamma_c)$ vs. $(1/\lambda)$.

The supersymmetry method may also be used to obtain various correlation functions, both static and dynamic. For example, the frequency-dependent density-density correlation function defined[11-12] in terms of both advanced and retarded electronic Green functions G^A and G^R may be found exactly in the $\omega \to 0$ limit and from its dependence on $|x-x'|$, the energy-dependent localization length may be found [16].

By using the same method we can also treat the diagonal disorder case exactly. By diagonal we mean that the random disorder is introduced in the diagonal matrix element in the Hamiltonian. Due to the space limit here, we will publish the results separately. We wish to thank Dr. J.S. Zmuidzinas for bringing the work of Efetov [11] to our attention and for helpful discussions.

1. W. P. Su, J. R. Schrieffer and A. J. Heeger, Phys. Rev. Lett. 42, 1698(1979); Phys. Rev. B 22, 2099(1980)
2. M.J. Rice, Phys.Lett. 71A,152 (1979)

3. For a review of solitons in polyacetylene see S. Kivelson, in Solitons, ed. by S. E.Trullinger, V.E. Zakharov and V.L. Pokrovskii (North-Holland, Amsterdam, 1986), in press
4. B. Horovitz, Phys. Rev. Lett. 46, 742(1981)
5. E.J. Mele and M.J. Rice, Phys. Rev. B 23, 5397(1981)
6. W.P.Su, S. Kivelson and J.R. Schrieffer, in Physics in One Dimension ,ed. by J. Bernasconi and T. Schneider (Springer-Verlag, Berlin, 1980)
7. D. Boyanovsky, Phys. Rev. B 27, 6863 (1983)
8. D.Baeriswyl, in Theoretical Aspects of Band Structures & Electronic Properties of Pseudo-One-Dimensional Solids, ed. by H.Kamimura Physics and Chemistry of Materials with Low-Dimensional Structures (D. Reidel Publ. Co.) 1986
9. C.T. White, M.L. Elert and J.W. Mintmire, J.de Phys.44, c3-481(1983)
10. G. Gomez-Santos and F.Yndurain, Phys.Rev.B 29, 4459(1984); Phys.Rev.B 31 5087(1985)
11. K.B. Efetov, Adv.Phys. 32, 53 (1983) and references therein
12. T. Bohr and K.B. Efetov, J. Phys. C 15, L249 (1982)
13. G. Parisi and N. Sourlas, Phys. Rev. Lett.43 ,744 (1979)
14. A.J. McKane , Phys. Lett. A 76, 23 (1980)
15. H.Takayama, Y.R. Lin-Liu and K. Maki, Phys.Rev. B 21, 2388 (1980)
16. Bing-Chang Xu and S.E. Trullinger, to be published
17. R.P. Feynman, Rev. Mod. Phys. 20, 367 (1948)
18. M. Abramowitz and I. A. Stegun, Handbook of Mathematical Functions (Dover, New York, 1970), Chpt. 10
19. A similar effect occurs in the case of dilute impurity " pairs " giving rise to symmetrically split states. See J. Callaway, Quantum Theory of the Solid State (Academic, New York, 1974), p. 438

Part IV

Experimental Techniques

Neutron Scattering Studies of Non-Linear Systems

R. Pynn

Institut Laue-Langevin, 156X Centre de Tri,
F-38042 Grenoble Cedex, France

This article describes a number of neutron scattering experiments which have been performed recently on systems which exhibit non-linear behaviour. Examples chosen include the incommensurably modulated insulator thiourea for which defect-induced effects mimic intrinsic non-linear behaviour, an aggregated gold colloid which displays fractal properties and two low-dimensional magnets. Experiments on the latter systems have been carried out with neutron polarisation analysis, a powerful technique which allows different components of the magnetic fluctuation spectrum to be measured independently. All experiments have been chosen to illustrate "a lesson from the past" or "a prospect for the future."

1. Introduction

Scattering techniques involving light, X-rays and neutrons have contributed widely to the study of the non-linear phenomena discussed at this conference. For various reasons, not least of which are the author's ignorance and restrictions on the length of this article, this paper will be concerned mainly with neutron scattering experiments. This probe allows both static and dynamic effects to be studied and, by judicious choice of instrumentation, the wavevector ($\vec{Q}$) and energy (E) resolution can often be chosen to match the phenomenon of interest. Since the neutron couples both to fluctuations of nuclear density and of electronic spin, a wide variety of interesting systems are accessible to study. The complementarity of these investigations with X-ray scattering (where the principal coupling is to electronic charge fluctuations) and to light scattering (where the coupling is to electric polarisation) should be clear. In general the choice of scattering probe is dictated either by the type of fluctuation (mass, spin, charge, etc.) or by the resolution available with a given probe. For example, the intense X-radiation available from synchrotron sources allows a wavevector resolution which is several orders of magnitude better than that conventionally achieved with neutrons. However there are some cases, notably those in which radiation partially destroys the system under study, in which the choice of scattering probe must be made carefully. An example of this will be given below. Many of the systems for which calculations of non-linear effects are available involve more or less esoteric spin dynamics. In this case neutron scattering is an appropriate tool, and recent advances in the technology of polarised neutron beams allow a detailed investigation of various components of the spin fluctuations. A large part of this article is devoted to these techniques in the hope that a documentation of recent experimental advances will encourage increased theoretical effort. The neutron spin can also be used in a novel way to improve the energy resolution available in neutron scattering by means of the neutron spin echo (NSE) technique. Measurements can now be made in a time domain which effectively extrapolates or overlaps that available with other techniques such as light scattering, resonance spectroscopy or dielectric relaxation.

2. Static Properties

Materials with incommensurably modulated structures have been studied by scattering techniques for over a decade. They range from layered charge-density-wave materials such as $NbSe_2$ [1] through charge-transfer systems such as TTF-TCNQ [2] to insulators such as K_2SeO_4 [3]. Although the physical mechanisms responsible for the lattice modulations in these architypical materials vary, the resulting symptom observed by X-ray or neutron diffraction is the same; a series of satellite reflections at wavevectors which are rational or irrational fractions of Bragg vectors of the high-temperature (parent) structure. The satellites signify the existence of a structural modulation which is more or less sinusoidal. In the non-linear case, the modulation is supposed to comprise regions of commensurate modulation separated by discommensurations [4] or solitons [5], within which there is a rapid change of the phase of the modulation with respect to the underlying lattice. To the author's knowledge there is no unambiguous evidence from scattering experiments with any system which supports in detail the discommensuration model, and there are some cases [6] in which the temperature dependence of satellite intensities remains a mystery. Nevertheless, evidence from other techniques indicates that this physical picture is substantially correct [34].

One of the lessons learnt from studies of structural phase transitions is that dirty samples behave in unpredictable ways. When a susceptibility diverges, as it does for order-parameters close to second-order phase transitions, coupling to impurities may produce extrinsic effects which imitate more interesting, intrinsic physical phenomena. The central peak observed at some structural phase transitions is an example for which the literature is extensive [7]. Similarly, in sinusoidally modulated structures, fluctuations of the phase of the modulation (phasons) are Goldstone modes and the phase susceptibilty is divergent. Coupling to lattice defects may then cause localisation or pinning of discommensurations with little or no cost in energy. This effect, which may be viewed as a conspiracy between a divergent susceptibilty and the reduced dimensionality of walls or discommensurations, may be difficult to distinguish from more interesting, intrinsic non-linear behaviour.

Examples of hysteresis and memory effects due to wall localisation appear in many systems; random-field Ising materials are a recent controversial example [8]. Less fraught with controversy are recent measurements by X-ray and neutron scattering on thiourea, a modulated insulator in which a frozen electric polarisation wave is observed. Commensurable phases of order 9, 7 and 3 have been observed in a neutron study of the (P,T) phase diagram [9] and a phase of order 8, predicted to occur under the influence of an applied electric field, has also been confirmed experimentally [10]. In a recent X-ray experiment [11], phases with high commensurable order (such as 2/15, 4/29 and 6/43) were reportedly seen in large applied fields and interpreted as intrinsic non-linear effects resulting from lattice discreteness. Neutron scattering experiments [32] failed to find this apparently beautiful example of Devil's staircase behaviour. In a recent experiment Denoyer and collaborators [12] revisited this phenomenon by carrying out neutron scattering experiments on samples which had previously been subjected to various doses of X-radiation. Their results for a sample which was subjected to a 100 MRad dose of X-rays about one week before the neutron experiment are shown in Fig. 1. Prior irradiation is found to cause enhanced overall hysteresis,

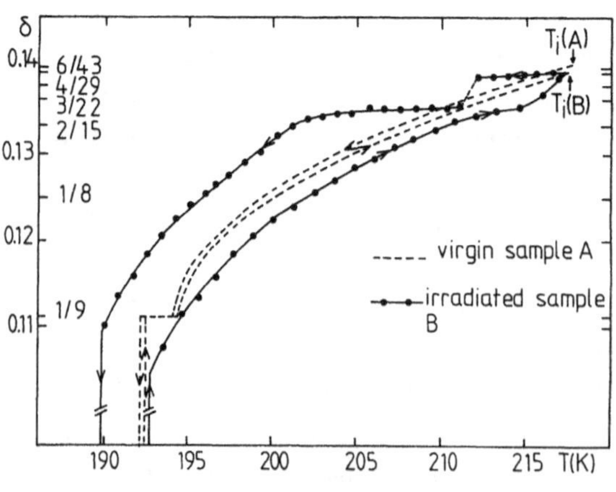

Fig. 1: Modulation wavevector δ as a function of temperature for deuterated thiourea. The solid curve refers to sample which had been exposed to 100 MRad of X-radiation while the dashed curve is for a virgin sample.

quasi-constant values of the modulation wavevector over various temperature regions within the incommensurable phase, and smearing of the lock-in transition of order 9 which depends on radiation dose. A possible explanation for the quasi-plateaux in Fig. 1, and perhaps for the earlier X-ray results [11], is that the incommensurable modulation is decorated by mobile defects which make it energetically unfavourable to change the modulation wavevector as temperature is further lowered. Credence is added to this explanation by the observation of a memory effect in Fig. 1: the warming curve exhibits a bend at the temperature at which a quasi-plateau was observed on cooling.

Although the data presented in Ref. 12 do not prove that the X-ray measurements of thiourea [11] observed defect-induced phenomena, they do make this explanation possible and indeed likely. The lesson to be learnt is that great care is needed to discriminate between intrinsic non-linear effects and defect-induced effects. The popular current belief that this problem occurs only for incommensurably modulated compounds containing hydrogen remains to be confirmed experimentally.

The neutron spectrometers used in the studies of incommensurable structures are principally diffractometers capable of measuring Bragg peaks at wavevectors up to ~20 $Å^{-1}$ with an accuracy of ~0.05%. A completely different type of instrument, the small-angle neutron scattering (SANS) spectrometer, probes wavevectors in the range $10^{-3} \lesssim Q \lesssim 10^{-1}$ $Å^{-1}$. An example of data taken with the D11 SANS instrument at ILL is shown in Fig. 2. The sample was a suspension of small (~72 Å diameter) gold particles which were persuaded to aggregate rapidly by a change in the pH of the

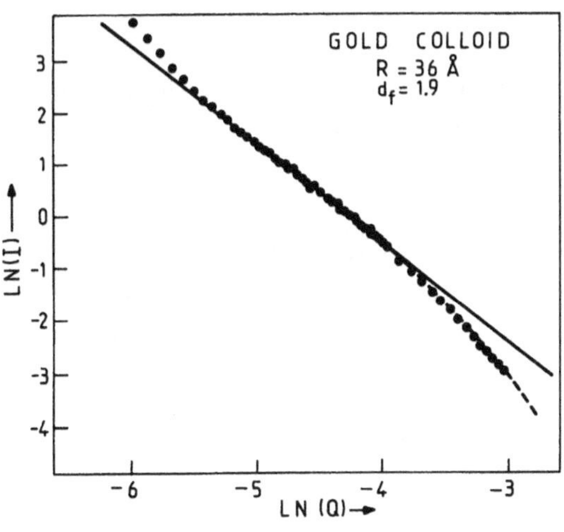

Fig. 2: Double logarithmic plot of small-angle scattering from a rapidly aggregated gold colloid whose individual particles have a radius R = 36 Å. Data were taken at a neutron wavelength of 10 Å with the detector at 20 m, 10 m, and 3 m from the sample. Instrumental resolution is about 10% (FWHM) at any value of wavevector Q.

supension [13]. Such growth processes are supposed to yield highly ramified fractal structures characterised by a fractal dimension, d_f. For such an aggregate it is easy to show by scaling arguments that the scattered intensity should behave as Q^{-d_f} provided one probes length scales which are larger than individual particle sizes and less than the aggregate size. To obtain the large dynamic range of wavevectors in Fig. 2, data for three positions of the neutron area detector were combined. The straight line through the data represents scattering from aggregates with $d_f = 1.9$. This value is within the range of values which have been reported for aggregated colloids but differs from the canonical value of $d_f = 1.75$ found for diffusion-limited cluster aggregation [13]. When a Guinier approximation to the particle form-factor is included, the suppression of the scattering at large Q is accounted for, as the broken line in Fig. 2 indicates. The up-turn at low Q, which is in the wrong sense to be explained by an upper length cut-off, seems to indicate some collapse of the outer arms of large aggregates [14]. However, there is no detailed explanation for this effect.

Various measurements of fractal dimensions for aggregates obtained by light, X-ray and neutron scattering have appeared in the literature [13,15,16] and sparked controversy because differing results were obtained for ostensibly similar systems. However, Weitz et al. [13] found distinctly different values of d_f for gold colloids aggregated slowly ($d_f = 2.05$) and rapidly ($d_f = 1.77$). Although these authors only identified two limiting cases, experiments with micron-sized polystyrene spheres [17] have shown that, at least in two dimensions, the fractal dimension of an aggregate can be tuned more-or-less continuously by varying the sticking probability of individual particles. The lesson to be drawn appears to be that it is futile to argue the merits of different techniques when there is no theoretical evidence of universality and when sample preparation and history are difficult to document in an unambiguous way.

3. Dynamical Properties

Although measurements of the dynamics of non-magnetic systems have been made both by light and neutron-scattering techniques, I will restrict my attention to studies of dynamics in low-dimensional spin systems. In this area, recent advances in techniques of neutron polarisation analysis have opened possibilities for more sophisticated and detailed measurements.

In most cases the simple treatment given by Moon, Riste and Koehler [18] is sufficient to determine the nature of the scattering when polarised neutrons interact with matter. Application of various combinations of these rules, which are recalled below, allows different components of the scattering to be discriminated. The rules for polarised neutron scattering [18] are expressed in terms of the scattering length, U, corresponding to neutron interaction with a single scattering centre. Within the Born approximation the scattered intensity is proportional to the space and time Fourier transforms of the correlation function $\langle U(0,0)\, U^*(\vec{r},t)\rangle$. We append to U one of the combinations (++), (--), (+-) and (-+), the first two of which represent non-spin-flip (n.s.f.) scattering and the last two spin-flip (s.f.) scattering. The combination (+-), for example, means that a neutron is scattered from an up (+) state in which its spin is parallel to the applied magnetic field $\vec{H}$ and to the quantisation axis z, to a down (-) state. One finds that:

$$U^{++} = b - pS_{\perp}^{z} + BI_{z} \tag{1a}$$

$$U^{--} = b + pS_{\perp}^{z} - BI_{z} \tag{1b}$$

$$U^{+-} = -p(S_{\perp}^{x} + iS_{\perp}^{y}) + B(I_{x} + iI_{y}) \tag{1c}$$

$$U^{-+} = -p(S_{\perp}^{x} - iS_{\perp}^{y}) + B(I_{x} - iI_{y}) \quad . \tag{1d}$$

In these equations b, B and p are constants related to the nuclear and magnetic scattering lengths of the scattering centre and $\vec{I}$ is the nuclear-spin operator. For our purposes the most important quantity is $\vec{S}_{\perp}$, the component of the atomic-spin operator perpendicular to the scattering vector $\vec{Q}$. A judicious choice of the relative orientation of $\vec{H}$ and $\vec{Q}$ and the measurement of partial n.s.f. and s.f. cross sections can often be used to isolate spin fluctuations of interest. Two examples in which such separation has recently been achieved are $CsNiF_3$ [19,20] and TMMC [21].

$CsNiF_3$ is a quasi-one-dimensional ferromagnet in which the Ni spins are restricted by crystal field anisotropy to lie in planes perpendicular to the crystallographic c axis. Soliton excitations are predicted to occur in this sine-Gordon system when a magnetic field is applied perpendicular to c [24]. These excitations have been observed and characterised in a series of neutron scattering experiments by Steiner and his collaborators [22]. A sketch of the ordered spin configurations obtained when an external field $\vec{H}$ is applied in, and perpendicular to, the (horizontal) scattering plane of a neutron experiment is shown in Figs. 3a and 3b. These figures also depict

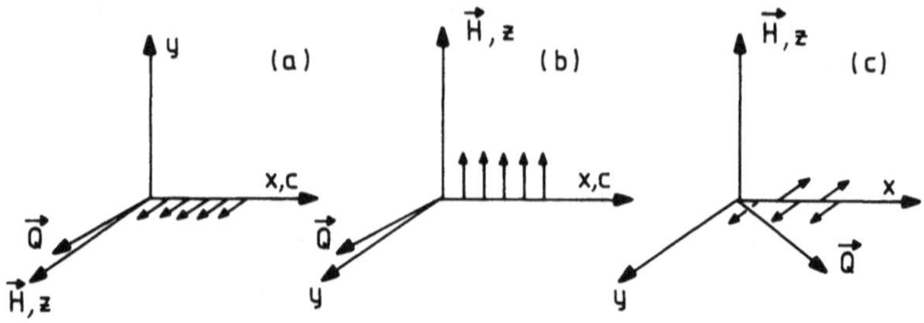

Fig. 3: Schematic representations of the scattering geometry used for the polarised-neutron experiments described in the text. In each figure the directions of the applied field $\vec{H}$ and the scattering vector $\vec{Q}$ are shown together with the ordered spin configurations of the magnetic chains. Parts (a) and (b) refer to experiments with $CsNiF_3$ while part (c) refers to the experiment with TMMC.

the scattering vectors $\vec{Q}$ at which measurements with polarised neutrons have been performed.

For the configuration given by Fig. 3a an experiment with <u>unpolarised</u> neutrons yields, after subtraction of incoherent background, a spectrum of the form shown in Fig. 4a. The intensity centred at zero-energy transfer is the soliton scattering while the two side bands correspond to magnon creation (positive energy) and annihilation. A similar experiment in which only the (+-) spin-flip scattering is measured yields the spectrum shown in Fig 4b. The magnon annihilation peak is largely suppressed and the soliton scattering is no longer symmetric about zero-energy transfer. Decrease of magnon annihilation scattering is expected and can be explained by linear

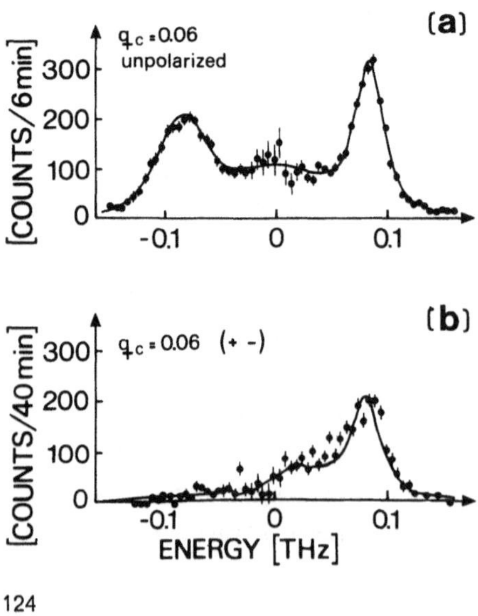

Fig. 4: Spectra recorded for $CsNiF_3$ at 12 K in an applied field of 1T with $\vec{Q}$ = (0.6, 0, -0.06): (a) using unpolarised neutrons; (b) (+-) spin-flip scattering.

spin-wave theory using the accepted Hamiltonian for $CsNiF_3$ [23]. The observed asymmetry of the soliton scattering in Fig. 4b is less straightforward to explain. In a naive picture, solitons comprise rotation of the spin direction in the (yz) plane of Fig. 3a. In view of the cross-section formulae given in (1), this model implies that only the y component of soliton spin fluctuations should be ovserved in (+-) s.f. scattering. However in this case, where out-of-plane fluctuations are assumed absent, (1c) and (1d) are identical and there should be no difference between the soliton scattering measured with and without polarisation analysis. The central peaks in Figs. 3a and 3b should be identical. The fact that they are not is a direct result of out-of-plane fluctuations associated with solitons. This experiment thus demonstrates unambiguously the existence of such fluctuations, which are indeed expected within the sine-Gordon theory [24]. Recently Osano [25] has calculated the spectrum depicted in Fig. 4b in terms of the $<S_yS_y>$ and $<S_xS_x>$ correlation functions. Using the previously measured $<S_yS_y>$ cross section [26] and the value for $<S_xS_x>$ deduced from the experiment described below, the spectrum of Fig. 4b can be calculated without adjustable parameters. The result, convolved with the experimental resolution function, gives the line in Fig. 4b.

In order to determine the out-of-plane component $<S_xS_x>$ of the soliton fluctuations directly, the scattering geometry depicted in Fig. 3b was used for a measurement of the spin-flip signal. A plot of the intensity measured at zero-energy transfer, from which incoherent background has been subtracted, is shown in Fig. 5. There is no evidence in this data for the soliton instability which is predicted to occur as the applied field, $\vec{H}$, is increased [35]. Indeed the sine-Gordon theory, scaled to the data at

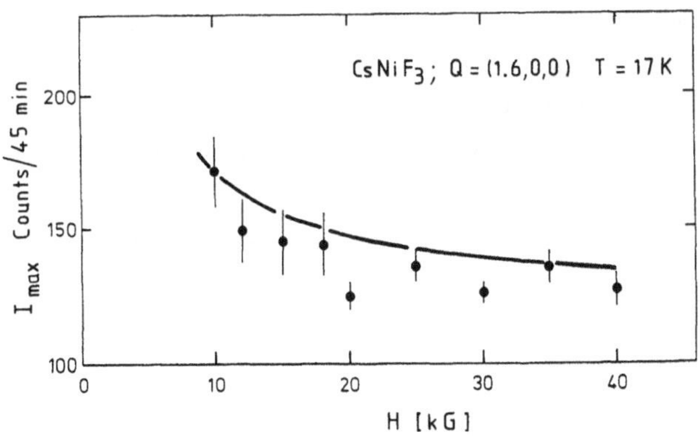

Fig. 5: Field dependence of out-of-plane soliton fluctuations in $CsNiF_3$ measured with polarised neutrons at $\vec{Q}$ = (0.6, 0, 0) and T = 17 K

H = 1T, provides a reasonable description of the measurement as the line in Fig. 5 shows. The mystery of $CsNiF_3$ is that the simple sine-Gordon theory seems to describe well the large body of neutron scattering data which has now been obtained with this compound.

Experiments with the antiferromagnetic counterpart of $CsNiF_3$, TMMC [$(CD_3)4NMnCl_3$] have been described at this conference by Boucher [27]. In a recent experiment with cold, polarised neutrons, the scattering geometry depicted in Fig. 3c was used for a measurement of both the n.s.f. and s.f. spectra. According to the rules given by (1), the magnetic contribution to these cross sections represent $\langle S_z S_z \rangle$ and a combination of $\langle S_x S_x \rangle$ and $\langle S_y S_y \rangle$ respectively. $\langle S_z S_z \rangle$ is the quantity denoted $S_\parallel(\vec{q},\omega)$ by Boucher [27]. Since the out-of-plane fluctuations are small compared to $\langle S_y S_y \rangle$, the s.f. scattering is related to Boucher's $S_\perp(\vec{q},\omega)$ by a simple geometric transformation. With unpolarised neutrons one measures a linear combination of $S_\parallel$ and $S_\perp$ which, since $S_\parallel$ is weak, is a fair approximation to $S_\perp$ itself. The latter, which is the Bragg scattering resulting from the antiferromagnetic order broadened by the presence of solitons, has been analysed previously [28]. Widths of the $S_\perp(\vec{q},\omega)$ spectrum, both in energy and wavevector, have been measured as a function of temperature and applied field and found to be in accord with theoretical expectations. The observation of $S_\parallel(\vec{q},\omega)$ is, however, a new result and represents the first measurement of the energy spectrum of soliton fluctuations (as opposed to the effect of solitons on antiferromagnetic order) in an antiferromagnetic material. As discussed by Boucher [27] a simple non-interacting soliton model (dashed curve in Fig. 6) is not adequate to explain the observed energy-width of $S_\parallel(q,\omega)$. The dot-dash line in Fig. 6 accounts phenomenologically for the dynamical effects of soliton-soliton and soliton-magnon collisions, which essentially introduce a soliton lifetime. This damping is not sufficient to account fully for the observed spectrum. To obtain the solid curve in Fig. 6 it is necessary to take into account the normalisation of the soliton amplitude which results from collision processes [21]. The theory for these collision effects is currently phenomenological and more work is necessary for a proper understanding (c.f., the article by Sasaki and Maki in this proceedings).

The lesson I want to draw from the experiments with $CsNiF_3$ and TMMC is that neutron scattering combined with polarisation analysis is now able to separate various components of spin fluctuations in a way which is often unambiguous. This method has obvious applications not only to low-dimensional spin systems but also, for example, to heavy fermion materials and random-field systems.

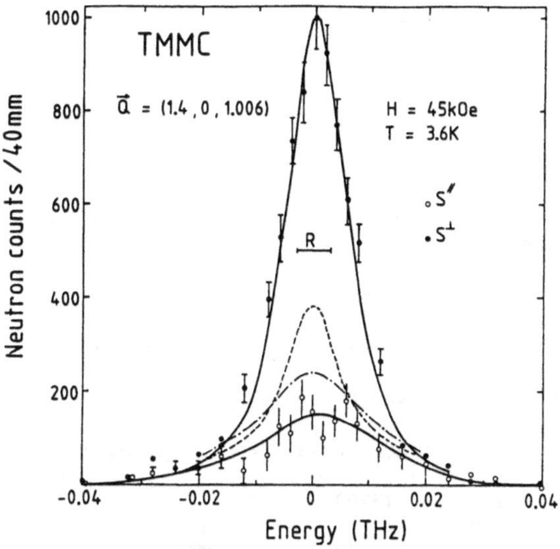

Fig. 6: Spectra of magnetic fluctuations observed in TMMC using polarised neutrons. $S_{\parallel}$ refers to fluctuations parallel to the applied field $\vec{H}$ while $S_{\perp}$ is the spectrum of fluctuations perpendicular to $\vec{H}$ and to the chain direction. R represents the instrumental resolution. The various lines through the points are described in the text.

The neutron spin may be used not only to permit polarisation analysis but also, via a novel method known as neutron spin echo (NSE), to improve the energy resolution available in neutron scattering. The technique [29] involves coding the velocities of neutrons before and after scattering by the number of Larmor precessions which the neutron spin makes in a suitable magnetic field. Energy resolution of a few nano-electron-volts can be obtained in this way. A further advantage of the NSE technique is that it naturally measures in the time domain rather than in the energy domain; that is, the quantity measured is the intermediate scattering function $S(\vec{q},t)$ rather than its Fourier transform $S(\vec{q},\omega)$. For phenomena which involve esoteric time dependence, NSE thus provides a more suitable probe than conventional neutron scattering. One example of this has already been presented by Boucher [27,33] at this meeting; the $\exp(-\sqrt{t})$ behaviour expected for diffusing solitons was easily distinguishable from the $\exp(-t)$ law which would pertain for ballistic solitons. Further examples relevant to this conference include measurements on CuMn spin glasses [30] and recent work by Durand et al. [31] on the dynamics of the incommensurable modulation in $NaNO_2$. In the latter case the time domain available to NSE ($\lesssim 10^{-8}$ sec) overlaps with that of dielectric relaxation measurements.

Acknowledgements

I would like to thank Roland Currat for illuminating discussions on incommensurable structures and the authors of reference [12] for allowing me to use the data in Fig. 1 which was unpublished at the time of this conference. My thanks are also due to those with whom I have collaborated on some of the experiments described here. In particular I thank J. P. Boucher, K. Kakurai, L. P. Regnault, S. K. Sinha, M. Steiner and D. Weitz in this context.

Dedication

This article is dedicated to the memory of Dr. W. C. Koehler, a pioneer in the field of neutron scattering.

References

1. D. E. Moncton, J. D. Axe and F. J. DiSalvo: Phys. Rev. Lett., 34, 734 (1975)
2. S. Megtert, R. Comès, C. Vettier, R. Pynn and A. F. Garito: Solid St. Commun., 37, 875 (1981)
3. C. F. Majkrzak, J. D. Axe and A. D. Bruce: Phys. Rev., B22, 5278 (1980)
4. W. L. McMillan: Phys. Rev., B14, 1496 (1976)
5. P. Bak and V. J. Emery: Phys. Rev. Lett., 36, 978 (1976)
6. S. R. Andrews and H. Mashiyama: J. Phys. C16, 4985 (1983)
7. B. I. Halperin and C. M. Varma: Phys. Rev., B14, 4030 (1976)
8. J. Villain: In Scaling Phenomena in Disordered Systems, ed. by R. Pynn and A. T. Skjeltorp (Plenum, New York, 1985)
9. F. Denoyer, A. H. Moudden, R. Currat, C. Vettier, A. Bellamy and M. Lambert: Phys. Rev., B25, 1697 (1982)
10. A. H. Moudden, E. C. Svensson and G. Shirane: Phys. Rev. Lett., 49, 557 (1982)
11. A. H. Moudden, D. E. Moncton and J. D. Axe: Phys. Rev. Lett., 51, 2390 (1983)
12. G. Andre, D. Durand, F. Denoyer, R. Currat and F. Moussa: to be published
13. D. A. Weitz, J. S. Huang, M. Y. Lin and J. Sung: Phys. Rev. Lett., 54, 1416 (1985)
14. Y. Kantor: In Scaling Phenomena in Disordered Systems, ed. by R. Pynn and A. T. Skjeltorp (Plenum, New York, 1985)
15. D. W. Schaefer, J. E. Martin, P. Wiltzius and D. S. Cannell: Phys. Rev. Lett., 52, 2371 (1984)
16. D. Johnston and G. Benedek: In Kinetics of Aggregation and Gelation, ed. by F. Family and D. P. Landau (North Holland, Amsterdam, 1984)
17. A. T. Skjeltorp: to be published
18. R. M. Moon, T. Riste and W. C. Koehler: Phys. Rev., 181, 920 (1969)
19. K. Kakurai, R. Pynn, B. Dorner and M. Steiner: J. Phys. C17, L123 (1984)
20. K. Kakurai, M. Steiner, R. Pynn and B. Dorner: J. Mag. and Mag. Mat., 54-57, 835 (1986)
21. J. P. Boucher, L. P. Regnault, R. Pynn, J. Bouillot and J. P. Renard: Europhys. Lett., 1, 415 (1986)
22. see references in [19] and [20]
23. M. Hood: J. Phys., C16, 4655 (1983)
24. H. J. Mikeska: J. Phys. C11, L29 (1978)
25. K. Osano: J. Phys. C17, L511 (1984)
26. J. K. Kjems and M. Steiner: Phys. Rev. Lett., 41, 1137 (1978)
27. J. P. Boucher: this proceedings
28. L. P. Regnault, J. P. Boucher, J. Rossat-Mignod, J. P. Renard, J. Bouillot and W. G. Stirling: J. Phys. C15, 1261 (1982)
29. Neutron Spin Echo, ed. by F. Mezei (Springer Verlag, New York, 1980)

30. F. Mezei and A. P. Murani: J. Mag. and Mag. Mat., 14, 211 (1980)
31. D. Durand, F. Mezei, R. Currat, L. Bernard and F. Denoyer: to be published
32. D. Durand, F. Denoyer, R. Currat and C. Vettier: Phys. Rev., B30, 1112 (1984)
33. J. P. Boucher, F. Mezei, L. P. Regnault, J. P. Renard, Phys. Rev. Lett. 55, 1778 (1985) and erratum Phys. Rev. Lett. 55, 2370 (1985)
34. J. C. Scott: this proceedings
35. P. Cumar: Physica 50, 359 (1982)

Novel Experimental Techniques and Realizations of Quasi-Periodicity

P.M. Chaikin[1,2], *A. Behrooz*[1], *M.J. Burns*[1,3], *D. Levine*[1,5], *D. Ou-Yang*[1,2], *B. Whitehead*[4], *and X. Yan*[1]

[1]Dept. of Physics and Laboratory for Research on the Structure of Matter, University of Pennsylvania, PA 19104, USA
[2]Exxon Research and Engineering Co. Rt. 22E, Annandale, NJ 08801, USA
[3]Dept. of Applied Sciences and Dept. of Physics, Harvard University, Cambridge, MA 02138, USA
[4]National Research and Resource Facility for Submicron Structures, Cornell University, Ithaca, NY 14853, USA
[5]After Sept. 1986, Dept. of Physics, UC Santa Barbara, CA 93103, USA

Although there has been a great deal of recent activity on possible structures and some of the static properties of quasi-periodic and quasi-crystalline phases, some fundamental questions still remain about the dynamics, possibilities of commensurability and pinning, and the nature of the states for such systems. Since three-dimensional realizations of such structures, with sufficient perfection and characterization to allow meaningful studies, are not yet available, we describe some experiments on artificial two-dimensional structures which answer some of these questions experimentally. The main results are attained on superconducting networks, where the application of a magnetic field allows us to see the generalization of commensurability pinning for a variety of quasi-crystalline, quasi-periodic and disordered structures.

1. Introduction

Although there are many examples of quasi-periodicity in nature, we are just beginning to understand some of its most fundamental consequences . In particular the electronic states that one might find in a quasi-periodic potential have been under investigation for quite some time, but the exact spectra for a few well-defined models has only been available over the past 10 years [1]. Quasi-periodic potentials have often been studied as 'intermediate' between disordered and periodic. Like disordered structures they lack translational symmetry. Like periodic structures the potential can be described throughout space by specifying a few variables locally.

In quasi-periodic potentials, such as the superposition of two incommensurate sine waves, the distance between two equivalent points can take on an infinite number of values, from zero to the shorter wavelength. Recently it has been shown that quasi- periodic structures with a finite number of lengths can be mathematically constructed in any number of dimensions [2]. We will take as the definition of 'quasi-crystals' quasi-periodic potentials with a finite number of lengths. This introduces a new step in going from disordered to crystalline: disorder -> quasi-periodic -> quasi-crystal -> crystal. Recently the discovery of an icosahedral phase of quenched AlMn [3] has led to a great deal of work on quasi-crystals.(Icosahedral symmetry is not allowed for conventional crystals, but can occur in quasi-crystals.)

Unfortunately very little is known about the spectra of quasi-periodic or quasi-crystalline systems. Theoretically the only problems which have been solved are one dimensional [4]. Experimentally there have been no detailed measurements of the spectra in any number of dimensions. There was hope that with the discovery of the icosahedral AlMn experiments could be performed on three-dimensional quasi-crystals, but as we shall see the samples are not yet of sufficient quality to allow detailed experiments. The 'novel' idea which is the nucleus of this talk is to artificially make quasi-periodic, quasi-crystalline, disordered and crystalline arrays so that the structures themselves are very well defined and then to study their physical properties. It is not easy to make large-scale structures (many basic units) of arbitrary symmetry in three dimensions, so we are limited to one or two dimensions. There already has been some work done on one dimensional quasi-crystals. Theoretically the 'Fibonacci' spectrum is known [1], and experimentally 'Fibonacci lattices' have been made by varying the spacings of multilayer structures [5]. Since we wanted to explore virgin territory we decided to work in two dimensions.

The easiest way to make two-dimensional structures of any pattern is to use a plotter connected to a computer. If you want to do more than diffraction experiments you must adapt the plotter to use other than paper and ink. In some experiments currently under way we use the plotting pen to apply epoxy dots to a rubber sheet. Then the sheet is mounted in a frame and we look at the 'drum' mode resonances. The epoxy mass loads the thin sheet and allows an evaluation of the phonon spectrum for the two-dimensional structure. It is not difficult to put down 10^3 - 10^6 epoxy plots on a 10cm x 10cm rubber sheet. It is also possible to substitute an electron beam and resist for ink and paper, and therefore microlithographically make a two-dimensional metal film. If the metal is superconducting we can perform flux quantization experiments on it and learn something about the electron spectrum using the macroscopic wavefunction of the superconductor.

2. Experiments on Icosahedrally Packed Al-Transition Metal Alloys

Before reporting on the artificial structures, we summarize what we have learned from the remarkable icosahedral phases discovered by Shechtman et al. Most of the experiments that have been published on these materials, $Al_{1-x}Mn_x$ (x~0.2), are structural studies using electron diffraction from small crystallites or x-ray powder patterns [3,6]. The scattering studies show sharp diffraction peaks. A 'usual' interpretation would lead to a translational coherence length of the order of several hundred Angstroms. The icosahedral orientational order, however, is much larger, namely on the scale of microns. Since one interpretation of the scattering experiments is that the system is quasi- crystalline, we decided to do a number of transport, elastic and magnetization measurements.

Our most interesting result is that the resistivity is extremely high, $\rho \sim 200\ \mu\Omega\text{-cm} \pm 50\%$ (and sometimes as high as 700μΩ-cm), indicating

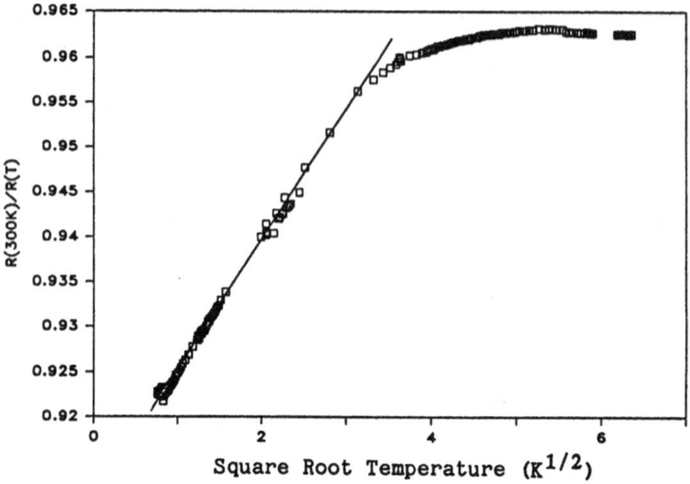

Figure 1. Low-Temperature resistivity of icosahedral phase $Al_{.8}Mn_{.2}$.

a mean free path of less than 5 Angstrom, comparable to that in a metallic glass. Moveover, from 1-10K, the conductivity decreases with decreasing temperature with a form, $\sigma = \sigma_o + AT^{1/2}$ characteristic of weak localization in three dimensions (Fig. 1). The conclusion we would like to draw from these studies is that the states in a three-dimensional quasi-crystal are weakly localized. However, the systems are sufficiently complex that it is not possible to prove this conjecture. The complications arise from a combination of the following factors: 1) Mn is magnetic in Al and in the present samples may show effects due to resonant d levels near the Fermi level E_f, spin fluctuations, the Kondo effect and the formation of a spin glass. 2) The 'stoichiometry' (x=.16-Al with some percentage of Mn, makes up an unknown (<<10%) of the samples. 3) The icosahedral phase may have substitutional disorder (e.g. as in brass). 4) The icosahedral phase may be a 'random' stacking of oriented icosahedral units rather than a quasi-crystal [7]. 5) The icosahedral phase may be a quasi-crystal with a large number of defects and/or phason strain fields.

We have eliminated many of the above objections through magnetization, thermopower and magnetoresistance measurements on a variety of samples (AlRe, AlRuMn, AlSiMn) and all of our results point to small mean free paths and weak localization. The question of substitutional disorder and the possibility of it being an icosahedral glass remain unresolved and require knowledge of the atomic positions. Nonetheless, a perturbative treatment of the low-temperature conductivity requires input simply from the structure factor S(q) and there does not appear to be sufficient diffuse scattering to account for the small mean free path.

3. Flux Quantization on Periodic Arrays

The idea of using superconductivity to probe solutions to Schroedinger's equation is an old one [8]. The superconducting order parameter has an amplitude and a phase similar to that of a wavefunction, and in mean field is the solution to the Ginzburg-Landau equations. The minimization of the Ginzburg-Landau free energy with respect to the order parameter gives two equations, known as the GL differential equations. After linearization (appropriate at the normal-superconducting phase boundary), one of these is formally the same as Schroedinger's equation, while the other is the quantum-mechanical expression for the current due to a wavefunction.

The wavefunction - order parameter must be single valued, thus the phase integral around any closed loop must be 2π times an integer. In the presence of a magnetic field this quantization criterion becomes [9]

$$\int \nabla\phi\cdot dl = \frac{4\pi}{c\Phi_0}\int \lambda^2 J\cdot dl + \frac{1}{\Phi_0}\int A\cdot dl = 2\pi n \qquad (1)$$

where Φ_0 is the flux quantum $(hc/2e) \approx 2\times10^{-7}$ gauss-cm^2, and λ is the penetration depth. The integral over the vector potential is just the flux through the loop and the phase integral (known as the fluxoid) must always be quantized. If the magnetic field H, through the loop is exactly an integral number of flux quanta eq. 3 is satisfied with no currents circulating the loop. Then the additional kinetic energy($E \propto J^2$) is a minimum and the transition temperature is a maximum. In the case of a single circular loop $E \sim \Delta T_c \approx (\xi/a)^2(n-\Phi/\Phi_0)^2$, where n is an integer giving the number of fluxoids in the loop, and the phase boundary is as shown in Fig. 2. This is the beautiful result demonstrated in the work of Little and Parks [9].

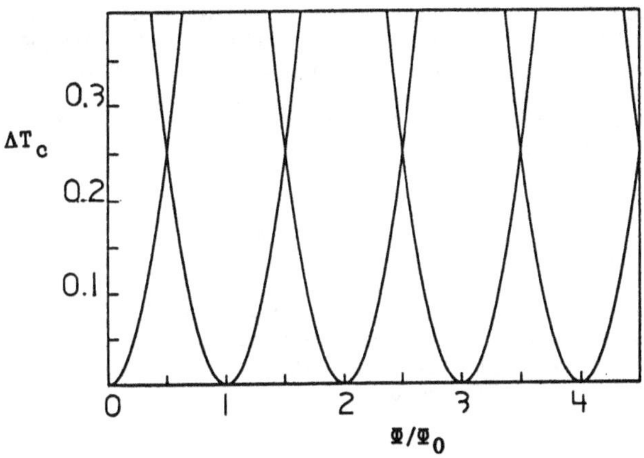

Figure 2. Decrease in transition temperature vs. flux through a single loop as in the experiment of Parks and Little[9].

The problem becomes considerably more complex when we deal with a periodic network [10]. The fluxoid is quantized on every closed path. To understand qualitatively what we might expect, it is convenient to imagine the limit where the wires which make up the network have a diameter which is much larger than the penetration depth but much smaller than the unit cell dimensions. In the center of the wire the currents are then negligible and the flux through any closed path (through the middle of the wires) is quantized. In this limit the currents flowing on the surfaces of the wire must produce a magnetic field which makes up the difference between the applied flux and an integral number of flux quanta. The 'rules' for this simplified calculation are then current conservation at each node, and an average field for the entire network which is equal to the applied field. For example, an applied field $H<\Phi_0/a^2$ (a^2 is the area of a unit cell or an elementary tile) would correspond to a fraction f of the elementary cells having a penetrating flux quantum and (1-f) having zero flux with $f\Phi_0/a^2 = H$. The currents required for this arrangement can be calculated and the kinetic energy and ΔTc evaluated. The experimental situation can be qualitatively recovered by realizing that the actual currents are much smaller than the ones calculated in the above approximation and the tiles containing integer flux quanta are to associated with tiles containing the same integer fluxoids. (The screening currents for most experimental situations are very small. They produce very little field modulation and to lowest order the magnetic field is uniform. However the actual currents are similar to those obtained by the recipe above and thus the T_c shift is qualitatively correct in its dependence on magnetic field.)

If we ignore correlations between occupied and unoccupied tiles (a sort of mean field treatment) then we can proceed to calculate the currents. In an applied field $H=f\Phi_0/a^2$, f cells or tiles require a current proportional to $\Phi_0(1-f)$ while (1-f) tiles require a current proportional to $\Phi_0 f$. In this situation all tiles will satisfy flux quantization. The average current is zero (as it must be to insure the average field as the applied field), however the average squared current circulating around each tile is $f(1-f)^2\Phi_0^2+(1-f)f^2\Phi_0^2=f(1-f)\Phi_0^2=\Phi(\Phi_0-\Phi)$ for f<1. The above arguments are easily generalized to any field and we have $\Delta T_c \sim(\Phi/\Phi_0-m)(m+1-\Phi/\Phi_0)$ for all f, (m is an integer). Thus for an infinite network ΔTc is zero for H=0 and for $H=n\Phi_0/a^2$, but it increases linearily as H varies from these points rather than quadratically (as for the single loop problem).

The periodicity is again in terms of an integral number of flux quanta per tile but we now have cusps instead of parabolas at each minimum. The cusps result from the fact that for an infinite system when we have a field where each tile is satisfied, a slightly different field corresponds to adding a few additional widely separated flux quanta The energy cost is thus linear with the deviation from the fully satisfied case in which there are no currents.

There are additional structures at nonintegral flux quanta per tile which result from the correlations which we have neglected above. The most straightforward example is the case where $H=\Phi_0/2a^2$. If we randomly fill half of the tiles the result is $\langle J^2\rangle=1/32$. However, if we correlate the tiles with flux quanta so that they only share edges with tiles without flux (as a checkerboard for the square lattice) then the kinetic energy is decreased to $\langle J^2\rangle=1/64$. Similarly, there is a favorable correlated arrangement of the flux quanta, forming a flux lattice, for any rational fraction of a flux quantum per tile[10]. This is equivalent to having a flux lattice commensurate with the underlying periodic pattern.

A more rigorous treatment for the phase boundary of a superconducting network is found in References 11 and 12, where the linearized Ginzburg-Landau equations are treated with the constraint of current conservation at each node.

$$- \Delta_i \sum_j \cot(\theta_{ij}) + \sum_j \Delta_j e^{i\gamma_{ij}} / \sin(\theta_{ij}) = 0 \quad . \tag{2}$$

For regular networks, with all lengths equal, the above eigenvalue problem is the same as that for an electron on a similar tight binding lattice in a magnetic field [11,12]. The eigenvalue which corresponds to the energy in Schroedinger's equation determines the coherence length, ξ, in the Ginzburg-Landau case, $\varepsilon = \cos(a/\xi)$. Since $\xi \sim \xi_o (T_{co}-T)^{-1/2}$ the maximum eigenvalue corresponds to the transition temperature. The best known example is the square lattice where the highest eigenvalue (corresponding to the Tc(H) phase boundary) is the top of the spectrum solved by Azbel and Hofstadter [13] . Again, this phase boundary shows cusps at every rational field.

We now want to see what happens when we try the same experiment on quasi-periodic or quasi-crystalline networks. In the quasi-crystalline case there are two fundamental differences from the regular periodic networks. There are at least two different tiles with irrationally related areas. It is therefore not possible to find a field which puts an integral number of flux quanta in each tile and this should dramatically affect the large-scale structure. However, the fine structure is most interesting since it relates to the 'commensurability' of the flux lattice with the network. With a quasi-periodic system it is not even clear what commensurability should mean.

4. Experiments on Quasicrystalline Arrays

The samples were fabricated at the National Research and Resource Facility for Submicron Structures at Cornell using electron beam lithography. A typical pattern consisted of 400x400 tiles with a tile area of 1-4 square microns. The network wires were aluminum, 500 Anstroms thick by 3000 Angstroms wide, deposited on a silicon wafer. The patterns were written under computer control using a Cambridge EBMF-2-150 Electron Beam Microfabricator.

Our first series of samples are 'one-dimensional' patterns consisting of a set of periodic parallel lines along y crossed by periodic, quasi-periodic, quasi-crystalline, random and several other patterns along x. The one-dimensional structures have two advantages: 1) they are considerably more variable (two or more dimensions puts constraints on many of the parameters in a quasi-crystal) and 2) it is possible to analyze the results theoretically, using the periodicity in the y direction. For the two-dimensional quasicrystal we used an eight fold generalization of a Penrose tiling. In Fig. 3 we show the optical micrographs along with the Fourier transforms of some of our networks .

In Fig. 3a is shown the 'Fibonacci' pattern which consists of two tiles with relative area $\tau = 1.618...$ (the golden mean) arranged in a Fibonacci sequence [2]. The number of large tiles is τ times the number of small tiles. The Fourier spectrum [2] consists of a dense set of points at $(n+m\tau)/a$ reflecting the quasi- periodicity with two frequencies related by τ . In the present Fibonacci pattern all of the incommensurate ratios are given by τ. In order to distinguish the effects of quasi- periodicity from that of the irrational area ratio we have made a 'random Fibonacci' pattern, shown in Fig. 3b. The ratio of the area of the tiles is τ and there are τ times as many large as small, as in the Fibonacci. However the tiles are randomly arranged. The optical transform reflects this in showing a broad diffuse scattering instead of distinct Bragg peaks. (But there is a rather sharp spot at $q=(5/3)(2\pi/a_s)$ which comes from the Hendricks-Teller effect.[14])

The phase boundary Tc(H) for the Fibonacci and random Fibonacci patterns are shown in Fig. 4. A quadratic background due to the finite thickness of the aluminum wires has been subtracted. It is immediately clear that neither pattern is periodic and that, although similar in

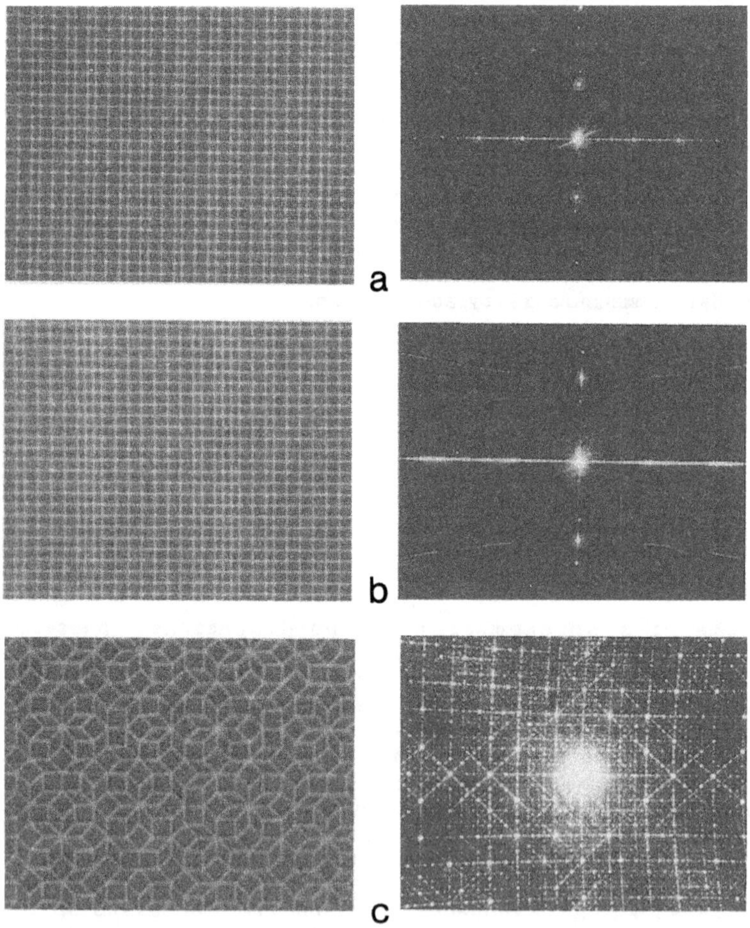

Figure 3. Optical Micrographs and Optical diffraction patterns for: a) 'Fibonacci Lattice', b)'Random Fibonacci, c)'Eight-fold Penrose Tiling'.

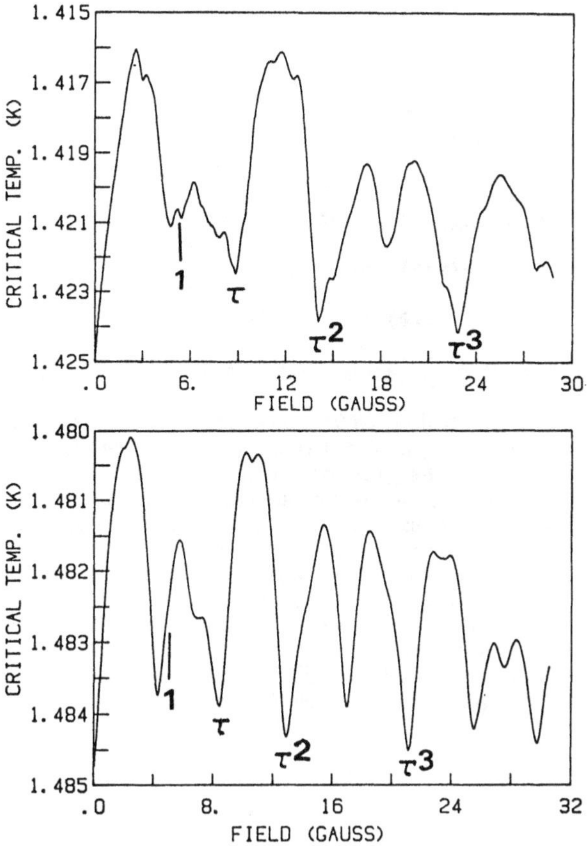

Figure 4. Experimental data on the phase boundary for a) the Fibonacci Lattice and b) the Random Fibonacci pattern.

gross features, they are different in detail. In particular there are large dips in both curves which, in analogy to the periodic network, should reflect the area or areas of the tiles. However there is no fine structure in the random Fibonacci and a lot of fine structure in the Fibonacci. This immediately tells us that the quasi- periodicity is important and that the flux lattice has found a way to be commensurate with the quasi- crystal.

The large-scale structure in both Fibonacci and random Fibonacci is easy to understand. The transition temperature for a given field would return to zero-field value if we could put an integral number of flux quanta in each tile [10]. In the infinite field limit we could do this. For finite fields we can approximate flux quantization at fields which correspond to rational approximates to τ (M/N=1,2,3/2,5/3,8/5,13/8, etc.). If we put M flux quanta in the small tiles and N in the large tiles then the average field is $H=(M+N\tau)/(1+\tau^2)$. For the low rational approximates (M,N) = (1,1), (1,2), (2,3), (3,5), (Fn,Fn+1), corresponding to applied fields of M/Ho = 1, τ, τ^2, τ^3 , ,,,τ^n where Ho = $\Phi_0/a_s{}^2)(1+\tau)/(1+\tau^2)$. H_0 is the field which corresponds to one flux quantum in each tile. Thus the largest structures which occur in $T_c(H)$ for both Fibonacci and random Fibonacci are again related to the properties of the elementary tiles.

In order to proceed further we turn to the network equations of references 11,12. It is a straightforward matter to write down the coupled equations for the Fibonacci sequence using its generating function $x_m=m+(1/\tau)\mathrm{INT}(m/\tau)$, defining $\Delta_m=x_{m+1}-x_m$, and using equation 2

$$\begin{aligned}&\Lambda_{mn}[\,2\sin(\tau a/\xi)\cos(a/\xi) + \cot(\Delta_{m-1}a/\xi)\sin(a/\xi)\sin(\tau a/\xi)\\ &\qquad + \cot(\Delta_m a/\xi)\sin(a/\xi)\sin(\tau a/\xi)\,]\\ &-\sin(a/\xi)\sin(\tau a/\xi)[\Lambda_{m-1,n}/\sin(\Delta_{m-1}a/\xi) + \Lambda_{m+1,n}/\sin(\Delta_m a/\xi)]\\ &-\sin(\tau a/\xi)[\Lambda_{m,n+1}\exp(i\gamma x_m a) + \Lambda_{m,n-1}\exp(-i\gamma x_m a)] = 0 \quad . \qquad (3)\end{aligned}$$

We are using a Landau gauge, (0,Hx), with $\gamma=eH/hc$. The only place where a 'y' dependence enters the equation is in the last terms on the right, involving the nodes m,n+1 and m,n-1. The form is such that the solution must merely contain a Floquet factor and hence the problem reduces to one dimension. However the eigenvalue problem still needs computer solutions. Evaluated on a 23x23 network we find the normalized phase boundary shown in Fig. 5. The computation reveals the major dips at factors of τ predicted above and some of the fine structure seen in the experiment. In fact the calculation fits the experiment remarkably well although the experiments have finer structure. An additional feature which appears in the calculation is a quasi mirror symmetry about the points $(\tau^n+\tau^{n+1})/2$ which extends from 0 to $\tau^n + \tau^{n+1}$ [15].

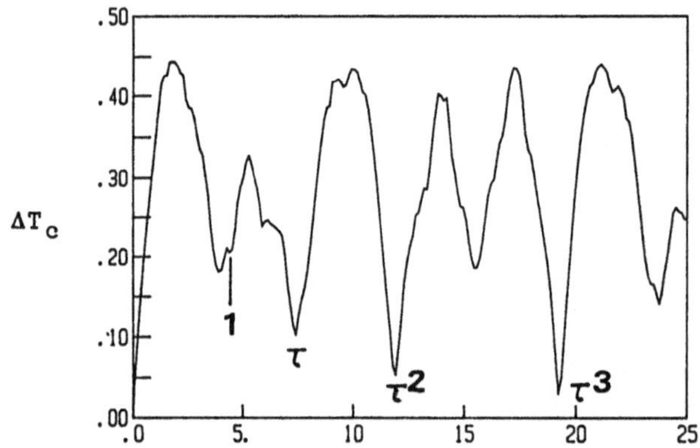

Figure 5. ΔT_c vs. Φ/Φ_0 calculated from the linearized GL equations for the Fibonacci Lattice.

The Fibonacci and the Random Fibonacci patterns tell us much that we want to know about flux quantization in quasi-periodic networks. The features due to the areas of the tiles dominate the phase boundary and the main dips which were at $n\Phi_0/a^2$ for the periodic case now vary as $(\Phi_0/a^2)\tau^n$. We have also seen that fine structure is present only in the ordered quasi-periodic network indicating the importance of the long range coherent structure for the flux lattice commensurability. This 'locking in' of the flux lattice is also indicated by the important observation that many dips in the Fibonacci lattice are 'cusp-like', while those in the random Fibonacci are all quadratic. Unfortunately the interpretation of each dip in the 'one-dimensional' patterns is difficult , because the periodicity along y actually complicates a simple labelling scheme.

Our two-dimensional quasi-crystal is the eight fold version of a Penrose tiling (a Penrose tile could not be fabricated since the E - Beam Fabricator cannot draw lines at 72° angles), fig. 3c. It consists of two tiles, a fat and a skinny, with area ratio $\Omega=2^{1/2}$. Unlike the Penrose tiling there are more skinnies than fats by the ratio Ω . However the quasi-periodicity is characterized by the number $\sigma=1+\Omega$, that is the pattern is generated by an inflation scheme with each inflation changing characteristics distances by σ. The average field for N_f (N_s) flux quantum in every fat (skinny) tile is $H=(\Omega N_s+N_f)\Phi_o/(2\Omega)A_s=(\sigma)\Phi_o/(2\Omega)A_s$. We expect the biggest effects for rational approximates to the area ratio $\Omega=2^{1/2}$. The lowest few are the pairs (N_f, N_s) = (1,1), (3,2), (7,5) which correspond to average fields $H/Ho=1,\sigma,\sigma^2$ as indicated in Fig. 6. It is interesting that the parameter that comes in when we are trying to rationalize the area ratio Ω, is the quasi-periodicity ratio σ. This is of course not accidental, but comes from the restrictions in forming quasi-crystals.

Aside from the 'major dips' there are smaller dips between the values $H/Ho=\sigma^n$. These dips correspond to the flux lattice finding favorable configurations on the quasi-crystalline substrate. There should of course be dips everywhere, but the largest ones in the present case can be readily indexed by $H(n,m) = Ho(n+m\sigma)$ with n,m positive and

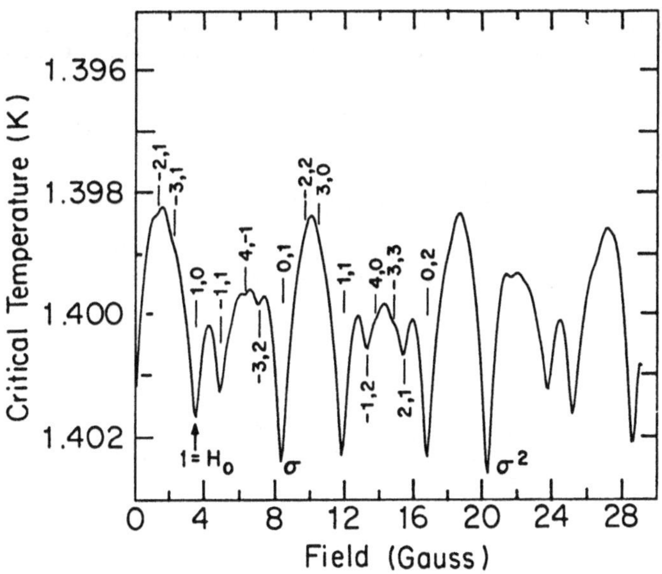

Figure 6. Experimental Data on the transition temperature vs. applied field for the 'Eight fold Penrose tiling'.

negative integers. The combination which yields fields less than σ^2 with n,m less than 4 are shown in Fig. 6. Most are readily identified. Thus the system which replaces the rational fields $H = (p/q)Ho$ for the periodic structure is $H=Ho(n+m\sigma)$ for the quasi-crystalline networks. It is worth pointing out that for quadratic irrationals, say β, p/q is in general not a member of $n+m\beta$, whereas $s\beta^t$ is a member (n,m,p,q,s,t integers). Thus some combinations of $n+m\sigma$ may be viewed as deflations of the quasicrystal. It is not unreasonable that a flux lattice may lock in or show epitaxy with a deflation of a quasi-crystalline network.

Our results on the 'Fibonacci' sample and the eight-fold Penrose tiling suggest the generalization of the concept of commensurability to include superstructures on whose Fourier transform the fourier transform of the original structure is completely found. This definition works equally well for the sparse points which are the reciprocal lattice for conventional crystals and the dense points which form the diffraction pattern of quasi-crystals.

Unfortunately we cannot calculate the expected phase boundary for the two-dimensional patterns. As discussed earlier the problem is equivalent to solving the electronic structure which is not yet possible. However we see that the main features are similar to what has been seen in the one-dimensional quasi- crystals: large dips at σ^n, fine structure and the remarkable quasi-reflection symmetry about $(\sigma^n+\sigma^{n+1})/2$.

The use of artificial structures and in particular superconducting networks can be useful for studying the properties of other patterns as well. Self-similar ordered fractals such as Sierpinski gaskets have been made [16] and random fractals such as percolation clusters are being studied along with quasi-periodic systems.

In conclusion,we have prepared a series of superconducting networks in the form of periodic, quasi- periodic , quasi- crystalline and disordered patterns. Our main result from this study is that a flux lattice can be commensurate with a quasi-crystal. This implies that it should be possible to grow epitaxially (at other than the trivial 1:1 case) on a quasi- crystalline substrate. We have also shown the utility of such artifical patterns for gaining insight into some unsolved problems in new structures with unusual symmetries and construction.

Part of this research is supported by NSF under grants DMR83-18060 (M. J. B.) and through the Laboratory for Research on the Structure of Matter, DMR85-19059 (A. B.)

References

1. S. Ostlund and R. Pandit: Phys.Rev B 29, 1394 (1984)
2. D. Levine and P.J. Steinhardt: Phys.Rev. Lett. 53, 2477 (1984)
3. D. Shechtman, I. Blech, D. Gratias, and J.W. Cahn: Phys. Rev. Lett. 53 , 1951 (1984)
4. B. Simon: Adv. Appl. Math. 3, 463 (1982).
5. R. Merlin, K. Bajema, R. Clarke, F.-Y. Juang, and P.K.Bhattacharya: Phys. Rev. Lett. 55, 1768 (1985)
6. P.A. Bancel, P.A. Heiney, P.W. Stephens, A.I. Goldman, and P.M. Horn: Phys. Rev. Lett. 54 ,2422 (1985)

7. P.W. Stephens and A.I. Goldman: Phys. Rev. Lett. 56, 1168 (1986)
8. V. L. Ginzburg and L. D. Landau, Zh. Eksperim. i Teor. Fiz 20, 1064 (1950).
9. R.D. Parks and W.A. Little: Phys. Rev. A133, 97 (1964)
10. B. Pannetier, J. Chaussy, R. Rammal, and J.C. Villegier: Phys.Rev.Lett. 53, 1845 (1984)
11. S. Alexander: Phys. Rev.B 27, 1541 (1983)
12. R. Rammal, T.C. Lubensky, and G. Toulouse: Phys. Rev.B 27, 2820 (1983)
13. M. Ya. Azbel: Zh. Eksp. Teor. Fiz. 46, 939 (1964) [Sov. Phys. -JETP 19, 634 (1964), D.R. Hofstadter: Phys. Rev. B14, 2239 (1976)
14. S. Hendricks and E. Teller, J. Chem. Phys. 10, 147 (1942).
15. A. Behrooz, M. J. Burns, H. Deckman, B. Whitehead, and P. M. Chaikin, Phys. Rev. Lett. 57, 368 (1986).
16. J. M. Gordon, A. M. Goldman, J. Maps, D. Costello, R. Tiberio and B. Whitehead, Phys. Rev. Lett. 56, 2280 (1986).

Heavy-Electron Materials

Z. Fisk, H.R. Ott, and J.L. Smith*

Los Alamos National Laboratory, Los Alamos, NM 87545, USA

1. Introduction

The earmark of the intermetallic compounds known as heavy-Fermion or heavy-electron materials is a low-temperature linear electronic specific-heat coefficient (γ) which shows an upturn on cooling to low temperatures, typically below 4 K (Fig. 1). It is clear from data on the gap anomaly of heavy-electron superconductors that this γ is of electronic origin and, further, that f-electrons must have something to do with it. Only the 4f and 5f elements, typically thought of as forming valence-fluctuation compounds (mainly Ce, Yb, U, and Np) seem to also form heavy-electron compounds. We list a number of these heavy electron compounds in Table 1.

All the heavy-electron compounds have magnetic susceptibilities which are Curie-Weiss-like at high temperature, with effective moments close to that of their expected f-configurations. At low temperature this susceptibility becomes Pauli-like, showing no tendency to saturate even for large H/T (magnetic field over temperature). One way to think about this is in Kondo terms: at high temperature we have local moment behavior, while on cooling

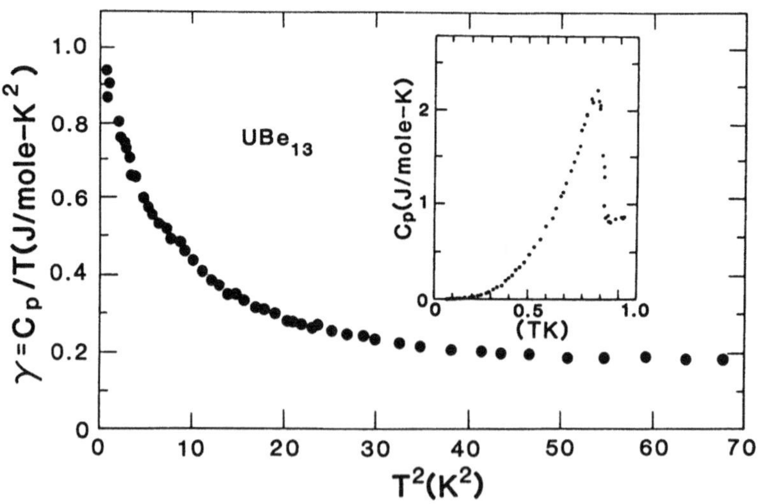

Fig. 1. C/T versus T^2 for UBe_{13}. Inset shows the anomaly in C at T_C.

*Permanent address: Laboratorium für Feskörperphysik, ETH-Hönggerberg 8093 Zürich, Switzerland

Table 1. Some Heavy-Electron Compounds

Compound	γ*(mJ/mole f-K^2)	ordering**
$CeCu_2Si_2$	1200	s, T_s = 0.6 K
$CeCu_6$	1600	n
$CeAl_3$	1500	n
UAl_2	150	n
UPt_3	450	s, T_s = 0.54 K
U_2Zn_{17}	535	m, T_N = 9.8 K
UCd_{11}	840	m, T_N = 5.0 K
$UAuPt_4$	725	n
UBe_{13}	1100	s, T_s = 0.9 K

*γ extrapolated from above T_N for magnetically ordering materials.
**s ≡ superconducting
m ≡ magnetically ordering
n ≡ not ordering

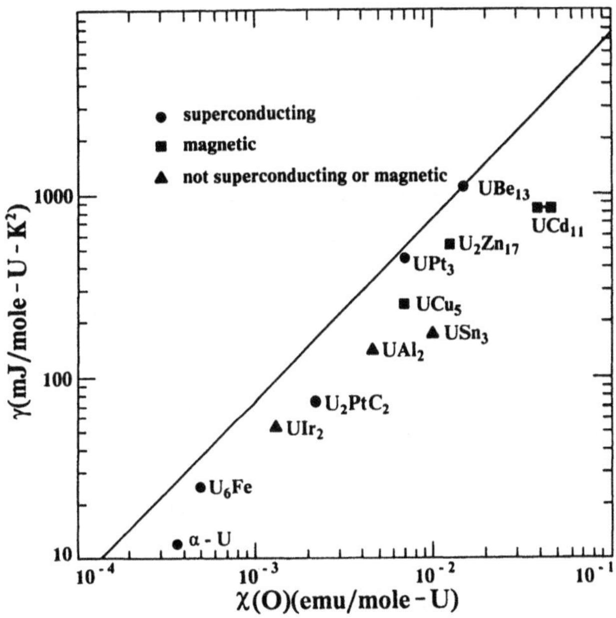

Fig. 2. Lnγ versus lnχ for U compounds. The line is the free electron relationship.

to low temperature these local moments are compensated by a screening cloud of conduction electrons, the resulting low-temperature state of the lattice being a Fermi liquid involving coherent, Bloch electron states. The plot of γ(o) versus χ(o) (Fig. 2) shows all the data below the Sommerfeld line, the line for which, loosely speaking, all the χ shows up in γ.

This Kondo viewpoint allows a qualitative estimate of γ. Suppose the crystal-field ground state of the f-level is a doublet. The kln2 entropy of

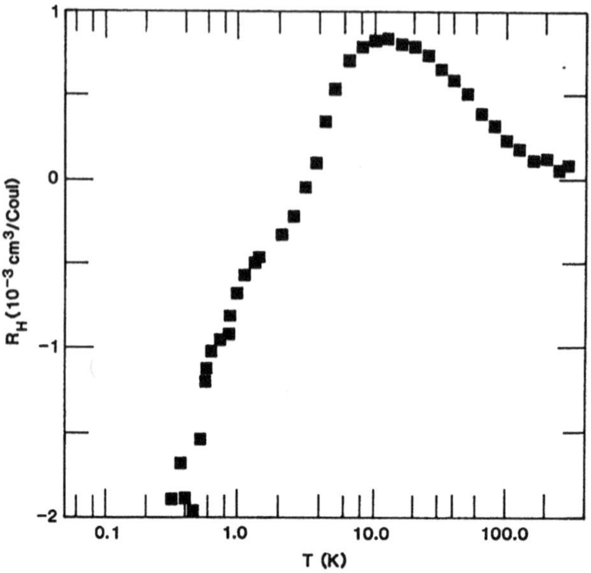

Fig. 3. Hall coefficient of $CeCu_6$. Data from ref. 1.

this doublet will disappear upon the establishment of the low-temperature coherent state and must therefore appear in the conduction electron entropy. If the compensation has a characteristic temperature T_o, then we estimate $\gamma \sim R\ln 2/T_o = 5.76/T_o$ J/mole-K^2.

Hall-effect data give further useful information. Studies of a number of heavy-electron compounds [1,2] show that at high temperatures there is a large, positive contribution to the Hall constant which is lost upon the establishment of coherence. While the temperature variation of the Hall-effect is not here understood in detail, the gross effects are [3,4]: incoherent skew scattering from the impurity-like f's at high temperature makes a large, positive contribution to the Hall constant. This scattering disappears upon the establishment of coherent Bloch states at low temperature (Fig. 3). The extra insight derived from the Hall data is that compounds such as $CePd_3$, whose γ = 37 mJ/mole Ce-K^2 [5] and is not normally thought of as a heavy-electron material, also show this temperature dependence in their Hall constants, only here T_o is much larger.

The Hall data then allows us to hope that there is possibly a whole continuum in the physics linking typical transition metal γ's of 10 mJ/mole-K^2 with the heaviest electron γ's over 1 J/mole-K^2. Thinking in band-structure terms, as the bands become narrower and γ increases, electron correlations lead to even narrower effective bands, where the Anderson Hamiltonian is appropriate. The simple viewpoint presented above is clearly not applicable over the whole range of variation of γ; our point is that the underlying physics may not change as drastically as γ does. We note that recent de Haas-van Alphen results on the compounds $CeCu_6$ [6] and UPt_3 [7] confirm the large masses expected, demonstrate the existence of a Fermi surface at sufficiently low temperature, and show that the entire Fermi surface involves heavy electrons.

2. Superconductivity

We turn now to the phase transitions in this heavy-electron state observed at low temperatures. These are either magnetic or superconducting. Some compounds (e.g. $CeAl_3$, $CeCu_6$, $UAuPt_4$) do not show a co-operative low-temperature phase transition, and, to date, no charge density wave transition has been identified in these materials. We deal first with the superconductors.

There are three known examples of heavy-electron superconductivity: $CeCu_2Si_2$ (T_c = 0.6 K) [8], UPt_3 (T_c = 0.54 K) [9] and UBe_{13} (T_c = 0.9 K) [10]. The specific heat plotted as C/T versus T^2 is shown in Fig. 1 for UBe_{13}. The anomaly at T_c is very much in the strong-coupling regime, and it demonstrates, in fact, that the superconducting gap opens in the heavy quasiparticle band. This shows that γ really does represent the electronic density of states. The specific heat anomalies at T_c for $CeCu_2Si_2$ and UPt_3 are generally somewhat smaller than weak coupling BCS theory and very dependent on sample quality. The accurate fit of the low-temperature normal state specific heat of UPt_3 to the spin fluctuation form $T^3 \ln T$ led to an early suggestion that paramagnon exchange might be the mechanism producing its superconductivity [9].

The special interest in the superconductivity of the heavy-electron materials is that both a new type of pairing and/or a new mechanism may be involved. The idea behind this is that in the highly correlated narrow band, s-wave pairing could well be repulsive. It is worth noting, further, that the superconductivity of these compounds must be thought of in different terms from that of the so-called re-entrant magnetic superconductors: for these latter, two separate weakly coupled electronic systems are in competition, while with the heavy-electron superconductors it is one and the same set of electrons which is deciding whether to be superconducting or magnetic.

In the presence of spin-orbit coupling, parity is the only obvious symmetry, and the superconducting state will have either even or odd parity. Among these states, the ones with essentially new properties will be those for which the gap possesses nodes on the Fermi surface, either points or lines where the gap vanishes. These should manifest themselves in non-exponential temperature variations in various properties below T_c, and a number of these have been seen. For example, $C = \gamma' T + \alpha T^3$ in UBe_{13} at very low temperatures [11]. The γ' term of 20 mJ/mole U-K^2, which is sample dependent, can be rationalized as coming from the smearing out of points of the gap into finite areas due to impurity scattering, and it is known that the residual resistivity of good samples of UBe_{13} is some 10 to 20 $\mu\Omega$-cm. We note in passing that the slope of the upper critical field of UBe_{13} at T_c is at least 40 T/K (and perhaps infinite) [12].

The really surprising extra feature connected with UBe_{13} is the effect of Th impurities (Fig. 4). There is a negative cusp in T_c near 1.8 a/o Th [13]. Between this concentration and approximately 4 a/o Th there are two phase transitions [14]: the superconducting one and one near 0.4 K of unknown origin - perhaps to a second kind of superconducting state. Neutron scattering experiments [15] have detected no ordered magnetic moment in this state, and ^{9}Be NMR [16] has also found no sign of magnetic ordering. No other impurity has been found to cause this behavior. It is interesting that the value extrapolated for γ approximately doubles between x = 0 and x = 3.3 a/o Th. A few experiments have also involved substituting for Be in UBe_{13}: for B substitution, depressions in T_c are of approximately

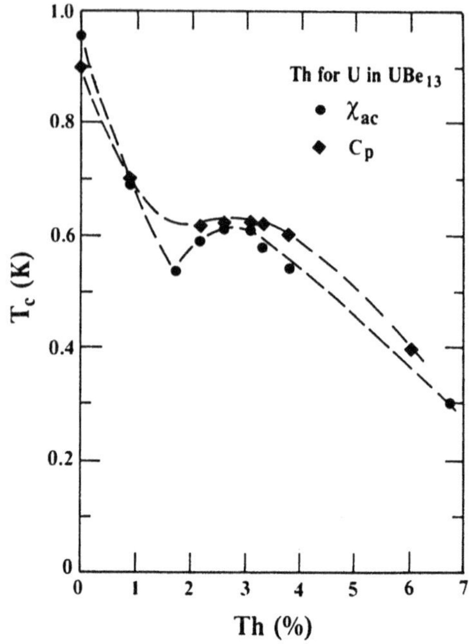

Fig. 4. Variation of T_C for $U_{1-x}\ Th_x\ Be_{13}$ alloys.

the same magnitude as seen for rare earth and Th impurities [17]. It also appears that there is no special extra depression associated with a magnetic moment on the impurity [13].

Further evidence for a new kind of superconducting state is the negative proximity-effect experiment of Han et al. (Fig. 5) [18]. In this experiment, conventional superconductivity is induced via the proximity effect in UBe_{13} with a Ta probe. Below the T_c of UBe_{13}, a depression of the Josephson critical current is found. The interpretation given is that the differing superconducting states of Ta and UBe_{13} are competing for electrons in a

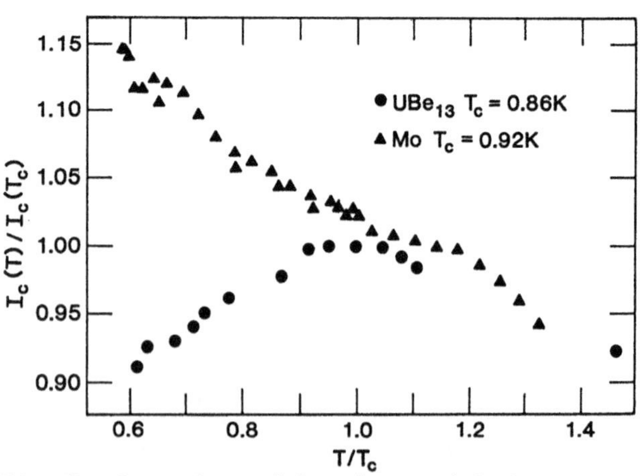

Fig. 5. Comparison of Josephson critical currents observed in Mo-Ta and UBe_{13} - Ta junctions. From ref. 18.

surface sheath. London penetration depth experiments [19] also provide evidence for a new superconducting state in UBe_{13}.

The present consensus, by no means unanimous, is that the superconducting states of UBe_{13} and UPt_3 are anisotropic with zeroes of the gap on the Fermi surface, and that either a p- or a d-type pair, loosely speaking, is involved. The possible mechanism is subject to considerable debate.

3. Magnetism

The low-temperature normal state has been the subject of several neutron studies, in particular in the compounds UPt_3 [20], UBe_{13} [21], $CeCu_6$ [22] and U_2Zn_{17} [23]. For the first three, strong magnetic fluctuations are found with characteristic energies of some tens of wavenumbers, with $\chi(q)$ peaking near a zone boundary for temperatures below the coherence temperature.

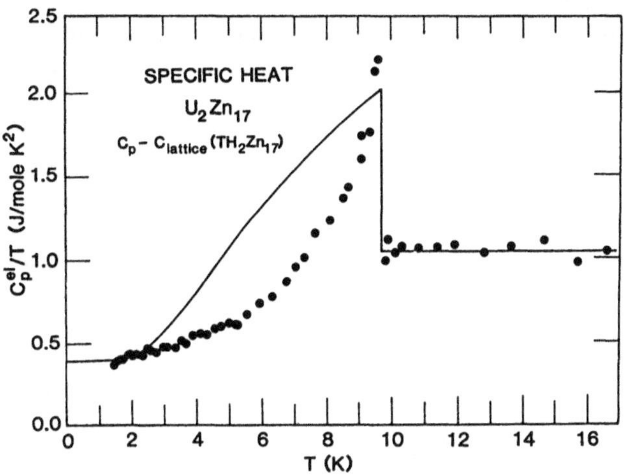

Fig. 6. Electronic specific heat of U_2Zn_{17} through T_N. C^{el}/T versus T is plotted. Solid line is curve calculated with a BCS-type model. C^{el}/T is per formula unit.

U_2Zn_{17} [24] and UCd_{11} [25] order magnetically at T_N = 9.8 K and 5 K respectively. Although the phase transition in U_2Zn_{17} bears some qualitative resemblance to a BCS-type transition (Fig. 6), this comparison does not hold up in detail, although entropy is approximately conserved through it. The ordered moment on U in U_2Zn_{17} is 0.8 μ_B, considerably smaller than the effective Curie-Weiss moment of approximately 3.3 μ_B. It is interesting that the square of the ratio of the ordered moment to the effective moment gives roughly the relative loss of γ upon ordering. Spin waves have been looked for below T_N in U_2Zn_{17} and not found [26] - this could possibly be similar to the Cr case where the spin waves are believed to travel with the Fermi velocity.

Another interesting and somewhat different heavy-electron antiferromagnet is UCu_5. A simple, antiferromagnetic order is observed below T_N = 15 K with ordered moment of 0.8 μ_B [27]. Below 3 K, however, C/T shows an upturn and a second phase transition occurs at 1.0 K [28]. Impurities suppress this lower transition, and it is clear that a heavy-electron state is developing within the ordered magnetic state below $T_N = 15$ K.

4. Comments

It might be expected from the high-temperature properties of the heavy-electron compounds that they would order magnetically upon cooling. In fact, only a few of them do. We show in Table 2 that the low-temperature phase transitions for U-compounds seem to be ranked according to γ_V, the γ per unit volume. It is clear that the superconductors have larger γ_V's than the magnetically ordering compounds. Experiments show that γ for UBe_{13} is strongly depressed by hydrostatic pressure, 30% at 6 kbar [29], while χ changes roughly 10% at this pressure [30]. This suggests that sufficiently high pressure could cause magnetic order to appear in UBe_{13}, a counter-intuitive notion.

Table 2. Electronic Specific Heats per Mole-U for Various U Compounds

Compound	γ(mJ/mole-K^2)	γ_V(mJ/cm^3-K^2)	Ordering*
α-U	12	0.96	s
UPt_5	85	1.38	n
URu_2Si_2	75	1.52	m, s
U_2PtC_2	75	1.59	s
UIr_2	52	1.63	n
URu_4B_4	170	1.63	n
U_6Fe	25	1.79	s
$U_2Co_3Si_5$	115	2.55	n
USn_3	169	2.84	n
UCu_5	210	4.03	m
UAl_2	150	4.25	n
U_2Zn_{17}	500	5.08	m
UCd_{11}	840	5.21	m
UPt_3	450	10.59	s
$UAuPt_4$	725	11.7	n
UBe_{13}	1180	13.55	s

*n = not ordering, m = magnetic ordering, s = superconducting.

The heavy-electron compounds which order magnetically lie further from the Sommerfeld free-electron line in Fig. 2 than the superconductors do. Speaking loosely, the magnets are less efficient than the superconductors in converting local moment entropy into conduction electron entropy. Experiments measuring the specific heat of U_2Zn_{17} and $CeCu_6$ [31] to fairly high temperature find, in this regard, that an integrated entropy for the electronic system of Rln6 per mole Ce is achieved by 70 K in $CeCu_6$, whereas U_2Zn_{17} has something quite a bit in excess of Rln10 per mole U by this temperature. The first number is exactly that expected for the J = 5/2 Ce Hund's Rule ground state; the U_2Zn_{17} value is anomalous. A possible source for this extra entropy could be, one imagines, an extra degree of freedom derived from the conduction electrons via hybridization with the U f-electrons.

Strong effects of impurities on magnetic ordering have been observed for Cu substitutions in U_2Zn_{17} [32] and Ni in UCu_5 [33]. T_N disappears with small additions in both cases. This is not a simple dirt sensitivity; it depends, we are now finding, on impurity valence, among other things. We also now know that a number of additions (Th, Au, Pd) to UPt_3 at the 5% level lead to magnetic ordering [34]. It is peculiar in this case that the ordering observed has a structure in k-space significantly different from the structure observed in $\chi(q)$ for pure UPt_3 [35].

Nothing has been said in this short summary concerning theoretical attempts to treat the ordered lattice of f-atoms in these compounds. Anderson lattice models have been investigated, but as yet there is very little certainty about what is happening in detail as coherence develops [36]. It is surprising that Oguchi and Freeman [37] seem to get results in rough agreement with the de Haas-van Alphen measurements on UPt_3. The effective masses, of course, turn out to be completely wrong, but the idea is that a proper many-body theory applied to the band structure results would give something reasonable. This approach has not been successful as yet.

UPt_3 has been quite extensively attacked via Fermi liquid theory [38], UPt_3 seeming to have the right low-temperature resistivity and specific heat behavior for the theory. The large anisotropy in χ of UPt_3 is clearly something of a problem for this theory. For UBe_{13}, the Fermi liquid regime is presumably at a much lower temperature, and the superconducting transition is reached before the coherent state is achieved. This makes the Fermi liquid approach to the superconductivity of UBe_{13} less appropriate.

A final point to make about the physics of the heavy-electron materials is that it appears to be dominated by spin fluctuations. In transition metal magnetic materials, both charge and spin fluctuations are important. The heavy-electron materials, therefore, represent a projected piece of the more general transition metal problem, and it is to be expected that what we learn from the heavy-electron compounds can throw new light on transition metal materials.

This work was performed under the auspices of the U.S. Department of Energy, Office of Basic Energy Sciences, Division of Materials Science.

References

1. T. Penney, J. Stankiewicz, S. von Molnar, Z. Fisk, J. L. Smith and H. R. Ott, J. Magn. Magn. Mater. 54-57, 370 (1985).
2. M. Hadzic-Leroux, A. Hamzic, A. Fert, P. Haen, F. Lapierre and O. Laborde, Europhys. Lett. 1, 579 (1986).
3. T. V. Ramakrischnam, P. Coleman and P. W. Anderson, J. Magn. Magn. Mater. 47-48, 493 (1985).
4. A. Fert, J. Phys. F 3, 2126 (1973).
5. W. E. Wallace, Rare Earth Intermetallics (New York, Academic), 1973.
6. P.H.P. Reinders, M. Springford, P. T. Coleridge, R. Boulet, and D. Ravot, Phys. Rev. Lett. 57, 1631 (1986).
7. L. Taillefer, R. Newbury, G. G. Lonzarich, Z. Fisk, and J. L. Smith, to appear J. Magn. Magn. Mater.
8. F. Steglich, J. Aarts, C. D. Bredl, W. Lieke, D. Meschede, W. Frantz and H. Schäfer, Phys. Rev. Lett. 43, 1892 (1979).
9. G. R. Stewart, Z. Fisk, J. O. Willis and J. L. Smith, Phys. Rev. Lett. 52, 679 (1984).
10. H. R. Ott, H. Rudigier, Z. Fisk and J. L. Smith, Phys. Rev. Lett. 50, 1595 (1983).

11. H. R. Ott, unpublished data.
12. M. B. Maple, J. W. Chen, S. E. Lambert, Z. Fisk, J. L. Smith, H. R. Ott, J. S. Brooks and M. J. Naughton, Phys. Rev. Lett. 54, 477 (1985).
13. J. L. Smith, Z. Fisk, J. O. Willis, A. L. Giorgi, R. B. Roof, H. R. Ott, H. Rudigier and E. Felder, Physica 135B, 3 (1985).
14. H. R. Ott, H. Rudigier, Z. Fisk and J. L. Smith, Phys. Rev. B31, 1651 (1985).
15. H. Mook, private communication.
16. D. E. MacLaughlin, C. Tien, W. G. Clark, M. D. Lan, Z. Fisk, J. L. Smith and H. R. Ott, Phys. Rev. Letter, 53, 1833 (1984).
17. Unpublished data.
18. S. Han, K. W. Ng, E. L. Wolf, A. Millis, J. L. Smith and Z. Fisk, Phys. Rev. Letter 57, 238 (1986).
19. D. Einzel, P. J. Hirschfeld, F. Gross, B. S. Chandrasekhar, K. Andres, H. R. Ott, J. Beuers, Z. Fisk and J. L. Smith, Phys. Rev. Lett. 56, 2513 (1986).
20. G. Aeppli, private communication.
21. G. Aeppli and H. Mook, private communication.
22. G. Aeppli, H. Yoshizawa, Y. Endoh, E. Bucher, J. Hufnagl, Y. Onuki and T. Komatsubara, Phys. Rev. Lett. 57, 122 (1986).
23. D. E. Cox, G. Shirane, S. M. Shapiro, G. Aeppli, Z. Fisk, J. L. Smith, J. K. Kjems and H. R. Ott, Phys. Rev. B 33, 3614 (1986).
24. H. R. Ott, H. Rudigier, P. Delsing and Z. Fisk, Phys. Rev. Lett. 52 1551 (1984).
25. Z. Fisk, G. R. Stewart, J. O. Willis, H. R. Ott and F. Hulliger, Phys. Rev. B 30, 6360 (1984).
26. G. Aeppli, private communication.
27. A. Murasik, S. Ligenza and A. Zygmunt, Phys. Status Solidi (a) 23, K163 (1974).
28. H. R. Ott, H. Rudigier, E. Felder, Z. Fisk and B. Batlogg, Phys. Rev. Lett. 55, 1595 (1985).
29. G. E. Brodale, R. A. Fisher, N. E. Phillips, G. R. Stewart and A. L. Giorgi, Phys. Rev. Lett. 57, 234 (1986).
30. M. McElfresh, private communication.
31. E. Swartz, private communication.
32. J. O. Willis, Z. Fisk, G. R. Stewart and H. R. Ott, J. Magn. Magn. Mat. 54-57, 395 (1986).
33. H. J. von Daal, K.H.J. Buschow, P. B. von Aken and M. H. van Maaren, Phys. Rev. Lett. 34, 1457 (1975).
34. A. P. Ramirez, B. Batlogg, E. Bucher and A. S. Cooper, Phys. Rev. Lett. 57, 1072 (1986).
35. G. Aeppli, private communication.
36. P. A. Lee, T. M. Rice, J. W. Serene, L. J. Sham and J. W. Wilkins, Comments on Condensed Matter Physics XII, 99 (1986).
37. T. Oguchi and A. J. Freeman, J. Magn. Magn. Mat. 52, 174 (1985).
38. See discussion in ref. 36.

Magnetic Resonance as a Probe of Nonlinearity in Condensed Matter Physics

W.G. Clark

Department of Physics and Solid State Science Center, University of California at Los Angeles, Los Angeles, CA 90066, USA

1. INTRODUCTION

The various forms of magnetic resonance have become ubiquitous tools for many kinds of scientific investigations. In this paper, I will review some of the basic ideas that govern the main applications of magnetic resonance, and present examples of their use in problems of current interest to the field of nonlinearities in condensed matter physics. At the end, prospects for future developments in magnetic resonance techniques and applications will be presented.

Before going into the more technical aspects of magnetic resonance, I would like to point out a few "statistical" aspects of magnetic resonance that express how much it has developed into a major scientific investigative tool. Even though magnetic resonance is considered to be about 40 years old (E.M. Purcell and F. Bloch received the Nobel Prize for the first demonstrations of nuclear magnetic resonance that were published in 1946), it is a mature subject. This is evident both from the numbers of published papers that make use of it, and from the different varieties of experimental techniques that have been developed. A simple exercise illustrates the first point. At the time of this conference, the hundred most recent papers in Physical Review Letters include five in which magnetic resonance plays a major role, from which we conclude that about 5% of the papers in the current physics literature make use of magnetic resonance. (The corresponding figure for this conference is 14%). If we apply the 5% figure to the 127,385 abstracts published in Physics Abstracts in 1985, an estimate of 6,400 is obtained for the annual production of papers in the physics literature with a significant content of magnetic resonance work. Over 10 years, that amounts to 64,000 papers, or about one for *each* professional physicist in the world!

Another demonstration of the maturity of the technique is the great variety of methods and techniques that have been developed. A partial listing on them (with acronyms) is: electron spin resonance (ESR), electron paramagnetic resonance (EPR), nuclear magnetic resonance (NMR), muon spin rotation (μSR), epin echoes, spin echo spectroscopy, electron nuclear double resonance (ENDOR), dynamic nuclear polarization (DNP), proton-enhanced nuclear induction spectroscopy (the acronym of this one is not usually printed in the physics literature), optical detection of magnetic resonance (ODMR), electron spin echo envelope modulation (ESEEM), ferromagnetic resonance (FMR), antiferromagnetic resonance (AFMR), electron-electron double resonance (ELDOR), proton decoupling, multiple pulse line narrowing, magic angle spinning, Fourier transform NMR, magnetic resonance imaging, solid state effect, Overhauser effect, etc.

An additional point worth noting here is the current cost required to purchase or construct a magnetic resonance apparatus. For nuclear magnetic resonance and nuclear quadrupolar resonance, the cost varies from about $1,000 to $500,000, depending on the complexity of the system. The same range applies to electron spin resonance systems. The cost to set up a μSR system is about $200,100,000. Of this, $200,000,000 is for the accelerator to produce the muons and $100,000 is for the additional apparatus to do the μSR experiments.

2. MAGNETIC RESONANCE PHENOMENA

In this section I will present a brief review of the main features of magnetic resonance, with special emphasis on those aspects that make it a useful tool in condensed matter science. For brevity, this discussion will concentrate mainly on NMR. The same or comparable principles apply, however, to other particles whose magnetic resonances are studied.

In an NMR experiment there are are two basic "handles" that are used to influence the motion of the nucleus: an external magnetic field couples to the nuclear magnetic moment, and an external electric field gradient (tensor) couples to the nuclear quadrupole moment. These are expressed in terms of the corresponding Hamiltonian operators. First, consider the case of a magnetic field whose z-component (H_0) is large and constant (or whose time dependence is negligible) and whose x- and y-components (H_x and H_y) are small, but can vary rapidly with time. The expression for the corresponding (Zeeman) interaction of a single spin (z-direction = quantization direction) is [1]

$$\mathcal{H}_Z = -\vec{\mu}\cdot\vec{H} = -\hbar\gamma_n\vec{I}\cdot\vec{H} = -\hbar\gamma_n[m_z H_0 + \tfrac{1}{2}(I_+H_- + I_-H_+)], \tag{1}$$

where $\mathcal{H}_Z$ is the Hamiltonian, $\vec{\mu} = \hbar\gamma_n\vec{I}$ is the nuclear magnetic moment, γ_n is the nuclear gyromagnetic ratio, $\vec{I}$ is the nuclear spin, and m_z is its projection on the z-axis (magnetic quantum number). The raising and lowering forms have been used for both the operator $\vec{I}$ ($I_\pm = I_x \pm iI_y$) and the transverse part of $\vec{H}$ ($H_\pm = H_x \pm iH_y$).

The term with m_z in (1) defines a set of 2I+1 equally spaced energy levels with an energy difference $\hbar\omega_0 = \hbar\gamma_n H_0$. It is normally the most important term in determining the energy level structure. The terms with $I_\pm$ correspond to transitions between the energy levels induced by an externally applied rf field or by fluctuations in the magnetic environment associated with the sample. All of the terms in (1) can be used to study the properties of condensed matter.

The basic magnetic resonance phenomenon is absorption of energy by these states from an rf magnetic field. The corresponding shape of the absorption reflects the static and dynamic interactions that determine the structure of the levels.

The second handle on the nucleus that can be exploited is the interaction of its electric quadrupole moment with an electric field gradient generated by its surroundings. It corresponds physically to the orientational interaction of a non-spherical nuclear charge distribution with the second derivative of the electrosatic potential (V_{xz} = second derivative with respect to x and z, etc.) formed by the charges external to the nucleus. This interaction is somewhat more complicated than the magnetic dipolar interaction, and only a few essential facts regarding it will be mentioned here. The interaction is non-zero only for nuclei with a non-spherical charge distribution; i.e., for nuclei with $I \geq 1$. In addition, it is non-zero only for an electrostatic potential whose symmetry is lower than cubic. There are six independent parameters needed to completely specify this interaction: I, the three angles that specify the orientation of the principal axes of the electric field gradient in space, and the product of the nuclear quadrupole moment (Q) with the principal value of the electric field gradient (V_{zz}) and with the asymmetry parameter [$\eta = (V_{xx} - V_{yy})/V_{zz}$]. Here we restrict the discussion to the case of axial symmetry for the field gradient ($\eta = 0$) and use the raising and lowering operators for I. By using the transformations $V_0 = V_{zz}$, $V_{\pm1} = V_{zx} \pm iV_{zy}$, and $V_{\pm2} = \pm iV_{xy}$ for the field gradients, the quadrupolar Hamiltonian $\mathcal{H}_Q$ becomes [2]

$$\mathcal{H}_Q = \frac{eQ}{4I(2I-1)}\left[V_0(3I_z^2 - I^2) + V_{+1}(I_-I_z + I_zI_-) + V_{-1}(I_+I_z + I_zI_+) + V_{+2}(I_-)^2 + V_{-2}(I_+)^2\right] \tag{2}$$

The first term in $\mathcal{H}_Q$ is a term added to the Zeeman levels, and the other terms form matrix elements that connect these energy levels. They can further modify them or, if they are time dependent, they can induce transitions among them.

Now we describe in Fig. 1 the effect of both $\mathcal{H}_Z$ and $\mathcal{H}_Q$ on the NMR spectrum using I=3/2. This discussion will be limited to the cases in which H_0 is held constant, the spectrometer frequency ω is swept slowly, other time variations are small, and $\mathcal{H}_Q$ is considered as a small, first order perturbation to $\mathcal{H}_Z$. The effect of $\mathcal{H}_Z$ is to give a single line (Fig. 1a) centered at ω_0 with a full width $2\delta\omega$. This width can be due to a spatially inhomogeneous H_0 or to lifetime broadening of the levels from various relaxation processes (described later). There are several effects of the first term in (2) that depend on the characteristics of the sample [3]. If the sample is a single crystal with a single equivalent site for all the resonant spins, V_0 is the same for all nuclei and each level is shifted by $V_0[3m_z^2 - I(I+1)]$. This means that the energy difference for the central energy levels is unchanged, but the other two transitions are shifted outwards to two satellites symmetrically placed by an amount $\pm\omega_Q$ about the central transition. It

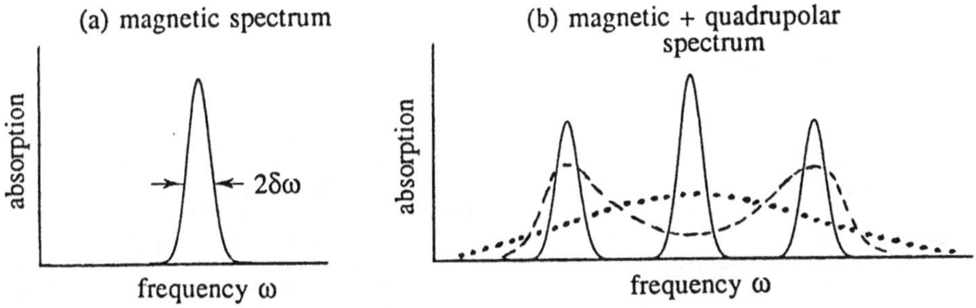

Fig. 1. Magnetic and quadrupolar absorption spectra (schematic)

is illustrated in Fig. 1b by the solid line. If the same sample were a powder with equal averaging over all orientations, an additional factor $(1/2)(3\cos^2\theta-1)$ multiplies the satellite splittings for each crystallite (θ is the angle between the principal axis of the field gradient and H_0). The result is a superposition of satellite lines into a "powder pattern" shown schematically by the dashed line on Fig. 1b. A further modification occurs if the magnitude of the field gradients is also distributed at random. In this case, the envelope of the satellite lines reflects the distribution, as shown schematically by the dotted line on Fig. 1b. Finally, if $\mathcal{H}_Q$ is large enough that one must go to second order in perturbation theory to adequately describe the absorption, there will be an additional shift (and perhaps broadening) to the central transition (illustrated in Sec. 4.1). Since these quadrupolar effects depend sensitively on the location of charges in the vicinity of the nuclei, they are useful for following any effect that produces ionic displacements, such as the static and dynamic behavior of structural phase transitions, critical phenomena, etc.

3. EXPERIMENTAL METHODS AND WHAT THEY REVEAL .

Before describing examples of the application of magnetic resonance to studies of nonlinear systems, it is useful to review briefly the two main methods of doing magnetic resonance experiments, and the kind of information that is obtained from them.

The simplest and most commonly used experimental setup is a steady-state absorption spectrometer [4]. It usually registers the magnetic resonance absorption signal at constant ω while H_0 is slowly swept through the resonance. In addition, low amplitude field modulation in the frequency range 10-10^5 Hz is often employed with lock-in detection to enhance the signal-to-noise ratio. This method normally generates the derivative of the absorption which, when integrated, gives results similar to those sketched in Fig. 1. In this case, the main information obtained from the experiment is the value of the applied field at the center of the line (H_r) and the linewidth. Since γ is usually known, one can often extract the local magnetic field produced by the material (H_{loc}) with the relationship

$$H_{loc} = H_r - \frac{\omega}{\gamma} . \tag{3}$$

This quantity is often of great interest, as it probes the magnetic field generated by the electronic structure of the material. It includes the order parameter in ordered magnetic phases, the Knight shift due to the hyperfine interaction with the conduction electrons in a metal, etc. In the absence of quadrupolar interactions, the linewidth $\delta H=\delta\omega/\gamma$ is used to study both the static and dynamic character of the local magnetic field.

The second major type of experimental setup is the pulsed magnetic resonance spectrometer [5,6]. In this case, a large rf magnetic field at or near the resonance frequency is used to quickly tip the spins through an angle ϕ from the direction of the externally applied field. After this has been done, the average component of the magnetization perpenducular to H_0, $M_\perp$, precesses freely at the Larmor frequency $\omega_L=\gamma H_0$. The amplitude of this precessing signal is detected in an rf coil (often the same one used to generate the large rf pulse). Because of the range of local fields that give the width to the absorption line, each spin packet precesses at a different frequency, with the result that even though they all start out in phase together, they gradually dephase so that the amplitude of $M_\perp$ decays in time. For this reason the signal is called a free induction decay (FID). It is the Fourier cosine transform of the absorption line shape [7] and, therefore, duplicates the lineshape information one would get from a steady-state absorption experiment. The characteristic time for the decay is called T_2^*, and is related to $\delta\omega$ by $T_2^*=1/\delta\omega$.

There are two advantages of the pulsed magnetic resonance apparatus over the steady state one that justify its greater complexity. The first is that it is usually much better adapted to measure another magnetic resonance quantity, the spin-lattice relaxation rate, $1/T_1$ [8]. This rate is the one at which the energy level populations of the spin system being studied return to their thermal equilibrium values when they have been disturbed from them. The most common way of preparing a non-equilibrium state is with a $\phi=\pi/2$ rf pulse. The free induction decay that follows this pulse has an amplitude that is proportional to the value of the z-component of the magnetization (M_Z) immediately before the pulse. Immediately after the pulse, $M_Z=0$. If the relaxation rate toward thermal equilibrium is proportional to the deviation from equilibrium, the solution of the corresponding rate equation is [8]

$$M_Z(t) = M_0\left(1 - \exp\left\{-\frac{t}{T_1}\right\}\right) \tag{4}$$

where M_0 is the equilibrium magnetization and t is the time after the $\pi/2$ pulse. Measurement of M_Z as a function of t then yields the value of T_1. The physical information obtained from T_1 is the power spectrum of local magnetic field and electric field gradient fluctuations at ω_L and $2\omega_L$. As we shall see, it is often very important for an understanding of the microscopic behavior of solids and liquids.

The second valuable feature of a pulsed spectrometer is its ability to measure the spin-phase memory time (T_2) with spin echoes. Roughly speaking, T_2 is the time a given spin "remembers" how much it has dephased relative to the average following a $\pi/2$ pulse. The simplest spin echo is generated by following the $\pi/2$ pulse after a time τ with a π pulse. At a time 2τ a signal appears (the spin echo) whose amplitude is proportional to the number of spins that have not undergone an irreversible variation of precession phase during the time 2τ. At 2τ, all of the packets that have not "forgotten" their phase arrive at the average phase to form the echo. This phenomenon itself is fundamentally nonlinear. If the response of the spins to the rf field were linear instead of a rotation through a large angle, no spin echo would be formed.

There are several ways in which the spin echo is useful for the study of condensed matter. The observed value of T_2 is determined mainly by dynamic fluctuations of the local field [9,10]. One contribution is from low frequencies on the order of $1/T_2$, and the rest comes from the same frequencies that determine T_1 (ω_L and $2\omega_L$). Since the detailed interpretation of T_2 measurements is often rather specialized, I will simply point out here some of the important ways it has been used in studies of condensed matter: (1) It is often important to know whether an observed magnetic resonance linewidth is due to inhomogeneous, static interactions ($T_2 \gg T_2^*$), or caused by dynamic fluctuations ($T_2^* \simeq T_2$) [11]. This question can be decided by comparing the two quantities. (2) The self-diffusion coefficient of ^{3}He and other materials is obtained from spin-echo decay measurements in an externally applied magnetic field gradient [12]. (3) Another example is that the introduction of an imhomogeneous field on a microscopic scale (associated with flux penetration in a superconductor or with a spin density wave transition) can be observed by its effect on spin echoes [13]. (4) The actual form of the spin echo envelope can be used to probe coherent spin-spin interactions that are not visible in the FID [14].

Finally, I would like to point out an important magnetic resonance effect that is used to study many different kinds of motion in condensed matter. It is "motional narrowing" of resonance lines [15], and it can be studied both by pulsed and steady-state methods. The main idea of motional narrowing can be made plausible by consideration of spatially fixed, non-interacting spins in a static, spatially inhomogeneous, magnetic field of width δH_i. Each spin resonates at the value of the local field, with the result that the measured linewidth is $\delta H_i = \delta\omega_i/\gamma$. Now let the spins move rapidly over the volume of the sample, so that each of them rapidly samples all of the field values. On the average, each spin will see the same field, so that a single, narrow line is observed. There must be some characteristic time for the spatial motion of the spins that corresponds to the crossover from a broad line with no motion and a narrow line with rapid motion. Theory (and common sense) tells us that this crossover occurs when the averaging time for the field (τ_c) is on the order of the inverse of the linewidth; $\tau_c=1/\delta\omega_i$. In the case of extreme motional narrowing (short τ_c limit), the narrowed linewidth $\delta\omega_n$ is given by [15]

$$\delta\omega_n \approx (\delta\omega_i)^2\tau_c. \tag{5}$$

The motion responsible for this narrowing can also affect T_1. That aspect will not be discussed here [16].

The applicability of the result in (5) is much wider than that suggested by the simple plausibility argument given above. In particular, the width δH_i need not be just an inhomogeneous magnetic field; it can be due to static spin-spin interactions or to quadrupolar interactions. As these interactions become sufficiently "dynamic," motional narrowing occurs. This averaging can be due to rapid motion of spatial coordinates, rapid motion of the purely spin coordiates, or both.

One common manifestation of motional narrowing is the narrow NMR lines seen in many liquids. In a magnetically dense solid, the dipolar interactions among the spins often lead to a broad resonance absorption (dipole-dipole broadened) line [17]. In ice, for example, the protons are at fixed positions and δH_i is on the order of a few Gauss. In pure water, however, the translational and rotational motion of the molecules is so fast that the dipolar interaction is averaged to $\delta H_n < 10^{-4}$ Gauss. This narrowing of the dipolar interaction in a liquid is what makes possible much of the extremely high-resolution NMR analysis that is done by chemists.

A different example of narrowing by spatial motion of sliding charge density waves is discussed in Sec. 4.1. Another commonly encountered situation is "exchange narrowing" [18]. In this case, isotropic exchange causes rapid enough orientational motion of the spins that there can be narrowing of other spin-spin interactions. The narrowing then becomes a way of investigating the exchange interactions responsible for it.

4. EXAMPLES OF THE APPLICATION OF MAGNETIC RESONANCE TO NONLINEAR SYSTEMS

In this section I will briefly review two recent examples of the application of magnetic resonance to nonlinear phenomena in condensed matter systems. The first is a beautiful experiment that detects directly the motion of charge density waves in a solid. It is clearly in the mainstream of nonlinear phenomena. The other is an investigation of quasiparticle excitations in a heavy fermion superconductor. Its selection is justified on the basis that all phase transitions represent a nonlinear response of matter. In both cases, the complete description of the work is too detailed to present here. Instead, I will attempt to present the qualitative flavor of the work to show how magnetic resonance methods play an important role in these physical problems. Those who hunger for a real explanation should consult the original articles.

4.1 Sliding charge density waves

One of the more interesting examples of nonlinear phenomena to appear recently in condensed matter systems is that of electrical transport by sliding charge density waves (CDW) [19]. A fundamental, and controversial question has been whether the charge density waves actually move in the bulk of the solid. Here, I will briefly describe a magnetic resonance experiment by SEGRANSAN et al. [20] which is a direct proof that the electron condensate moves as a whole.

The material they studied was a single crystal of $Rb_{0.30}MoO_3$ ("blue bronze") at 77 K, which has the unusual characteristics of electrical transport by sliding charge density waves. The phenomonon they exploited was the effect of CDW motion on the NMR spectrum of the ^{87}Rb nuclei. In their non-cubic environment, these nuclei experience a very large quadrupolar interaction. The satellite lines are so widely split that their NMR absorption is not seen. The central transition, which is observed, has a substantial frequency shift due to the second order quadrupolar interaction. In the absence of a charge density wave, all of the quadrupolar shifts are the same, and a single, narrow line would be seen. The effect of a static, incommensurate CDW is to introduce a lattice distortion that induces a distribution of electric field gradients at the ^{87}Rb nuclei. This, in turn, creates a superposition of NMR lines whose shape can be calculated in terms of the CDW parameters [20]. The result is shown on the top curve of Fig. 2. The solid line is the calculated shape for the chosen parameters, and the wiggly line is the measured absorption when no electric field is applied to move the CDW. A very good fit is obtained for the static case.

The lower two curves show the lineshape for progressively larger currents above the threshold for CDW conduction. Two major features are evident in the spectrum: the outer peaks are reduced in amplitude, and a third peak emerges in the center of the spectrum. The latter is due to motional

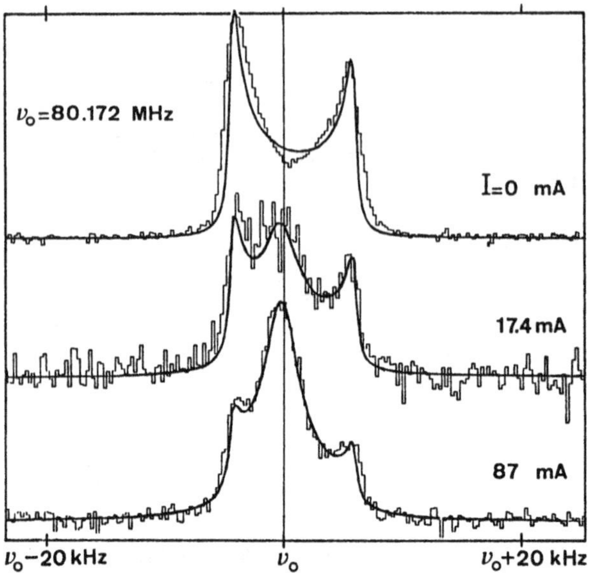

Fig. 2. ^{87}Rb NMR spectrum for the central transition in blue bronze at 77 K

narrowing of the two outer peaks. Motional narrowing occurs because as the CDW moves, it varies the electric field gradient so that a given nuclear spin will spend part of its time in the outer peak on the right and part of its time in the one on the left. When the CDW motion is rapid enough to meet the motional narrowing criterion described in Sec. 3, the structure collapses to the central peak.

The significance of the remaining outer peaks is that not all of the nuclei experience CDW motion. This fact is interpreted as evidence that the CDW does not move uniformly throughout the sample, but remains pinned in a substantial fraction of its volume. By following the intensity of these outer wings as a function of the current, SEGRANSAN et al [20]. are able to tell how much of the sample has sliding CDW's at each value of the current. They also point out that if the CDW were to slide in a perfectly rigid manner, the central line would be much narrower. The large width actually observed is taken as evidence of a finite correlation time of 200 μs corresponding to fluctuations of the CDW velocity in individual domains.

The assertion that these results are proof of CDW motion in the sample rests on the fact that the nuclear spins are a truly microscopic probe of the electric field gradients responsible for the spectrum. Motional narrowing is a demonstration that the motion occurs on a microscopic distance scale.

4.2 Quasiparticle excitations in a heavy fermion superconductor

One of the more actively investigated problems recently is the superconducting state of systems with heavy fermions. A substantial part of the motivation for this activity is that the unusual properties observed suggest the possibility of a new mechanism for superconductivity as well as the speculation that there is pairing that is different from the conventional BCS singlet type. In this section I will briefly describe an experiment of MACLAUGHLIN et al. [21] that uses NMR to probe the spectrum of quasiparticle excitations in the heavy fermion superconductor UBe_{13}. The ^{9}Be NMR signal is observed.

The quantity measured in this work is the spin-lattice relaxation rate $1/T_1$ in the superconducting state. The data are shown in Fig. 3.

The first point to be noticed in this work is that in the normal state, there is a huge enhancement of $1/T_1$. In comparison to the isostructural material $ThBe_{13}$, which does not display heavy fermion effects,

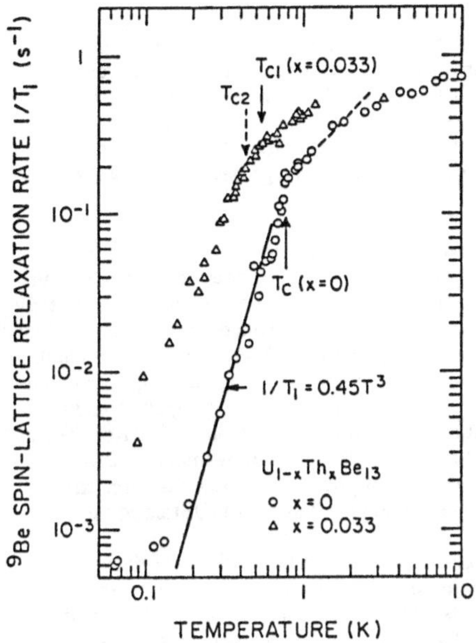

Fig. 3. Spin-latice relaxation rate of ^{9}Be in the heavy fermion superconductor UBe_{13}.

the enhancement is a factor of about 1000 [21]. This fact shows that one is sensing the heavy fermions in such measurements. The next thing to notice is that somewhat below the transition temperature T_C, the relaxation rate is proportional to T^3. For this situation, $1/T_1$ can be thought of as proportional to the number electron spins that can relax the nuclei. In the superconducting state, however, this means it is proportional to the number of thermally excited quasiparticles, as the paired electrons in the superconducting state can not relax them. At any temperature, the number of excited quasiparticles is determined by the dependence of their density of states on energy. By using the model for relaxation that has been successful before with superconductors, one finds that the observed $1/T_1 \propto T^3$ implies a linear dependence of the quasiparticle density of states on energy [21]. Another way of characterizing the result is that it implies a highly anisotropic gap with a line of nodes on the fermi surface.

It has been tempting to compare these results with apparently similar ones for the superfluid phases of liquid ^{3}He, where there is clearcut evidence of an exotic order parameter and of triplet pairing. This is, however, dangerous because of the major role expected of the spin orbit interaction in UBe_{13} and because the low temperature heat capacity indicates an anisotropic gap with point nodes on the fermi surface [22]. More progress on the theoretical side is needed before a confident interpretation of the situation in UBe_{13} can be given. It is clear, however, that magnetic resonance has a major role to play in the understanding of heavy fermion superconductivity.

5. PROSPECTS FOR THE FUTURE

The prospects for the future of magnetic resonance applications to problems in condensed matter science fall into three rough categories: new techniques, new instrumentation, and new physical problems. Judging from recent history, it looks like new techniques will advance at a rate of about one new acronym per year, but it is difficult to predict what they will be.

It is easier to see two important lines of instrumentation development. One of them is the thrust to higher magnetic fields. The availability of higher fields is needed for two reasons. In most cases, a higher field provides greater sensitivity, so that one can use smaller samples, investigate nuclei with smaller moments, make measurements on materials whose spin concentration is dilute, or obtain improved measurement accuracy. The second reason is to investigate field-dependent phenomena over a wider range of field.

Following current technology, laboratory electromagnets going up to 1-2 Tesla are common. Current laboratory superconducting magnets range up to 7-14 T, and hybrid superconducting-Bitter magnets can provide fields up to about 30 T. This latter figure appears difficult to exceed by any substantial margin for magnets that operate on the steady state basis that is needed for most magnetic resonance measurements. At present, pulsed, ultrahigh field magnets spend such a short time at high field that it is unlikely that they will come into use for magnetic resonance except for some of the most specialized cases. In the foreseeable future 15-30 T should be the practical upper limit. To go beyond that will require a new technological breakthrough.

It appears that there is room for a substantial improvement in sensitivity through the use of low temperature, low noise electronic circuits. Low-temperature operation is needed because in current, well designed NMR instrumentation operating near room temperature,. the noise of the system is already rather close to the minimum dictated by thermodynamic fluctuations of the resonant circuit that receives the magnetic resonance signal. The situation for microwave ESR is not far behind. For experiments that are carried out at low temperatures, however, the thermal noise in the first receiving elements can be one or more orders of magnitude smaller than that obtained when they are at room temperature. In this situation, the amplifying electronics used in most apparatus becomes the major source of noise. For this reason, low temperature, low noise amplifiers are needed to obtain a major improvement in sensitivity. Some progress along this line has been made already, and it is reasonable to expect that over the next decade a substantial improvement in sensitivity will become widely available through this route.

Finally it can be asked what major new physics insights will be gained over the next decade using magnetic resonance? I am sure that they will occur, but regret to say that I can't predict what they will be. The main reason is that magnetic resonance experiments are usually a small scale effort and can therefore be done quickly. Investigators can respond rapidly to new opportunities, to new ideas, to new developments, and to new materials. Many of those presently doing the most significant work using magnetic resonance would not have been able to predict what it would be five to ten years ago. We will simply have to hope that the best researchers are supported and wait to see what they accomplish.

I thank the organizers of this conference for the opportunity to participate and express my gratitude to the NSF for its support in preparing the manuscript with the help of Grant DMR-8409390.

1. A. Abragam: *The Pinciples of Nuclear Magnetism* (Oxford University Press, London, 1962) p. 22
2. C.P. Slichter: *Pinciples of Magnetic Resonance* (Springer, Berlin, Heidelberg, 1978) p. 288
3. M.H. Cohen and F. Reif: In *Solid State Physics, Advances in Research Applications*, ed. by F. Seitz and D. Turnbull, Vol. 5 (Academic, New York, 1957) pp. 336-339
4. Ref. 1, p. 40-44
5. Ref. 2, p. 23
6. W.G. Clark, Rev. Sci. Instr. 35, 316 (1964)
7. Ref. 2, pp. 43-55
8. Ref. 1, pp. 264-304
9. Ref. 2, pp. 38-43
10. J.R. Klauder and P.W. Anderson, Phys. Rev. 125, 912 (1962)
11. Ref. 1, p. 58
12. Ref. 2, pp. 377-82
13. B.G. Silbernagel, M. Weger, W.G. Clark, and J.H. Wernick, Phys. Rev. 153, 525 (1967). A. Andrieux, D. Jerome, and K. Bechgaard, J. Phys. (Paris) Lett. 42, 187 (1981)
14. Ref. 1, pp. 497-501
15. Ref. 1, pp. 180-2
16. N. Bloembergen, E.M. Purcell, and R.V. Pound, Phys. Rev. 73, 679 (1946)
17. Ref. 2, pp. 56-76
18. Ref. 1, pp. 435-40
19. G. Gruner, Physica 269D (1983)
20. P. Segransan, A. Janossy, C. Berthier, J. Marcus, and P. Butaud, Phys. Rev. Lett. 56, 1854 (1986)
21. D.E. MacLaughlin, Cheng Tien, W.G. Clark, M.D. Lan, Z. Fisk, J.L. Smith, and H.R. Ott, Phys. Rev. Lett. 53, 1833 (1984)
22. H.R. Ott, H. Rudigier, T.M. Rice, K. Ueda, Z. Fisk, and J.L. Smith, Phys. Rev. Lett. 52, 1915 (1984)

High-Resolution He Beam Scattering as a Tool for the Investigation of the Structural and Dynamical Properties of Surface Soliton Dislocations

M. El-Batanouny and K.M. Martini

Department of Physics, Boston University, Boston MA 02215, USA

Abstract

We discuss the applicability of high-resolution-*He*-beam/surface scattering to the investigation of the structural and dynamic properties of soliton-like surface misfit dislocations and associated phase transitions. We present evidence, based on recent *He* diffraction measurements, for the existence of double-sine-Gordon soliton-like dislocations on the reconstructed Au(111) surface.

Investigations of the structure of clean and adsorbate-covered single crystal surfaces of metals and semiconductors have revealed that under certain conditions the topmost layer incorporates some degree of mismatch with respect to the underlying bulk structure.[1,2] In some systems the mismatch is accommodated into commensurate structures, in others through the creation of misfit dislocations, yet, in others it is manifest in some degree of disorder.[1,2] In the case of clean metal surfaces such reconstructions have been observed on the (100) surfaces of Ir, Pt, Au, Cr, Mo, and W, the (110) surfaces of Ir, Pt, and Au, and the (111) surface of Au.[2] A large number of investigations of the structure of adlayers, both metallic and nonmetallic, have reported similar findings.[1,2] These results have led to an extensive revival of the Frank and van der Merwe(FvdM) theory of soliton-like misfit dislocations in surface physics.[3] The FvdM work was based on the Frenkel-Kontorova (FK) chain model.[2] A large part of the work, both theoretical and experimental, has dealt mainly with the nature of the critical behavior of incipient phase transitions, especially the commensurate-incommensurate (CI) and the melting transitions.[1,2] Studies of the dynamical aspects have been relatively few and mostly theoretical. The experimental investigations of surfaces were conducted, primarily, with LEED and X-ray diffraction; the latter was extensively used in studies involving noble gas overlayers on graphite.[1] Because of the nature and characteristics of these probes the dynamical properties of these transitions, as well as the details of the misfit dislocations have hardly been explored. However, microscopic studies of the dynamics of soliton excitations in quasi-one-dimensional magnetic materials have been studied very recently using neutron scattering techniques, and the results are reported as part of this proceedings.[4] These investigations are confined to the case of very low soliton concentrations dictated by the very nature of the magnetic solitons being excitations above the system's ground state where solitons and antisolitons are degenerate and usually are generated in pairs. The dynamic aspects reported concern the nature of the soliton mobility, diffusive or ballistic. Studies of surface soliton dislocations provide microscopic information concerning the high soliton concentration regime which is of a complementing nature. The reason for the high concentration is the presence of mismatch which removes the degeneracy of the soliton-antisoliton pair and favors the presence of one or the other as ground state configuration. The dynamical information obtainable in this case involves the small oscillations about the ground state soliton configuration[5] and, possibly, the nature of the soliton meandering motion during the CI transition.[2]

A relatively novel and promising experimental technique that has gone through considerable development over the past ten years , and is applicable to this problem, is that of neutral atom/surface scattering. Its competitive edge lies in its nondestructive nature ($10-100meV$ beam energy), extreme surface sensitivity (changes in surface corrugation of $\sim 0.01Å$ can be detected), and the achievement of high intensity ($10^{15}-10^{16}$ atoms/cm^{2}sec), as well as, angular and energy resolution ($\Delta k \sim 0.01Å^{-1}, \Delta E \sim 0.2meV$ the former corresponds to a coherence length$\sim 300-500Å$).[6-9,16] Furthermore, the scattering process is not surface atom specific since it depends solely on the local electronic surface charge density;[16] which also qualifies this technique for the investigation of surface charge density waves. Finally, the He atom momentum, ($k_i \sim 5.4-11Å^{-1}$ for nozzle stagnation temperatures in the range $77-300°K$), is adequate to probe the entire phonon Brillouin Zone (BZ). Consequently, this technique has been used in the past four years, with great success, to measure the surface-phonon dispersion of several transition and noble metal surfaces, in addition to providing detailed structural information about clean surfaces and overlayers.[16]

The current momentum resolution attainable with He beams is adequate for investigating the structural properties of soliton-like surface misfit dislocations. The coherence length cited above matches the expected surface-island sizes. However, in the dynamical studies of the small oscillations about the surface solitons the instrument momentum resolution imposes an upper limit on the soliton-lattice periodicity that can be investigated. This limit is based on the availability of a satisfactory number of sampling points in the corresponding BZ. Such investigations are, therefore, feasible for soliton length l_0 smaller than $10-15$ atoms in order that some $10-20$ points can be sampled in the superlattice BZ. Furthermore, the energy resolution available $\sim 0.2meV$ puts a more stringent limit on the feasibility. Measuring the soliton (or phason)-branch requires that ω_m is at least one order of magnitude higher than the instrument energy resolution, i.e. $\geq 2meV$. Consequently, such measurements will bias sample selection to materials with high surface Debye temperature and a superlattice periodicity d_0, which is determined by the natural mismatch, $(\pi d_0/2l_0) \leq 2$. This criterion is based on the dependence of ω_m on the inter-soliton distance, i.e. concentration, and on l_0, and is given by[5]

$$\omega_m \simeq \frac{\pi\sqrt{1-\kappa^2}}{\kappa l_0}\omega_0$$

where ω_0 is the maximum phonon frequency , $d_0 = 2\pi\kappa l_0 K(\kappa)$ where $K(\kappa)$ is the complete elliptic integral of the first kind.

Reports of He beam measurements concerning either soliton dislocations or the CI transition have been published recently.[6,7,9] Among these is the work of Harten *et al*[6] who reported detailed diffraction measurements on the reconstructed Au(111) surface using a high-resolution He beam facility. Their work shall serve as an example of the potential of this probe for such investigations. Harten *et al* interpreted the diffraction pattern as a manifestation of quasi-one-dimensional sine-Gordon (SG) soliton-like misfit structures. However, instead of fitting a soliton-like atomic arrangement model to the diffraction data, they accounted for their presence by a model containing equivalent Gaussian corrugations with full-width-half-maxima equal to the soliton width. The adoption of such a procedure was dictated by the inadequacy of the simple SG soliton to simultaneously account for the extra packing of half an atom per soliton dislocation and justify the unequal areas of the 'A' and 'C' regions. We propose an interpretation of their data using a different soliton model which accounts *a priori* for all observed features. The model is based on Double-Sine-Gordon (DSG)-like[10,11] soliton dislocations. It incorporates the new concept of competing sites in addition to that of competing periodicities.

It is well known[11] that surfaces with hexagonal structure have two equivalent adatom sites which are conventionally labeled A and C; A is associated with the hcp stacking and C with fcc stacking. The two sites are not energetically degenerate when one type of stacking is favored. In the case of the Au(111) surface layer the C-site should be favored if a continuation of the bulk structure occurs. However, it has been proposed[12] that a redistribution of *sp* electrons in the surface layer may favor a reduction of the interatomic separation from the bulk value. The competition between occupying the C-sites and reducing the mean interatomic separation in the surface layer can be resolved by accommodating both A- and C-sites, thus reducing the average interatomic spacing through the concept of defectability, i.e. the creation of local defects or, in this case, solitons. The topology of the underlying substrate potential is such that the minimum energy path for the transition from one site type to the other is effected across the shared bridge site. Potential energy minima are located at the A- and C-sites, while a saddle point is found at the bridge site. The resulting zigzag shape of the underlying potential can be unfolded into a one-dimensional potential. The nondegeneracy of the A- and C-site energy minima is suggestive of a Double-Sine-Gordon-like substrate potential that can be written

$$V_{DSG}(x) = \frac{W}{2} V_0(x),$$

where W is the peak-to-peak amplitude, and

$$V_0(x) = \frac{16}{(R+4)^2}[R(1-cos(\frac{2\pi x}{a})) + 1 - cos(\frac{4\pi x}{a})],$$

and $0.0 \leq R \leq 4.0$. The parameter R determines the A-site well depth relative to that of the C-site. It varies from W at $R = 0$, to zero at $R = 4.0$. The former limit corresponds to an incommensurability $p = 2$ sine-Gordon system,[3] while the latter corresponds to $p = 1$. The Hamiltonian of the corresponding finite Frenkel-Kontorova-like chain is

$$H = \frac{\mu}{2}\sum_{i=0}^{N}(x_{i+1} - x_i - (b-a))^2 + \frac{W}{2}\frac{16}{(4+R)^2}\sum_{i=0}^{N}[R\{1-cos(\frac{2\pi x_i}{a})\} + 1 - cos(\frac{4\pi x_i}{a})],$$

where x_i is the displacement of the ith particle from the bottom of the ith C-well, a is the C-well periodicity, b and μ are the equilibrium interparticle spacing and the elastic constant of the surface layer, respectively, and N is the number of particles comprising the chain. This model is represented schematically in figure 1a. By substituting $u_i = \frac{x_i}{a}$, H becomes

$$H = \frac{\mu a^2}{2}\sum_{i=0}^{N}(u_{i+1} - u_i - \frac{1}{p_0})^2 + \frac{W}{2}\frac{16}{(4+R)^2}\sum_{i=0}^{N}[R\{1-cos(2\pi u_i)\} + 1 - cos(4\pi u_i)],$$

where $p_0 = a/(b-a)$. The corresponding equations of motion are

$$u_{i+1} - 2u_i + u_{i-1} = (\frac{\pi}{l_0})^2[R\, sin(2\pi u_i) + 2\, sin(4\pi u_i)],$$

where $l_0^2 = \mu a^2/2W$. The continuum limit of this equation has stable solutions of the form

$$tan(\pi u) = \sqrt{1+\frac{4}{R}}cosech(\frac{4\pi n}{\sqrt{R+4}l_0}).$$

This solution can be viewed as a configuration of two bound π-solitons, each with an effective length of $l_0/2$, and center-to-center separation $L(l_0, R)$. For $R \ll 1.0$, it leads to $L \simeq (l_0/4\pi)\, ln(1/R)$.

To obtain the diffraction intensities from the DSG model we fixed the mismatch to the value reported by Harten *et al*, namely, $b/a = 22/23$. The choice of the other parameters of the DSG model was based on the information given in ref. 9, namely, $l_0 \simeq 8a$ and $R = 0.01$ were found to agree with a DSG subsoliton size of about $4a \simeq 12\mathring{A}$ and a ratio of A to C regions of 0.70.

We used molecular dynamics procedures outlined in earlier publications[13-15] to determine the equilibrium ground-state and metastable configurations of a harmonic chain with DSG solitons. A typical equilibrium configuration, u_i, containing DSG solitons is shown in fig.1.b, where the dots indicate the positions of the atomic centers. Subsequently, we used the configuration of the central soliton in the relaxed chain con-

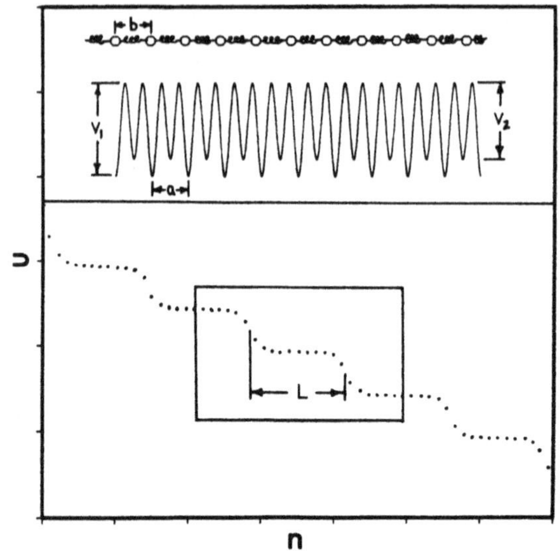

figure 1. (a.) A schematic diagram of a DSG-like Frenkel Kontorova model where $V_1 = W$ and $V_2 = (1 - R)W$. (b.) Profile of a DSG soliton chain showing the displacement u_n of the nth particle from the nth C-type well. The inset isolates an individual DSg soliton.

figuration to construct the $(23\text{x}\sqrt{3})$ unit cell of the reconstructed Au(111) surface. The equilibrium positions of the particles constituting the DSG soliton were used as the x-coordinates of the atomic centers along the < 110 >-direction, the y- and z-coordinates, along the < 112 > and surface normal respectively, were then determined using a hard sphere stacking model, where $y_i = x_i/\sqrt{3}$,

$$z_i = \left[\sqrt{\frac{3d^2}{4} - \frac{\left(\frac{d}{2} - 2x_i\right)^2}{3}} - d\sqrt{\frac{2}{3}}\right],$$

and d is the interatomic separation in the bulk. The y-coordinate was reduced by a factor of 0.98 in order to account for the shift of the center spot of the < 112 > diffraction group to higher k values. A typical planar view of the atomic arrangement is shown in fig.2.a. The soliton stacking pattern in the < 112 > direction is dictated by the doubling of periodicity in that direction, which corresponds to having two inequivalent sites in the unit cell. This arrangement results in a zigzag-like soliton arrangement along the < 112 > direction, which is also indicated in fig.2.a. The last step in simulating the reconstructed surface was to superimpose Gaussian charge distributions characterized by a width σ and an amplitude A on the atomic center positions in order to reproduce the associated surface electron corrugation manifested in the He scattering. Since the purpose of the present paper is to introduce and motivate the DSG model as possible candidate for surface misfit dislocations, we opted not to carry out a systematic and thorough search over the parameter space (l_0, R, σ, A) for a best fit to the diffraction data. Instead we simply chose $\sigma = 0.4a$ and $A = 1.5$ in order to reproduce the peak-to-peak atomic corrugation of $0.07\AA$ and a height of the envelope corrugation of $0.15\AA$ cited in ref. 9. The resulting simulation of the unit cell corrugation is shown in fig.2.b. We computed the atomic beam diffraction intensities associated with the predicted corrugation using the hard wall and the eikonal approximations.[16]

The diffraction intensities computed from the DSG model are given in table-I where they are compared with the experimental values and with the fitted intensities reported for the model of Harten *et al*. Furthermore, in table I we have identified the observed

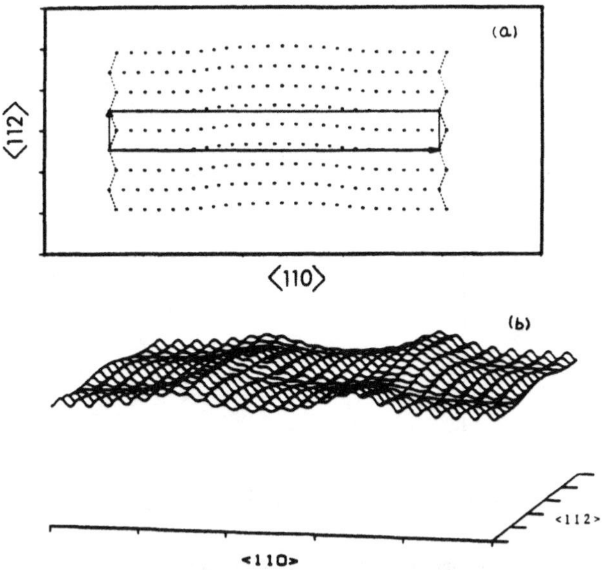

figure 2. (a.) Planar view of the atomic arrangement involving DSG solitons. Notice the atomic displacements in the < 110 > and the zigzag stacking of solitons along the < 112 > directions. (b.) unit cell corrugation.

Table I
Diffraction Peaks of
Reconstructed Au(111) Surface

Peak	Intensity		
	Experiment	DSG	Harten *et al*
1(1, 0)	4.0	7.5	7.4
2(2, 0)	26.0	28.0	24.7
3(3, 0)	5.5	5.0	5.5
4(4, 0)	—	0.2	0.5
5(5, 0)	0.5	1.0	0.7
A(24,-1)	1.3	0.8	1.0
B(-24,-1)	1.0	0.8	1.0
C(± 1,∓2)	0.5	0.7	0.5
D(0,±2)	0.7	1.7	1.2
E(± 1,±2)	0.5	0.5	0.5
F(±22,±1)	0.2	0.1	0.2
G(±22,∓1)	0.2	0.05	0.1
H(±22,∓1)	0.1	0.05	0.1
I(±22,±1)	0.1	0.05	0.1
J(± 1,±2)	—	—	0.2
K(0,±2)	2.0	2.4	2.0
L(± 1,∓2)	0.05	—	0.2
M(25, 1)	0.9	0.1	0.8
N(-25, 1)	0.6	0.08	0.8

diffraction peaks in terms of the reciprocal lattice vectors of the new rectangular unit cell.

The apparent six-fold symmetry manifest in the diffraction spots is generated by the presence of three domains with 120° relative rotations. To further facilitate the

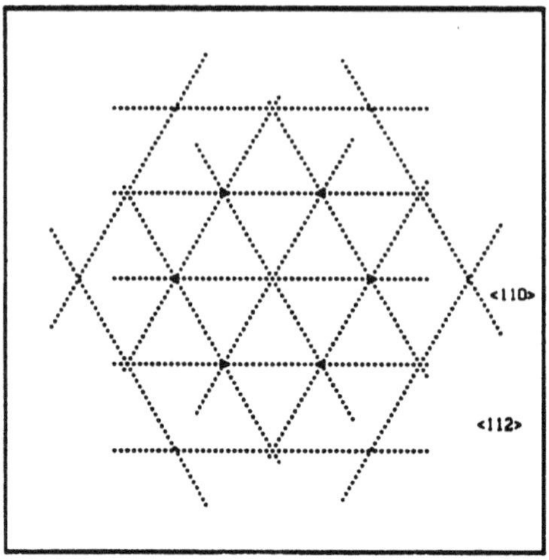

figure 3. Reciprocal lattice view of the superposition of the three equivalent domains.

identification of the diffraction pattern we provide in fig.3 a schematic diagram of reciprocal space where we show the superimposed contributions of the three equivalent reconstructed domains. A comparison of this diagram with figure 2 of ref.9 will prove helpful in the identification. Furthermore, it will underline the uniqueness of the proposed soliton zigzag stacking in the $< 112 >$-direction, since the orientation of the $< 112 >$ hexagonal group of diffraction spots can only be reproduced by the proposed stacking.

Inelastic *He* scattering measurements of the phonon dispersion associated with the reconstructed Au(111) have been also reported recently by Toennies and his collaborators.[9] Although the manifestation of the unit cell doubling can be noticed along the $< 112 >$-direction in a change of the slope of the Rayleigh mode branch at the mid k-point, the phason branch is absent along the $< 110 >$-direction. This is to be expected since a crude estimate of ω_m, based on the above arguments and an $\omega_0 \simeq 10meV$ for the Au(111) surface, yields a value of$\omega_m \leq 0.05\omega_0 \simeq .5meV$; too small to provide unambiguous and meaningful results. However, systems incorporating soliton misfit dislocations which satisfy the constraints outlined above can easily be found among the multitude of mismatched overlayers known to date.

The success of Harten *et al* in recognizing the presence of soliton structures on the Au(111) surface, and our effort to isolate and interpret the salient features of these structures, demonstrates the potential of *He* scattering as a tool for the investigation of soliton surface dislocations. It even provides information detailed enough to allow the introduction and testing of a more complex DSG soliton model which is also of interest because of its internal dynamical structure.[17,18] Furthermore, we have discussed the current limitations that will be encountered in extending these studies to the related dynamics. However, in doing so we have established a prescription for proper choice of surface systems that make such studies feasible. The realization of such experiments provides an arena that allows existing concepts and theories to be put to a fine test, and may lead to uncovering new and unexpected concepts.

Acknowledgements

This work is supported by the U.S. Department of Energy under contract # DE-FG02-85ER45222.

References

1. *Ordering in Two Dimensions*, ed. by S.K. Sinha, (North-Holland, 1981).
2. Per Bak, Rep. Prog. Phys. **45**, 587 (1982), and references therein.
3. F. C. Frank and J. H. van der Merwe, Proc. R. Soc. London, Ser. **A198**, 205 (1949).
4. J.P. Boucher, this proceedings.
5. V.L. Pokrovsky and A.L. Talapov, Theory of Incommensurate Crystals, (Hardwood Academic publishers,1985).
6. U. Harten, A.M. Lahee, J.P. Toennies, and Ch. Wöll, Phys. Rev. Lett. **54**, 2619 (1985).
7. K. Kern, R. David, R.L. Palmer, and G. Comsa, Phys. Rev. Lett. **56**, 620 (1986).
8. K.M. Martini, W. Franzen, and M. El-Batanouny, to be published in Rev. Sci. Instrum.
9. U. Harten, J.P. Toennies, and Ch. Wöll, Faraday Discuss. Chem. Soc.**80**, 1 (1985).
9. P. Sodano, Proc. Conf.:*Quantum Field Theory: Past, Present and Future*, June 1985. (North-Holland, Amsterdam 1986).
10. D. Campbell, M. Peyrard, and P. Sodano, Physica **19D**, 165 (1986).
11. C. Kittel, *Introduction to Solid State Physics*, (John Wiley, 1976).
12. L. D. Marks, V. Heine, and David J. Smith, Phys. Rev. Lett. **52**, 656 (1984); and V. Heine and David J. Smith, to be published.
13. K. M. Martini, S. Burdick, M. El-Batanouny and G. Kirczenow, Phys. Rev. **B30**, 492 (1984).
14. K. M. Martini, S. Burdick, M. El-Batanouny and G. Kirczenow, *The Structure of Surfaces*, ed. by M.A. Van Hove and S.Y. Tong, in Springer Series in Surface Sciences, Vol. 2, 347 (Springer Verlag, New York 1985).
15. K. M. Martini, S. Burdick, M. El-Batanouny and G. Kirczenow, *Dynamical Phenomena at Surfaces, Interfaces and Superlattices*, ed. by F. Nizzoli, K.-H. Rieder and R. F. Willis, in Springer Series in Surface Sciences, Vol. 3, 159 (Springer Verlag, New York 1985).
16. K.H. Rieder and T. Engel, in *Structural Studies of Surfaces*, Springer Tracts in Modern Physics, Vol. 91, 55 (Springer 1982).
17. P. Sodano, M. El-Batanouny, and C. Willis, to appear in Phys. Rev. **A34**.
18. C. Willis, M. El-Batanouny, S. Burdick, R. Boesch, and P. Sodano, to be published in Phys. Rev. **B34**.

Part V

Theoretical Techniques

Quantum Mechanics at the Macroscopic Level

A.J. Leggett

Department of Physics, University of Illinois,
1110 West Green Street, Urbana, IL 61801, USA

Since most participants at this conference come from sub-areas of physics different from the one in which most of the technical work associated with the subject of this talk has been done (namely superconducting devices exploiting the Josephson effect), I thought that rather than talking about the technical details it might be more interesting to review the general motivation for the work and the philosophy associated with it. An additional reason for doing so is that I believe that the sorts of effects we are looking for in the superconducting-device area may also occur in other systems, and are most likely to occur in those which show the strongly nonlinear behavior which is the subject of this conference.

The fundamental motivation underlying attempts to test quantum mechanics at the macroscopic level - at least in my own case, though perhaps not for everyone working in this area - is a belief in the severity of the notorious quantum measurement paradox. As you know, the paradox typically arises in a situation when the behavior of a macroscopic system such as an electron or an atom is used to trigger a (macroscopic) change in a macroscopic system, such as a Geiger counter or, in the best-known formulation of the paradox, Schrödinger's famous cat. Suppose that a microstate ψ_1 of the atom (say) triggers a state 1 of the macrosystem and a microstate ψ_2 similarly triggers a state 2. Generally speaking, the "states" of this system must be characterized in thermodynamic terms - for example, in the case of a counter one might specify the voltage across it, which will be different according to whether it has or has not fired. But in principle it should be possible to characterize this state by a quantum-mechanical description - if not by a pure-state wave function, at least by a density matrix. Suppose now that a particular measurement apparatus is exposed to a single atom drawn from an ensemble in a pure state described by the linear-superposition wave function $\alpha\psi_1 + \beta\psi_2$, when $|\alpha|^2 + |\beta|^2 = 1$. According to the standard interpretation of the quantum formalism, it is incorrect to interpret this description as saying that the electron, before the measurement, was in one or the other of the states ψ_1, ψ_2; if one accepts the standard measurement axioms, this follows directly from the observation that such a hypothesis would give incorrect predictions for the ensemble expectation value of an operator which has nonzero matrix elements between ψ_1 and ψ_2, while if one objects (reasonably) that to rely on the measurement axioms begs the whole question, then Bell's theorem and the related experiments show that (in a situation where ψ_1, ψ_2 describe two-particle product states) the hypothesis can be maintained only at the cost of sacrificing one or more other assumptions such as local causality which most of us would dearly like to keep. Anyway, the question is: after a particular microsystem has interacted with the measuring apparatus, is the "universe" - i.e. the combined system formed by the microsystem plus apparatus -- is a definite state or not? In particular, is the apparatus in one and only one of the macroscopically different states corresponding, e.g., to different values of pointer position or voltage (with fluctuations around the mean value of at best a microscopic order of magnitude)? A straight forward extrapolation of the linear formalism of quantum mechanics, and its interpretation, suggests that it is not: the "common sense" we normally apply to the macroscopic world seems to require that it is.

One standard "resolution" of the dilemma is to say in effect: If we insist on the quantum-mechanically correct description, then the universe is indeed technically not describable as being in a single definite macroscopic state; however, the results of any possible experiment will turn out exactly as if it was, and hence it "really" is so. For simplicity of notation (only!) let us consider the physically somewhat unrealistic case in which not only is the original microsystem absorbed in the counter, but we know enough about the latter to ascribe to it a pure state initially and hence also a pure state for each possible definite macroscopic outcome of the measurement. In that case, an initial state ψ_1 of the microsystem will lead to a final pure state Ψ_1 of the counter, an initial state ψ_2 will lead to a final pure state Ψ_2 (macroscopically different from Ψ_1), and the linear conbination $\alpha\psi_1 + \beta\psi_2$ leads to the linear combination $\alpha\Psi_1 + \beta\Psi_2$ of macroscopically distinct states. The "resolution" just referred to then holds (a) that no measurement could reveal the interference terms, and (b) that since (as we all agree) inspection of the apparatus will reveal it to be in either state Ψ_1 or state Ψ_2, and since no interference between them is observable, it "really is" in one state or the other even before our inspection.

Now this is a rather peculiar argument. It says, in effect, that once we know for sure that observation of the interference terms is impossible, the state of the system changes from the linear combination $\alpha\Psi_1 + \beta\Psi_2$ to one of its branches, Ψ_1 or Ψ_2. But one is tempted to ask: How does the system "know" that further observation of the interference will be impossible? What features are relevant to this? Generally speaking, in the literature the relevant features are taken to be the <u>complexity</u> and <u>irreversibility</u> characteristic of a macroscopic system. Unfortunately, in a quantum-mechanical context irreversibility is a somewhat ambiguous concept. Suppose for example that a single atom emits a photon. We are tempted to think of this as an irreversible process. But if the system is contained in a suitably engineered cavity, it may not be: the amplitude for the photon to be reflected from the walls and reabsorbed by the emitting atom may not be zero. So for present purposes let's concentrate on the aspect of complexity. This is clearly a rather qualitative concept, but as a zeroth approximation it may be reasonable to try to define it in terms of the number of particles (as usually defined in a nonrelativistic context) which are behaving in different ways in the two branches of the superposition: when the states in question are macroscopically different, this is typically of the order of 10^{23}. (I have tried, elsewhere [1], to give a more quantitative definition of the property we are seeking in terms of a quantity christened the "disconnectivity" of the superposition). Given a suitable definition of the property which corresponds roughly to our intuitive version of "complexity", we can then ask: Is sufficient "complexity" in fact guaranteed to destroy the interference effects? Or could we, given sufficient ingenuity, build a state whose complexity is by our criterion very high, but where we can still see the interefence effects occurring? In other words, can we verify experimentallly that linear superpositions of complex macroscopic states do occur in nature and are not, always, equivalent even in their experimental properties to classical probabilistic mixtures of the same states, i.e. to the hypothesis that the superposition has been converted into one of its branches but we do not know which? If we can, then it seems that much less reasonable to assume that in the usual case, where the interference terms cannot be observed, the superpositions have somewhat been converted into one of its branches.

I will not discuss here the attempts to demonstrate in the context of superconducting devices that such superpositions do occur: see for example ref. [2]. What I would like to emphasise is that it is quite conceivable that we may be able to find evidence for them in other areas of physics, and probably the more nonlinear the better!

Is it in fact obvious that such superpositions do indeed occur (whether or not we can in practice observe the interference terms)? Within the framework of orthodox quantum mechanics the answer is clearly yes. But who says that orthodox quantum mechanics is the whole story? Is it not possible that "complexity" is itself a relevant variable, and that systems with a large value, let us say, of the "disconnectivity' defined in ref. [1] behave quite differently from those with a small value, such as electrons and neutrons, which to the best of our current knowledge are certainly quite adequately described by quantum mechanics? This, surely, is a matter for experiment rather than theoretical dogma, and this is why I would encourage people working in all areas of nonlinear physics to keep an eye open not only for positive evidence that quantum mechanics works under the conditions referred to above, but also for hints that it may not.

One obvious objection needs to be addressed. The notion of "disconnectivity" defined in ref. [1], and many possible variants of it, place heavy reliance on as it were counting the degrees of freedom of a system in terms of the familiar stable (or quasi-stable) "elementary" particles (electron, proton, neutron, etc.): crudely speaking, states have high or low disconnectivity according as a large number of elementary particles are behaving in a different way in the two branches of the superposition. But what is "fundamental" about this classification? Even if the electron is truly elementary, we now know that the nucleon is composed of quarks and moreover contains a vast tangle of virtual gluons, quark-antiquark pairs and other things. Thus, if we consider a neutron passing through a neutron inteferometer, and even if we neglect for the present argument the coupling of the spin to the radiation field, does its state not have just as much "complexity" as some of the superpositions of macroscopically distinct states which we ascribe to counters, etc? And if despite this quantum mechanics works for the neutron, why not for the counter?

The only satisfactory answer to this objection is: yes, <u>within</u> the framework of orthodox quantum mechanics of course the neutron is as "complex" as the counter, and there is no good reason to distinguish the familiar stable elementary particles as in any way special. (For that matter, as noted in ref. [1], there is no good reason to regard the coordinates of individual particles as any more fundamental than combinations of them such as center-of-mass and relative coordinates). But the whole suggestion I am making is that the quantum-mechanical "paradigm" may not be the whole truth, and it is typical of the breakdown of a paradigm that it cannot be anticipated by, as it were, inspecting the paradigm "from the inside" (cf. KUHN/3/). To take a possible historical parallel: there was simply no way, in 1895, of deducing from the then accepted framework of classical mechanics that it had to break down when the characteristic scale of the action involved became comparable to 10^{-34} joule secs; and indeed any such suggestion would have been obviously incompatible with the general framework of classical mechanics, which clearly cannot tolerate a characteristic scale of this kind. In the end, this apparently outlandish concept was forced on the physics community not by any internal inconsistency in classical mechanics itself, but by experiment, and at that point the significance of all the long-established concepts of classical mechanics had to be re-examined. I am suggesting that it is not inconceivable that experiment may one day force us to look again at all the things we have taken for granted in the quantum-mechanical description - including, in particular, its "complexity-invariance". At least, while we will all no doubt have our prejudices about the probability of such an outcome, it would be unwise to dismiss on a priori grounds any experiment which fails to give agreement with the predictions of quantum mechanics.

This work was supported through the MacArthur Chair established by the John D. and Catherine T. MacArthur Foundation at the University of Illinois. I am grateful to Gordon Baym for discussions which helped me to clarify my thoughts about the question discussed in the last two paragraphs.

References

1. A. J. Leggett, Prog. Theor. Phys. supp. 64, 80 (1980).
2. A. J. Leggett, in Directions in Condensed Matter Physics, ed. G. Grinstein and G. Mazenko, World Scientific, Singapore 1986.
3. T. S. Kuhn, The Structure of Scientific Revolutions, University of Chicago Press, Chicago 1962.

Soluble Models

M. Fowler

Department of Physics, University of Virginia,
Charlottesville, VA 22901, USA

To give an adequate survey of the field of soluble models in fifty minutes is manifestly impossible. What I shall do instead here is to trace a particular thread or set of threads through the sixty years or so of historical development of the subject, examining the evolution of a particular series of models. I believe this exercise offers some Lessons from the Past, thus at least partially fulfilling the guideline requirements for talks at this conference.

I take it that a "soluble model" in the present context means an exactly soluble model of a system with many degrees of freedom. Thus I exclude mean field-type models, for example, where the many degrees of freedom are reduced to one or a few. The first soluble model by the above definition was the Ising model of a ferromagnetic system, formulated by Lenz (Ising's thesis supervisor) in 1920.[1] This model has spins on the sites of a square lattice which can take values ±1, and a nearest neighbor only interaction JS_iS_j (as first formulated) and, possibly, an external magnetic field B adding energy BS_i to site i. In 1925, Ising solved the one-dimensional version of the model, an elementary exercise in the transfer matrix method. For the particular case of zero external field, if one changes variables from the S_i to a set $\sigma_i = S_iS_{i+1}$ the system is found to be equivalent to N independent spins (for periodic boundary conditions) so there is obviously no phase transition at nonzero temperature.

One might wonder why such a simple model as Ising's had not been discussed before - after all, the Weiss mean field analysis of ferromagnetism had been carried out twenty years earlier. But in Weiss' model, an atomic spin can orient in any direction - the Ising model is more tractable because the spin direction is restricted to $\pm\hat{z}$, and Lenz' rationale for this was quantum mechanical. At the time Ising was working on his thesis with Lenz, Stern and Gerlach were working in the same institute in Hamburg on their classic experiment on direction quantization. Thus in this case, at least, the advent of quantum mechanics led to a new model which was simpler as well as being more realistic (at least for some materials).

Unfortunately, after he found that his one-dimensional model exhibited no phase transition, Ising went on to argue that this must also be the case for similar systems in two or three dimensions. This erroneous conclusion delayed investigation of his model for many years.

Meanwhile, Heisenberg was developing a theory of ferromagnetism based on new insights into the nature of the many-electron wavefunction, and in particular the rôle of exchange in spin alignment. This led him to a nearest neighbor interaction Hamiltonian $J\vec{S}_i\cdot\vec{S}_j$, the $\vec{S}_i$ now being quantum mechanical spin operators. In the late twenties and early thirties, attention focussed on this Hamiltonian. Bloch presented a simple spin wave picture for a one-dimensional Heisenberg chain in 1930, and in 1931 Bethe gave the exact many-spin wave eigenstates - the Bethe Ansatz. This Ansatz, still practically the only exact solution available for a true many-body system, has made possible precise calculations of thermodynamic properties for Kondo systems, spin chains, etc., as well as providing much insight into their excitation spectra. However, Bethe's Ansatz is notorious for its mathematical complexity. In 1931 Bloch grumbled "it appears to me that Bethe's tedious algebraic manipulation is somewhat academic in character...". This was possibly partly in response to Bethe's pointing out that Bloch's simpler spin wave theory failed to give the correct number of eigenstates for a given total spin.

As Bethe's Ansatz plays a central rôle in the models to be discussed here, it is important to understand the form of the wavefunction Bethe found. We examine first the simplest interacting system soluble by Bethe's technique, the "quantum billiard ball" problem.[2] The trivial classical version has N equal mass perfectly elastic billiard balls on a line, so that the initial set of momenta $k_1 \ldots k_N$ is conserved, but the momenta keep getting transferred to different balls. We consider the quantum problem with a finite strength repulsive delta function potential (so tunnelling can also occur)

$$\left(-\sum_{i=1}^{N} \frac{\partial^2}{\partial x_i^2} + 2c\sum \delta(x_i - x_j)\right)\psi = E\psi \tag{1}$$

where $\hbar = 2m = 1$.
The $N = 2$ particle system is equivalent to a single particle interacting with a static delta function potential, and the solution (apart from a normalization factor) is

$$\psi(x_1,x_2) = e^{i(k_1x_1 + k_2x_2)} + e^{i\theta_{12}}\, e^{i(k_2x_1 + k_1x_2)} \tag{2}$$

where the phase shift

$$e^{i\theta_{12}} = -\frac{c - i(k_1 - k_2)}{c + i(k_1 - k_2)} \quad . \tag{3}$$

This is a Bethe Ansatz wavefunction.
For N bosons, it is straightforward to verify that the solution to Schrodinger's equation is a generalization of (2),

$$\psi(x_1 \ldots x_N) = \sum_P a(P) e^{i\sum_{j=1}^{N} k_{Pj}x_j} \quad \text{for} \quad x_1 < x_2 < \ldots < x_N \tag{4}$$

with other x_i orderings given by the boson symmetry. Here a(P) is a product of factors $e^{i\theta_{ij}}$ for each transposition of k_i's needed to get the ordering P from 1 ... N. The energy and momentum are given by

$$E = \Sigma k_i^2 \quad , \quad K = \Sigma k_i \tag{5}$$

but the appearance of a free particle system is deceptive. Although the particles propagate as plane waves between collisions, the phase shifting at each collision implies that the individual k_i's may not be noticeable in the Fourier transform, that is, in scattering of external waves by the system. Another important point about the above wavefunction is that it would not be correct for three equal mass but distinguishable particles interacting with repulsive delta function potentials of different strengths. A nice discussion of this point has been given by McGUIRE.[3]

The boson wavefunction above has a fermionlike property - the k_i's must all be different. This is most easily seen for the two-particle case by noticing that $e^{i\theta_{\text{"}}} = -1$, so ψ vanishes identically if $k_1 = k_2$. This is true in general. The actual allowed k_i values for a system of length L with periodic boundary conditions are given by a simple extension of the free particle rule,

$$e^{ik_jL} = 1, \quad k_jL = 0, \pm 2\pi, \ldots \text{ (free particles)} \quad . \tag{6}$$

For the Bethe Ansatz wavefunction, one must add to the kinetic phase changing term e^{ik_jx} the phase shifts occurring at collisions, to give

$$k_jL + \sum_{\ell \neq j} \theta_{j\ell} = 2\pi I_j, \quad e^{i\theta_{j\ell}} = -\frac{c - i(k_j - k_\ell)}{c + i(k_j - k_\ell)} \quad . \tag{7}$$

The integers I_j are the quantum numbers of the system. For a given set of quantum numbers $\{I_j\}$ these coupled nonlinear equations are in general difficult to solve for the k_j's (and hence the wavefunction and energy) but LIEB and LINIGER[2] gave the exact solution in the $L \to \infty$ finite density limit for the ground state, which corresponds to choosing the lowest possible $\{I_j\}$, i.e., 0, ±1, ±2, ... For large L, the k_j's are closely spaced so from (7), with $I_{j+1} = I_j + 1$,

$$(k_{j+1} - k_j)L = \Sigma(\theta_{j+1,\ell} - \theta_{j,\ell}) + 2\pi$$

$$\cong -2c(k_{j+1} - k_j) \sum_e \frac{1}{c^2 + (k_j - k_\ell)^2} + 2\pi \quad . \tag{8}$$

Defining f(k) by

$$k_{j+1} - k_j = \frac{1}{Lf(k_j)} \tag{9}$$

so that Lf(k)dk is the number of k_j's in (k, k+dk), (8) becomes

$$2f(k) - 1 = 2c \int_{-K}^{K} \frac{f(p)}{c^2 + (p-k)^2}\, dp, \qquad L\int_{-k}^{k} f(k)dk = N \tag{10}$$

a nonsingular integral equation soluble by iteration. Having found f(k), the ground state energy is

$$E_0 = L\int_{-K}^{K} k^2 f(k)dk \quad . \tag{11}$$

Thus in the ground state the bosons fill a "fermi sea" up to a "fermi momentum" K, but the density of states varies with momentum, and the density of states at zero momentum f(o), for example, itself is a function of K, because it depends on phase shifting by the other particles to the sea.

The elementary excitations of the system are analogous to those of the free electron gas - a particle hole pair is generated by raising a boson from $k < k_F$ to $k > k_F$. The novel feature is that as a result of moving the boson in momentum space, all the other boson momenta and energies must shift slightly because the allowed k_i values, including those in the sea, are determined by the momentum-dependent phase shifts. This "backflow" term must be included in computing the energy and momentum of an elementary excitation, and it is not in general small. The particle and hole dispersion curves were worked out by LIEB.[2]

We now turn to the Heisenberg spin one half antiferromagnetic chain, the system actually analyzed by Bethe in 1931.[4] The Hamiltonian is

$$H = J\Sigma\vec{S}_i\cdot\vec{S}_{i+1} + B\Sigma S_i^{\,z} \tag{12}$$

The simplest way to describe this system is in terms of a gas of interacting magnons. A single magnon of wavenumber k is created from the ferromagnetic (all spins down) state by

$$|k\rangle = \sum_n e^{ikn} S_n^{\,+}|\text{ferro}\rangle \quad . \tag{13}$$

The two-magnon state $|k_1,k_2\rangle$ is not simply a superposition of $|k_1\rangle$, $|k_2\rangle$ because $(S_n^{\,+})^2 = 0$, that is, there is an effective local repulsion between magnons. Bethe proved that the lattice wavefunction analogous to (2) above,

$$\psi(n_1,n_2) = e^{i(k_1n_1 + k_2n_2)} + e^{i\theta_{12}}\, e^{i(k_1n_2 + k_2n_1)} \tag{14}$$

where n_1, n_2 label sites of the lattice, is an eigenstate of the Hamiltonian (12). Here θ_{12}, the magnon scattering phase shift, is given by

$$2\cot\left(\frac{\theta_{12}}{2}\right) = \cot\left(\frac{k_1}{2}\right) - \cot\left(\frac{k_2}{2}\right) \quad . \tag{15}$$

Bethe went on to show that the N-particle generalization corresponding to (4) above was also an exact eigenstate of the system.

For the antiferromagnetic chain, the ground state (in zero external magnetic field) has zero total spin, so corresponds to a "fermi sea" of N/2 magnons over the ferromagnetic state. The ground state energy density of this system for $N\to\infty$ was found by HULTHÉN in 1938,[5] using the integral equation method outlined above for the boson gas. Further understanding of spin chain properties came slowly. In an attempt to connect the two known models, the Heisenberg-Ising or XXZ system was formulated,

$$H = \sum_i \left(J_xS_i^xS_{i+1}^x + J_xS_i^yS_{i+1}^y + J_zS_z^zS_{i+1}^z\right) \quad . \tag{16}$$

It was conjectured on the basis of variational calculations that this system would undergo a discontinuous change from Heisenberg to Ising behavior as a function of J_z/J_x.[6] It was thought that at a critical value, long-range order would appear and the ground state energy $E(J_z/J_x)$ would have a discontinuous second derivative. In 1958, however, ORBACH [7] demonstrated that numerically there was no evidence of such sudden change, and in 1959 WALKER [8] solved for the ground state energy exactly, finding $E(J_z/J_x)$ to be analytic on the real line between one and infinity.

The Bethe Ansatz elementary excitation spectrum for the isotropic spin one half Heisenberg chain was worked out in 1962 by DES CLOIZEAUX and PEARSON,[9] who computed the total energy and momentum, including backflow, for a hole in the magnon fermi sea. (The sea just fills a Brillouin zone so there are no "particle" excitations.) The dispersion curve they found was

$$E = \frac{\pi}{2} J \cos k \quad . \tag{17}$$

Simple spin wave theory (that is, ignoring magnon-magnon interactions so there is no backflow term) give $E = J \cos k$. Neutron scattering experiments [10] on an essentially one-dimensional Heisenberg spin system, CPC, gives $E = \pi/2\ J \cos k \pm 5\%$. (J can be found accurately from the susceptibility.) The lesson here is that Bethe's tedious algebraic manipulations describe nature better than the more straightforward spin wave theory. Actually, the details of the picture are even more complicated - turning over a single spin in fact creates two holes in the sea, one generated by the backflow, and the des Cloizeaux-Pearson analysis rather arbitrarily fixes one of these at the zone boundary. (This was not appreciated at the time.) The true excitation spectrum is therefore a two-hole continuum (range of allowed energies for a given momentum), but its lower bound in E, k space is fortunately just $E = \pi/2\ J \cos k$, and most of the spectral weight is in that vicinity.[11] This more precise Bethe Ansatz excitation picture is in excellent agreement with experiment. It must be added, however, that the theoretical analysis of the problem involved a great deal of effort, including substantial numerical work. Despite explicit knowledge of the wavefunctions, the matrix elements needed to find scattering amplitudes have not been found analytically except in certain simple cases, because the Bethe Ansatz wavefunction is so complicated. Evaluation of these matrix elements, or, equivalently, the spin-spin correlation functions, is perhaps the major unsolved problem in the field of soluble models.[12]

To pick up another thread, we backtrack to 1930 and review the increasing understanding of the Ising model in two or more dimensions since Ising's erroneous conclusion that it would never exhibit ferromagnetism. As mentioned earlier, Heisenberg was skeptical, as was Pauli, who in 1930 conjectured that an extension of the model to three dimensions would in fact be ferromagnetic at low temperatures. Neither of them pursued these ideas further, however, and the Ising model might have been forgotten, had not the rather successful Bragg-Williams theory of alloys been developed in the early thirties. This theory was recognized by Fowler and Peierls to be mathematically equivalent to the Ising model, sparking a new interest in possible Ising phase transitions. In 1936, PEIERLS [13] gave a simple argument why the two and three-dimensional Ising models should order ferromagnetically below a critical temperature. He counted the possible boundary configurations between up and down spins to calculate the entropy and claimed that at low enough temperatures the free energy minimum would correspond to "small" total boundary lengths, so the system would be predominantly in one of the two phases (most spins up). Peierls' argument actually had some technical flaws,[1] but the basic idea was sound, and in 1941 KRAMERS and WANNIER [14] found the transition temperature for the two-dimensional Ising model by a symmetry (duality) argument. In 1942, ONSAGER [15] found the exact expression for the free energy (in zero magnetic field), perhaps the most important single achievement in the field of soluble models. Unfortunately the very elegant mathematics he used is also very difficult to follow. The work was simplified by KAUFMAN,[16] and a lucid, accessible account was presented in 1964 by SCHULTZ, MATTIS and LIEB.[17] The basic technique is the transfer matrix method, which provides a way of summing over the 2^{N^2} possible configurations of an N x N matrix M whose rows are labelled by particular configurations of the rth row of the N x N Ising lattice, and the columns of the matrix are labelled by the 2^N possible configurations of the (r+1)th row of the Ising lattice. The m, n element of this $2^N \times 2^N$ transfer matrix M is $e^{-E_{mn}/kT}$ where E_{mn} is essentially the energy associated with the m th configuration of the rth row in conjunction with the nth configuration of the (r+1)th row. Taking the trace of a product of N such identical matrices, then, gives the free energy as a sum over all possible configurations of the entire lattice. (Such a sum would actually count once the energy associated with vertical bonds between the rth row and the (r+1)th row, but count twice the energy in horizontal bonds in the rth row, say. The transfer matrix is easily adjusted to eliminate this double counting.) A great simplification occurs in the $N \to \infty$ limit - considers TrM^N for large N. If M is diagonalized, its elements are its eigenvalues $\lambda_1, \lambda_2 \ldots (\lambda_1 \geq \lambda_2 \geq \ldots)$. Provided it is nondegenerate, $TrM^N = \lambda_1{}^N$ as $N \to \infty$, so the free energy is given directly by the largest eigenvalue of the transfer matrix, which Onsager was able to calculate.

Onsager's method has been generalized to other two-dimensional classical lattice models, and a fascinating connection to the one-dimensional spin one half Heisenberg chains has emerged. In 1967, LIEB and SUTHERLAND [18] solved square ice models, in which arrows are placed on each bond of a two-dimensional square lattice, constrained by the ice rule, that is, having two and only two arrows pointing into each vertex. The transfer matrix for this problem can be taken to act on a (horizontal) row of N vertical bonds, each of which carries one up or down arrow. The space spanned by such configurations is the same as the Hilbert space of a spin one half Heisenberg XXZ chain. It turns out, remarkably, that the transfer matrix, viewed as an operator on this space, commutes with an XXZ Heisenberg spin chain Hamiltonian (for appropriate J_z/J_x) so the eigenstates of the ice transfer matrix are just the Bethe Ansatz states of the spin chain. This makes possible the computation of the largest eigenvalue and hence the free energy of the ice model. The method was further generalized by BAXTER [19] in 1971 to the eight vertex model, in which 0, 2 or 4 arrows can point into each vertex (thus allowing eight possible vertex configurations). The transfer matrix in this case commutes with the fully anisotropic (XYZ) spin chain, and Baxter found the appropriate generalization of Bethe's Ansatz to this system. Over the last fifteen years, the field of two-dimensional statistical models has continued to develop rapidly. A notable triumph was BAXTER's solution of the hard hexagon model in 1980,[20] which describes, for example, phase transitions in gases adsorbed on graphite. Very recently, it has been emphasized that many of the two-dimensional models are conformally invariant at the critical point. The algebraic structure of the conformal group severely constrains critical indices, and connects these models with string theories in elementary particle physics.[21] The prospects for future development along these lines appear very good, and this route may lead to much wider knowledge of correlation functions, for the one-dimensional quantum as well as the two-dimensional classical systems.

Let us now return to the development of Bethe Ansatz methods for the one-dimensional quantum systems. So far, we have only discussed single quantum states of such systems. In 1969, YANG and YANG [22] generalized the Bethe Ansatz analysis to finite temperatures, working with the repulsive boson gas. Away from T=0, the fermi surface, or, rather, fermi points, become blurred. The filled "fermi sea" quantum ground state is replaced by a mixed state, described by a local particle density $L\rho(k)$ and a local hole (unoccupied state) density $L\tilde{\rho}(k)$. The total local density of states $L(\rho(k)+\tilde{\rho}(k))$ depends on the overall particle distribution for the same reason as at zero temperature, so ρ, $\tilde{\rho}$ obey an equation like (9) with $\rho+\tilde{\rho}$ on the left and ρ on the right. Another integral equation for ρ, $\tilde{\rho}$ is given by minimizing the total free energy F as a functional of the local density. Writing $F = E - TS$, the entropy is an integral over a local entropy density which corresponds to the number of possible ways of distributing the bosons among available states,

$$\text{local entropy density} \approx (\rho + \tilde{\rho})\ln(\rho + \tilde{\rho}) - \rho \ln \rho - \tilde{\rho} \ln \tilde{\rho} \quad . \tag{18}$$

Thus we arrive at two coupled integral equations for the two densities ρ, $\tilde{\rho}$. These can be solved and the free energy calculated as a function of overall (macroscopic) density.

Another major contribution by YANG [23] (and, independently, GAUDIN [24]) was the realization that the boson gas analysis could be extended to spin one half fermions. The eigenstates still have the form

$$\psi = \sum_P a(P,Q) e^{i \sum_i k_{Pi} x_{Qi}} \quad , \quad x_{Q_1} < x_{Q_2} < \ldots < x_{Q_N} \quad , \tag{19}$$

but now necessarily the a(P,Q) depend on the ordering of the particles in configuration space, reflecting the nontrivial permutation symmetry of ψ. (For M spin up, N-M spins down, and total spin (N/2)-M, for example, the spatial ψ has symmetry $[2^M 1^{N-2M}]$). The equation for allowed k_i's is now more complicated than (7), and Yang introduced a new set of parameters $\Lambda_1, \ldots, \Lambda_M$, which essentially correspond to spin excitation wavenumbers, just as the $k_1 \ldots k_N$ describe particle or charge excitations. A discrete version is also exactly soluble - the Hubbard model of a one-dimensional metal,[25]

$$H = T \sum_i (c^+_{i\sigma} c_{i+1\sigma} + c^+_{i+1\sigma} c_{i\sigma}) + U \sum_i c^+_{i\uparrow} c_{i\uparrow} c^+_{i\downarrow} c_{i\downarrow} \quad .$$

For large U at a density of one electron per site, the system becomes effectively an antiferromagnetic chain, and the Λ_i turn out to be the k_i of Bethe's spin chain analysis. The Yang delta function fermion gas is actually the low-density continuum limit of the Hubbard model. Another interesting limit of the model is the set of massless fermions given by considering only excitations near its fermi points, a set of right movers and left movers. This gives the field theoretical Gross-Neveu model.[26] It was noticed by ANDREI [27] that this model is strikingly similar to the Kondo model of a single magnetic impurity in a metal, described by the interaction Hamiltonian

$$H = -J\vec{S}\cdot\vec{\sigma} \tag{21}$$

between a local impurity spin S and the (s-wave) conduction electron spin density at the impurity. In fact, if a Gross Neveu model is constructed having only one left mover and many right movers, and the kinetic term for the left mover is dropped from the Hamiltonian, we have essentially a Kondo system - the delta function interaction between fermions can be cast in the $J\vec{S}\cdot\vec{\sigma}$ form, because from the exclusion principle it only operates between fermions having opposite spins. A right mover passing and interacting with the left mover corresponds to an s wave conduction electron coming in, interacting with the impurity and going out again. Thus the Kondo model is a Bethe Ansatz system, and its thermodynamic properties can be found in standard fashion. The results for specific heat and susceptibility are in excellent agreement with experiment.[28] Furthermore, one can find analytically the Wilson number, the ratio of energy scales for low-temperature (strong coupling) phenomena and high-temperature behavior (for which perturbation theory can be used). Wilson found the number by calculating literally hundreds of eigenvalues of a renormalization group model.[29] In the Bethe Ansatz method, it takes a couple of pages of analysis. More recently, the Bethe Ansatz has been extended to the more general Anderson model of a magnetic impurity [30], and certain models of mixed valence impurities. This area is still under active development. An unsolved problem here is calculating the resistivity of the Kondo system - the original "Kondo effect". This requires computation of matrix elements between Bethe Ansatz eigenstates.

In another line of development of exactly soluble one-dimensional quantum systems, WALKER's [8] analytic solution of the XXZ spin chain was generalized to the fully anisotropic XYZ spin chain by BAXTER,[19] as mentioned above, in connection with the eight vertex model. The thermodynamics of these systems was formulated by GAUDIN [30] (XXZ) and TAKAHASHI and SUZUKI [32] (XYZ) in the early seventies. There are physical examples of XXZ spin chains ($CsCoC\ell_3$, for example) which are well described by the theory. Although there are presently no known examples of XYZ spin chains, LUTHER and COLEMAN [33] showed that in a particular continuum limit the XYZ spin chain becomes the quantum sine Gordon Hamiltonian, a model which has been widely used to describe various one-dimensional systems, including spin chains with spins greater than one half. This implies that one can use Bethe Ansatz thermodynamics to evaluate the specific heat of a quantum sine Gordon system. The sine Gordon system has a rich spectrum of excitations - phonons, solitons and soliton-antisoliton bound states call breathers. It has been appreciated for a long time from simple physical arguments that the soliton contribution to the specific heat would be concentrated in a peak centered at around one quarter the soliton mass, where the rate of soliton production with rising temperature is a maximum. It had been argued from this that certain physical systems, such as TMMC in a magnetic field, were quantum sine Gordon systems - the peak was there as predicted. However, using the Bethe Ansatz, and feeding in the known sine Gordon coupling constant for TMMC, the specific heat curve is given without further adjustable parameters, and found to be too low by a factor of two.[34] Thus TMMC cannot be a sine Gordon system - the spins cannot be constrained to a plane as originally thought, and a nonlinear sigma model [35] is a more appropriate description.

An unusual feature of quantum sine Gordon Bethe Ansatz thermodynamics is that one can take the classical limit, and, therefore, use the Bethe Ansatz to compute the specific heat of a purely classical system. The Bethe Ansatz analysis is quite complicated for the sine Gordon system, because there are separate particle and hole densities, ρ and $\tilde{\rho}$ for each distinct type of excitation, leading to a set of coupled integral equations for all these densities. Furthermore, in the classical limit the number of allowed breather (soliton-antisoliton bound state) energies diverges as the quantization becomes finer, giving a diverging number of coupled integral equations. However, it has been found possible to solve this set of equations to give an expression for the specific heat as a double series in temperature t (in soliton mass units) and a soliton density parameter $e^{-1/t}$, which is in exact agreement with the result of purely classical transfer integral techniques.[36] As both methods involve very nontrivial and apparently quite different renormalization procedures, this is a reassuring result. It also puts the phenomenological approach to the sine Gordon model, viewing it as a gas of interacting phonons, solitons, etc., on a firm basis, although, as usual with the Bethe Ansatz techniques, it has not quite led to evaluation of correlation functions at the present time.

SOME SOLUBLE MODELS

1920	Lenz formulates Lenz-Ising model: $H = \Sigma S_i S_{i+1} + B\Sigma S_i \quad (S_i = \pm 1)$	
1925	Ising solves model in 1-D: no phase transition. Concludes none in 2-D, 3-D	Heisenberg takes note; Formulates his model: $H = \Sigma \vec{S}_i \cdot \vec{S}_{i+1} + B\Sigma S_i{}^z$
1930		
		(1931) Bethe solves 1-D: Bethe Ansatz
	(1935) Peierls says maybe there <u>is</u> 2D phase transition	
		(1938) Hulthén - ground state energy
1940		
	Kramers, Wannier find T_c	
	Onsager finds free energy	(Variational methods)
	Onsager finds $\beta = 1/8$	
1950		
	Kaufman simplifies Onsager work	
		(1958) Orbach, Walker: XXZ spin chain
	((1964) Schultz Mattis Lieb)	$H = \sum_i J_x S_i^x S_{i+1}^x + J_x S_i^y S_{i+1}^y + J_z S_i^z S_{i+1}^z$

(1958) Orbach: XXZ spin chain ground state energy smooth in $J_z = (1, \infty)$

(1959) Walker: analytic solution

(1962) des Cloizeaux and Pearson: elem excitations of Heisenberg chain

(1963) Lieb and Liniger: bosons on line, δ function repulsion

(1964) Schultz Mattis Lieb easy formulation of Onsager's work

(1967) Lieb, Sutherland Ice models and —— XXZ spin chains

(1967) Yang: fermions on line with δ function potential between them

Lieb, Wu Hubbard model

(1969) Yang and Yang: BA thermodynamics

(1970) Baxter 8-vertex model and —— XYZ spin chain

(1976) Luther, Coleman lim XYZ ≡ SG ≡ MTM

Baxter's book: hard hexagons, etc.

(1980) Bergknoff, Thacker Korepin: SG BA

SG BA thermodynamics

(1980) Andrei, Weigmann: Kondo problem Anderson model

(surface phase transitions, etc)

(quasi 1-D systems) spin chains

(magnetic impurities, etc.)

1. For the history of soluble models up to 1955, we have relied heavily on History of the Lenz-Ising Model, S.G. Brush, Rev. Mod. Phys. 39, 883 (1967) and The Development of the Quantum Mechanical Electron Theory of Metals II: 1928-33 by L. Hoddeson, G. Baym and M. Eckert, University of Illinois Preprint. These papers contain all the earlier references.
2. E.H. lieb and W. Liniger, Phys. Rev. 130, 1605 (1963), E.H. Lieb, Phys. Rev. 130, 1616 (1963).
3. J.B. McGuire, J. Math. Phys. 5, 622 (1964).
4. H.A. Bethe, Z. Phys. 71, 205 (1931).
5. L. Hulthén, Arkiv. Mat. Astr. Fysik 26A, No. 1 (1938).
6. P.W. Kasteleijn, Physica 18, 104 (1952), W. Marshall, Proc. Roy. Soc. A232, 48 (1955).
7. R. Orbach, Phys. Rev. 112, 309 (1958).
8. L.R. Walker, Phys. Rev. 116, 1089 (1959).
9. J. des Cloizeaux and J.J. Pearson, Phys. Rev. 128, 2131 (1962).
10. R.J. Birgeneau and G. Shirane, Physics Today, p. 32, December 1978.
11. G. Müller, H. Thomas, H. Beck and J.C. Bonner, Phys. Rev. Lett. 43, 75 (1979).
12. Significant progress in calculating correlations functions for certain quantum spin systems has been made by McCoy and coworkers, e.g. B.M. McCoy, J.H.H. Perk and R.E. Schrock, Nucl. Phys. B220 [FS8], 35 (1983). This is partly based on an important series of papers by Jimbo and co-workers, e.g. M. Jimbo and T. Miwa, Physica 2D, 407 (1981).
13. R.E. Peierls, Proc. Camb. Phil. Soc. 32, 471 (1936).
14. H.A. Kramers and G.H. Wannier, Phys. Rev. 60, 252 (1941).
15. L. Onsager, Phys. Rev. 65, 117 (1944).
16. B. Kaufman, Phys. Rev. 76, 1272 (1949).
17. T.D. Schultz, D.C. Mattis and E.H. Lieb, Rev. Mod. Phys. 36, 856 (1964).
18. E.H. Lieb, Phys. Rev. Lett. 18, 692, 1046 (1967), B. Sutherland, Phys. Rev. Lett. 19, 103 (1967).
19. R.J. Baxter, Ann. Phys. NY 70, 193, 212 (1972).
20. R.J. Baxter, Exactly Solved Models in Statistical Mechanics, Academic Press, London (1982).
21. J.L. Cardy, Nucl. Phys. B240 [FS12] 514 (1984).
22. C.N. Yang and C.P. Yang, J. Math. Phys. 10, 1115 (1969).
23. C.N. Yang, Phys. Rev. Lett. 19, 1312 (1967).
24. M. Gaudin, Phys. Lett. 24A, 55 (1967) and thesis, University of Paris (1968).
25. E.H. Lieb and F.Y. Wu, Phys. Rev. Lett. 20, 1445 (1968).
26. N. Andrei and J.H. Lowenstein, Phys. Rev. Lett. 43, 1698 (1979).
27. N. Andrei, Phys. Rev. Lett. 45, 379 (1980); P.B. Wiegmann, J. Phys. C 14, 1463 (1981).
28. N. Andrei, K. Furuya and J.H. Lowenstein, Rev. Mod. Phys. 55, 331 (1983).
29. K.G. Wilson, Rev. Mod. Phys. 47, 773 (1975).
30. A.M. Tsvelick and P.B. Wiegmann, Adv. Phys. 32, 453 (1983).
31. M. Gaudin, Phys. Rev. Lett. 26, 1301 (1971).
32. M. Takahashi and M. Suzuki, Prog. Theor. Phys. 48, 2187 (1972).
33. A. Luther, Phys. Rev. B 14, 2153 (1976); S. Coleman, Phys. Rev. D 11, 2088 (1975); H. Bergknoff and H.B. Thacker, Phys. Rev. D 19, 3666 (1979); V.E. Korepin, Theor. Math. Phys. 41, 953 (1979); H.B. Thacker, Rev. Mod. Phys. 53, 253 (1981).
34. M. Fowler and X. Zotos, Phys. Rev. B 25, 5806 (1982); N.F. Wright and M.D. Johnson, Phys. Rev. B 32, 5798 (1985).
35. F.D.M. Haldane, Phys. Lett. 93A, 464 (1983); I. Affleck, Phys. Rev. Lett. 56, 2763 (1986).
36. N.N. Chen, M.D. Johnson and M. Fowler, Phys. Rev. Lett. 56, 904 (1986), J. Timonen, M. Stirland, D.J. Pilling, Yi Cheng and R.K. Bullough, Phys. Rev. Lett. 56, 2233 (1986).

Variational Schemes for Many-Electron Systems

*D. Baeriswyl**

NORDITA, Blegdamsvej 17, DK-2100 Copenhagen, Denmark

A brief summary of variational methods for many-electron systems is given with main emphasis on approaches à la Gutzwiller. A simple ansatz is proposed for the large U limit of the Hubbard model and shown to agree extremely well with the exact solution in one dimension. Variational schemes can be useful in different contexts like the derivation of Landau parameters for heavy-fermion systems, the influence of U on lattice dimerization in conjugated polymers or even the spectra of excited states in semiconductors.

1. Introduction

Variational methods can be very useful in dealing with strongly interacting many-body systems. They often allow one to describe states of matter which cannot be treated perturbatively such as classical liquids, strongly interacting Bose systems or nuclear matter. Many-fermion systems require a special treatment since they have to satisfy the Pauli exclusion principle. Most approaches (except those based on simple product wavefunctions of the Hartree type) fulfil this symmetry requirement, at least approximately.

This contribution gives a brief and certainly incomplete account of variational methods for many-fermion systems. All approaches described in the following use second quantization and therefore satisfy automatically the Pauli principle. The general variational principle of quantum mechanics is recalled in Section 2. The Hubbard model is presented in Section 3 and treated in Hartree-Fock (HF) approximation. The Gutzwiller ansatz is introduced as a natural device to repair the deficiencies of the HF method. In Section 4 three different ways for calculating the ground state energy using the Gutzwiller ansatz are discussed: the small U expansion, the Gutzwiller approximation and a new way for treating the large U limit. Section 5 presents the results of various variational methods for the one-dimensional Hubbard model and compares them with the Bethe-ansatz solution. In Section 6 several applications, mainly of the Gutzwiller scheme and of related methods, are briefly summarized. A connection to Fermi liquid theory is discussed and extensions of the method to more complicated cases like the periodic Anderson model and the Peierls-Hubbard model are described. Section 7 contains a few final remarks.

*Permanent address: Institut für theoretische Physik, ETH-Hönggerberg, CH-8093 Zürich, Switzerland.

2. The Variational Principle of Quantum Mechanics

Let the states $|\nu>$ form a complete orthonormal set of eigenstates of the Hamiltonian H with energies $E_\nu \leq E_{\nu+1}$ ($\nu=0,1,2,\ldots$). Any state $|\psi>$ can be written in the form

$$|\psi> = \sum_\nu a_\nu |\nu> \tag{1}$$

and it follows that the energy expectation value

$$E[\psi] = <\psi|H|\psi>/<\psi|\psi> = \sum_\nu E_\nu |a_\nu|^2 / \sum_\nu |a_\nu|^2 \tag{2}$$

fulfils the inequality

$$E[\psi] \geq E_0 \ . \tag{3}$$

In the variational approach a trial wavefunction $|\psi_t>$ is constructed which depends on one or several parameters η_i and has the appropriate symmetry. The minimization of the energy,

$$\frac{\partial}{\partial \eta_i} E[\psi_t] = 0 \ , \tag{4}$$

yields the best state within the family of states defined by the variational ansatz.

It is often difficult to estimate how close a variational solution comes to the exact ground state. Even if one knows that the variational result for the energy provides an excellent approximation for the exact ground state energy one cannot be sure that the same is true for the corresponding wavefunction. One has to be particularly suspicious if several states are nearly degenerate with the ground state. A striking example is the Hubbard model in the large U limit where a large number of configurations, namely all those without doubly occupied sites, have nearly the same energy.

One has also to notice that in most but some simple cases (like the Unrestricted Hartree-Fock (UHF) method applied to the Hubbard model) $E[\psi_t]$ cannot be calculated without further approximations. If these can be identified with a restricted set of basis states they will not invalidate the variational character of the ansatz. In most other cases, however, there is no guarantee that $E[\psi_t]$ still represents an upper bound for the ground state energy. Nevertheless, if the approximations involve an expansion in powers of some parameter they can still provide good solutions, at least for an appropriate parameter range.

Sometimes exact solutions are available for certain particular cases, e.g. for a finite number of sites or for reduced dimensionality. They can of course provide very useful touchstones for testing the validity of a given variational procedure. A famous example is the Bethe-ansatz solution for the one-dimensional Hubbard model [1].

3. The Hubbard Model

The Hubbard Hamiltonian is defined as

$$H = -t \sum_{<ij>\sigma} (c^+_{i\sigma} c_{j\sigma} + c^+_{j\sigma} c_{i\sigma}) + U \sum_i n_{i\uparrow} n_{i\downarrow} \tag{5}$$

where $c^+_{i\sigma}$ and $c_{i\sigma}$ are, respectively, creation and annihilation operators for an electron at site i and with spin projection σ, $n_{i\sigma} = c^+_{i\sigma} c_{i\sigma}$ and the summation is restricted to nearest neighbours. Both the $U \to 0$ and the $U \to \infty$ limits are simple to visualize. For $U = 0$ the ground state is given by

$$|\psi_0\rangle = \prod_{k<k_F} c^+_{\vec{k}\uparrow} c^+_{\vec{k}\downarrow} |0\rangle , \qquad (6)$$

corresponding to a Slater determinant of extended Bloch functions. For $U \to \infty$ only those configurations survive which do not have doubly occupied sites. There is an important difference between the two limits. For $U \to 0$ the ground state is not degenerate. For $U \to \infty$ there is a huge degeneracy both due to the distribution of particles and due to the spin configurations. The former type of degeneracy is absent for the particular case of an average density of one particle per site ("half-filled band"). The remaining problem of spin degeneracy can be mapped onto a Heisenberg model with antiferromagnetic exchange, as will be discussed later.

The HF method, where the interaction term is replaced by a self-consistent field, becomes particularly simple for the Hubbard model. The exchange contribution vanishes and the Hartree term produces a constant energy shift in the single-particle spectrum. The ground state energy is obtained as

$$E_{HF} = \langle\psi_0|H|\psi_0\rangle = E_0 + NU/4 \qquad (7)$$

where E_0 is the energy for $U = 0$ and N is the number of electrons. One notices that there remains nothing to be varied in this case.

In the limit $t \to 0$ the ground state energy is simply the minimum potential energy with $n_{i\sigma} = 0$ or 1. This energy vanishes for $N \leq L$ where L is the number of lattice sites. Therefore the HF approximation badly fails for small t (or large U). The reason is that in the HF approximation, configurations including doubly occupied sites have essentially the same weight as those without. In fact, there exists a unique expansion of the HF state (6)

$$|\psi_0\rangle = |\phi_0\rangle + |\phi_1\rangle + \ldots + |\phi_{N/2}\rangle \qquad (8)$$

where the states $|\phi_D\rangle$ have D doubly occupied sites, i.e. they are eigenstates of the operator

$$\hat{D} = \sum_i n_{i\uparrow} n_{i\downarrow} \qquad (9)$$

with eigenvalues $D = 1,2, \ldots , N/2$. For the particular case $N = L = 2$ one easily finds

$$|\phi_0\rangle = \frac{1}{2} (c^+_{1\uparrow} c^+_{2\downarrow} + c^+_{2\uparrow} c^+_{1\downarrow}) |0\rangle , \qquad (10a)$$

$$|\phi_1\rangle = \frac{1}{2} (c^+_{1\uparrow} c^+_{1\downarrow} + c^+_{2\uparrow} c^+_{2\downarrow}) |0\rangle . \qquad (10b)$$

There are

$$N_D(L,N) = L!/\{D!(L+D-N)!\ [(N/2-D)!]^2\} \qquad (11)$$

different configurations contained in $|\phi_D\rangle$ for given L,N. The amplitude for a particular configuration is given by a product of two $(N/2) \times (N/2)$ Slater determinants and exhibits the usual "exchange hole": the amplitudes become small if the spin-up (or spin-down) particles are close to each other but are quite indifferent to the relative arrangement of spin-up and spin-down particles and thus to the number D of doubly occupied sites.

In order to reduce the weight of configurations with large D Gutzwiller [2] proposed the following ansatz

$$|\psi\rangle = \exp(-\frac{1}{2}\eta\hat{D})\ |\psi_0\rangle = \prod_i\ [1-(1-g)n_{i\uparrow}n_{i\downarrow}]\ |\psi_0\rangle \qquad (12)$$

where $g = \exp(-\frac{1}{2}\eta)$. The variational parameter η (or g) is determined by minimizing

$$E(\eta) = \langle\psi_0|e^S\ H\ e^S\ |\psi_0\rangle/\langle\psi_0|e^{2S}\ |\psi_0\rangle\ , \qquad (13)$$

where $S = -\frac{1}{2}\eta\hat{D}$, and is expected to increase monotonously with U and tend to infinity for $U \to \infty$. In this limit the Gutzwiller ansatz tends to $|\phi_0\rangle$. It is worthwhile to mention that it gives the exact ground state for $L=N=2$. In general the evaluation of (13) still represents a very hard task and approximations will be unavoidable. This will be further discussed in the next section.

4. Gutzwiller's Approximation and the Small and Large U Limits

The name Gutzwiller is not only connected with the ansatz (12) but also with a particular way to calculate approximately the expectation value (13) [3]. A nice interpretation of his approach has been given by Ogawa et al. [4]. One first notices that the parameter U acts like a chemical potential for the number of doubly occupied sites D. In the thermodynamic limit the ground state will be characterized by a particular fraction $d = D/L$ of doubly occupied sites, namely that for which the energy is a minimum. The potential energy simply becomes UD. Gutzwiller's approximation for the kinetic energy consists in neglecting spatial correlations in the evaluation of matrix elements except for the constraints imposed by the hopping term [5]. Thus the number of doubly occupied sites is simply determined by the maximum of $g^{2D}\ N_D(L,N)$. Using (11) one finds

$$g^2 = d(1+d-n)/(\tfrac{1}{2}n-d)^2 \qquad (14)$$

where $n = N/L$ is the electron density. The total energy turns out to have the following simple form

$$E = q(d,n)\ E_0 + UD \qquad (15)$$

where the "loss factor" of the kinetic energy $q(d,n)$ is found to be ($n \leq 1$)

$$q(d,n) = \frac{2(n-2d)}{n(2-n)}\ [(1-n+d)^{\frac{1}{2}} + d^{\frac{1}{2}}]^2\ . \qquad (16)$$

This number determines at the same time the discontinuity of the momentum distribution function at the Fermi surface [3]. The minimization of the energy with respect to d becomes particularly simple for $n = 1$ [6] where

$$q(d) = 8d\,(1-2d)\ . \tag{17}$$

One finds

$$d = (1 - U/U_c)/4 \tag{18}$$

where

$$U_c = 8|E_0|/L \tag{19}$$

corresponds to the critical value where the kinetic energy vanishes. Brinkman and Rice [6] interpreted this result in terms of a metal-insulator transition to a completely localized state. On the other hand, starting from a paramagnetic state (6) one cannot exclude the possibility of magnetic ordering. Allowing for antiferromagnetism in $|\psi_0\rangle$ Ogawa et al. [4] indeed found a first-order transition to a state with antiferromagnetic long-range order at a smaller value of U than the critical value for the metal-insulator transition.

The results given above do not depend on lattice dimensionality and therefore represent kind of a mean-field treatment of the Gutzwiller ansatz. In one dimension there is, however, neither a Mott-Hubbard transition at finite U [1] nor long-range antiferromagnetic order, in contradiction with the general conclusions based on the Gutzwiller approximation. This failure is not inherent in the ansatz (12) but rather produced by neglecting spatial correlations. To show this we have a closer look at the one-dimensional case for $n = 1$ both for $U \ll t$ and for $U \gg t$.

For $U \ll t$ the correlation parameter η will be small and therefore one can expand (13) in powers of η. Due to the form of the Gutzwiller operator e^S one arrives at an expansion in terms of connected diagrams [7]

$$E(\eta) = \langle\psi_0|e^S\ H\ e^S|\psi_0\rangle_c \approx \langle\psi_0|H|\psi_0\rangle - \frac{1}{2}\eta\langle\psi_0|\{\hat{D},H\}|\psi_0\rangle_c + \frac{1}{8}\eta^2\langle\psi_0|\{\hat{D},\{\hat{D},H\}\}|\psi_0\rangle_c \tag{20}$$

where $\{A,B\} = AB + BA$. Wick's theorem yields a decomposition of these expectation values in terms of correlation functions

$$P_{nm} = \langle\psi_0|\ c^+_{n\sigma}\ c_{m\sigma}\ |\psi_0\rangle \tag{21}$$

which have a simple form in the half-filled band case of the one-dimensional Hubbard model,

$$P_{n,n+2m} = \frac{1}{2}\ \delta_{m,0} \tag{22a}$$

$$P_{n,n+2m+1} = (-1)^m/[\pi(2m+1)]\ . \tag{22b}$$

The energy becomes [8]

$$E(\eta) = E_{HF} - 2t\eta^2 \sum_i \left(\frac{1}{8}P_{i,i+1} + \frac{1}{2}P^3_{i,i+1} - 2\sum_{j\ell} P^3_{j,\ell} P_{j,i} P_{\ell,i+1}\right)$$

$$- U\eta \sum_{ij} P^4_{i,j} \ . \tag{23}$$

Minimizing with respect to η and performing the summations one finds $\eta \sim U/t$ and [8,9]

$$(E-E_{HF})/(Lt) = -\ 0.2484\ (U/4t)^2\ . \tag{24}$$

This represents more than 90% of the correlation energy which has been deduced from the Lieb-Wu solution (see Table 1 of Section 5). One can reduce still further the energy by allowing for a gap in the single-particle spectrum [10]. This gap is non-zero for finite U and the system is insulating, in qualitative agreement with the exact solution. It is important to notice that this gap is associated with charge transport between neighbouring sites and thus can be identified with an optical gap. On the other hand a spin-density wave state which would be predicted to be stable within the UHF approximation and to produce both long-range magnetic order and a single-particle gap is found to be suppressed by the correlation term e^S, and, correspondingly, the spin-spin correlation function decays algebraically [11].

For $U \gg t$ the Gutzwiller wavefunction is reduced to $|\phi_0\rangle$ in the expansion (8), i.e. to a configuration without doubly occupied sites. This is, however, not done in an arbitrary way but the resulting state has a particular symmetry. This is obvious for L=N=2 where $|\psi_0\rangle$, given by (10a), corresponds to a spin singlet. In the general case the pairs of operators in (6) become

$$c^+_{k\uparrow} c^+_{k\downarrow} \rightarrow \frac{1}{N} \sum_{m<n} e^{ik(m+n)} \left(c^+_{m\uparrow} c^+_{n\downarrow} - c^+_{m\downarrow} c^+_{n\uparrow}\right) \tag{25}$$

where the term with two electrons on a single site has already been omitted. The state $|\phi_0\rangle$ is therefore a superposition of states each of which is a product of singlet pairs. But these are precisely the states which contribute to the exact ground state of the Heisenberg chain with antiferromagnetic exchange [12]. The total spin is zero, in agreement with a theorem of Lieb and Mathis [13]. Therefore it is not surprising that the state $|\phi_0\rangle$ yields spin correlation functions which agree very well with the corresponding results for the Heisenberg model [14]. In particular the numerical calculations for finite chains are consistent with the asymptotic behaviour

$$\langle\phi_0|s^z_i\, s^z_{i+\ell}|\phi_0\rangle \sim \ell^{-1} \qquad \text{for } \ell \rightarrow \infty\ , \tag{26}$$

in agreement with analytical [15] and numerical results [16] for the Heisenberg chain.

The energy expectation value with respect to $|\phi_0\rangle$ vanishes (for N=L). This corresponds to the limit $U \rightarrow \infty$. For large but finite U states involving doubly occupied sites will also contribute and it is tempting to use the state $|\phi_0\rangle + |\phi_1\rangle$ to calculate the first corrections to the $U \rightarrow \infty$ result. However, the numerical study of Kaplan et al. [14] indicates that polarization processes involving doubly occupied sites are poorly described by $|\phi_1\rangle$ where they are spread out over large distances (due to the extended nature of Bloch functions). For large U the charge susceptibility is expected to decay extremely fast with distance due to the large gap for charge excitations. Therefore, while the Gutzwiller ansatz turns out to describe very well the spin degrees of freedom even for $U \to \infty$, it fails with respect to the charge degrees of freedom. Kaplan et al. [14] propose a modified version of the Gutzwiller ansatz where all polarizations except those among nearest neighbours are projected out.

An alternative way for dealing with the large U region could be in terms of a variational ansatz where the roles of kinetic and potential energies are interchanged, for instance in the form

$$|\psi\rangle = \exp(-\tfrac{1}{2}\kappa\hat{T})\ |\phi\rangle \tag{27}$$

where

$$\hat{T} = \sum_{i\sigma}(c^+_{i\sigma}c_{i+1\sigma} + c^+_{i+1\sigma}c_{i\sigma}) , \tag{28}$$

and κ is a variational parameter. $|\phi\rangle$ can be chosen equal to $|\phi_0\rangle$ or determined variationally as a superposition of the 2^N states involving only singly occupied sites. Expanding the expectation value of the Hamiltonian (5) up to second order in κ one finds

$$E[\psi] \approx (\kappa t + \tfrac{1}{4}\kappa^2 U)\ \langle\phi|\hat{T}^2|\phi\rangle/\langle\phi|\phi\rangle \quad . \tag{29}$$

The minimizations with respect to κ and $|\phi\rangle$ completely decouple giving $\kappa = -2t/U$. It is well known that within a restricted space of states with exactly one electron per site the operator $\hat{T}^2$ can be simply expressed in terms of spin ½ operators [17], leading to

$$E[\psi] = (4t^2/U)\sum_i \langle\phi|(\vec{s}_i\cdot\vec{s}_{i+1} - \tfrac{1}{4})|\phi\rangle/\langle\phi|\phi\rangle \quad . \tag{30}$$

Thus the problem is reduced to finding the ground state of the one-dimensional Heisenberg model with antiferromagnetic exchange $J = -4t^2/U$. Its ground state energy is known [12], giving

$$E/N = -4\ \ell n2\ (t^2/U) . \tag{31}$$

It is instructive to compare this result with the value one obtains using the Gutzwiller wavefunction, $|\phi\rangle = |\phi_0\rangle$, in (30). The numerical results for the nearest-neighbour spin correlation function [14] give

$$E/N = -2.7688\ (t^2/U) \tag{32}$$

which is extremely close to the exact value (31). This confirms that the Gutzwiller state $|\phi_0\rangle$ is an excellent starting point for large U expansions.

5. Comparison of Different Variational Schemes

The previous two Sections have been mainly concerned with the Gutzwiller ansatz. Here a short summary of other approaches will be given. The ground state energies for the one-dimensional Hubbard model as obtained from the various methods are presented in Table 1 and compared to the exact Bethe-ansatz solution.

The most simple generalization of the HF method, the UHF method, consists in introducing a staggered magnetic field which leads to a spin-density wave ground state the amplitude of which increases with increasing U. For $U \to \infty$ only a single purely antiferromagnetic configuration (Néel state) survives. The UHF ground state exhibits long-range antiferromagnetic order for all $U > 0$ in contrast to what is known from the $U \to \infty$ limit [15,16] and Monte Carlo calculations for finite U [18]. The energy also disagrees appreciably with the exact result (Table 1).

In Ukrainskii's method of "varying localized geminals" (VLG) two types of Wannier states are introduced and the ground state is written in terms of pairs of these Wannier functions [19]. This method is analytically tractable and agrees qualitatively with exactly known properties. Charge excitations have a gap whereas there exist also gapless neutral excitations. The latter are responsible for a power-law decay of the spin-spin correlation function.

A combination of a variational ansatz similar to the Gutzwiller wavefunction and a cluster expansion has been studied by Bernasconi [20]. In its simplest version, the one-site cluster approximation, this method gives the same energy as the Gutzwiller approximation for small U (Table 1) but it fails in tending to the proper large U limit. This shortcoming can be removed by starting from a spin-density wave ground state. In this case the energy approaches zero for $U \to \infty$. At the same time this modification produces the artifact of long-range antiferromagnetic order, although it is strongly reduced as compared to the UHF approach [20]. As shown by Razafimandimby [21], the calculation of the energy expectation value for the Gutzwiller wavefunction using a two-site cluster approximation essentially reproduces the results of the Gutzwiller approximation.

Table 1 shows that the small and large U expansions of the Gutzwiller ansatz come very close to the exact results. It is worthwhile to add that an energy gap in the single-particle spectrum due to charge excitations produces an additional term which is non-analytic in U for $U \to 0$ [10]. Such a term is indeed also found in the asymptotic expansion of the exact ground state energy [22].

Table 1 Ground state energy of the one-dimensional Hubbard model as a function of $u = U/(4t)$. The different methods are explained in the text.

Method	$(E-E_{HF})/(Nt)$, $u \ll 1$	$E/(Nt)$, $u \gg 1$	References
UHF	$-16\,\pi^{-1}\exp(-\pi/u)$	$-0.5/u$	e.g. [20]
VLG	$-0.1745\,u^2$	$-0.5/u$	[19]
1-Cluster	$-0.1963\,u^2$	1.273	[20]
Gutzwiller approximation	$-0.1963\,u^2$	0	[3,6]
Gutzwiller, expansions	$-0.2484\,u^2$	$-0.6922/u$	[8,9], Section 4
Bethe ansatz	$-0.2719\,u^2$	$-0.6931/u$	[1,22]

6. Applications

The variational approach allows, at least in principle, to calculate any static correlation function at zero temperature and thus various types of response functions like the magnetic susceptibility or the dielectric constant. It is usually much harder to deal with dynamic correlation functions since these involve excites states. Fortunately, the frequency-moments can often be reduced to simple static correlation functions. The conductivity sum rule for instance is proportional to the kinetic energy which can be easily obtained as a derivative of the ground state energy [23]. Since the kinetic energy decreases monotonously with U, the experimental determination of the conductivity sum rule can give useful information about the "effective U" of a given material [24].

Variational schemes are also well suited for providing the necessary link between microscopic model Hamiltonians and the phenomenological Fermi liquid theory where the strongly interacting bare particles are replaced by (hopefully) weakly interacting quasi-particles. In particular one can deduce values for the effective mass m^* and the various Landau parameters describing the interaction between quasi-particles. This connection turns out to be particularly illuminating for the Gutzwiller approximation [5,6]. Identifying the step in the momentum distribution function with the mass enhancement one finds $m/m^* = q$. Essentially the same enhancement is found for the magnetic susceptibility. For $n = 1$ charge transport can only occur through doubly occupied sites ("particles" or empty sites ("holes"). Following an argument by Mott [25] one can describe the current on a "microscopic" time scale as being due to a fraction $2d$ of carriers moving freely, i.e. with the bare mass m. Fermi liquid theory is concerned with low-energy excitations and therefore with long times as compared to the time scale of a single hop. On these "macroscopic" time scales every electron gets the chance to hop. Averaging out all the fast but rare events results in a slow motion of quasi-particles. This means that, while microscopically only a small fraction $2d$ of the electrons move with their bare mass m, on a longer time scale all particles move, but with a heavier mass m_c. Identifying the two currents gives $m_c/m = (2d)^{-1}$. In view of (17) the effective mass for transport m_c has the same order of magnitude as the effective mass m^* associated with the specific heat and the magnetic susceptibility. Similar arguments can be given for $n < 1$ and they can be made more precise on the basis of the conductivity sum rule [26]. The argument presented above does not apply to a quantum liquid like He^3 where also on a microscopic time scale all particles move. In fact, for a translationally invariant system the mass associated with transport is equal to the bare mass [27].

At first sight it seems surprising that Gutzwiller's approach, which has been derived for fermions on a lattice, should be able to describe a quantum liquid. However, as discussed by Vollhardt [5], the essential features of this system are well represented in terms of a contact interaction. Furthermore, the Gutzwiller approximation (15) does not depend on details of the lattice structure and should therefore be applicable to systems without any particular point symmetry. Indeed, the theory yields Landau parameters which are in good agreement with those deduced from experiments. The values of the parameter U deduced from the data lead to the conclusion that the normal phase of liquid He^3 is an almost localized Fermi liquid (not in the sense of transport but in the sense of a strong tendency of the particles to avoid each other) [5].

The Gutzwiller approximation has recently been extended to deal with the periodic Anderson Hamiltonian [28,29] which is of great importance in the context of heavy-electron systems [30]. In the large U limit a paramagnetic heavy Fermi-liquid is found to be stable if both the orbital degeneracy of the f-level and its hybridization with the conduction electrons are large enough. This state can then be viewed as an almost localized Fermi liquid with n close to 1 [28]. The variational approach thus allows to build a bridge between normal liquid He^3 and heavy-electron metals, systems which at first sight look extremely different but experimentally are found to exhibit many common features.

Variational studies have also been made for models of conjugated polymers, in particular the "Hubbard-Peierls" chain which incorporates both electron-electron (U) and electron-phonon interaction (parameter λ) [8,9,10,31,32]. Mainly the half-filled band case ($n=1$) has been studied so far. Exact solutions are available for $U = 0$, $\lambda > 0$ where the system is a dimerized Peierls semiconductor and for $U > 0$, $\lambda = 0$ where it corresponds to a Mott-Hubbard insulator [1]. Both the VLG approach [31,32] and the small U expansion of the Gutzwiller ansatz [8,9,10] produce the surprising result that the lattice dimerization (for small λ and small U) increases with U. This has since been confirmed by other methods. Unfortunately, there is no general agreement about the appropriate values of U and λ. Several experiments indicate that U is of the order of the bandwidth, or maybe slightly less. Properties of small organic molecules indicate $\lambda \approx 0.1$. In this case the optical gap is mainly determined by correlation effects [10] and the conjugated polymers have to be considered as Mott-Hubbard insulators rather than Peierls semiconductors. An opposite point of view has been advocated recently by Wu and Kivelson [33]. It is worthwhile to add that the variational procedure yields very reasonable results in regions of parameter space where perturbation theory breaks down.

The Gutzwiller scheme has been generalized by Fulde and coworkers to what they call the "local ansatz" (for a review see [34]). Realistic Hamiltonians for atoms, molecules and solids are investigated by means of an ansatz of the form $\exp(S)\,|\psi_0\rangle$ where $|\psi_0\rangle$ is the Hartree-Fock ground state and the operator S not only reduces the double occupancy for individual orbitals (or hybrids) but also includes correlations between different orbitals. S contains various variational parameters η_ν and the energy is expanded to second order in these parameters in a similar way as in the small U expansion for the Hubbard model. Therefore this approach is limited to systems with weak or at most intermediate correlation. Finally it should be mentioned that the method has been generalized to study excited states in semiconductors [35].

7. Concluding Remarks

The Gutzwiller ansatz and its generalizations are of the form

$$|\psi\rangle = \exp(S)\,|\psi_0\rangle \tag{33}$$

where $|\psi_0\rangle$ is the Hartree-Fock ground state or some other state. One should notice that the exact ground state can always be written in this form (if $\langle\psi|\psi_0\rangle \neq 0$) and that, at least in principle, the operator S can be determined. The way to achieve this is known as the "Coupled Cluster Method" which has been widely used by nuclear theorists [36] and quantum chemists [37]. This method might also be useful in combination with a variational approach since it could give a hint how to construct the variational ansatz.

Recently Kotliar and Ruckenstein devised a functional integral approach both for the Hubbard and the periodic Anderson model [38]. Within the simplest saddle-point approximation they recover the Gutzwiller approximation [6,28] and thus confirm its role of a mean-field solution.

References

1. E.H. Lieb, F.Y. Wu: Phys. Rev. Lett. 20, 1445 (1968)
2. M.C. Gutzwiller: Phys. Rev. Lett. 10, 159 (1963)
3. M.C. Gutzwiller: Phys. Rev. B 137, A1726 (1965)
4. T. Ogawa, K. Kanda, T. Matsubara: Progr. Theor. Phys. 53, 614 (1975)
5. D. Vollhardt: Rev. Mod. Phys. 56, 99 (1984)
6. W.F. Brinkman, T.M. Rice: Phys. Rev. B 2, 4302 (1970)
7. P. Horsch, P. Fulde: Z. Phys. B 36, 23 (1979)
8. D. Baeriswyl, K. Maki: Phys. Rev. B31, 6633 (1985)
9. P. Horsch: Phys. Rev. B 24. 7351 (1981)
10. D. Baeriswyl, K. Maki: Synth. Met., to be published
11. D. Baeriswyl, K. Maki: unpublished
12. L. Hulthén: Arkiv Mat. Astron. Fysik: 26A, 1 (1938)
13. E.H. Lieb, D.C. Mattis: Phys. Rev. 125, 164 (1962)
14. T.A. Kaplan, P. Horsch, P. Fulde: Phys. Rev. Lett. 49, 889 (1982)
15. A. Luther, I. Peschel: Phys. Rev. B 12, 3908 (1975)
16. J. Bonner, M.E. Fisher: Phys. Rev. 135, A640 (1964)
17. V.J. Emery: Phys. Rev. B 14, 2989 (1976)
18. J.E. Hirsch, R.L. Sugar, D.J. Scalapino, R. Blankenbecler: Phys. Rev. B 26, 5033 (1982)
19. I.I. Ukrainskii: Teor. Mat. Fiz. 32, 392 (1977) (Theor. Math. Phys. 32, 816 (1977))
20. J. Bernasconi: Phys. Kondens. Mat. 14, 225 (1972)
21. H. Razafimandimby: Z. Phys. B 49, 33 (1982)
22. E.N. Economou, P.N. Poulopoulos: Phys. Rev. B 36, 23 (1979)
23. D. Baeriswyl, J. Carmelo, A. Luther: Phys. Rev. B 33, 7247 (1986)
24. C.S. Jacobsen, Ib Johannsen, K. Bechgaard: Phys. Rev. Lett. 53, 194 (1984)
25. N.F. Mott: Metal-Insulator Transitions (Taylor and Francis, London 1974) p. 139
26. D. Baeriswyl, C. Gros, T.M. Rice: submitted to Phys. Rev. B
27. D. Pines, P. Nozières: The Theory of Quantum Liquids (Benjamin, New York 1966)
28. T.M. Rice, K. Ueda: Phys. Rev. Lett. 55, 995; 2093 (E) (1985)
29. C. Varma, W. Weber, L.J. Randall: Phys. Rev. B 33, 1015 (1986)
30. P.A. Lee, T.M. Rice, J.W. Serene, L.J. Sham, J.W. Wilkins: Comm. Cond. Matt. Phys. 12, 99 (1986)
31. I.I. Ukrainskii: Zh. Eksp. Teor. Fiz. 76, 760 (1979) [Sov. Phys. JETP 49, 381 (1979)]
32. I.I. Ukrainskii: phys. stat. sol. (b) 106, 55 (1981)
33. W.-K. Wu, S. Kivelson: Phys. Rev. B 33, 8546 (1986)
34. P. Fulde: Progr. Theor. Phys. Suppl. 80, 47 (1984)
35. S. Horsch, P. Horsch, P. Fulde: Phys. Rev. B 28, 5977 (1983); Phys. Rev. B 29, 1870 (1984)
36. H. Kümmel, K.H. Lührmann, J.G. Zabolitzky: Phys. Rep. 36, 1 (1978)
37. V. Kvasnicka, V. Laurinc, S. Biskupic: Phys. Rep. 90, 159 (1982)
38. G. Kotliar, A.E. Ruckenstein: preprint

Quantum Monte Carlo

D.J. Scalapino

Department of Physics, University of California,
Santa Barbara, CA 93106, USA

Abstract

Monte Carlo methods allow us to explore the properties of interacting many-body quantum systems. Here we discuss what we have learned about (1) the 2-D XXZ spin-1/2 system and (2) superconductivity near a Van Hove singularity from simulations; and comment on future needs and prospects.

The development of stochastic methods for the numerical simulation of interacting many-body quantum systems provides a new method for studying the non-linear properties of these systems. Here I'll review two examples which illustrate the types of things which have been done and discuss some future prospects. Just as Monte Carlo studies [1] have provided information on the phase diagram of the classical Heisenberg-Ising XXZ model, similar studies can now be carried out for the quantum spin-1/2 XXZ model [2-7]. As with all simulations of phase transitions, it is necessary to carry out some type of finite size scaling analysis. For the quantum systems, the lattice size is usually more restricted so that at present one may only conclude that the scaling properties of the simulation are consistent with a given type of transition. However, with the evolution of both new algorithms and new machines, our ability to extend these initial efforts is being greatly enhanced. Below we describe some results obtained from simulations of the two-dimensional quantum XXZ spin-1/2 model [7]. In the area of electronic properties, simulations offer the possibility of treating band structure and many-body effects on the same footing. As an example, consider the question of what a peak in the band density of states does to the superconducting transition. From the BCS expression, $T_c = \omega_c \exp(-1/N(0)|V|)$, one expects that an increase in the single particle density of states at the fermi energy, $N(0)$, will enhance T_c. However, the effect of $N(0)$ on the screening, self-energy and vertex corrections and their resultant effect on T_c remain open questions. This is particularly acute in two dimensions, where saddle points in ϵ_k versus k lead to logarithmic Van Hove singularities in the single particle density of states. Here we will discuss what simulation tells us regarding the enhancement of the pairing susceptibility for a 2-D system with a Van Hove singularity at the Fermi energy [8].

For some one-dimensional electron systems, such as the Hubbard model, and for certain spin systems such as the XXZ model in any dimension, one can formulate the simulation directly in terms of a sum over configurations of the primary electron (occupation) or spin (S_z) variables. For example, for the 2-D XXZ model

$$H = -\sum_{\langle ij \rangle} (S_i^x S_j^x + S_i^y S_j^y + \lambda S_i^z S_j^z) \tag{1}$$

we can write

$$Z = Tre^{-\beta H} = \sum_{\alpha_1, \alpha_2 \cdots \alpha_L} \langle \alpha_1 | e^{-\Delta\tau H} | \alpha_2 \rangle \langle \alpha_2 | e^{-\Delta\tau H} | \alpha_3 \rangle \cdots \langle \alpha_L | e^{-\Delta\tau H} | \alpha_1 \rangle \tag{2}$$

Here $\beta = \Delta\tau \cdot L$ and the states are specified by listing the $\{S_i^z\}$ quantum numbers of all the spins on the lattice

$$|\alpha_i\rangle = |\{S_1^z, S_2^z, \cdots S_N^z\}_i\rangle \tag{3}$$

To evaluate the matrix elements in Eq. (2), one takes $\Delta\tau$ (and $\Delta\tau\lambda$) small compared to 1 and breaks up H into solvable pieces $H = H_1 + H_2$

$$e^{-\Delta\tau H} \simeq e^{-\Delta\tau H_1} \cdot e^{-\Delta\tau H_2} \tag{4}$$

Suzuki has labeled this approximation the generalized Trotter formula and discussed various forms for the breakup [2]. For the 2-D XXZ model we use a checkerboard decomposition [9] in which H_1 and H_2 are each composed of sums of nonoverlapping four-site cell Hamiltonians as indicated in Fig. 1. With this breakup, the matrix elements of $e^{-\Delta\tau H_1}$ and $e^{-\Delta\tau H_2}$ can be separated into exponentials of the four-site Hamiltonians without further approximation. The matrix elements of these exponentials are then simply evaluated numerically. For spin Hamiltonians for which one can arrange for all the matrix elements to be positive, one can interpret their product as a probability. This is the case for the XXZ model in a basis where the spins are quantized along the

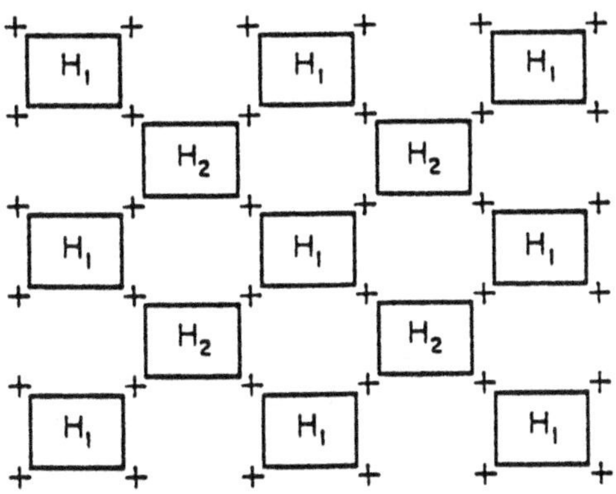

Fig. 1. Checkerboard breakup of the Hamiltonian for the two-dimensional lattice. Terms in H are grouped into two parts, H_1 and H_2, each of which is composed of four-site cell sub-Hamiltonians which commute among themselves.

z-direction. Then the resulting sum over configurations, Eq. (2), can be carried out using Monte Carlo importance sampling. In this case a configuration consists of the L τ–configurations so that a d–dimensional quantum system has been mapped to a $d+1$-dimensional classical problem.

In order to have a fast algorithm it is important to be able to make local moves. This is accomplished for the present problem by making four-site updates. As discussed in detail in Ref. 7, these moves span only a part of the configuration space. In particular, the total magnetization is taken to be zero, as well as a quantity called the *winding number*. These restrictions, however, become unimportant in the thermodynamic limit. For spins quantized in the z-direction, we have measured the energy, specific heat, vortex density and various spin-spin correlation functions such as the staggered susceptibility

$$\chi = \int_0^\beta d\tau \sum_l (-1)^{l_x+l_y} \langle S^z_{i+l}(\tau) S^z_i(0) \rangle \tag{5}$$

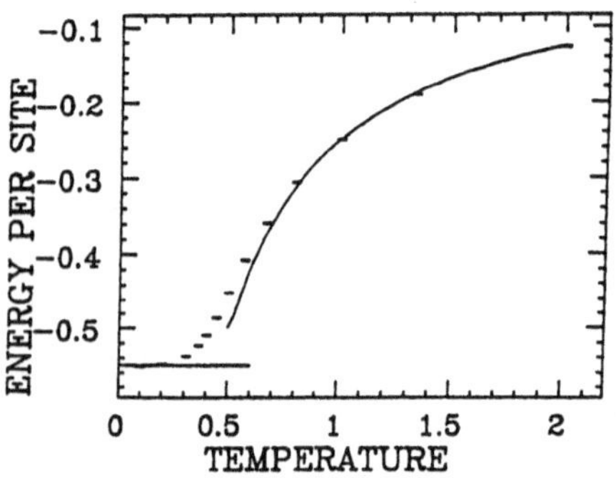

Fig. 2. Energy per site versus T for an 8×8 lattice with $\Delta\tau = 0.25$. Results for the ground state energy and the leading terms in a high temperature expansion are shown for comparison.

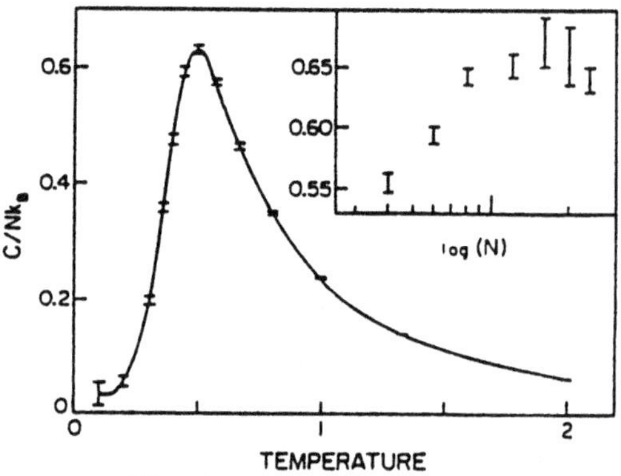

Fig. 3. Specific heat per site versus T. The inset shows the dependence of the peak height on size for 4×4 through 24×24 lattices.

For $\lambda = 0$, the Hamiltonian, Eq. (1), is the XY model. Results [7] for the energy per site and the specific heat are shown in Figures 2 and 3. At high temperatures, the ground state energy varies as $1/4T$ and at low temperature it approaches -0.543 ± 0.002 which is in good agreement with estimates from both variational and exact diagonalization calculations. The inset in Figure 3 shows the dependence of the specific heat peak per site versus the log of the site number. For comparison, Figure 4 shows the behavior of the peak height obtained from a Monte Carlo calculation [10] for the classical XY model with $H = -\sum_{\langle ij\rangle} \cos(\theta_i - \theta_j)$. From this data, we concluded that the

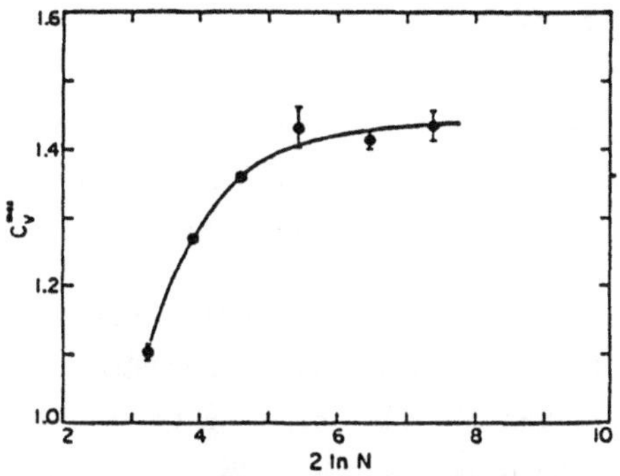

Fig. 4. Height of the specific heat peak for the classical XY model versus $2\ln(N)$ for 5×5 through 40×40 lattices (from Ref. 10).

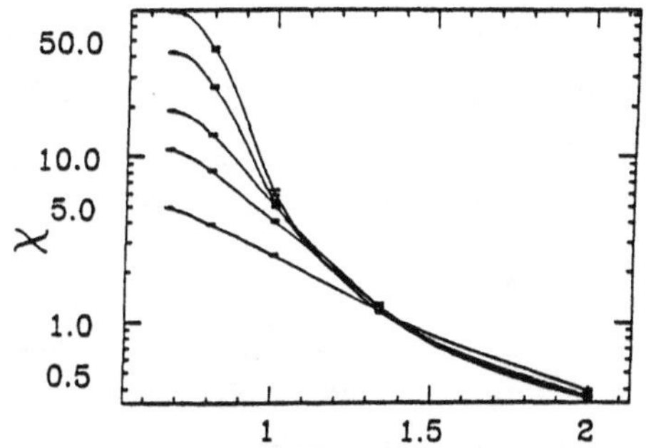

Fig. 5. Staggered susceptibility, Eq. (5), versus T for the XXZ model with $\lambda = -1.6$. Data is shown for 4×4, 6×6, 8×8, 12×12 and 16×16 lattices. The lines are drawn as a guide to the eye.

specific heat peak of the quantum $S = 1/2$ model saturates at a finite value as the lattice size increases. This is in contrast to the results obtained by de Raedt, et al., [6] who claimed that the specific heat peak grew as $\ln(N)$. Our specific heat results along with measurements of the helicity modulus [7] lead us to conclude that the XY model has a phase transition which is consistent with the Kosterlitz-Thouless [11] transition.

For $-1 < \lambda < 0$, a similar behavior was observed. For $\lambda < -1$, the behavior changed to Ising like. Figure 5 shows the staggered susceptibility for $\lambda = -1.6$ for various sized lattices. Using finite size scaling we obtained the T_c versus λ data shown in Fig. 6. Our conclusions were that our results were consistent with the quantum spin-1/2 XXZ model being in the same universality class as the classical XXZ model [7].

These simulations were carried out on an IBM 3081. For $\Delta\tau = 0.25$, $L = 8$ time slices corresponded to a β of 2 and a temperature $T = 0.5$ near the peak in the specific heat of the XY model. A run consisting of 100,000 lattice sweeps on a 16×16 lattice

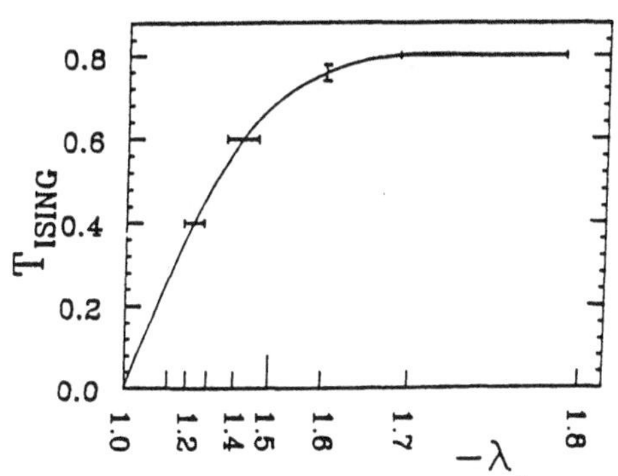

Fig. 6. The temperature T_{Ising} obtained from finite size scaling for the Ising-like transition plotted against $-1/\log(-\lambda - 1)$.

with $L = 8$ time slices took of order an hour. With L fixed, the time for a sweep of the lattice scales with the number of spatial lattice sites. With the factor of 50 to 100 increase in speed presently available with today's supercomputers one could certainly do 100×100 lattices even allowing for critical slowing down near T_c, putting the quantum calculations in the same N-region as typical classical results.

Unfortunately, the situation is at present quite different for fermions in higher dimensions, where interchanges of position give rise to negative signs which frustrate the type of direct sum over configurations discussed above. In this case, present algorithms involve the introduction of a Hubbard-Stratonovich or possibly naturally occurring boson field (phonons, excitons) so that the fermion fields enter only quadratically. Then formally tracing over the fermion fields one obtains an effective non-local action involving configurations of the space-imaginary-time boson field. In some cases of interest (e.g. where there is particle-hole symmetry), this can provide a positive probability measure for Monte Carlo sampling. In the more general case where the weight can become negative for certain configurations, problems remain. For temperatures which are not too low and for systems which are not too large, it has been possible to treat this by introducing a weight defined by an absolute value and keeping separate track of the signs.

However, even when a suitable weight can be found, the Monte Carlo updating is much more time consuming than the procedure described for the XXZ model because of the non-local nature of the action. One method which has been used on small two and three-dimensional lattices involves an exact updating of the lattice Green's function at each change of configuration [12]. With this procedure, the time for one sweep of the lattice varies as the cube of the space volume times the number of time slices and a central problem for the future involves the construction of algorithms where the computational time varies linearly with the space-time lattice volume.

To illustrate some of these features, consider a 2-D Hubbard model with an attractive on-site interaction [8]

$$H = -\sum_{\langle ij\sigma\rangle} t_{ij}(C^+_{i\sigma}C_{j\sigma} + C^+_{j\sigma}C_{i\sigma}) - \mu\sum_{i\sigma} n_{i\sigma} - |U|\sum_i n_{i\uparrow}n_{i\downarrow} \tag{6}$$

Here $t_{ij} = t$ for (i,j) near neighbors and $t_{ij} = t_2$ for (i,j) next-nearest neighbors which gives rise to a band structure $\epsilon_k = -2t(\cos k_x + \cos k_y) - 4t_2 \cos k_x \cos k_y$. For $t > t_2/2$ there are saddle points at $(k_x, k_y) = (\pm\pi, 0)$ and $(0, \pm\pi)$ which produce a logarithmic singularity in the single particle density of states $N(\epsilon) \approx \frac{1}{t}\ln(t/|\epsilon - 4t_2|)$ at $\epsilon = 4t_2$. Figure 7 shows $N(\epsilon)$ as a dashed curve for $t = 1$ and $t_2 = 0$ and as a solid curve for $t = 1$ and $t_2 = -0.4$. For $\mu = -1.6$, the fermi energy is at the Van Hove singularity for

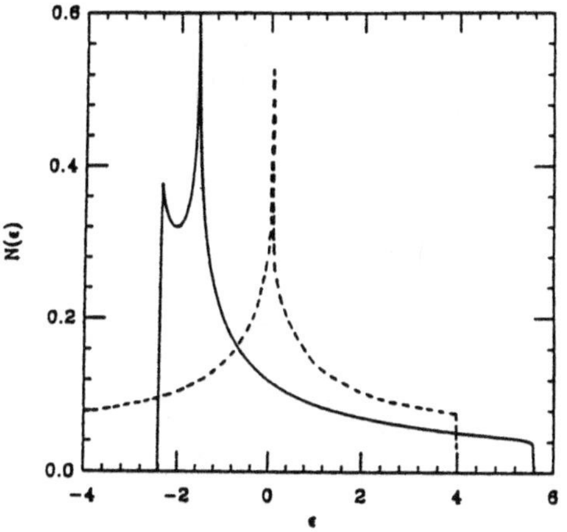

Fig. 7. Single electron density of states $N(\epsilon)$ versus ϵ for $t = 1$, $t_2 = 0.0$ (dashed) and $t = 1$, $t_2 = -0.4$ (solid). For $\epsilon_F = -1.6$, $\rho = 0.58$ and the fermi energy is at the Van Hove singularity for $t = 1$ and $t_2 = -0.4$.

$t_2 = -0.4$ and well away from it for $t_2 = 0.0$. Here we are interested in exploring the temperature dependence of the pair field susceptibility

$$P = \int_0^\beta d\tau \langle \Delta(\tau) \Delta^+(0) \rangle \tag{7}$$

with $\Delta^+ = \sum_l C_{l\uparrow}^+ C_{l\downarrow}^+$, for these two cases.

In order to construct a simulation, $e^{-\beta H}$ is again written as a product of L factors $e^{-\Delta\tau H}$ with $\beta = L\Delta\tau$. Then using the small parameters $\Delta\tau t$ and $\Delta\tau|U|$, the interaction term is separated from the kinetic term, and a discrete Hubbard-Stratonovich transformation [13] used to write

$$e^{\Delta\tau |U| n_\uparrow n_\downarrow} = \frac{1}{2} \sum_{\sigma=\pm 1} e^{\Delta\tau[\frac{1}{2}\,|U| + 2a\sigma](n_\uparrow + n_\downarrow - 1)} \tag{8}$$

with $\tanh^2(\Delta\tau a) = \tanh(\Delta\tau|U|/4)$. Carrying this out at each time slice $\tau_l (l = 1$ to $L)$ and at each site leads to a partition function

$$Z = Tr_\sigma (\det 0_\uparrow \det 0_\downarrow) \tag{9}$$

with $\det 0_\uparrow = \det 0_\downarrow$ and

$$\begin{aligned} \det 0_\uparrow &= Tr \Pi_l e^{\Delta\tau \sum_{ij} (C_{i\uparrow}^+ t_{ij} C_{j\uparrow})} e^{+\Delta\tau \sum_i (\mu + \frac{|U|}{2} + 2a\sigma_i(\tau_l)) n_{i\uparrow}} \\ &= \det(1 + B_L \cdots B_1) \end{aligned} \tag{10}$$

Here, after tracing over the fermion degrees of freedom

$$B_l = e^{\Delta\tau t} e^{\Delta\tau V(\tau_l)} \tag{11}$$

with t and $V(\tau_l)$ matrices on the spatial lattice $t = (t_{ij})$ and $V(\tau_l) = \delta_{ij}(\mu + \frac{|U|}{2} +$

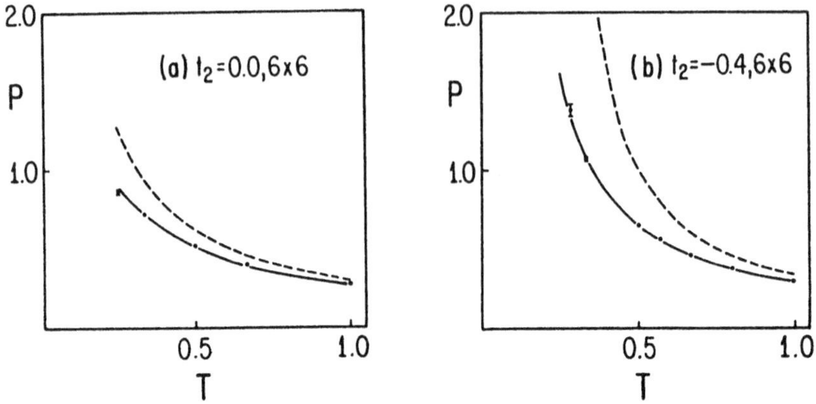

Fig. 8. Singlet pair susceptibility P, Eq. (7), versus T on a 6×6 lattice with $U = -2.0$. (a) $t_2 = 0, \rho = -0.5$; (b) $t_2 = -0.4, \rho = 0.58$. The dashed line is the RPA prediction while the full line includes terms shown in Fig. 9. The points were obtained from the Monte Carlo simulation.

$2a\sigma_i(\tau_l))$. Note that since $\det \mathrm{O}_\uparrow = \det \mathrm{O}_\downarrow$, the factor $\det \mathrm{O}_\uparrow \det \mathrm{O}_\downarrow$ provides a positive weight for a given $\{\sigma_i(\tau_l)\}$ configuration and can serve as a probability for the Monte Carlo calculation.

In the Monte Carlo procedure, a given configuration is changed by flipping a spin $\sigma_i(\tau_l) \to -\sigma_i(\tau_l)$. Using a heat-bath algorithm, the new configuration is accepted with a probability $P = \det \mathrm{O}_\uparrow \det \mathrm{O}_\downarrow / (1 + \det \mathrm{O}_\uparrow \det \mathrm{O}_\downarrow)$. Since a local change has been made, one only needs the local Green's function $(1 + B_L B_{L-1} \cdots B_1)^{-1}_{ii}$ to determine P. However, when a change is accepted, the problem of changing the Green's function arises. One procedure which we have used involves the exact updating [12] of G which requires N^2 operations per update with N the number of spatial sites. However, as previously noted, this type of algorithm requires LN^3 operations for one sweep of the space-time lattice. In addition, round off errors lead to low temperature instabilities, limiting the size of β. Nevertheless, this algorithm is the one which has been most extensively used for these types of models.

With this procedure, the singlet pair susceptibility P, Eq. (7), was measured as a function of temperature on 6×6 lattices and results for $U = -2$ are shown in Figs. 8a and 8b. One clearly sees that in this case the pairing is enhanced when the Van Hove singularity occurs near the Fermi energy. The dashed curves in Fig. 8 represent the RPA ladder approximation for P, while the solid curves represent the result of a resummation of the higher order contributions shown in Fig. 9 to both the self-energy of the propagators and the interaction. The agreement of the Monte Carlo simulations with this higher order approximation encouraged us to use this approximation to predict T_c. The results for T_c obtained in this manner are shown in Fig. 10.

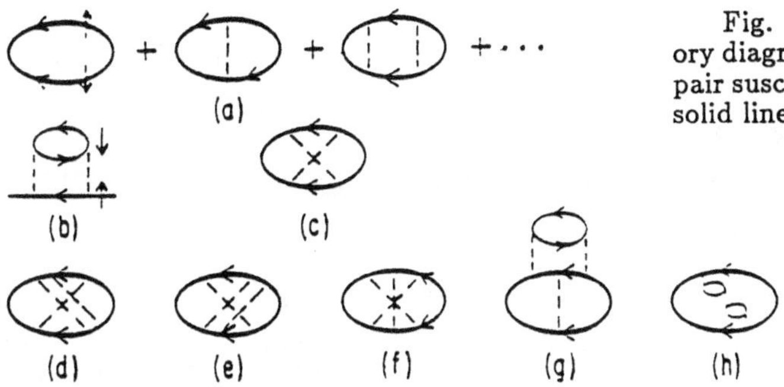

Fig. 9. Perturbation theory diagrams associated with the pair susceptibility shown as the solid line in Fig. 8.

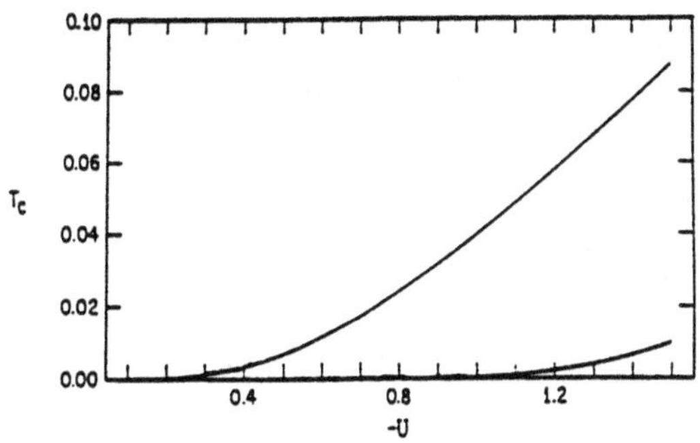

Fig. 10. T_c versus U in units where $t = 1$. The upper curve corresponds to the case in which the Van Hove singularity is at the Fermi energy ($t = 1, t_2 = -0.4, \rho = 0.58$) and the lower curve to a case ($t = 1, t_2 = 0.0, \rho = 0.5$) where it is away from the Fermi energy.

Clearly, one would like to push the simulations into the low temperature regime shown in Fig. 10. However, since the finite energy level spacing on a lattice restricts β to vary as some fraction of the linear lattice dimension, one would need lattices of order 100×100 with $L = 100$ to adequately scale into the low temperature T_c regime. For such lattices, new algorithms which scale as the space-time volume are needed as well as methods which allow larger values of β to be achieved. It seems likely that the role of fermion quantum Monte Carlo will remain that of a bridge between analytic weak and strong-coupling results. Future algorithms and faster machines should allow us to reach lattices of sufficient size to make this bridge meaningful for a variety of condensed matter models of current interest such as the periodic Anderson model and the degenerate Hubbard model.

Acknowledgment

The work reviewed here has been supported in part by the National Science Foundation under Grant DMR83-20481 and the Department of Energy under Grant No. DE-FG03-85ER45197. An ST-100 Array Processor was purchased under NSF-DMR83-20423. We gratefully acknowledge E.I. du Pont de Nemours and Company and the Xerox Corporation for their support.

References

1. *Monte Carlo Methods in Statistical Physics*, edited by K. Binder (Springer Verlag, Berlin 1979).
2. M. Suzuki, *Prog. Theor. Phys.* **56**, 1454, (1976).
3. M. Suzuki, S. Miyashita and A. Kuroda, *Prog. Theor. Phys.* **58**, 1377, (1977).
4. J.E. Hirsch, D.J. Scalapino, R.L. Sugar and R. Blankenbecler, *Phys. Rev. Lett.* **47**, 1628, (1981).
5. J.J. Cullen and D.P. Landau, *Phys. Rev. B* **27**, 297, (1983).
6. H. de Raedt, B. de Raedt, J. Fivez and A. Lagendijk, *Phys. Lett.* **104A**, 430, (1984); H. de Raedt, B. de Raedt and A. Lagendijk, *Z. Phys.* **B57**, 109, (1984); H. de Raedt and A. Lagendijk, *Phys. Rev. B* **33**, 5102, (1986).
7. E.Loh, D.J. Scalapino and P.M. Grant, *Phys. Rev. B* **31**, 4712, (1985); *Physica Scripta* **32**, 327, (1985); *Phys. Rev. B* **33**, 5104, (1986).
8. J.E. Hirsch and D.J. Scalapino, *Phys. Rev. Lett.* **56**, 2732, (1986).
9. M. Barma and B.S. Shastry, *Phys. Rev. B* **18**, 3351, (1978).
10. J.E. Van Himbergen and S. Chakravarty, *Phys. Rev. B* **23**, 359, (1981).
11. J.M. Kosterlitz and D.J. Thouless, *J. Phys. C* **6**, 1181, (1973).
12. R. Blankenbecler, D.J. Scalapino and R.L. Sugar, *Phys. Rev. D* **24**, 2278, (1981); D.J. Scalapino and R.L. Sugar, *Phys. Rev. B* **24**, 4295, (1981).
13. J.E. Hirsch, *Phys. Rev. B* **28**, 4059, (1983).

Quantum Spin Chains and Conformally Invariant Non-Linear σ-Models *

I. Affleck *

Joseph Henry Laboratories, Princeton University,
Princeton, NJ 08544, USA

ABSTRACT

The critical behavior of quantum antiferromagnetic chains is related to certain recently discovered conformally-invariant non-linear σ-models.

This paper concerns the critical behavior of isotropic antiferromagnetic quantum spin chains, with Hamiltonian

$$H = J\Sigma \mathbf{S}\cdot\mathbf{S}_{+1}\ ,\ \mathbf{S}^2 = s(s+1),\ J>0 \qquad (1)$$

This model has been the subject of much theoretical work over the last half-century beginning with the exact solution for $s = 1/2$ by Bethe in 1931. Experimental compounds have been found which are quite well described by this one-dimensional model down to low temperatures where three-dimensional effects become important.[1]

The Bethe-ansatz solution (for s = 1/2) as well as experimental data and linearized theory suggest that the low-energy excitations are spin-waves with dispersion relation

$$E = v\ |sink|$$

Thus there are two linear branches near $k = 0$ or π. These low-energy modes can be described by quantum field theory as shown by Luther and Peschel[2] for $s = 1/2$, in 1974. These authors gave an approximate mapping of the system onto a massless free boson (phonon) quantum field theory. Fixing one free parameter from the Bethe ansatz solution they were able to predict all critical exponents. It was argued relatively recently by Haldane[3] that this massless critical behavior holds for arbitrary half-integer s but that for integer s there is a gap in the spectrum (indeed for large integer s the low energy spectrum is that of a massive quantum field theory).

The Heisenberg Hamiltonian of Eq. (1) is not Bethe-ansatz integrable for $s > 1/2$. However, if this Hamiltonian is generalized to a degree $2s$ polynomial in $\mathbf{S}_i\cdot\mathbf{S}_{i+1}$ then an integrable Hamiltonian can be found for each s.[4] It is found to have a gapless, sinusoidal dispersion relation as for $s = 1/2$. The Bethe ansatz results obtained so far for $s > 1/2$ do not give the critical exponents. (For $s = 1/2$ the Bethe ansatz solution for the anisotropic xyz model gives the scaling dimension of one operator.) They do, however, determine the low-temperature (linear) specific heat and (finite) susceptibility.

⋆ Supported by NSF Grant PHY80-19754 and by the A.P.Sloan Foundation.

There has recently been considerable progress in the study of scale-invariant, gapless quantum field theories in($1+1$) space-time dimensions.[5] This has been motivated in part by developments in string theory. In particular an infinite discrete series of such theories with $SU(2)$ symmetry has been found by Witten.[6] These theories are known as the Wess-Zumino-Witten non-linear σ-models (WZW models). All critical exponents were found exactly for these models.[7]

In this paper, I wish to summarize briefly some contact that has been made between these recent field theory developments and quantum spin chains. Since isotropic spin chains have an $SU(2)$ symmetry, one might hope that they were described at low energies by the WZW models. It will be argued that this is the case and operator relations will be derived. These will allow predictions of exact critical numbers which can be checked against Bethe ansatz results, finite chain numerical results and experiment.

Let me begin with the case $s = 1/2$. A convenient starting point is the representation of the spin chain in terms of the infinite-coupling Hubbard model.[8] Electrons with two spin states are introduced on a one-dimensional lattice with a nearest neighbour hopping term. An on-site coupling favors states with exactly one electron on each site. These states form a basis for the $s = 1/2$ representation of $SU(2)$.

$$H = \Sigma[-\sqrt{JU}(\psi_i^{+\alpha}\psi_{i+1\alpha} + h.c.) + U(\psi_i^{+\alpha}\psi_{i\alpha} - 1)] \tag{2}$$

The repeated spin-index, $\alpha = 1,2$ is summed over; "h. c." stands for hermitian conjugate. At $U \to \infty$ the spin $-1/2$ antiferromagnet is recovered, as can be seen, for example, by considering perturbation theory in the hopping term (only second-order perturbation theory survives). The spin variables are represented as

$$\mathbf{S}_i = (1/2)\psi_i^{+\alpha}\,\vec{\sigma}_\alpha^\beta\,\psi_{i\beta}\,.$$

In order to extract the low-energy sector it is convenient to imagine, at first, that U is small and then pass to the large-U limit later. This procedure is reasonable because we will discover a massless sector of the Hilbert space which decouples from the massive sector that has a gap which grows with U. At $U = 0$ the free fermions have a dispersion relation

$$E = \sqrt{JU} cosk\,.$$

Thus the ground-state has all levels filled with $|k| < \pi/2$, a half-filled band. Low energy excitations have electrons raised from just below to just above the Fermi surface. Thus an effective Hamiltonian for the low-energy sector keeps only Fourier modes of ψ_i close to $\pm\pi/2$, which we refer to as ψ_L and ψ_R respectively. ψ_L annihilates left-moving electrons and creates left-moving holes.

$$a^{-1/2}\psi_n \approx i^n\psi_L + (-i)^n\psi_R\,. \tag{3}$$

Here we have introduced the lattice spacing, "a" to give $\psi_{L,R}$ the conventional field theory dimensions of (length)$^{-1/2}$. The continuum limit of the spin operators then contains both uniform and alternating parts:

$$2a^1\mathbf{S}_n \approx (\psi_L^+\vec{\sigma}\psi_L + \psi_R^+\vec{\sigma}\psi_R) + (-1)^n(\psi_L^+\vec{\sigma}\psi_R + h.c.)$$

(spin indices have been suppressed). Linearizing the dispersion relation near the Fermi surface we find

$$H = iv\int dx(\psi_L^{+\alpha}(d/dx)\psi_{L\alpha} - \psi_R^{+}\alpha(d/dx)\psi_{R\alpha})$$

This is the Hamiltonian for a free massless Dirac fermion, with Lagrangian density

$$L = i(\psi_L^{+\alpha}\partial_-\psi_L\alpha + \psi_R^{+\alpha}\partial_+\psi_{R\alpha})$$

where $\partial_\pm = (\partial_0 \pm \partial_1)$ are light-cone derivatives and we have set $v = 1$. This Lagrangian density is invariant under Lorentz transformations $x_\pm \to x_\pm e^{\pm\phi}$, $\psi_{L,R} \to \psi_{L,R}e^{\pm\phi/2}$. In order to understand the effect of the interaction term it is convenient to bosonize. Ordinary, abelian bosonization has been applied to models of this type by many authors. However, in order to make the isotropy manifest I prefer to use non-abelian bosonization.[6] This scheme is based on the observation that the free theory has both chiral $U(1)$ and chiral $U(2)$ symmetries. The left and right chiral groups act on the left- and right-moving fermions only. The $U(1)$ currents are represented in terms of a free boson:

$$J_L \equiv :\psi_L^{+\alpha}\psi_{L\alpha}: = \partial_-\varphi/\sqrt{4\pi}\,, \quad J_R \equiv :\psi_R^{+\alpha}\psi_{R\alpha}: = -\partial_+\varphi/\sqrt{4\pi}\,.$$

Here : : represents normal-ordering (subtraction of the vacuum expectation value). Note that by the equations of motion $\partial_\pm J_{L,R} = 0$. On the other hand, the $SU(2)$ currents are represented by an $SU(2)$ matrix field, g_β^α

$$\begin{aligned} 2\mathbf{J}_L &\equiv \psi_L^+\sigma\psi_L = trg^+\partial_- g\vec{\sigma} \\ 2\mathbf{J}_R &\equiv \psi_R^+\sigma\psi_R = tr\partial_+ g^+ g\vec{\sigma} \end{aligned}$$

The field g has a non-linear σ-model action

$$S = \lambda^{-1}\int d^2x tr\partial_\mu g\partial^\mu g^+ + k\Gamma_{WZ}$$

Here the Wess-Zumino term Γ_{WZ} has a topological definition and is only uniquely defined mod 2π. Thus the corresponding coupling constant, k must be an integer in order for the path integral weight e^{-iS} to be well-defined. This model is *not* a free theory. It was argued[6] that it has a conformally-invariant fixed point at $\lambda = 4\pi/k$. The bosonization formulas hold at this fixed point for $k = 1$. The free fermion Hamiltonian can be written in a form quadratic in these currents. Combining the Hamiltonian density $\mathcal{H}$ and momentum density $\mathcal{P}$ to form the light-cone components of the energy-momentum tensor, $T_{L,R} \equiv \mathcal{H} \pm \mathcal{P}$ we find

$$(4\pi)^{-1}T_L = J_L J_L + \mathbf{J}_L \cdot \mathbf{J}_L/3$$

and similarly for T_R.

We now turn to a discussion of the interaction term in the Hamiltonian of Eq.(2). Passing to the continuum representation of Eq. (3) we obtain a variety of four-fermion terms. Those that involve four left fermions or four right fermions can be written in current-current form. Thus they simply renormalize the velocity of "light" (differently for φ and g) and will not be considered further. Those that contain two left and two right fermions correspond to marginal Lorentz-invariant interaction terms in the Lagrangian. (In addition there are higher derivative irrelevant terms.) There are three such terms permitted by the diagonal $SU(2)$ and $U(1)$ symmetries (which simultaneously rotate left and right fermions):

$$L_{int} = \lambda_1 J_L J_R + \lambda_2 \mathbf{J}_L \cdot \mathbf{J}_R + \lambda_3(\epsilon_{\alpha\beta}\psi_L^{+\alpha}\psi_L^{+}{}^{\beta}\psi_{R\gamma}\psi_{R\delta}\epsilon^{\gamma\delta} + h.c.)$$

The first two terms can be bosonized as discussed above. The third term can also be bosonized using the formula

$$\psi_L^{+\alpha}\psi_{R\beta} \propto g_\beta^\alpha \, exp\,(i\sqrt{2\pi\varphi})\,. \tag{4}$$

The third interaction term is a singlet under both chiral $SU(2)$ groups. Correspondingly, its bosonized form involves only the determinant of g which is one giving

$$\lambda_3 \, cos\, \sqrt{8\pi}\varphi\,.$$

Thus bosonization has completely separated the theory into a sine-Gordon model, with Lagrangian density:

$$L = (1/2)\partial_\mu\varphi\partial^\mu\varphi(1-\lambda_1) + \lambda_2 \, cos\, \sqrt{8\pi\varphi}$$

and a WZW model with a current-current interaction. The behavior of the sine-Gordon model is well-known. There is a gap in the spectrum of order $e^{-1/\lambda} \approx exp(-\sqrt{U/J})$. The renormalization of the coupling λ_3 in the WZW model can easily be calculated,

$$d\lambda_3/d(lnL) = -(\lambda_3)^2\,. \tag{5}$$

Thus the interaction is marginally irrelevant. The effective Lagrangian at low energy scales is just the WZW model at its critical point. The representation of the spin operators in the low energy, long-wavelength regime becomes (after replacing $exp(i\sqrt{2\pi\varphi})$ by its expectation value)

$$a^{-1}\mathbf{S}_n \approx (\mathbf{J}_L + \mathbf{J}_R) + constant \cdot (-1)^n tr g\vec{\sigma}\,.$$

The (zero-temperature) correlation exponents exponents are then determined by the scaling dimensions of J and g.

$$< \mathbf{S}_n \cdot \mathbf{S}_0 > \approx \; n^{-2} + constant \cdot (-1)^n |n|^{-1}$$

The coefficient in front of the term above also determines the zero-temperature susceptibility, $\chi = 1/2\pi v$ (reinstating v but setting $\hbar$ and the Bohr magneton to one). This is given by the normalization of the two-point function of the current, $\mathbf{J}$. In general it is proportional to the coupling constant, k, in front of the Wess-Zumino term. Another critical parameter is the slope of the specific heat at low temperature, T. Conformal field theories have linear specific heats (this follows from scale invariance) with slope[9] $(\pi/3)c$, where the dimensionless parameter, c, measures the density of states. It also measures the two-point function of T_L (and T_R) and consequently the anomalous response of the theory to curving the underlying space. It is known as the conformal anomaly parameter. It has the value $c = 1$ for the $k = 1$ WZW model. This is the value for a free boson, as established for the $s = 1/2$ chain by Luther and Peschel. (The other critical numbers are also in agreement.) These field theory predictions for the susceptibility and specific heat are in exact agreement with the Bethe ansatz result for $s = 1/2$. They also agree well with experimental data[10] on $CuCl_2 \cdot 2(N_5H_5)$.

I would now like to touch briefly on the situation for higher spin. A generalized Hubbard model representation exists. Now the fermions carry an additional index i, called color, with $i = 1, 2, 3, \cdots n_c = 2s$. The various colors can be thought of as the different orbitals in an atomic shell of orbital angular momentum 1, with $21 + 1 = n_c$. We again have a half-filled band and introduce a very large Hund's rule coupling which projects out spin-s states.

$$U\Sigma\, (\psi_n^{+i\alpha}\, \psi_{nj\alpha} - \delta^i_j)\, (\psi_n^{+j\beta}\, \psi_{ni\beta} - \delta^j_i)$$

Low energy states have n_c fermions per site in a color singlet state which is totally anti-symmetric with respect to color and therefore (by Fermi statistics) totally symmetric with respect to spin, hence spin $n_c/2$. We may pass to the continuum limit by introducing left and right fermions as before. The free Dirac fermion theory now has conserved chiral charge $[U(1)]$, spin $[SU(2)]$ and color $[SU(n_c)]$ currents.

$$\begin{aligned} J_L &\equiv \, : \psi_L^{+\alpha i}\, \psi_{L_{\alpha i}} : \\ \mathbf{J}_L &\equiv (1/2)\psi_L^{+\alpha i}\, \vec{\sigma}^{\beta}_{\alpha} \psi_{L_{\beta i}} \\ J_L^A &\equiv \psi_L^{+\alpha i}\, \tau^A{}_i{}^j\, \psi_{L_{\alpha j}} \end{aligned}$$

where the τ^A are generators of $SU(n_c)$. The energy-momentum tensor can be written in a form quadratic in charge, spin and color currents. Thus it is tempting to conclude that the theory can be bosonized in terms of a charge boson, an $SU(2)$ WZW model and an $SU(n_c)$ WZW model. It can be shown that all multi-current Green's functions are correctly reproduced by this representation,[11] if the $SU(2)$ WZW model has topological coupling constant, $k = n_c$ and the $SU(n_c)$ WZW model has $k = 2$ (the number of *spin* degrees of freedom of the fermions). However, local bosonization formulas do not exist for left-right fermion bilinears as in Eq. (4). (It remains an open question whether some type of non-local representation exists, perhaps involving twisted boundary conditions.) Nevertheless, it is reasonable to expect that the critical theory for arbitrary spin will be an $SU(2)$ WZW models with some values of k. This can be seen by first arguing[12] that a conformally invariant critical theory that has a diagonal $SU(2)$ symmetry will have a chiral $SU(2)$ symmetry. Given the existence of conserved diagonal $SU(2)$ currents, $\mathbf{J}_\mu$, we may form the light-cone component, $\mathbf{J}_L \equiv \mathbf{J}_0 - \mathbf{J}_1$. Lorentz invariance dictates that the two-point function is of the form

$$< \mathbf{J}_L(x) \cdot \mathbf{J}_L(0) > \propto \; x_-^{-2} \cdot G(x_\mu x^\mu)$$

but the function G must in fact be a constant since J has the scaling dimension one. Thus the scalar field $\partial_+ J_L$ has vanishing two-point function. This implies that it is identically zero. Likewise for $\partial_- J_R$. Given the existence of these chiral currents $J_\pm$, we may ask what commutation relations they obey. The most general possibility consistent with conformal invariance, is the Kac-Moody algebra,

$$[J^a(x_-), J^b(0)] \;=\; i\epsilon^{abc} J_c \delta(x_-) \,+\, k\delta^{ab}\delta'(x_-)\,.$$

This algebra is known to have unitary representations only for the central charge, k, integer. The WZW model obeys this algebra with the central charge equal to the topological coupling constant. In fact the WZW model represents the "minimal" theory for a given value of the central charge. It has the smallest possible value of c and hence the smallest possible number of massless particles. Non-minimal theories can always be constructed by appending additional massless fields which commute with the currents [i.e. are chiral $SU(2)$ singlets]. Taking the usual point of view of critical phenomena, that all terms allowed by symmetry are present in the effective Lagrangian, we expect that assuming a chiral $SU(2)$ symmetry with a particular value of k, only the minimal set of massless particles will be present.

The generalized Hubbard model gives six different marginal interactions in the continuum limit. The first three are current-current interactions for charge, spin and color currents. The last three (Umklapp) terms are

$$\begin{aligned} \lambda_4 \left(\psi_L^+ \sigma\tau^A \psi_L\right) \cdot \left(\psi_R^+ \sigma\tau^A \psi_R\right) \,+\, & \lambda_5(\epsilon_{\alpha\beta}\psi_L^{+\alpha\{i}\,\psi_L^{\beta j\}}\psi_{R\gamma\{i}\psi_{R\delta j\}}\epsilon^{\gamma\delta} + h.c.) \\ +\, & \lambda_6(\psi_L^{+\{\alpha[i}\,\psi_L^{+\beta\}j]}\,\psi_{R\{\alpha[i}\,\psi_{R\beta\}j]} \,+\, hc) \end{aligned}$$

The notation { } and [ ] denotes symmetrization and anti-symmetrization respectively. While the fifth term is invariant under chiral $SU(2)$, the fourth and sixth terms are not. We may represent the various current-current interactions in terms of a charge boson and spin and color WZW fields as discussed above. It has been shown that the fifth term can be represented entirely in terms of the charge boson and color WZW field (as might be expected since it is a chiral $SU(2)$ invariant). On the other hand, there is no known representation for the fourth and sixth terms, which appear to mix all the fields together in some very complicated way. If we add additional interactions to our original spin Hamiltonian, such as the higher order couplings necessary to make the model integrable then we should expect the six coupling constants to change in a more or less arbitrary way. In particular we might expect to find special points where λ_4 and λ_6 vanish. In this case the charge and color current-current interactions and the first Umklapp term together produce a gap in the charge and color sector of the Hilbert space. However the only interaction in the spin WZW sector is the marginally irrelevant current-current interaction. Thus we conclude that the critical theory for these special Hamiltonians where λ_4 and λ_6 vanish should be the $k = 2s$ $SU(2)$ WZW model. The predicted susceptibility and specific heat are then $\chi = s/\pi v$ and $C = T\pi s/(1+s)v$. These appear to agree exactly with the Bethe ansatz results, suggesting that the integrable models are special multi-critical points where λ_4 and λ_6 vanish in the coninuum limit.

An ordinary biquadric Hamiltonian (or for that matter, any generic Hamiltonian) should have $\lambda_4, \lambda_6 \neq 0$. The chiral symmetry is broken by relevant operators with complicated couplings between the charge, color and spin sectors. After integrating out the color and charge fields, we should expect to generate all relevant operators permitted by symmetry in the spin sector. There are no such operators for $k = 1$, and an increasing number for larger k. (The number is the largest integer s such that $s(s+1) < 2+k$ and $s \leq k/2$.) Thus the larger k WZW models represent increasingly multicritical points. Only the $k = 1$ fixed point is stable (it is only destabilized by the marginal operator if the coupling constant λ_3 in Eq. (5) is negative).

This analysis leaves it unclear whether the system develops a gap or else flows to a different conformally invariant fixed point. (In the latter case it should be another WZW model by the argument given above.) It can be argued[13] that for integer spin the system develops a gap in general, whereas for half-integer spin the generic behavior within some finite range of parameters is flow to the $k = 1$ fixed point. As s is increased the critical behavior is expected to set in at lower and lower T. Experimental data on $CsVCl_3$ and $(CH_3)_4NM_nC\ell_3$ with $s = 3/2$ and $5/2$ do not seem to probe the critical region (the limitations are the anisotropy and the interchain couplings which lead to three-dimensional ordering below some temperature).

REFERENCES

1. For a review see M. Steiner, J. Villain and C.G. Windsor, *Adv. Phys.* **25** (1976), 87.

2. A. Luther and I. Peschel, *Phys. Rev.* **B9** (1974), 2911.

3. F.D.M. Haldane, *Phys. Lett.* **93A** (1983), 464;
   Phys. Rev. Lett. **50** (1983) 1153;
   J. Appl. Phys. **57** (1985) 3359.
   See also I. Affleck, *Nucl. Phys.* **B257** (1985), 397;
   Phys. Rev. Lett. **54** (1985) 966;
   ibid **56** (1986) 408;
   I. Affleck and E.H. Lieb, *Lett. Math. Phys.* **12** (1986), 57.

4. P. Kulish and E. Sklyanin, *Lecture Notes in Physics* **151** (1982), 61;
   P. Kulish, N. Yu. Reshetikhin and E. Sklyanin, *Lett. Math. Phys.* **5** (1981), 393;

L. Takhtajan, *Phys. Lett.* **90A** (1982), 479;
J. Babudjian, *Phys. Lett.* **90A** (1982), 479;
Nucl. Phys. **B215** (1983) 317.

5. For a review see J.L. Cardy, in *Phase Transitions and Critical Phenomena*, vol. 11, ed. C. Domb and J.L. Lebowitz.

6. E. Witten, *Comm. Math. Phys.* **92** (1984), 455.

7. V. Knizhnick and A. Zamalodchikov, *Nucl. Phys.* **B247** (1984), 83.
See also D. Gepner and E. Witten, Princeton preprint (1986).

8. P.W. Anderson, in *Solid State Physics*, ed. F. Seitz and D. Turnbull (Academic Press, N. Y. 1963) Vol. 14, 99.
For an application of this mapping in the one-dimensional case using abelian bosonization see F.D.M. Haldane, in *Electron Correlation and Magnetism in Narrow Band Systems*, ed. T. Moriya, 150 (1980).

9. H. Blote, J.L. Cardy and M. Nightingale, *Phys. Rev. Lett* **56** (1986), 742;
I. Affleck, *ibid* (1986), 746.

10. K. Takeda, S. Matsukawa and T. Haseda, *J. Phys. Soc. Jap.* **30** (1971), 1330.

11. I. Affleck, *Nucl. Phys.* **B265** (1986), 409.

12. R. Dashen and Y. Frishman, *Phys. Rev.* **D11** (1975), 2781.

13. To appear.

Collective Coordinates and Linear Modes of the Double-Sine-Gordon Kink

C.R. Willis [1], *M. El-Batanouny* [1], *and P. Sodano* [2,*]

[1]Department of Physics, Boston University, Boston, MA 02215, USA
[2]Center for Theoretical Physics, Laboratory for Nuclear Science and Department of Physics, Massachusetts Institute of Technology, Cambridge, MA 02139, USA

Abstract

We present a complete Hamiltonian treatment of a kink with an internal degree of freedom, namely the double-sine-Gordon (DSG) kink. In this formalism we assign two canonical coordinates and their associated momenta to describe the motion of the center of mass of the DSG kink and the relative motion of its two sub-kinks. We show that the canonical coordinate representing the separation of the two sub-kinks describes a nonlinear oscillatory degree of freedom. We have also used the method of supersymmetry to obtain, for the first time,the complete set of eigenfunctions of small oscillations about the 4π DSG kink. This analysis was motivated by the need for an accurate expression of the shape mode in the Hamiltonian formalism.

Over the past years two-dimensional models have played a crucial role in field theory and statistical mechanics. Recently, in the realm of condensed matter physics, two dimensions became an experimental world in its own right. Systems such as ferromagnets, polymer chains, organic superconductors, reconstructed surfaces, etc. have been rather successfully modeled by two-dimensional field theory models. Many of these theories support kink-like excitations, whose properties have been object of recent investigations.

A remarkable theory is the double sine-Gordon model (DSG). This is a widely studied [1] classically non-integrable model of a single scalar field in two dimensions whose dynamics are described in the continuum limit by

$$(4\pi)^2 L = (2\pi)^2 \int dx\mathcal{L} = \int dx(\frac{1}{2}\dot{\phi}^2 - \frac{1}{2}(\frac{\partial\phi}{\partial x})^2 - (\frac{2\pi}{l_0})^2 V(\phi)) \tag{1.a}$$

with

$$-V(\phi) \equiv 4(\cosh R)^{-2}\left[\frac{1}{4}\sinh^2 R(\cos\phi - 1) - \left(1 + \cos\frac{\phi}{2}\right)\right] \tag{1.b}$$

Here, $l_0^2 = \frac{\mu a^2}{2W}$ is a dimensionless coupling constant. The parameter μ is the spring constant, W is the depth and $2\pi a$ is the period of the potential.

The theory admits [2]as a solution a static topological kink which can be expressed, for all values of R, as a superposition of two sine-Gordon solitons separated by a distance 2R. Due to this structure the linearized spectrum about the DSG kink admits [2,3] a bound state whose frequency depends upon the parameter R. The presence of this shape mode is responsible for many non-trivial properties [4] of the DSG kinks; this is reminiscent of the situation arising in the study of polaron[5] solutions in polymer models.

Here we report on the results of our recent analysis [6] of the complete Hamiltonian dynamics for the DSG kink. Our approach follows Tomboulis' work [7] and promotes both the translational mode and internal mode to dynamical variables. The field ϕ is expressed as

$$\phi(x,t) \equiv \sigma(x, X(t), R(t)) + \chi(x,t) \tag{2.a}$$

---

* Permanent Address: Dipartimento di Fisica, Universitá di Perugia, Via Elce di Sotto, 10. Perugia, 06100, ITALY.

where

$$\sigma(x,X,R)=\sigma_{SG}[\frac{2\pi}{l_o}(x-X)+R]-\sigma_{SG}[R-\frac{2\pi}{l_0}(x-X)] \tag{2.b}$$

and

$$\sigma_{SG}(x)=4\tan^{-1}\exp(x) \quad . \tag{3}$$

Next, we introduce the center of mass variable X(t) and the relative coordinate R(t) (where 2R(t) is the separation of the two solitons making up the DSG equation) as canonical coordinates. for ease of presentation, we will carry out the derivation of the canonical equations in the center of mass system so that we can $X(t)=0$, for all t, and concentrate on the variable $R(t)$. We use the method of Dirac brackets developed in Ref.(8) to introduce $R(t)$ and its conjugate momentum P_R as canonical variables. Since $R(t)$ and P_R increase the number of canonical variables by two, it becomes necessary to introduce two constraints on the variables $\chi(x,t)$ and its conjugate momenta $\pi(x,t)$. These constraints are:

$$C_1\equiv\int dx\frac{\partial\sigma}{\partial R}\chi(x,t)=0;\quad C_2\equiv\int dx\frac{\partial\sigma}{\partial R}\pi(x,t)=0 \quad . \tag{4}$$

The constraints mean that the motion of R(t) is such that the "shape mode" $\frac{\partial\sigma}{\partial R}$ must be orthogonal to $\chi(x,t)$ and $\pi(x,t)$. With the imposition of the constraints C_1 and C_2 the variables R, P_R, χ, and π form a complete canonical set of equations. The effect of the constraints is to modify the Poisson brackets involving χ and π to Dirac brackets. Here we present only the final equations of motion for χ and R which are:

$$\ddot{\chi}=\frac{\partial^2\chi}{\partial x^2}+\frac{\partial^2\sigma}{\partial x^2}-\frac{\partial V_{\mathcal{R}}(\sigma+\chi)}{\partial\chi}$$

$$-\frac{1}{M_R}\frac{\partial\sigma(x)}{\partial R}\int dx'\frac{\partial\sigma(x')}{\partial R}\left(\frac{\partial^2\chi}{\partial x'^2}+\frac{\partial^2\sigma}{\partial x'^2}-\frac{\partial V_{\mathcal{R}}(x')}{\partial\chi(x')}\right) \tag{5.a}$$

$$\ddot{R}=-\frac{1}{2}\dot{R}^2\frac{\partial \ln M_R}{\partial R}+\frac{1}{M_R}\int dx'\frac{\partial\sigma(x')}{\partial R}\left(\frac{\partial^2\chi}{\partial x'^2}+\frac{\partial^2\sigma}{\partial x'^2}-\frac{\partial V_{\mathcal{R}}(x')}{\partial\chi(x')}\right) \tag{5.b}$$

and

$$M_R\equiv\int dx\left(\frac{\partial\sigma}{\partial R}\right)^2=\frac{8l_0}{\pi}\left[1-\frac{1}{\sinh 2R}\ln\left(\frac{1+\cosh 2R+\sinh 2R}{1+\cosh 2R-\sinh 2R}\right)\right] \quad . \tag{6}$$

The notation $\mathcal{R}$ indicates the fixed parameter value of R in the potential $V(\phi)$, Eq. (1,b), and the symbol R(t) denotes the canonical variable R(t). Equations (5.a) and (5.b) are the complete, coupled nonlinear canonical equations of motion for $R(t)$ and $\chi(x,t)$.

When we linearize the equation of motion, (5.a), for χ we obtain

$$\ddot{\chi}=(1-\mathcal{P}_R)\Lambda\chi+(1-\mathcal{P}_R)S \tag{7}$$

where the projection operator P_R is

$$\mathcal{P}_R O(x)\equiv\frac{\partial\sigma(x)}{\partial R}\frac{1}{M_R}\int dx'\frac{\partial\sigma(x')}{\partial R}O(x') \tag{8}$$

the operator S is

$$S\equiv V'_R[\sigma(x,R)]-V'_{\mathcal{R}}[\sigma(x,R)] \tag{9}$$

and the prime denotes differentiation with respect to σ. The linear operator Λ is defined as

$$\Lambda\equiv\frac{\partial^2}{\partial x^2}-V''_{\mathcal{R}}[\sigma(x,R)]\equiv-\Lambda_{SCH}+\delta\Lambda(t) \tag{10.a}$$

where

$$\Lambda_{SCH}=\frac{-\partial^2}{\partial x^2}+V''_{\mathcal{R}}\Big[\sigma(\mathcal{R})\Big] \tag{10.b}$$

and

$$\delta\Lambda(t)=V''_{\mathcal{R}}\Big[\sigma(\mathcal{R})\Big]-V''_{\mathcal{R}}\Big[\sigma(R(t))\Big] \tag{10.c}$$

The time dependence in $\delta\Lambda(t)$ results from the fact that the DSG kink moves in time as a result of the internal motion $R(t)$. To find the linear modes around the DSG kink, we make use [3] of the supersymmetry which naturally arises in any one-dimensional Schrödinger equation for which the ground state wave function is exactly known. The spectrum of Λ_{SCH} consists of the two discrete eigenvalues corresponding to the Goldstone mode in addition to the continuum which starts at $\omega = (\frac{2\pi}{l_o})$. The shape mode is

$$\psi_b(x) = \left[s\alpha tanh\left(\frac{2\pi}{l_o}\alpha x\right) + tanh\left(\frac{2\pi}{l_o}x + R\right) + tanh\left(\frac{2\pi}{l_o}x - R\right) - tanh\left(\frac{2\pi}{l_o}x\right)\right]$$

$$\times \left[cosh\left(\frac{2\pi}{l_o}\alpha x\right)\right]^{-s} \qquad (11)$$

where

$$\alpha \equiv tanh^2 R\frac{sinh2R}{sinh2R - 2R}; \qquad s \equiv \frac{1}{2}\left[-1 + (1 + \alpha^{-2}8tanh^2 R)^{\frac{1}{2}}\right]$$

and the eigenfrequency

$$\omega_b = \frac{2\pi}{l_0}\left(1 - s^2\right)^{\frac{1}{2}} \ . \qquad (12)$$

The explicit form of the continuum eigenfunctions,as well as the discussion of the R dependence of the completeness condition,are given in ref. (3). Here, we stress the fact that for $R \geq 2$ $\psi_b = \frac{\partial\sigma}{\partial R}$. Since in this range of R we have

$$\int[(\frac{\partial\sigma}{\partial x})^2 + (\frac{\partial\sigma}{\partial R})^2]dx = 16(\frac{2\pi}{l_0})^2 \ . \qquad (13)$$

There is no R-dependence in the normalization of the continuum eigenstates. This situation is drastically changed when $R \leq 2$ since in this case ψ_b is given by (11) and the sum of the normalization constants of the discrete states is R-dependent. Thus, the result of the linear analysis suggest that for $R \leq 2$ the phonon modes are vigorously influenced by the presence of the kink. This expectation is confirmed and it contents are enriched by the non-linear analysis of ref. (6) where the effects of the field χ on the shape mode have been discussed in some detail.

In the absence of external radiation or spontaneous emission we can set $\chi = 0$ in Eq. (2.a) and the nonlinear solution ϕ is

$$\phi(x, R(t)) = \sigma(x, R(t))$$

where for convenience we use the center of mass system in order to set $X = 0$. Consequently, to find the nonlinear solution of the original field equation all we need is to determine the solution for $R(t)$ from the ordinary differential equation

$$\ddot{R} + \frac{1}{2}\dot{R}^2\frac{\partial lnM_R}{\partial R} = \frac{1}{M_R}\int dx'\frac{\partial\sigma(x')}{\partial R}S(x') \equiv -\frac{1}{M_R}\frac{\partial v_{\mathcal{R}}}{\partial R} \qquad (14)$$

where

$$v_{\mathcal{R}}(R) \equiv \int dx\left[\frac{1}{2}\left(\frac{\partial\sigma(x,R)}{\partial x}\right)^2 + \left(\frac{2\pi}{l_o}\right)^2 V_{\mathcal{R}}\left[\sigma(x,R)\right]\right]$$

$$= 8\left(\frac{2\pi}{l_0}\right)\left[1 + \frac{tanh^2\mathcal{R}}{tanh^2 R} + 2R\left(\frac{1}{sinh2R} + \frac{cothR}{cosh^2\mathcal{R}} - \frac{tanh^2\mathcal{R}coshR}{2sinh^2 R}\right)\right] \ . \qquad (15)$$

The potential appearing in Eq. (15) is asymmetric about $R = \mathcal{R}$, and has a single minimum at that point. For $R < \mathcal{R}$ the $v_{\mathcal{R}}(R)$ is repulsive, resisting compression of the sub-kinks; while for $R > \mathcal{R}$ it is attractive with almost linear dependence on R which can be roughly viewed as a contribution from the underlying potential $\sim (V(2\pi) - V(4\pi)) \times R$. Increasing the parameter $\mathcal{R}$ leads to a decrease in the magnitude of potential without change in its shape. $v_{\mathcal{R}}(R)$ roughly resembles the potential of a diatomic molecule. For very small deviations of R from $\mathcal{R}$ we have harmonic oscillations about the potential minimum with a frequency given by

$$\omega_0^2 = M_{\mathcal{R}}^{-1}\frac{\partial^2}{\partial R^2}v_{\mathcal{R}}(R)|_{R=\mathcal{R}} = \left(\frac{2\pi}{l_0}\right)M_{\mathcal{R}}^{-1}16\left[\frac{1}{sinh^2\mathcal{R}}\left(3 - \mathcal{R}coth\mathcal{R}\right)\right.$$

$$\left. + \frac{1}{cosh^2\mathcal{R}}(\mathcal{R}tanh\mathcal{R} - 1) - \frac{2\mathcal{R}coth\mathcal{R}}{cosh^2\mathcal{R}sinh^2\mathcal{R}}\right] \quad . \tag{16}$$

Concluding Remarks:

In this paper we developed for the first time a complete Hamiltonian dynamics for a kink with an internal degree of freedom. We introduced, in addition to the sine-Gordon field, two particle variables and their conjugate momenta as canonical variables. We solved the ensuing equations of motion in some limiting and physically interesting cases. Our treatment of the kink motion has been non-relativistic.

Our analysis can be regarded as a canonical and fully non-linear treatment of the motion of a wobbling [1] kink in a model which exhibits an exact topological soliton with an internal mode. This is relevant not only for the dynamical properties of kinks in the numerous physical systems modeled by the DSG equation but could reveal interesting applications in the study of polaron-like solutions [5] in polymer models.

An interesting extension of our analysis would be to include the effect of stochastic noise on the kink dynamics of this model. This will allow for a more detailed study of the thermodynamical properties of the DSG equation [3]. It will also provide the tools for the investigation of a stochastic activation mechanism of the DSG-kink shape mode. This is subject of ongoing research and a forthcoming publication.

Acknowledgements

This work is supported by the U.S. Department of Energy under contract # DE-FG02-85ER45222. One of the authors would like to thank NATO for financial support.

References

1. R. K. Bullough, P. J. Caudrey and H.M. Gibbs, in *Solitons*, R.K. Bullough and P.J. Caudry eds., (Springer, 1980).

2. D. Campbell, M. Peyrard, and P. Sodano, Physica **19D**, 165 (1986).

3. P. Sodano, M. El-Batanouny, and C. Willis, to appear in Phys. Rev. **B34** (Rap. Comm.), (1986).

4. P. Sodano, Proc. Conf.: *Quantum Field Theory, Present and Future*, June 1985, (North-Holand, Amsterdam 1986).

5. K. Fesser, A.R. Bishop, and D.K. Campbell, Phys. Rev. **B27**, 4804 (1983).

6. C. Willis, M. El-Batanouny, S. Burdick, R. Boesch, and P. Sodano, to be published in Phys. Rev. **B**.

7. E. Tomboulis, Phys. Rev. **D12**, 1678 (1975).

8. P.A.M. Dirac, *Lectures on Quantum Mechanics*, (Academic, New York, 1964).

A Collective Coordinate Method for Classical Dynamics of Nonlinear Klein-Gordon Kinks

R.J. Flesch[1], *S.E. Trullinger*[1], *and B. Horovitz*[2]

[1]Center for Nonlinear Studies, Los Alamos National Laboratory, Los Alamos, NM 87545, USA and
Department of Physics*, University of Southern California, Los Angeles, CA 90089, USA

[2]Department of Physics, Ben-Gurion University of the Negev, Beer-Sheva, Israel

Abstract A collective coordinate method is used to study the motion of a nonlinear Klein-Gordon (NKG) kink [1] in the presence of a weak, localized perturbation. An equation of motion is derived for the kink "center of mass" position which includes the effects of phonons. A perturbation expansion of these equations shows that through second order, no extended phonons are generated by the "collision" of the kink with a static perturbation. As a consequence, the kink recovers its initial velocity after passing through the perturbation region.

The study of kink dynamics in NKG models has been greatly facilitated by the introduction of a collective coordinate describing the "center of mass" (CM) motion of the kink [2-5]. As the kink is a coherent, extended object, it seems natural to assign a coordinate which describes the motion of its center. Separating out this degree of freedom also removes secular terms caused by the zero-frequency Goldstone mode (translation mode) [6].

Although many times introduced purely as an ansatz, a canonical transformation which utilizes a collective coordinate has been discovered for the NKG class of field theories [7]. This canonical structure allows us to easily derive the equations of motion and leads to a well-defined quantization procedure [2,3,7,8]. We extend this canonical formalism to include the effects of a spatially localized perturbation $\lambda v(x,t)$ which couples linearly to the field $\Phi(x,t)$. With λ as a small parameter measuring the strength of the perturbation, we consider Lagrangians of the form:

$$L=\int_{-\infty}^{\infty} dx \left\{ \frac{1}{2}\partial_\mu \Phi\, \partial^\mu \Phi - U(\Phi) + \lambda\, v(x,t)\, \Phi \right\} \quad . \tag{1}$$

The equation of motion for the Lagrangian in (1) is:

$$\Phi_{tt} - \Phi_{xx} + U'(\Phi) - \lambda\, v(x,t) = 0 \quad . \tag{2}$$

The unperturbed equation (λ=0) is assumed to have a static (classical) kink solution $\phi_c(x)$. Solutions to (2) are further studied via the canonical transformation:

$$\Phi(x,t) = \phi_c(x-X) + \psi(x-X,t) + \chi_0(x,t) \quad , \tag{3a}$$

$$\Pi_0(x,t) = \pi(x-X,t) - (M_0+\xi)^{-1} \left\{ p + \int_{-\infty}^{\infty} dx\, \pi(x,t)\, \psi'(x,t) \right\} \phi_c'(x-X) - \dot{\chi}_0(x,t) \quad , \tag{3b}$$

$$\xi = \int_{-\infty}^{\infty} dx\, \phi_c'(x)\, \psi'(x,t) \ , \quad M_0 = \int_{-\infty}^{\infty} dx\, \phi_c'(x)\, \phi_c'(x) \ , \quad \int_{-\infty}^{\infty} dx\, \phi_c'(x)\, \psi(x,t) = 0 \ , \quad \int_{-\infty}^{\infty} dx\, \phi_c'(x)\, \pi(x,t) = 0 \quad . \tag{3c}$$

*Permanent address

The first term on the right-hand side of (3a) represents a kink whose CM moves according to the dynamical variable X(t). The ψ field will account for the interaction of the kink with the perturbation. The major contribution to ψ will be localized about the kink center and will have an appreciable amplitude only when the kink is in the region of the perturbation. In addition, $\psi(x,t)$ must account for any extended phonons radiated in the "collision", however, as will be shown later, for static perturbations, such phonons first appear in the third order terms of a perturbation series for ψ in powers of the strength parameter λ. The final contribution to the field is the "background" response χ_0, that is, the response of the system in the absence of the kink. The equation which determines χ_0 is (2) with Φ replaced by χ_0. We approximate $\chi_0(x,t)$ by linearizing (2) to obtain

$$\chi_{0_{tt}} - \chi_{0_{xx}} + \chi_0 - \lambda v(x,t) = 0 \qquad . \tag{4}$$

We explicitly account for this background or vacuum response of the field because unless the perturbation is turned on adiabatically, this response will be present long before and after the kink interacts with the perturbation (a situation often realized in physical systems [6]). When the kink is far from the impurity, ψ is zero and the ansatz of a translating kink ϕ_c plus the background χ_0 is a solution to the perturbed equation of motion to lowest order in χ_0. As the kink approaches the perturbation, this ansatz breaks down and the ψ field begins to contribute.

The momentum, Π_0, conjugate to the field Φ, is expressed in terms of the new canonically conjugate pairs (X,p) and (ψ,π) in (3b). The transformation,

$$\{\Phi(x,t), \Pi_0(x,t)\} \longrightarrow \{X(t), p(t), \psi(x,t), \pi(x,t)\} \qquad , \tag{5}$$

does not conserve the number of degrees of freedom, hence the last two equations in (3c) are introduced as constraints. The first of these constraints has the interpretation that the ψ field may not account for any translation of the kink since ϕ_c' is the translation mode. Using the Dirac formalism for constrained systems [9], (3) may be shown to form a canonical transformation [10]. With a canonical transformation in hand, we may derive the equations of motion using the standard rules of Hamilton-Jacobi theory. Details of this will be presented elsewhere [10] and here we simply state the equation of motion for the CM collective coordinate X(t):

$$M_0\ddot{X} = \frac{1}{1+\frac{\xi}{M_0}} \Big\{ \int_{-\infty}^{\infty} dx\, \phi_c'(x-X)\, [\, U'[\Phi(x,t)] - U'[\chi_0(x,t)]\,] + (1+\dot{X}^2)\int_{-\infty}^{\infty} dx\, \psi'(x,t)\, \phi_c''(x)$$
$$- 2\dot{X}\int_{-\infty}^{\infty} dx\, \phi_c'(x)\, [\pi'(x,t) - \dot{\chi}_0'(x+X,t)\,] \Big\} \qquad . \tag{6}$$

The right-hand side of (6) has what appear to be dissipative terms. However, in a system such as ours in which there is no coupling to other degrees of freedom such as a heat bath, these "dissipative" terms can only represent a transfer of energy between the degrees of freedom. In our case, the energy transfer is from the kink CM motion to the "phonon field" $\psi(x,t)$. Below we carry out a perturbation expansion in which we show that to lowest order, no energy transfer occurs, and in second order, the energy given to the phonon field during the collision is ultimately given back to the kink's translational motion.

Carrying out this expansion, we have for the first-order equation of motion for the CM variable:

$$M_0\ddot{X} = -\frac{\partial V(X,t)}{\partial X} \quad , \qquad V(X,t) \equiv \int_{-\infty}^{\infty} dx\, \chi_0(x+X,t)\, [\phi_c''(x) - \phi_c(x)\,] \qquad . \tag{7}$$

Equation (7) simply states that to first order the kink behaves as a Newtonian particle of mass M_0 (see (3c)) moving in the effective potential V(X,t). Proceeding to second order we have,

$$M_0 \ddot{X} = - \left[1 - \frac{\xi}{M_0}\right] \frac{\partial V(X,t)}{\partial X} + \frac{1}{2}\int_{-\infty}^{\infty} dx\, U'''[\phi_c(x)]\, [\psi(x,t) + \chi_0[x+X,t]]^2$$

$$- \frac{1}{2}\int_{-\infty}^{\infty} dx\, U'''(0)\, \chi_0^2(x+X,t) - 2\dot{X}\int_{-\infty}^{\infty} dx\, \phi_c'(x)\, \dot{\psi}'(x,t) \quad . \tag{8}$$

From (8), we see that to obtain X(t) through second order, we need ψ to first order. The first order equation for ψ may be written as [10]:

$$\ddot{\psi}(x,t) - \psi''(x,t) + U''[\phi_c(x)]\,\psi(x,t) = \chi_0(x+X,t)\,\{1 - U''[\phi_c(x)]\}$$

$$- \frac{\phi_c'(x)}{M_0}\int_{-\infty}^{\infty} dx\, \phi_c'(x)\, \chi_0(x+X,t)\,\{1 - U''[\phi_c(x)]\} \quad . \tag{9}$$

Denoting the (inhomogeneous) terms on the right-hand side by I(x,t), we have the following integral expression for $\psi(x,t)$:

$$\psi(x,t) = \int_{-\infty}^{\infty} dx' \int_{-\infty}^{\infty} dt'\; G(x,x',t,t')\, I(x',t') \quad , \tag{10}$$

where G(x,x',t,t') is the appropriate Green function [10,11]. Analytic expressions for G(x,x',t,t') are available in terms of modified Lommel functions of two variables for the sine-Gordon, ϕ^4, and double-quadratic potentials [11].

Using the localized nature of χ_0 and the assumption that $U(\phi_c)$ is scaled so that $U''[\phi_c(x)] \to 1$ as $x \to \pm\infty$, one can show [10] that $\psi(x,t)$ is localized in both space and time. Since all second order terms in (8) are proportional to ψ, ψ^2, or spatial derivatives of ψ, all second order terms in (8) are localized in time, therefore as $t \to \pm\infty$, we have $M_0\ddot{X} \to -\partial V(X,t)/\partial X$. Furthermore, for static perturbations or perturbations which are turned on and off adiabatically we have $V(X,t) \to 0$ as $|X| \to \infty$. Thus we see that through second order, the kink behaves as a free particle as $t \to \pm\infty$ (recall that in zeroth order, $X(t)=X_0 + V_0 t$). Since energy is conserved, all energy is returned to the translational motion of the kink.

To illustrate the methods outlined, we present an example in which a sine-Gordon kink, initially traveling to the right, encounters the time-independent perturbation

$$v(x) = e^{-(x-x_0)^2} - e^{-(x+x_0)^2} \quad . \tag{11}$$

In our simulation, the following initial conditions and parameters were used:

$$X(t=0) \equiv X_0 = -20 \quad , \quad \dot{X}(t=0) \equiv \dot{X}_0 = .3 \quad , \quad \lambda = .04 \quad , \quad x_0 = 5 \quad . \tag{12}$$

Plots of v(x) along with the linear response χ_0 it generates are given in Fig. 1. The background $\chi_0(x)$ is localized as it should be for localized v(x). Figure 2 shows the effective potential V(X) experienced by the kink CM in first order. As expected, the effective potential is localized in X. Since to first order the kink behaves as a Newtonian particle moving in the potential well V(X), we expect the kink's velocity to increase upon entering the region of the perturbation, then return to its original value when leaving. The first-order motion for the kink CM is shown in Fig. 3 and confirms this potential energy analysis.

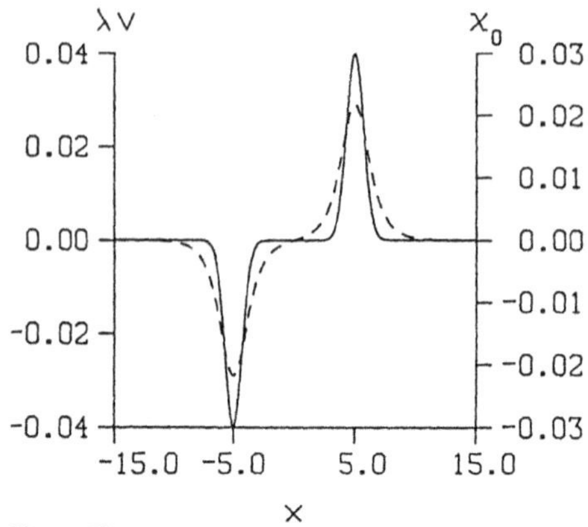

Fig. 1 The perturbation $\lambda v(x)$ and the linear response $\chi_0(x)$ (dashed curve) it generates

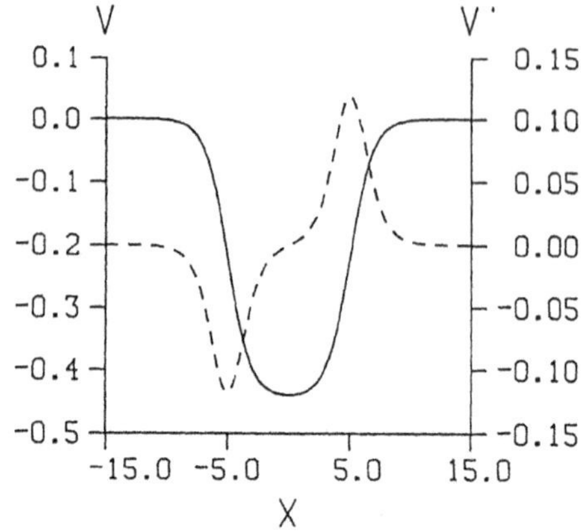

Fig. 2 The effective potential V(X) and its derivative (dashed curve)

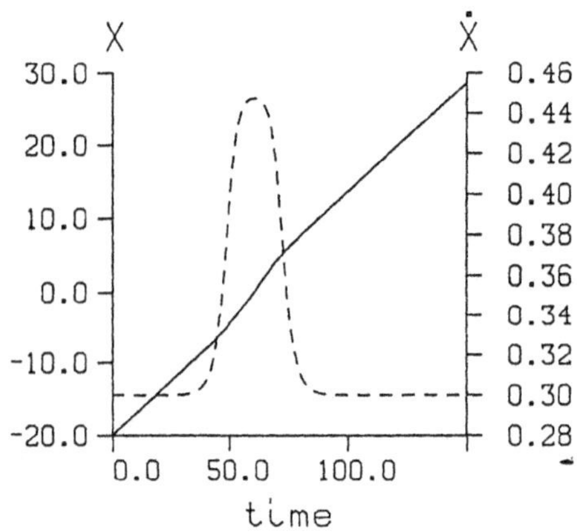

Fig. 3 The first-order kink position and velocity (dashed curve) as a function of time

The second-order motion for the CM variable X(t) deviates from the first order motion only in that the maximum velocity (see Fig. 3) attained by the kink is slightly smaller since some energy is given to the phonon field $\psi(x,t)$. However, since the final kink velocity equals the initial velocity, this energy is given back to the kink when it leaves the perturbation region. Since the phonon field ψ remains localized (through second order) about the kink CM, it can be regarded as a (temporary) shape change of the kink during the collision. It is possible that in third and higher orders, in addition to a phase shift, the kink will be accompanied by extended phonons and the kink's final velocity will not equal its initial velocity. However, before proceeding to higher order, one must examine the approximations made to see if they are still valid in higher order. One point of concern is the approximation of the response of the field to the perturbation by the background χ_0, which is obtained by linearization. These questions along with applications of this method to different classes of perturbations shall be addressed in further publications.

REFERENCES

1. A. R. Bishop, J. A. Krumhansl, and S. E. Trullinger, Physica 1D, 1 (1980)
2. J. L. Gervais, A. Jevicki, and B. Sakita, Phys. Rev. D12,1038 (1975), Physics Reports, 23C, 281 (1976)
3. J. L. Gervais and A. Jevicki, Nuclear Phys. B110, 93 (1976)
4. M. J. Rice and E. J. Mele, Solid St. Commun. 35, 487 (1980); T. Holstein, Proc. Int. Conf. Low-Dim. Cond., Boulder, Col., August 1981 (Gordon and Breach, Science Publishers, Inc. 1981)
5. D. K. Campbell, J. F. Schonfeld, and C. A. Wingate, Physica 9D, 1 (1983)
6. M. B. Fogel, S. E. Trullinger, A. R. Bishop, and J. A. Krumhansl, Phys. Rev. Lett. 36,1411 (1976); Phys. Rev. B15, 1578 (1977)
7. E. Tomboulis, Phys. Rev. D12, 1678 (1975)
8. S. E. Trullinger, Solid St. Commun. 29, 27 (1979)
9. P. A. M. Dirac, Lectures on Quantum Mechanics (Academic, New York, 1964)
10. B. Horovitz, R. J. Flesch, and S. E. Trullinger, to be published
11. R. J. Flesch and S. E. Trullinger, to be published

Quantum Coherence and the Nonlinear Schrödinger Equation

D.W. Brown, K. Lindenberg, and B.J. West *

Department of Chemistry B-014, University of California, San Diego, La Jolla, CA 92093, USA

We show how the validity of Hamilton equation methods for determining the time evolution of trial state vectors in quantum mechanics may be tested. We show how an ansatz state vector consisting of a product of coherent states allows a differential operator to be constructed under which a scalar Hamilton function must be invariant. Since the Hamilton equations for the coherent state amplitudes are derived without approximation from the exact Heisenberg equations of motion for creation and annihilation operators, the differential invariance condition provides information about the admissibility of coherent state products as state vectors and the validity of the equations of motion subsequently derived.

1. INTRODUCTION

In the study of quasiparticle transport in deformable solids, coherent state products have proven useful in describing the dynamic organization of vibrational normal modes about an immobile excitation introduced into the solid at the initial time[1]. In this transportless limit of the general problem based on the Froelich Hamiltonian[2], two distinct lines of investigation converge. The first, based on the concept of the polaron[3-6], makes extensive use of the fact that the eigenstates of the transportless problem can be reached through a unitary transformation which associates with the particle a deformation of the medium describable through a product of coherent states in each vibrational mode. The transport problem is then analyzed using these "dressed" or "polaron" states as zeroth order states in perturbation theory. The second approach assumes that the state of the system in the general transport problem is qualitatively similar to that in the transportless problem in that the state of the medium is assumed to be describable as a product of coherent states in each vibrational mode at all times[7]. The parameters determining the organization of the coherent states are assumed to be time dependent, and equations of motion are developed to determine the required time dependences. In the analysis of Davydov[7], and in recent improvements of that analysis by Kerr and Lomdahl[8], the "coherence" of the phonon system allows the development of a nonlinear Schroedinger equation for the wave function amplitudes of the mobile particle, introducing the "soliton" as an alternative to the polaron. Other approaches based on effective Hamiltonians have employed coherent states in the description of the state of the "bare" excitation as well as that of the lattice[9].

In this paper we present a straightforward illustration of how the applicability of coherent states may be overstepped to yield attractive but erroneous results. As with many mathematical illusions, the fallacy encountered is subtle. We show how the essential assumptions of the coherent state approach to transport may be tested in a particularly simple way.

2. THE DIFFERENTIAL INVARIANCE CONDITION

Throughout this section we will be concerned with a particular kind of state vector or wave function, and with the determination of this state vector's time evolution by methods grounded in the equations of motion for operators. This association involves frequent interchanges between the Schroedinger and Heisenberg pictures of the quantum evolution of a dynamical system. For clarity, we consistently indicate the picture in use by the quantities on which the time dependence is displayed; thus, $a_n |A(t)>$ is understood to indicate the use of the Schroedinger picture, while $a_n(t)|A>$ is understood to indicate the use of the Heisenberg picture. We let $\{a, a^\dagger\}$ denote the set $\{a_1, a^\dagger_1, \ldots, a_N, a^\dagger_N\}$ of Boson creation and annihilation operators appearing in the Hamiltonian operator $H\{a, a^\dagger\}$ of a given system. We consider state vectors $|A(t)>$ defined by

$$|A(t)> \equiv |\alpha_1(t)> \otimes |\alpha_2(t)> \otimes \cdots \otimes |\alpha_N(t)>, \tag{2.1}$$

wherein $|\alpha_n(t)>$ is the pure coherent state defined by [10]

$$|\alpha_n(t)> \equiv exp[-\frac{1}{2}|\alpha_n(t)|^2] exp[\alpha_n(t) a^\dagger_n]|0>. \tag{2.2}$$

The complex scalar $\alpha_n(t)$ is the coherent state amplitude, which may take on all values in the complex plane. The product state $|A(t)>$ may be defined by the property that

$$a_n |A(t)> = \alpha_n(t)|A(t)> \tag{2.3}$$

for all of the a_n. The expectation value of a Hamiltonian operator $H\{a, a^\dagger\}$ in the state $|A(t)>$ is therefore a real scalar function $H\{\alpha(t), \alpha^*(t)\}$ of all the $\alpha_n(t)$ and their complex conjugates. We presume that our starting Hamiltonian operator is in normal ordered form so that there is no ambiguity in the relationship between $H\{a, a^\dagger\}$ and $H\{\alpha(t), \alpha^*(t)\}$.

The following relationships between commutators of the Hamiltonian operator and derivatives of the scalar Hamilton function follow from the properties of the coherent state $|A(t)>$:

$$<A|[a_n(t), H(t)]|A> = <A(t)|[a_n, H]|A(t)>$$

$$= \frac{\partial}{\partial \alpha^*_n(t)} H\{\alpha(t), \alpha^*(t)\}, \tag{2.4a}$$

$$<A|[a^\dagger_n(t), H(t)]|A> = <A(t)|[a^\dagger_n, H]|A(t)>$$

$$= \frac{-\partial}{\partial \alpha_n(t)} H\{\alpha(t), \alpha^*(t)\}. \tag{2.4b}$$

Applying these relations to the expectation value of the corresponding Heisenberg equations of motion for the creation and annihilation operators yields

$$i\hbar \dot{\alpha}_n(t) = \frac{\partial}{\partial \alpha^*_n(t)} H\{\alpha(t), \alpha^*(t)\}, \tag{2.5a}$$

$$i\hbar \dot{\alpha}^*_n(t) = \frac{-\partial}{\partial \alpha_n(t)} H\{\alpha(t), \alpha^*(t)\}, \tag{2.5b}$$

where we have used the fact that $<A|\dot{a}_n(t)|A> = \frac{d}{dt}<A|a_n(t)|A> = \dot{\alpha}_n(t)$. The relations (2.5) are the classical Hamilton equations for generalized coordinates $\alpha_n(t)$ and momenta $i\hbar\alpha^*_n(t)$ with respect to a Hamiltonian function $H\{\alpha(t), \alpha^*(t)\}$, and follow without approximation directly from the definition of the coherent state $|A(t)>$, the Heisenberg equations of motion, and the Bose commutation relations for the operators $\{a(t),a^*(t)\}$. The set of coherent state amplitudes $\{\alpha(t),\alpha^*(t)\}$, and therefore $|A(t)\rangle$, may be determined for any Hamilton function $H\{\alpha(t),\alpha^*(t)\}$ for which the classical Hamilton equations (3.5) are solvable. The procedure outlined appears attractive as a means of determining particular solutions of the Schroedinger equation corresponding to the less tractable Heisenberg equations of motion for the operators $\{a(t),a^\dagger(t)\}$. We find, however, that for most Hamiltonians the vectors $|A(t)\rangle$ so determined are *not* solutions of the corresponding Schroedinger equation, due to the existence of a constraint on the form of the Hamilton function $H\{\alpha(t),\alpha^*(t)\}$ which fails to be satisfied in most cases.

Consider the Schroedinger equation for the state vector $|A(t)\rangle$

$$i\hbar\frac{d}{dt}|A(t)\rangle = H|A(t)\rangle\,. \tag{2.6}$$

Since $|A(t)\rangle$ is a trial state vector, the equality in (2.6) is really the statement of the basic assumption on which the coherent state analysis rests; namely, that there exist sets of coherent state amplitudes $\{\alpha(t),\alpha^*(t)\}$ which solve (2.6) and thus determine particular solutions of the Schroedinger equation. Rather than proceeding from (2.6) toward its solution, we will derive from it a condition which will indicate limitations in the underlying assumptions of the approach. Inserting into (2.6) the form of $|A(t)\rangle$ given by (2.1) and (2.2) and carrying out the differentiation with respect to time, we find

$$i\hbar\sum_n\{-\frac{1}{2}[\dot{\alpha}^*_n(t)\alpha_n(t)+\alpha^*_n(t)\dot{\alpha}_n(t)]+\dot{\alpha}_n(t)a^\dagger_n\}\,|A(t)\rangle = H|A(t)\rangle\,. \tag{2.7}$$

Left-multiplying (2.7) with the bra vector $\langle A(t)|$, we obtain the scalar relationship

$$\frac{i\hbar}{2}\sum_n[\dot{\alpha}^*_n(t)\alpha_n(t)-\alpha^*_n(t)\dot{\alpha}_n(t)] = H\{\alpha(t),\alpha^*(t)\}\,. \tag{2.8}$$

Using the Hamilton equations (2.5) to eliminate the time derivatives in (2.8), we obtain an invariance condition on the Hamilton function,

$$\frac{1}{2}\sum_n[\,\alpha_n(t)\frac{\partial}{\partial\alpha_n(t)}+\alpha^*_n(t)\frac{\partial}{\partial\alpha^*_n(t)}\,]\,H\{\alpha(t),\alpha^*(t)\} = H\{\alpha(t),\alpha^*(t)\}\,. \tag{2.9}$$

This condition may be cast in a simpler form by transforming to the "action-angle" variables defined by $\alpha_n \equiv J_n^{1/2}exp(i\theta_n)$, for which the partial derivative operators take the form

$$\frac{\partial}{\partial\alpha_n} = J_n^{1/2}exp(-i\theta_n)[\,\frac{\partial}{\partial J_n}+\frac{1}{2iJ_n}\frac{\partial}{\partial\theta_n}\,]\,, \tag{2.10}$$

whereupon the condition (2.9) becomes

$$\sum_n[\,J_n(t)\frac{\partial}{\partial J_n(t)}\,]H\{J(t),\theta(t)\} = H\{J(t),\theta(t)\}\,. \tag{2.11}$$

The requirement that the Hamilton function be invariant under the action of a specific differential operator strongly limits the class of Hamiltonians admitting the state vectors $|A(t)\rangle$ as solutions of their Schroedinger equations. The invariant Hamilton functions $H\{J,\theta\}$ satisfying (2.11) have the form

$$H\{J,\theta\} = \sum_m f_m(\theta_1,\theta_2,\ldots,\theta_N)J_1^{x_{m1}}J_2^{x_{m2}}\ldots J_N^{x_{mN}}\,, \tag{2.12}$$

wherein the functions f_m are arbitrary real functions of the phases only and the x_{mi} are real numbers subject to the restriction that $x_{m1}+x_{m2}+\cdots+x_{mN}=1$.

The scalar Hamilton function to which the invariance condition is applied is derived from a Hamiltonian operator represented in a given basis. A Hamiltonian operator admitting coherent state Schroedinger solutions in a given basis may not admit solutions of the same form constructed in another basis and vice versa, even if the bases are related by a canonical transformation. This observation underscores the fact that what is being tested by the condition (2.11) is the admissibility of a class of trial state vectors having a specified form, the form being represented in a basis-specific way.

Hamiltonians used to describe unperturbed excitons or vibrons often take forms which yield Hamilton functions satisfying condition (2.11). The Hamiltonian for pure Frenkel excitons is often represented as

$$H_{ex} = \sum_m E_m a^\dagger_m a_m + \sum_{m,n} V_{mn} a^\dagger_m a_n \ , \tag{2.13}$$

in which $a^\dagger_m$ (a_m) creates (annihilates) an exciton in the site state labeled by m. Forming $<A|H_{ex}|A>$ and making the transformation to action-angle variables yields

$$H_{ex}\{J,\theta\} = \sum_m E_m J_m + \sum_{m,n} V_{mn} (J_m J_n)^{1/2} exp[-i(\theta_m - \theta_n)] \ . \tag{2.14}$$

Since the degree of the action polynomial is unity (degree 2 in the square roots of the action variables), (2.11) is satisfied.

The Hamiltonian for pure vibrons is sometimes written

$$H_{vib} = \sum_n \hbar\omega_n (b^\dagger_n b_n + \frac{1}{2}) + \sum_{m,n} L_{mn} :(b^\dagger_m + b_m)(b^\dagger_n + b_n): , \tag{2.15}$$

in which $b^\dagger_n$ (b_n) creates (annihilates) a quantum of vibrational energy in a molecular mode labeled by n. Casting H_{vib} into explicit normal-ordered form as indicated by $: \cdots :$ and neglecting the zero-point energy gives

$$H_{vib} = \sum_n \hbar\omega_n b^\dagger_n b_n + \sum_{m,n} L_{mn} (b^\dagger_m b^\dagger_n + b^\dagger_m b_n + b^\dagger_n b_m + b_m b_n) \ . \tag{2.16}$$

Forming $<A|H_{vib}|A>$ and making the transformation to action-angle variables yields

$$H_{vib}\{J,\theta\} = \sum_n \hbar\omega_n J_n + 4\sum_{m,n} L_{mn} (J_m J_n)^{1/2} \cos\theta_m \cos\theta_n \ . \tag{2.17}$$

Again, the degree of the action polynomial is unity, so that (2.11) is satisfied.

3. VIOLATION

For an example of a Hamiltonian violating the condition (2.11) one need look no further than the general two-body interaction

$$H^{(2)} = \sum_{i,j,k,l} H^{(2)}_{ijkl} a^\dagger_j a^\dagger_i a_k a_l \ . \tag{3.1}$$

Such two-body operators have appeared in effective Hamiltonians obtained as approximations to the Froelich Hamiltonian [9,11]. When transformed into the action-angle representation, $H^{(2)}$ takes the form

$$H^{(2)}\{J,\theta\} = \sum_{i,j,k,l} H^{(2)}_{ijkl} (J_j J_i J_k J_l)^{1/2} exp[-i(\theta_j + \theta_i - \theta_k - \theta_l)] \ . \tag{3.2}$$

The degree of the action polynomial is 2, not 1 as required for $H^{(2)}\{J,\theta\}$ to satisfy the invariance condition. Since the Hamilton equations (2.3) are exact for the coherent states, the necessary conclusion is that *the Schroedinger equation for a Hamiltonian containing the two-body interaction* $H^{(2)}$ does not admit pure coherent state products as solutions .

If $H^{(2)}$ (or its continuum analog) is added to the Hamiltonian of an otherwise free particle, the Hamilton equations for the coherent state amplitudes are of the nonlinear Schroedinger type. Carrying out the manipulations of section 2, we find that the coherent state amplitudes follow equations of motion

$$i\hbar\dot{\alpha}_r(t) = \sum_n H_{rn}^{(1)}\alpha_n(t) + \sum_{j,k,l}(H_{jrkl}^{(2)} + H_{rjkl}^{(2)})\alpha^*_j(t)\alpha_k(t)\alpha_l(t) . \tag{3.3}$$

If the state labels are spatial coordinates and the two-body interaction is site-diagonal so that only the elements $H_{iiii}^{(2)}$ are non-vanishing, then the continuum limit of this equation is the usual nonlinear Schroedinger equation

$$i\hbar\alpha(x,t) = -\frac{\hbar^2}{2m^*}\frac{\partial^2}{\partial x^2}\alpha(x,t) + V(x)\alpha(x,t) - G(x)|\alpha(x,t)|^2\alpha(x,t) . \tag{3.4}$$

The violation of the invariance condition implicit in (3.3) and (3.4) implies that these nonlinear equations for the coherent state amplitudes do *not* describe the quantum evolution of a system with two-body interactions. The conclusion is easily generalized to interaction Hamiltonians for arbitrary numbers of particles, allowing us to conclude broadly that the evolutions of coherent state products embodied in the usual nonlinear Schroedinger equation (3.4) and the generalization of this evolution embodied in (3.3) *do not* correspond to the quantum evolution of a system state vector in *any* quantum system.

To verify the failure of the coherent state analysis for a simple two-body system, consider the trivial Hamiltonian operator

$$H^{(2)} = \epsilon_1 a^\dagger_1 a^\dagger_1 a_1 a_1 + \epsilon_2 a^\dagger_2 a^\dagger_2 a_2 a_2 . \tag{3.5}$$

Applying the coherent state analysis using equations (2.5), we find the solutions for the coherent state amplitudes

$$\alpha_n(t) = exp[i2\epsilon_n|\alpha_n(0)|^2 t/\hbar]\alpha_n(0) ; \tag{3.6}$$

however, when the product state $|\alpha_1(t)\rangle \otimes |\alpha_2(t)\rangle$ is inserted into the Schroedinger equation we find the necessary condition

$$\epsilon_1|\alpha_1(0)|^4 + \epsilon_2|\alpha_2(0)|^4 = 0 . \tag{3.7}$$

Thus we find that only a fortuitous few of the coherent state solutions (3.6) can have the precise combination of energies and amplitudes to escape definitive exclusion as possible quantum evolutions.

4. CONCLUSION

In this paper we have shown how manipulations of a system Hamiltonian can provide (negative) information about the solutions of the corresponding Schroedinger equation. These results are motivated by a method often used for analyzing the evolution of a quantum system in which the classical Hamilton equations are used to describe the time dependence of parameters appearing explicitly in a trial state vector. Given a trial state vector, it proves possible to construct an invariance condition which is a necessary (but not sufficient) condition for the set of Hamilton equations to be equivalent to the Schrodinger equation. The differential operator and scalar Hamilton function are simply constructed without any knowledge, exact or approximate, of the solutions of the Schroedinger equation.

The trial state vector for which we have illustrated the analysis is unique in that the Hamilton equations for the coherent state amplitudes are derived exactly from the Heisenberg equations for the corresponding creation and annihilation operators. This leaves as the only assumption being tested the most fundamental one; namely, the assumption that the trial state vector solves the Schroedinger equation. We find that any Hamiltonian containing interactions such as (3.1) between two or more particles leads to a violation of the invariance condition, implying that the Schroedinger equations of such Hamiltonians do not admit coherent state products as solutions. Since the two-body interaction is necessary to obtain equations of motion for the coherent state amplitudes which are of the nonlinear Schroedinger type, we conclude that the nonlinear Schroedinger equations obtained from the coherent state ansatz do not describe the evolution of a state vector in *any* quantum system.

The construction of a differential invariance condition is not limited to trial state vectors which are products of coherent states, but may be carried out whenever state vector parameters are assumed to satisfy the classical Hamilton equations. Unfortunately, the condition one obtains does not provide a measure of the deviation of Hamilton equation solutions from corresponding Schroedinger equation solutions having the same initial form. The condition should thus be viewed as a qualitative guide toward more quantitative analysis[1].

5. ACKNOWLEDGEMENTS

This work was supported in part by NSF Grant No. DMR-8315950, by U. S. Defense Advanced Research Projects Agency (DARPA) Contract No. DAAG-29-K-0246, and by the Independent Research Funds of the La Jolla Institute.

* Division of Applied Nonlinear Problems, La Jolla Institute, 3252 Holiday Court, Suite 208, La Jolla, California 92038, USA.

1. David W. Brown, Katja Lindenberg and Bruce J. West, J. Chem. Phys. *84* , 1574 (1986); Phys. Rev. A *33* , 4104 (1986); 4110 (1986).
2. H. Froehlich, Proc. R. Soc. London Ser. A *215* , 291 (1952).
3. T. Holstein, Ann. Phys. (N.Y.) *8* , 325 (1959); 343 (1959).
4. M. Grover and R. Silbey, J. Chem. Phys. *54* ,4843 (1971); S. Rackovsky and R. Silbey, Mol. Phys. *25* , 61 (1973); D. R. Yarkony and R. Silbey, J. Chem. Phys. *65* , 1042 (1976); *67* , 5818 (1977).
5. V. M. Kenkre, Phys. Rev. B *11* , 1741 (1975); *12* , 2150 (1975); V. M. Kenkre and T. S. Rahman, Phys. Lett. *50A* , 170 (1974).
6. V. M. Kenkre and P. Reineker, *Exciton Dynamics in Molecular Crystals and Aggregates* , Vol. 94 of *Springer Tracts in Modern Physics* , edited by G. Hoehler (Springer-Verlag, Berlin, 1982).
7. A. S. Davydov, Phys. Scr. *20* , 387 (1979); Zh. Eksp. Teor. Fiz. *78* , 789 (1980) [Sov. Phys.-JETP *78* , 397 (1980)]; Usp. Fiz. Nauk *138* , 603 (1982) [Sov. Phys.-Usp. *25* , 898 (1982)].
8. W. C. Kerr and P. S. Lomdahl, preprint entitled "Quantum-Mechanical Derivation of the Equations of Motion for Davydov Solitons," (1986).
9. J. A. Tuszynski, R. Paul, R. Chatterjee and S. R. Sreenivasan, Phys. Rev. A *30* , 2666 (1984).
10. R. J. Glauber, Phys. Rev. *131* , 2766 (1963).
11. A. C. Scott, Philos. Trans. R. Soc. London Ser. A *315* , 423 (1985).

Energy Transfer, Self-Trapping, and Solitons on a Nonlinear Dimer*

V.M. Kenkre[1], *G.P. Tsironis*[1], *and D.K. Campbell*[2]

[1]Department of Physics and Astronomy, University of New Mexico, Albuquerque, NM 87131, USA
[2]Center for Nonlinear Studies, Los Alamos National Laboratory, Los Alamos, NM 87545, USA

Abstract

A summary is presented of several results obtained recently concerning the time evolution of a soliton on a lattice. They include the derivation of a Liouville-Von Neumann equation for the density matrix from the discrete nonlinear Schrödinger equation and, for the two-site system, e.g. a molecular dimer, the derivation and explicit solution of a closed nonlinear evolution equation obeyed by the difference of the probabilities of occupation of the two sites. The probabilities are seen to evolve in the form of Jacobian elliptic functions, and a transition from free to self-trapped behavior is seen at appropriate relative values of the intersite interaction and the nonlinearity parameter. Features such as the polaron (or soliton) bandwidth reduction conjectured for the general polaron transport problem are seen to emerge as a consequence. Exact expressions for observables such as the scattering function of relevance to quasi-elastic neutron diffraction experiments are displayed and shown to contain explicit nonlinear characteristics.

The discrete nonlinear Schrödinger equation, which describes a number of phenomena in various fields of physics, may be written as

$$dc_m(t)/dt = -iV(c_{m+1} + c_{m-1}) + i\chi|c_m|^2 c_m \tag{1}$$

where c_m is the amplitude for the system to be in state $|m>$, V is the interstate matrix element (assumed "nearest-neighbor" in (1)) describing the linear evolution among the states $|m>$, and χ is the nonlinearity parameter. In the specific system of an electron or exciton in a crystal interacting strongly with phonons or vibrations, $|m>$ denotes the (localized) Wannier state centered on site m, V is proportional to the bandwidth of the bare electron or exciton, and χ is the energy lowering due to polaronic effects, often written as a sum of the products of the vibrational energies of the participating modes and the square of their coupling constants with the electron or exciton. In (1), we have set $\hbar = 1$.

Equation (1) or similar equations have been written down or derived by a variety of people in many different contexts [1-7]. While solutions for its continuum limit are well known [7], analytical solutions for the discrete chain are not available. We have recently found [8] exact time-dependent solutions of (1) for the case of a dimer, i.e. when m can take on values 1 and 2 only, and have applied them [9] to the calculation of experimental observables such as the quasi-elastic neutron scattering function and fluorescence depolarization. A brief description of those solutions and applications follows, our reasons for displaying these results being two. First, they show, in a completely explicit

*Work supported in part by the National Science Foundation under grant no. DMR-850638 and by the Department of Energy under grant nos. 7405/ENG-36 and BES/DMS-86-03-02.

manner, the phenomenon of self-trapping, energy mismatch behavior, polaron/soliton bandwidth reduction and other expected and conjectured features of the evolution described by the equation for the chain of arbitrary length, or generally for a crystal. And second, far from being an overidealized system, the dimer is interesting in its own right and attainable experimentally [10-12]. Calculations for dimer observables are therefore not merely pedagogical in nature but have direct significance in experiments.

Before proceeding further, a comment concerning terminology might prove helpful. Some current usage [2,13] refers to a "soliton" as simply a quasiparticle whose time evolution and motion are described by a nonlinear equation such as (1), independent of the total size of the system. In this usage, phrases like a "soliton in a Wannier or Bloch state" and the idea of a soliton on a dimer are clearly acceptable. In contrast, another usage reserves the term "soliton" for localized, nonlinear solutions only in spatially extended systems (i.e., much larger than a dimer). Those readers who prefer the former usage should look upon the present analysis as that of soliton motion in a dimer. Those who favor the latter terminology should interpret it simply as an investigation of energy transfer in a nonlinear dimer.

The derivation of the Liouville-Von Neumann equation obeyed by the density matrix ρ_{mn} from (1) is straightforward. Equation (1) and its complex conjugate can be combined to give

$$d\rho_{mn}/dt = iV(\rho_{m+1n} + \rho_{m-1n} - \rho_{mn+1} - \rho_{mn-1}) + i\chi(\rho_{mm} - \rho_{nn})\rho_{mn} \tag{2}$$

provided χ is assumed real. We point out that a linear chain with varying site energies would obey (2) with the replacement of the second term $i\chi(\rho_{mm}-\rho_{nn})\rho_{mn}$ by the energy mismatch term $-i(E_m-E_n)\rho_{mn}$, where E_m is the site energy at m. The energy lowering $-\chi\rho_{mm}$ which equals the product of χ and the probability of occupation of site m is thus completely evident in (2). For the dimer, (2) yields

$$d\rho_{11}/dt = -iV(\rho_{21} - \rho_{12}) \tag{3}$$

$$d\rho_{12}/dt = -iV(\rho_{22} - \rho_{11}) + i\chi(\rho_{11} - \rho_{22})\rho_{12} \tag{4}$$

with corresponding equations for ρ_{22} and ρ_{21}. The combination of (3) and (4) and a straightforward calculation involving the elimination of the off-diagonal elements of the density matrix leads to the following closed nonlinear equation for the probability difference $p(t) = \rho_{11} - \rho_{22}$:

$$d^2p/dt^2 = Ap - Bp^3 \ . \tag{5}$$

In the absence of the nonlinearity, i.e. for $\chi = 0$, the coefficient A equals $-4V^2$, B vanishes, and the probability difference p oscillates sinusoidally. In the presence of the nonlinearity, for arbitrary initial conditions, A and B are given by

$$A = (\chi^2/2)p_0^2 - 4V^2 - 2V\chi(\rho_{21} + \rho_{12})_0 \ ; \ B = (\chi^2/2) \tag{6}$$

where the subscript 0 denotes the initial (i.e., $t = 0$) value.

The general solution of (5) and (6) may be written as

$$p(t) = Ccn[(C\chi/2k)(t - t_0) \mid k] = Cdn[(C\chi/2)(t - t_0) \mid 1/k] \tag{7}$$

$$1/k^2 = 2 + (1/C^2)[(4V/\chi)^2 + (8V/\chi)(\rho_{21} + \rho_{12})_0 - 2p_0^2] \tag{8}$$

where C and t_0 are arbitrary constants to be determined from the initial conditions, e.g. the values of p and dp/dt at $t = 0$. One can verify directly the general solution (7) by substituting it in (5).

Three consequences of (7) are interesting to explore. The first involves the extraction of the specific form of the stationary states of the system by considering the situation in which all the elements of ρ, and therefore also p, are independent of t. The second is relevant to the initial condition that only one of the two sites is occupied, and shows a transition from free to self-trapped behavior. And the third is an application of (7) to the calculation of an explicit observable, the scattering function.

To extract the stationary states from (7), we differentiate it twice and equate both dp/dt and d^2p/dt^2 to 0 for all t. This leads to the results that $C = p_0$ and $t_0 = 0$ and to the condition $1/k = 0$. When substituted in (8), this results in

$$(4V/\chi)^2 + (8V/\chi)(\rho_{21} + \rho_{12})_0 = 0, \tag{9}$$

and thence to explicit expressions for the initial values of the density matrix elements. These initial values are, however, the values for all times, since the state under consideration is stationary. Rewriting the elements of ρ in terms of the coefficients of the eigenstates in the site representation, i.e., c_1 and c_2, we arrive at

$$|c_1| = (1/2)^{1/2}\{1 + [1 - (2V/\chi)^2]^{1/2}\}^{1/2} \tag{10}$$

$$|c_2| = (1/2)^{1/2}\{1 - [1 - (2V/\chi)^2]^{1/2}\}^{1/2} \tag{11}$$

which specifies the stationary states precisely from the general time-dependent solution (7). This exercise demonstrates that from our general solution for all times one may recover the stationary dimer results given earlier by Eilbeck et al. [6]. One can obtain (10) and (11) directly, and immediately, from our Eq. (5) by equating its right-hand side to zero for $p_0 \neq 0$.

As the second consequence of (7), consider the initial condition that only one of the two sites is occupied initially. The initial probability difference is then $p_0 = 1$ (or -1), and the off-diagonal elements of the density matrix are zero initially. Then,

$$A = (\chi^2/2) - 4V^2 \ ; \ B = (\chi^2/2) \tag{12}$$

The probability difference is given by

$$p(t) = cn(2Vt \mid k = \chi/4V) \ for \ \chi/4V < 1 \tag{13}$$

if the nonlinear parameter is small enough, but by

$$p(t) = dn(\frac{1}{2}\chi t \mid k = 4V/\chi) \ for \ \chi/4V > 1 \tag{14}$$

if it is large enough. In the former case the oscillations of the particle are such that it resides equally on both sites but in the latter case it resides more on the initially occupied site. The former case describes free behavior while the latter represents incomplete intersite transfer, i.e. relative trapping as would be characteristic of an energy mismatch. There is thus a transition at $\chi = 4V$. As a function of χ, the detailed behavior is as follows.

For $\chi = 0$, we see that the particle oscillates between the two sites with period π/V. As χ increases, the oscillations of the particle begin to depart from trigonometric behavior. The period of the oscillations of the particle is $T = 2K(k)/V$. It increases as $\chi/4V$ increases. As χ approaches very close to 4V, the oscillations of the particle show a marked departure from trigonometric behavior: The period becomes enormous as the particle oscillates sluggishly between the two sites. The logarithmic approximation to $K(k)$ is appropriate here and the period of the oscillations of the particle is given by

$$T = (1/V)ln\{16/[1 - (\chi^2/16V^2)]\} \tag{15}$$

The free particle motion of the particle between the two sites for $\chi < 4V$ described above is depicted in curves (a) and (b) of Fig. 1. When χ equals $4V$, a transition

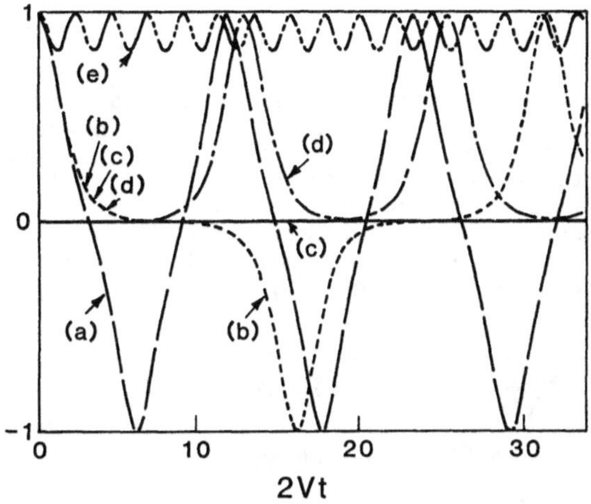

Figure 1. The difference in the probabilities of occupation of the two sites in a dimer plotted as a function of time t (in units of 2V) for various values of $(\chi/4V)$: (a) 0.95, (b) 0.9995, (c) 1, (d) 1.0001, (e) 1.75. Curves a and b are indicative of free-particle motion, c describes the transition, and d and e represent self-trapping behavior.

occurs between this free motion and self-trapped behavior. At the transition, the equalization of the probabilities of the two sites follows exactly a "sech" evolution:

$$p(t) = sech(2Vt) = sech(\chi t/2) \tag{16}$$

and the period of the oscillation is infinite. This is shown in curve (c) of Fig. 1.

As χ increases beyond the transition, the oscillations no longer follow the cn function but are described by the dn function (see (14)). In this region the probability difference never crosses 0. Increasing χ has now two effects: the amplitude of the oscillations decreases and the period also decreases. Far away from the transition, as $\chi/4V$ approaches infinity, the oscillations follow a trigonometric evolution, i.e., can be approximated by

$$p(t) = 1 - (8V^2/\chi^2)sin^2(\chi t/2) \tag{17}$$

This is precisely the evolution of the probability difference in a linear dimer with a true energy mismatch: the amplitude decreases with an increase in the amount of the energy mismatch. The period of the oscillations is simply $2\pi/\chi$.

This self-trapped particle behavior is depicted in curves (d) and (e) of Fig. 1. In the limit as $\chi/4V$ tends to infinity, the particle is entirely self-trapped as the amplitude (and also the period) of the oscillations becomes zero.

A consequence of our analysis is the clear demonstration that a polaronic/solitonic bandwidth reduction, which represents a lowering of the velocity of the carrier, is a natural outcome of the discrete nonlinear Schrödinger equation in the context of the dimer. Being inversely proportional to the period of oscillation, the reduced (or effective) bandwidth V_{eff} drops off sharply at the transition as the reciprocal of a logarithm:

$$V_{eff} = V[(2/\pi)K(\chi/4V)]^{-1} \approx V[\pi/ln\{16/1 - (\chi^2/16V^2)\}] \;. \tag{18}$$

the second equality in (18) being valid close to the transition. Equation (18) establishes unambiguously that a polaronic/solitonic bandwidth reduction does indeed occur.

Finally, as the third consequence of (7), we display the scattering function $S(\omega)$ which describes the intensity of probe particles such as neutrons which would lose or gain energy $\hbar\omega$ when scattered off the quasiparticle (the "moving soliton") as it moves between the dimer sites [9]. The general scattering function $S(q,\omega)$, which involves the (dimensionless) wavevector difference q as well as the frequency difference ω, is, in the case of a dimer, a weighted sum of an elastic contribution and the contribution $S(\omega)$, the weights being determined by the angle of the scattered beam to the line joining the two sites of the dimer [9,14]. The contribution $S(\omega)$ is identical to the Fourier transform of the Van Hove correlation function, which, at high temperatures, becomes identical to $p(t)$, the difference in the probabilities of the two sites for initial single-site occupation [15,16]. The Fourier transform of (7) for the initial single site occupation thus gives [9]

$$S(\omega) = \pi[2kK(k)]^{-1} \sum_{n=0}^{\infty} (sech\omega R_<)[\delta(\omega - \omega_n) + \delta(\omega + \omega_n)] \tag{19}$$

$$S(\omega) = (1/4V)sech(\omega\pi/4V) = (1/\chi)sech(\omega\pi/\chi) \tag{20}$$

$$S(\omega) = \pi[2K(1/k)]^{-1}\{\delta(\omega) + \sum_{n=1}^{\infty} (sech\ \omega R_>)[\delta(\omega - \tilde{\omega}_n) + \delta(\omega + \tilde{\omega}_n)]\} \tag{21}$$

for, respectively, the three cases $\chi < 4V, \chi = 4V$, and $\chi > 4V$. In (19)-(21), $R_<$ is given by $R_<(k) = K[(1-k^2)^{1/2}]/2V$, and $R_>(k) = (1/k)R_<(1/k)$. In (19) the summation runs from $n = 0$ to ∞ and $\omega_n = 2V_{eff}(2n+1)$. In (21), the summation is from $n = 1$ to ∞ and $\tilde{\omega}_n = (n\pi\chi)/2K(\frac{1}{k})$. While further discussion of these scattering results and an extension for arbitrary temperatures may be found elsewhere [9], we point out here that, as the nonlinearity is increased, (19) exhibits a "march" of the spectral lines towards the origin, (20) represents a collapse of the δ-function lines onto a finite spectrum at the transition, and (21) shows an interesting reemergence of the δ-functions followed by a strengthening of the line at the origin and a weakening, and outward "march", of the other lines. This behavior of the spectrum of the nonlinear undamped dimer as the nonlinearity is changed, resembles very closely, but is not identical to, the phenomenon of motional narrowing exhibited by the spectrum of the linear damped dimer as the damping is changed.

References

1. T. D. Holstein, Ann. Phys. 8, 325, 343 (1959).
2. A. S. Davydov, J. Theor. Biol. 38, 559 (1973); Sov. Phys. Usp. 25, 898 (1982) and refs. therein.
3. D. Emin, Physics Today 35, 34 (1982) and refs. therein.
4. Y. Toyozawa, in Organic Molecular Aggregates, eds. P. Reineker, H. Haken and H. C. Wolf, (Springer-Verlag, Berlin 1983), and refs. therein.
5. D. K. Campbell, A. R. Bishop, and K. Fesser, Phys. Rev. B 26, 6862 (1982).
6. J. C. Eilbeck, A. C. Scott, and P. S. Lomdahl, Chem. Phys. Lett. 113, 29 (1985); Physica 16D, 318 (1985).
7. A. C. Scott, F. Y. Chu, and D. W. McLaughlin, Proc. I.E.E.E. 61, 1443 (1973).
8. V. M. Kenkre and D. K. Campbell, Phys. Rev. B, to be published.
9. V. M. Kenkre and G. P. Tsironis, University of New Mexico preprint.
10. S. Stryer and R. P. Haugland, Proc. Nat. Acad. Sc. (US) 58, 719 (1967).

11. A. H. Zewail and C. B. Harris, Phys. Rev. B 11, 935, 952 (1975).

12. H. Wipf, A. Magerl, S. M. Shapiro, S. K. Satija, and W. Thomlinson, Phys. Rev. Lett. 46, 947 (1981); A. Magerl, A. J. Dianous, H. Wipf, K. Neumaier, and I. S. Anderson, Phys. Rev. Lett. 56, 159 (1986).

13. D. W. Brown, K. Lindenberg, and B. J. West, Phys. Rev. A33, 4104, 4110 (1986); J. Chem. Phys. 84, 1574 (1986).

14. D. W. Brown and V. M. Kenkre, J. Phys. Chem. Solids, to be published.

15. L. Van Hove, Phys. Rev. 95, 249 (1954).

16. V. M. Kenkre and D. W. Brown, Phys. Rev. B31, 2479 (1985).

Part VI

Structural Phase Transitions

Microscopic Probing of $BaTiO_3$ Ferroelectric Phase Transitions by EPR

K.A. Müller

IBM Zurich Research Laboratory, CH-8803 Rüschlikon, Switzerland

The displacive (soft-mode) versus order-disorder (relaxator) behavior is assessed. First, static and dynamic scattering experiments are reviewed, followed by a summary of the two most recent experiments probing locally the octahedral Ti site with paramagnetic resonance of Fe^{3+} and Mn^{4+}. These two experiments prove substantial order-disorder behavior.

(a) Fe^{3+} on a Ti^{4+} site senses the *local fourth-order potential*. The measured site, pressure and temperature dependences of the cubic EPR splitting parameter $\underline{a}$ prove a 34% enhanced local anharmonic potential as compared to inert cubic MgO or typical underdamped ferroelectric soft phonon displacive cases such as $SrTiO_3$ or $KTaO_3$.

(b) EPR of Mn^{4+} on a Ti^{4+} site is detected in the low-temperature rhombohedral phase. Its axial crystalline splitting of $D = +0.65\,cm^{-1}$ shows that Mn^{4+} is sited on Ti^{4+} off-center $\langle 111 \rangle$ positions. Absence of the Mn^{4+} spectrum in the high-temperature phases of $BaTiO_3$ indicates a reorientation of the Ti^{4+} ions between equivalent $\langle 111 \rangle$ off-center sites with a time constant of $\tau \leq (D \times c)^{-1} \simeq 10^{-10}$ sec. Such slow reorientation times can be expected for the one-dimensional $\langle 100 \rangle$ strands of $\langle 111 \rangle$ displaced and correlated Ti ions in the high-temperature phases as deduced from x-ray streaks.

1. Historical Introduction

The soft mode $\omega_s(T)$ was introduced by COCHRAN regarding the ferroelectricity occurring in $BaTiO_3$ [1]. Underdamped ferroelectric soft modes have been reported early in $SrTiO_3$ and $KTaO_3$ [2]. However, anharmonicity is so small in these two crystals that their soft ferroelectric modes do not freeze out. Quantum-mechanical fluctuations stabilize their paraelectric phases in the absence of uniaxial stress [3]. In contrast, in structural phase transitions in which certain units, like octahedra in $LaAlO_3$ or $SrTiO_3$, rotate, phase transitions occur and underdamped soft modes exist which almost freeze out [4,5]. Near T_c, critical phenomena or coupling to impurities hinder a complete freeze-out and a central peak phenomenon is found at $\omega = 0$ [5,6]. Apart from the examples cited, many cases of soft modes have been found [4,5].

$BaTiO_3$ is the first oxide ferroelectric discovered [4], and on cooling undergoes three consecutive phase transitions: Cubic – tetragonal – orthorhombic and rhombohedral, as does $KNbO_3$. The debate as to whether these transitions

Table 1: History

Order-disorder:		MASON and MATHIAS 1948	[7]
Soft mode:	*infrared*	SPITZER *et al.* 1962	[8]
		BARKER and HOPFIELD 1964	[9]
Order-disorder:	*x-rays*	COMES *et al.* 1968	[10]
Soft mode:	*neutron scattering*	HARADA *et al.*1971	[11]
Order-disorder:	*Raman scattering* $T > T_c$	GUITTET and LAMBERT 1973	[12]

are more displacive or order-disorder-like has a history of more than 20 years. Here, we concentrate on $BaTiO_3$. Table 1 lists earlier experiments and conclusions from 1948 to 1973: They alternate between order-disorder and displacive as a function of time [6-12]: Infrared [8,9] and neutron-scattering [11] experiments clearly revealed a transverse optical mode indicative of displacive behavior, the latter however becoming highly overdamped near the cubic-tetragonal phase transition. First-order-like Raman features observed by the Orsay group in 1973 were more indicative of order-disorder behavior [12]. Here, we emphasize the 1968 discovery by this group of diffuse x-ray streaks in $BaTiO_3$ and $KNbO_3$ cubic, tetragonal and orthorhombic phases [10]. The interpretation of these streaks was in terms of ⟨100⟩ correlated strands of 10 to 50 ⟨111⟩ displaced Ti (or Nb) atoms as shown for the orthorhombic phase in Fig. 1. In this phase, [111] displaced Ti ions reorient in 1D, [010] and [0$\bar{1}$0] strands, i.e., [111] and [11$\bar{1}$] Ti displacements, to yield an average [101] polarization. In the tetragonal phase, this reorientation occurs between four ⟨111⟩ displacements to obtain a ⟨100⟩ polarization, and in the cubic phase between all eight equivalent ⟨111⟩ displacements. Although it was pointed out that the x-ray streaks can result from the anisotropy of the soft-mode eigenvectors [11], LAMBERT and COMES earlier accounted *quantitatively* for the entropy change and Curie constant at the cubic-tetragonal phase transition of $BaTiO_3$ (and $KNbO_3$) with the large chain dipoles of strands of size $10 < n < 25(50)$, depending on the purity of the crystals [13].

After an interval of a few years, interest in $BaTiO_3$ has been resumed, and Table 2 lists recent experiments published between 1980 and 1985. They are grouped into those bearing on dynamic aspects [14-17], and those yielding instantaneous near static results [18-21]. The former bear on the highly overdamped transverse soft mode, and BWO techniques characterize it as "relaxator" [17]. In the "static" group, index of refraction measurements [18] indicate a nonvanishing $\langle P^2 \rangle$ component [18,19] in the cubic phase, as one may expect from the afore-mentioned polarized strands deduced from x-rays [10]. The ⟨111⟩ displacements were confirmed in Japan in 1985 [21]. X-rays also prove the presence of a strong order-disorder component in the cubic phase [20].

From the review presented, it is motivating to obtain information on the local cubic potential $V(\vec{R})$ in which the Ti is sited in $BaTiO_3$, and which reorientation times are involved. In a cubic environment, $V(\vec{R})$ can be written to lowest order by [22]

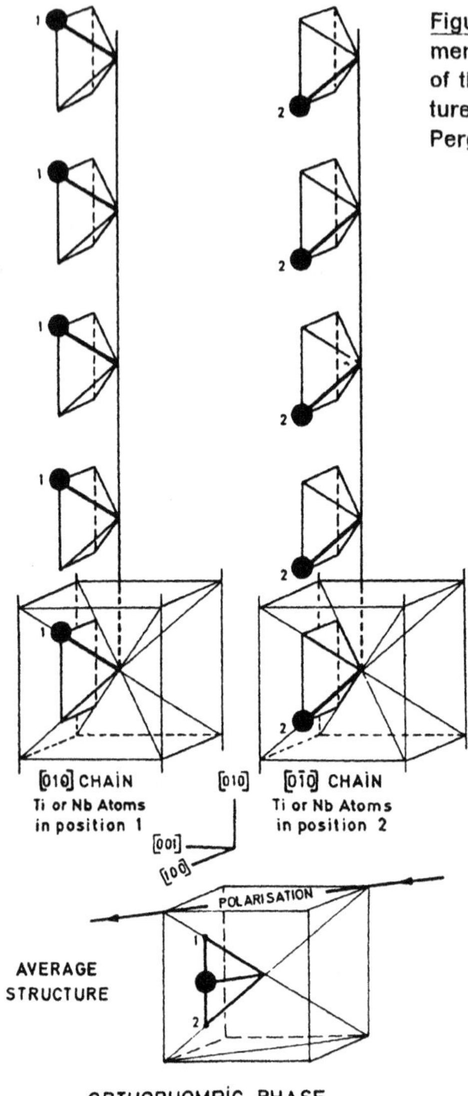

Figure 1 The Ti or Nb displacements in the [010] and [0$\bar{1}$0] chains of the average orthorhombic structure. From [10], Copyright © 1968 Pergamon Press

$$V(\vec{R}) = C_q\left(x^4 + y^4 + z^4 - \frac{3}{5}R^4\right). \tag{1}$$

EPR experiments of Fe^{3+} in $BaTiO_3$ probe $V(\vec{R})$, and show it is flatter in $BaTiO_3$ than for example in $SrTiO_3$ [22]. This one might expect as in $SrTiO_3$ the soft mode is underdamped and the Ti moves in more harmonic potential [4]. Section 2 describes earlier and more recent confirming results.

Table 2: Recent experiments

Dynamic			
Saturating soft mode	*infrared*	LUSPIN *et al.* 1980	[14]
Overdamped soft mode	*hyper-Raman*	VOGT *et al.* 1982/85	[15]
Soft mode→order-disorder crossover	*hyper-Raman*	INOUE *et al.* 1981/85	[16]
At T_c: relaxator	*BWO*	KOZLOV 1984/85	[17]
"Static"			
$\langle P^2 \rangle \neq 0$ for $T > T_c$	*n(T)*	BURNS 1981	[18]
Strong order-disorder component	*x-rays*	EHSES *et al.* 1981	[20]
$\langle 111 \rangle$ off-center Ti: $T > T_c$	*x-rays*	ITOH *et al.* 1985	[21]

The most recent advance in local probing by EPR is the finding of a paramagnetic impurity which follows the Ti^{4+} motion cooperatively. This is the Mn^{4+} ion. Although these experiments are not completed, one can already infer important insights [23]. The data so far obtained indicate that the Ti $\langle 111 \rangle$ off-center reorientation starts in the range of 10^{-10} sec, an order of magnitude lower than what the dynamic experiments listed in Table 2 could reach. A preliminary summary of the Mn^{4+} EPR is given in Section 3.

2. Probing Order-Disorder versus Displacive Behavior in $BaTiO_3$ by Fe^{3+} EPR

A paramagnetic singlet orbital ground state with spin $S \geq 2$ is split by (1) via spin-orbit and spin-spin interaction. The splitting is described empirically by a spin Hamiltonian of the form [24]

$$\mathcal{H} = \frac{a}{6}\left(S_x^4 + S_y^4 + S_z^4 - c\right). \tag{2}$$

For a Mn^{2+} or Fe^{3+} ion with half-filled 3d shell $3d^5$, $S = 5/2$, the $^6S_{5/2}$ ground state is split into a Γ_8 quartet and a Γ_6 doublet by $3\underline{a}$ which can be determined by EPR.

The dependence of $\underline{a}$ for Fe^{3+} in various cubic oxides as a function of lattice spacing 2d was found to follow an empirical curve $\underline{a} = a_0/d^n$ with $n = 6$ to 7, except for $SrTiO_3$, $BaTiO_3$ and $KNbO_3$. Whereas for $SrTiO_3$ the EPR parameter $\underline{a}$ is only slightly lower than the empirical curve, in the latter two crystals $\underline{a}$ is a factor of 2.5 smaller. This fact was pointed out a decade ago by Müller [22]. Figure 2a shows the measured $\underline{a}$ parameters scaled by $(4.2/2d)^6$ of seven cubic oxides where 4.2 Å is the lattice constant of MgO, and 2d are those of the others. The low $\underline{a}$ parameter correlates, as then recognized, with the strongly overdamped and anisotropic soft modes in $BaTiO_3$ and $KNbO_3$ and the consecutive tetragonal, orthorhombic and rhombohedral phases not observed in

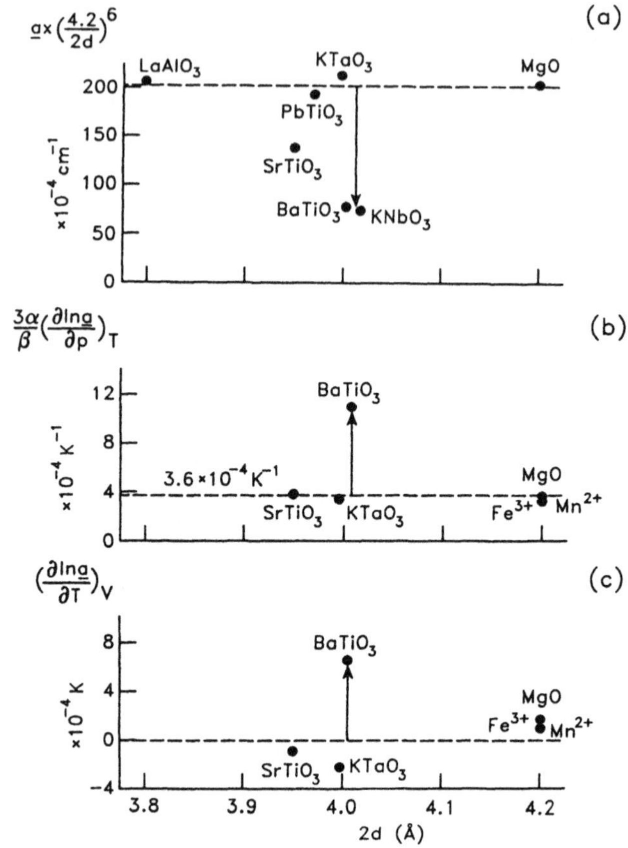

Figure 2 Comparison of scaled cubic field splitting parameter a, relative explicit volume and temperature effects in various cubic oxides as a function of lattice parameter 2d. From [28]

other perovskites. One of the reasons for the restricted interest at the time was the unclear theoretical relation of a to $V(\vec{R})$ for Fe^{3+}, and the position of this ion in the octahedral cell was not well understood, either.

Understanding of the position of Fe^{3+} in its octahedral cage was considerably improved with an analysis of SIEGEL and MÜLLER [25]. Using the superposition model to analyze some 20-year old EPR fine-structure splittings measured in the three ferroelectric phases of $BaTiO_3$ [26,27], it was shown that the Fe^{3+} remains *centered* in the octahedron. Thus, it is also clear why parameter a varies by no more than 10% in all phases of $BaTiO_3$ [26,27]. The reason why the Fe^{3+} is centered while the Ti^{4+} is not will be addressed in Section 3.

To obtain information on $V(\vec{R})$, we can, with restrictions, take advantage of the empirical a(d) dependence in the following way. We first compute, from

$\underline{a}(d)$, how much larger d_{eff} in $BaTiO_3$ has to be than its actual lattice constant d to observe $\underline{a}$ reduced by a factor of 2.5. From the power law of $\underline{a}$ on d, we get $d_{eff}/d = 2.5^{1/6} = 1.17$, i.e., the probing Fe^{3+} sees the oxygens in $BaTiO_3$ at a *distance 17% larger* than what it should be for an inert oxide. Consider the anharmonicity of $V(\vec{R})$ with $\vec{R}$ parallel to [111] with $|\vec{R}| \equiv R$. It can be parametrized with

$$V(R) = -AR^2 + BR^4, \tag{3}$$

with the constants A and B positive. This potential has minima at $R_m = \pm\sqrt{A/2B}$ with energy $V_m = (A/2)R_m^2$. Now, we assume the minimum R_m also at a distance larger by the same proportion $R_m(BaTiO_3)/R(I\,O) = d_{eff}/d$, where I O stands for inert cubic oxide. Of course, $R_m \neq d$, but to lowest order their variation is proportional. With this, we calculate for $V_m = (A/2)R_m^2$ in $BaTiO_3$, $V_m(BaTiO_3) = 1.34\,V_m(I\,O)$, a 34% enhanced anharmonicity $\propto 1/B$. The same enhanced anharmonicity has also to be present in $KNbO_3$. The distinction between the limiting cases of displacive versus order-disorder behavior at the transition temperature T_c is determined by whether $V_m \ll kT_c$: displacive, or $V_m \gg kT_c$: order-disorder [5].

The question arises at this point whether additional experiments can confirm the conclusion reached above. This has indeed been the case most recently by measurements of the pressure p and temperature T dependences of the cubic crystalline splitting parameter $\underline{a}(p,T)$ [28] whose total differential is given by

$$d\underline{a} = \left(\frac{\partial \underline{a}}{\partial p}\right)_T dp + \left(\frac{\partial \underline{a}}{\partial T}\right)_p dT. \tag{4}$$

$(\partial\underline{a}/\partial p)_T$ and $(\partial\underline{a}/\partial T)_p$ were first measured for Fe^{3+} and Mn^{2+}; then using the differentiated form of the equation of state $V = V(p,T)$ where V is the volume, WALSH *et al.* [29] obtained the relation

$$(\partial\underline{a}/\partial T)_p = -(3\alpha/\beta)(\partial\underline{a}/\partial p)_T + (\partial\underline{a}/\partial T)_V, \tag{5}$$

where $\alpha = (1/d)(\partial d/\partial T)_p$ is the coefficient of linear thermal expansion, and $\beta = -(3/d)(\partial d/\partial p)_T$ the volume compressibility. The first term on the right side is the explicit volume effect, and the second the explicit temperature effect. $(\partial\underline{a}/\partial p)_T$ and $(\partial\underline{a}/\partial T)_p$ have also been measured for Fe^{3+} in $SrTiO_3$ [30], and more recently for $KTaO_3$ [31]. In the latter publication, RYTZ *et al.* compared the values obtained for Fe^{3+} and Mn^{2+} in MgO as well as Fe^{3+} in $SrTiO_3$ and $KTaO_3$. Two very interesting properties of $\underline{a}(p,T)$ were noticed. The explicit volume effect relative to $\underline{a}$, reve $-(3\alpha/\beta)(\partial \ln \underline{a}/\partial p)_T$, was within 6% the same for all four measurements, see Fig. 2b. The optic modes for either MgO, $SrTiO_3$, or $KTaO_3$, being underdamped, indicate quite harmonic potentials. The explicit temperature effect is negative in $KTaO_3$ and $SrTiO_3$, whereas it is positive for Mn^{2+} and Fe^{3+} in MgO. Figure 2c shows the relative explicit temperature effect, rete, $(\partial \ln \underline{a}/\partial T)_V$. It was pointed out in Ref. 31 that there are two contributions to the explicit temperature effect

$$(\partial \underline{a}/\partial T)_V \sim +f_1 - f_2 T_c/(T - T_c)^2, \tag{6}$$

a positive Debye contribution $+f_1$ and a negative soft-mode contribution $-f_2 T_c/(T - T_c)^2$. The retes, i.e., $(1/a)(\partial \underline{a}/\partial T)_V$ shown in Fig. 2c reflect this in a clearer way. The values of Mn^{2+} and Fe^{3+} in MgO are closely the same. In this crystal, the soft-mode term is absent, whereas it is present in $KTaO_3$ and $SrTiO_3$. The two rete values of $KTaO_3$ and $SrTiO_3$ are negative and comparable to each other.

In Figs. 2b and 2c, the recent reve and rete values for $BaTiO_3$ are included and marked by arrows from the average and zero lines, respectively. They were obtained from the pressure and temperature dependences of $\underline{a}$ in the cubic and tetragonal phases [28]. They show the following remarkable results: The *explicit relative volume effect is larger by a factor of 3.0* than for the other oxides, and confirms the 34% enhanced anharmonicity of the Ti ion deduced from its absolute 2.5 times reduced $\underline{a}$ value as compared to other oxides. The *relative explicit temperature effect* $(\partial \ln \underline{a}/\partial T)_V$ *is positive, 4.5 times* that in MgO, whereas it is negative in $SrTiO_3$ and $KTaO_3$ with underdamped soft modes. This giant positive rete effect masks a negative soft-mode contribution of -2.0 possibly present, and proves the substantial order-disorder character of $BaTiO_3$ with its more anharmonic potential.

Anharmonic dynamical lattice theory explains the soft mode and dielectric constants in the incipient ferroelectrics $SrTiO_3$, $KTaO_3$ and possibly ferroelectric $PbTiO_3$, for example, in the work of COWLEY and BRUCE [5], or the more recent one of MIGONI *et al.*[32]. However, the Fe^{3+} EPR findings make it questionable whether such a theory can quantitatively account for the classic three ferroelectric phase transitions in $BaTiO_3$ or $KNbO_3$. In fact, some dynamic results indicate that in $BaTiO_3$ and more clearly in $KNbO_3$ the soft mode rather freezes out towards the orthorhombic-*trigonal* transition [33], i.e., $\langle 111 \rangle$ displaced Ti(Nb) ions are involved. The Fe^{3+} EPR results, also those on Mn^{4+} in the next section, make the theories resulting from electronic bands and densities [34,35] appear at least as promising [36]. In fact, from such an approach, in 1966 BERSUKER [37,38] predicted a strong anharmonicity in $BaTiO_3$ with the minimum position of the Ti ion in all phases to lie along the $\langle 111 \rangle$ directions.

3. Local Position and Dynamics of Ti^{4+} Probed by Mn^{4+} EPR

In the previous section, we postponed mentioning why the Fe^{3+} remains centered in the octahedral cage. Understanding this effect has led to a new paramagnetic probe which follows the cooperative motion of the Ti^{4+}, namely, the Mn^{4+} ion. The reason why the Fe^{3+} remains centered is two-fold. Fe^{3+} is nominally one unit negatively charged with respect to the Ti^{4+} it replaces, and therefore repels the negative nearest-neighbor O^{2-} ions as follows from a recent theory of SANGSTER [39]. Furthermore, the 3d shell of Fe^{3+} ($3d^5$) is half-filled with configuration $(t_{2g})^3,(e_g)^2$. The e_g orbitals are strongly σ − anti-bonding and add repulsive forces between the Fe^{3+} and O^{2-}, whereas the t_{2g} are π more nonbonding. Cr^{3+} with configuration $(t_{2g})^3$ is still centered because of its negative effective charge but less than Fe^{3+}, owing to the two missing e_g electrons [40]. This is borne out more quantitatively by the recent self-consistent tight binding calculations of SELME and PÉCHEUR [41] in $SrTiO_3$. From these calculations, the electronic charge of the $(e_g)^2$ orbitals is 38% on the d wave function of Fe^{3+}, but the $(t_{2g})^3$ only 18%.

The Cr^{3+} has three d-electrons. The half-filled t_{2g} shell carries no orbital angular momentum, and the magnetic state is described by S = 3/2. Such a spin state is not split by the cubic term (4), and consequently only second-order ligand-field splittings are observed in the lower than cubic-symmetry ferroelectric phases of $BaTiO_3$. The appropriate Hamiltonian is now,

$$\mathscr{H} = D\left(S_z^2 - (1/3)S(S+1)\right) + E\left(S_x^2 - S_y^2\right). \quad (7)$$

EPR spectra in the three ferroelectric phases have recently been reported [40]. The axial ligand parameter D(T) consists of two terms

$$D(T) = D^P(T) + D^T(T). \quad (8)$$

$D^P(T)$ is proportional to the square of the polarization P(T), and D^T is linear in T above 100 K and quite large. The existence of D^P shows that the Cr^{3+} remains centered in the octahedral cell. The existence of D^T, not observed for Fe^{3+}, points to large thermal fluctuations of the Cr^{3+} owing to the absence of repelling e_g electrons. Saturation of $D^T(T)$ for low T (below 100 K) is accounted for by a Debye model with energy θ_D = 237 K, proving independently a flat ionic potential for the Cr^{3+}. The Debye energy for Fe^{3+} is by comparison θ_D = 450 K, close to the one of $BaTiO_3$ [40].

The Mn^{4+}, $(t_{2g})^3$, is isoelectronic to Cr^{3+}. Its charge is the same as the substitutional Ti^{4+} ion. Thus, the Sangster effect is absent. In fact, the self-consistent tight binding calculations in $SrTiO_3$ place the $(t_{2g})^3$ orbitals of Mn^{4+} only 0.36 eV above the valence band, 1.25 eV lower than those of the Cr^{3+} [41,42]. This indicates that the Mn^{4+} ground state is similar to that of Ti^{4+}.

Axial EPR spectra of the Mn^{4+} have been observed in the rhombohedral phase with ligand-field axes z parallel to the $\langle 111\rangle$ domain axes [28]. Figure 3 shows the ±1/2 resonance magnetic fields taken at 19.2 GHz and 4.2 K in a $(1\bar{1}0)$ plane. The insert shows that in this plane the $\{111\}$ and $\{11\bar{1}\}$ domains each yield one set of the transitions observed, whereas the $\{1\bar{1}1\}$ and $\{1\bar{1}\bar{1}\}$ domains a degenerate set. The ligand-field splitting D deduced from the transitions perpendicular to [111] and $[\bar{1}\bar{1}2]$ measured at 19.2 and 9.2 GHz, respectively, is some 100 times larger than that of the isoelectronic Cr^{3+} despite the superposition model parameters of the Mn^{4+} being closely the

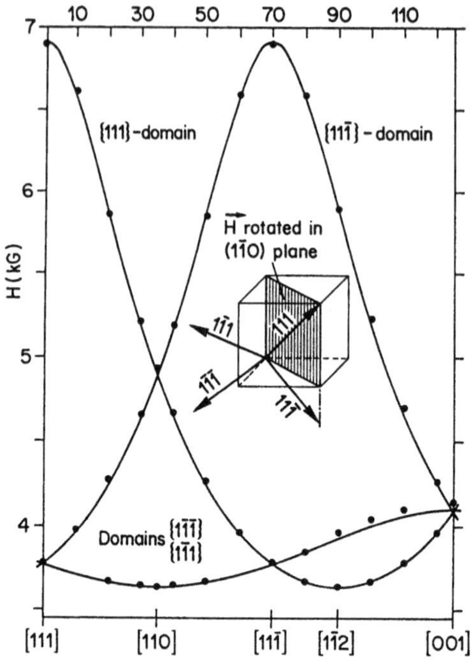

Figure 3 Center of paramagnetic resonance fields of Mn^{4+} in rhombohedral $BaTiO_3$ at 4.2 K and 19.2 GHz. Lines from the various domains are indicated. Note that EPR does *not* distinguish between {111} and reversely polarized {$\overline{111}$} domains, etc. Full lines are calculated theoretically.

same [23,40]. An analysis in terms of the superposition model for the Mn^{4+} gave:

$$\begin{array}{lll} & Ti^{4+}\text{ coordinate:} & \text{Centered in octahedron:} \\ D = & -0.44\pm0.23\ cm^{-1} & +0.0350\pm0.0088\ cm^{-1} \end{array} \tag{9}$$

as compared to the one observed of

$$|D| = 0.65\pm0.01\ cm^{-1}. \tag{10}$$

Comparing the model and experimental values of D, it is clear that the Mn^{4+} has quite the same coordinates as Ti^{4+} in the rhombohedral phase, i.e., displaced from the octahedral center by $\simeq 0.14$ Å along the $\langle 111\rangle$ directions. We recall here that two decades ago, BERSUKER, from vibronic theory, predicted the Ti^{4+} ions to lie off-center along $\langle 111\rangle$ directions [37].

From the above, it follows that the Mn^{4+} with nearly the same atomic mass as Ti^{4+} follows its cooperative motion and allows probing of its dynamics. No EPR spectrum of Mn^{4+} has so far been observed in the orthorhombic, tetragonal or cubic phases of $BaTiO_3$ along ⟨111⟩ directions of the magnetic field applied. This is only possible when the Mn^{4+} spectra are motionally smeared out. For this to be the case, a component of the movement of the Mn^{4+} and, therefore, Ti^{4+} is present with a frequency of

$$\nu = \frac{1}{h} g\mu_B(H_h - H_l) \sim 0.9 \times 10^{10}\ \text{sec}^{-1}, \tag{11}$$

where H_h and H_l are the high and low-resonance magnetic fields for $\vec{H} \parallel [111]$ in the rhombohedral phase. If the motion were faster than ν, a dynamically averaged Mn^{4+} should be visible. A local mode of the Mn^{4+} can be excluded with certainty because in the non-ferroelectric $SrTiO_3$, the Mn^{4+} spectrum is visible between 4 and 300 K [43].

The EPR of Mn^{4+} being a reliable probe of the Ti^{4+} position and motion resolves an important aspect of the statics and dynamics of the ferroelectric phase transition in $BaTiO_3$. In their important x-ray work, COMES *et al.* [10] had concluded that the Ti ion sits off-center along equivalent ⟨111⟩ directions in *all* phases: at rest in the low-temperature phase, motionally averaging between two equivalent ⟨111⟩ directions in the orthorhombic, four in the tetragonal, and all eight ⟨111⟩ in the cubic phase, as reviewed in Section 1, see Fig. 1. The particular [111] displaced Ti ions are correlated along ⟨100⟩ equivalent directions. Owing to these rather long correlated chains, the reorientation has to be slow as deduced from the new Mn^{4+} EPR investigations [23]. Because this motion is so slow, of the order of 10^{-10} sec, it escaped observation by neutron, Raman or equivalent scattering experiments but appears static to x-rays [10,21]. However, the neutron-scattering spin-echo technique invented by MEZEI [44] appears well suited to measure the slow reorientation times in the range of 10^{-10} sec. The reorientation times cannot be slower than the time scale set by the cubic crystalline splitting of Fe^{3+}, $1/5\,\underline{a}c \simeq 10^{-9}$ sec, because the EPR of Fe^{3+} is observed in *all* phases of $BaTiO_3$ [26,27]. This is shown schematically in Fig. 4.

ION	CHARACTERISTIC TIME				
Mn^{4+}	10^{-10} sec	OBSERVED	NOT OBSERVED	NOT OBSERVED	NOT OBSERVED
Fe^{3+}	10^{-9} sec	OBSERVED	OBSERVED	OBSERVED	OBSERVED
PHASE		RHOMBOHEDRAL	ORTHORHOMBIC	TETRAGONAL	CUBIC → T

Figure 4 Schematic representation of observed and unobserved EPR of Mn^{4+} and Fe^{3+} in the four phases of $BaTiO_3$

1. W. Cochran: Adv. Phys. **9**, 387 (1960)
2. P. A. Fleury, J. M. Worlock: Phys. Rev. Lett. **18**, 665 (1967); G. Shirane, R. Nathans, V. J. Minkiewicz: Phys. Rev. **157**, 396 (1967)
3. K. A. Müller: Jpn. J. Appl. Phys. **24**, Suppl. 2, 89 (1985)

4. M. E. Lines, A. M. Glass: In *Principles and Applications of Ferroelectrics and Related Materials* (Clarendon, Oxford 1977)
5. See the two recent texts with the title 'Structural Phase Transitions' by A. D. Bruce and R. A. Cowley, (Taylor and Francis Ltd., London 1981), and *Current Topics in Physics* **23** (Springer, Berlin 1981), ed. by K. A. Müller and H. Thomas.
6. K. A. Müller: In *NATO 1981 School on Nonlinear Phenomena at Phase Transitions and Instabilities*, ed. by T. Riste (Plenum, New York 1982) p. 1; and *Statics and Dynamics of Nonlinear Systems*, ed. by G. Benedek, H. Bilz and R. Zeyer, Solid-State Sciences, Vol. 47 (Springer, Berlin 1983) p. 68.
7. W. P. Mason, B. T. Mathias: Phys. Rev. **74**, 1622 (1948)
8. W. G. Spitzer, R. C. Miller, D. A. Kleinman, L. E. Howarth: Phys. Rev. **126**, 1710 (1962)
9. A. S. Barker, J. J. Hopfield: Phys. Rev. **135**, A 1732 (1964)
10. R. Comes, M. Lambert, A. Guinier: Solid State Commun. **6**, 715 (1968)
11. J. Harada, J. D. Axe, G. Shirane: Phys. Rev. B **4**, 155 (1971)
12. A. M. Guittet, M. Lambert: Commun. Solid St. Phys. **12**, 1053 (1973)
13. M. Lambert, R. Comes: Solid State Commun. **7**, 305 (1969)
14. Y. Luspin, J. L. Servoin, F. Gervais: J. Phys. C **13**, 3761 (1980); K. A. Müller, Y. Luspin, J. L. Servoin, F. Gervais: J. Phys. (Paris) **43**, L-537 (1982)
15. H. Vogt, J. A. Sanjurjo, G. Rossbroich: Phys. Rev. B **26**, 5904 (1982); H. Vogt: Jpn. J. Appl. Phys. **24**, Suppl. 2, 112 (1985)
16. K. Inoue, N. Asai: J. Phys. (Paris) **42**, C6-430 (1981); K. Inoue: Jpn. J. Appl. Phys. **24**, Suppl 2, 107 (1985)
17. G. V. Kozlov: In *Abstract Book Sixth International Meeting Ferroelectricity*, Kobe, Japan, (1985) p. 16
18. G. Burns, F. H. Dacol: Ferroelectrics **37**, 661 (1981)
19. R. Hofmann, S. H. Wemple, H. Gränicher: J. Phys. Soc. Jpn. Suppl. **28**, 265 (1970)
20. K. H. Ehses, H. Bock, K. Fischer: Ferroelectrics **37**, 507 (1981)
21. K. Itoh, L. Z. Zeng, E. Nakamura, N. Mishima: Ferroelectrics **63**, 29 (1985)
22. K. A. Müller: Ferroelectrics **13**, 381 (1976); *idem*: Phys. Rev. B **13**, 3209 (1976)
23. K. A. Müller, W. Berlinger, K. W. Blazey, J. Albers: in preparation
24. A. Abragam, B. Bleaney: In *Electron Paramagnetic Resonance of Transition Ions* (Clarendon, Oxford 1970).
25. E. Siegel, K. A. Müller: Phys. Rev. B **19**, 109 (1979)
26. A. W. Hornig, R. C. Rempel, H. E. Weaver: J. Phys. Chem. Solids **10**, 1 (1959)
27. T. Sakudo: J. Phys. Soc. Jpn. **18**, 1626 (1963); T. Sakudo, H. Unoki: *ibid*. **19**, 2109 (1964)
28. K. A. Müller, W. Berlinger: submitted for publication (1986)
29. W. M. Walsh, J. Jeener, N. Bloembergen: Phys. Rev. **139** A, 1338 (1965)
30. T. Revai: Sov. Phys. Solid State **10**, 2984 (1969)
31. D. Rytz, U. T. Höchli, K. A. Müller, W. Berlinger, L. A. Boatner: J. Phys. C **15**, 3371 (1982)

32. R. Migoni, H. Bilz, D. Bäuerle: Phys. Rev. Lett. **37**, 1155 (1976)
33. M. D. Fontana, G. Métrat, J. L. Servoin, F. Gervais: J. Phys. C **16**, 483 (1984); M. D. Fontana, G. E. Kugel: Jpn. J. Appl. Phys. **24**, Suppl. 2, 223 (1985)
34. L. Castet-Mejean, F. M. Michel-Calendini: Ferroelectrics **37**, 503 (1981)
35. K. H. Weyrich, R. Siems: Z. Phys. B **61**, 63 (1986)
36. K. A. Müller: to appear in Helv. Phys. Acta (1986)
37. I. B. Bersuker: Phys. Lett. **20**, 589 (1966)
38. I. B. Bersuker, B. G. Vekhter, A. A. Muzalevskii: J. Phys. (Paris) Suppl. 4, **33**, C2-139 (1972) and literature therein
39. M. J. L. Sangster: J. Phys. C **14**, 2889 (1981)
40. K. A. Müller, W. Berlinger, J. Albers: Phys. Rev. B **32**, 5837 (1985)
41. M. O. Selme, P. Pecheur, G. Toussaint: J. Phys. C **17**, 5185 (1985); M. O. Selme, P. Pecheur: *ibid.* **18**, 551 (1985); and to appear in J. Phys. C (1986)
42. F. M. Michel-Calendini, K. A. Müller: Solid State Commun. **40**, 255 (1981)
43. K. A. Müller: Phys. Rev. Lett. **2**, 341 (1959)
44 F. Mezei:. Z. Phys. **255**, 146 (1972); in: *Neutron Spin Echo*, ed. by F. Mezei (Springer-Verlag, Heidelberg, 1980)

Structural Transformations in DNA

S.M. Lindsay

Physics Department, Arizona State University, Tempe, AZ 85282, USA

DNA undergoes displacive transitions between the A, B and Z forms, and order-disorder transitions such as the strand separation. The displacive transitions may be controlled by the highly charged phosphate backbone of the polymer interacting via the surrounding electrolyte. Interactions between neighboring molecules dominate the A to B transition since A-DNA appears only in concentrated (aggregated) solutions or in fibers, and the degree to which adjacent molecules are aligned parallel controls the transition rate. Modes of GHz. frequencies are probably involved in the dynamical pathways of these transitions and this is a frequency region in which the dynamics of the surrounding hydration shell needs to be considered explicitly. This is because the water close to the double helix (primary hydration) undergoes a viscoelastic transition at about 4 GHz. while the more remote water (secondary hydration) undergoes this transition at about 80 GHz. Melting can be brought about by an increase in the phonon population in the small bandwidth occupied by modes which contribute stretch amplitude to the bridging hydrogen bonds.

1. Introduction

The DNA double helix can undergo remarkable changes in its spatial arrangement while retaining its basic covalent structure. For example, raising the salt concentration of the surrounding electrolyte can cause a transition from a right-handed (B) helix to a left-handed (Z) helix in certain polymers. The biological role of these conformational changes is not fully understood, but what little is known of the structure of DNA-protein complexes suggests that DNA is rather floppy, deforming considerably on binding, whereas proteins are somewhat more rigid. Thus the dynamics of DNA should be of particular importance. The molecule is particularly interesting because of its approximate screw symmetry and charged and polarazible backbone.

The three basic families of structure are outlined in Table 1. The most studied transition is between the A form and the more loosely wound B form. Both soft mode [1] and soliton models [2] of this transition have been proposed, but these theories deal with ideal isolated molecules (except in as much as force constants are renormalized). However, various 'external' influences are important in these transitions. Soumpasis [3,4] has demonstrated the contribution of electrolyte-mediated interactions between the charged phosphates to free energy differences between the various conformations. In the case of the

Table 1. Helix parameters for the A, B and Z structures.

	A-DNA	B-DNA	Z-DNA
Handedness	Right	Right	Left
Base-pairs/turn	11	10	6 dimers
Diameter	20Å	20Å	18Å
Pitch	28Å	34Å	45Å
Rotation per basepair	+33°	+36°	-60°/dimer

A-B transition, we have shown that interactions between phosphates of neighboring double helices control the transition [5,6] (we will describe these interactions more fully here). The role of the surrounding electrolyte is not only static, for we have demonstrated strong coupling between the hydration shell and gigahertz phonons [6,7]. We end this contribution with a brief description of the self-consistent phonon theory of the hydrogen bond melting of the double helix.

2. A Soft mode for the A to B transition?

Prohofsky and his coworkers have exploited the screw symmetry of DNA in carrying out a lattice dynamics analysis of the polymer [8]. Using a model of the long-range interactions adjusted to fit the measured speed of sound, the lattice dynamics calculations have proved extraordinarily successful in predicting the frequencies and polarizations of low-lying Raman bands. Fig. 1 shows the calculated bands below 100 cm^{-1} (note the one-dimensional Brillouin zone). The dots mark observations [8 - 11] located at the appropriate point in the zone by the polarization selection rules [8]. The eigenvector of the lowest zone center optic mode (~ 12 cm^{-1}) has almost exactly the geometry required for an order parameter for the A to B transition [8], and calculations show that this mode is particularly sensitive to the effective dielectric constant within the double helix, softening dramatically for small changes in this parameter [1]. Although the A-B transition is indeed accompanied by mode softening [8], we have shown that this is a consequence of the breaking up of crystallinity as water is added to DNA fibers [5,6,9]. Presumably the success of the lattice dynamics calculations is a consequence of the effective incorporation of these interhelical forces through the use of the experimentally determined sound speed in refining the model of non-bonded interactions (however, the eigenvectors will be modified considerably by the interhelical contacts).

3. A Mechanism for Phosphate-Phosphate Interactions.

Soumpasis [3,4] has identified the key role played by phosphate-phosphate interactions in conformational stability. The negative charges are treated as point anions interacting via the electrolyte through an effective pairwise interaction (calculated from a true many-body model based on ion-ion correlation functions). At low salt concentrations, this effective potential

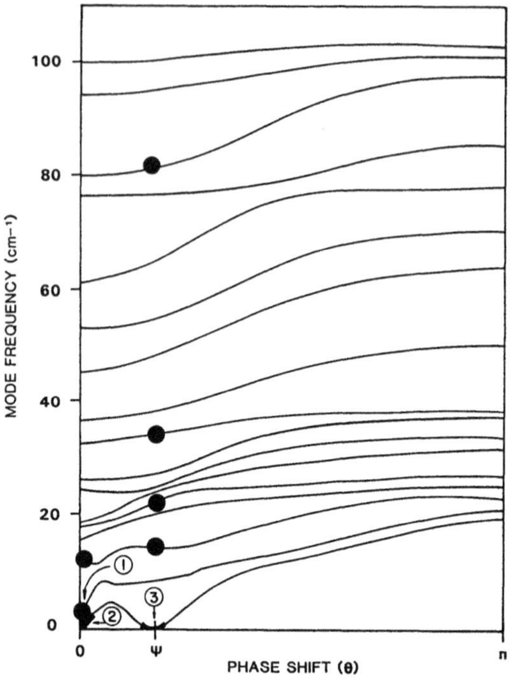

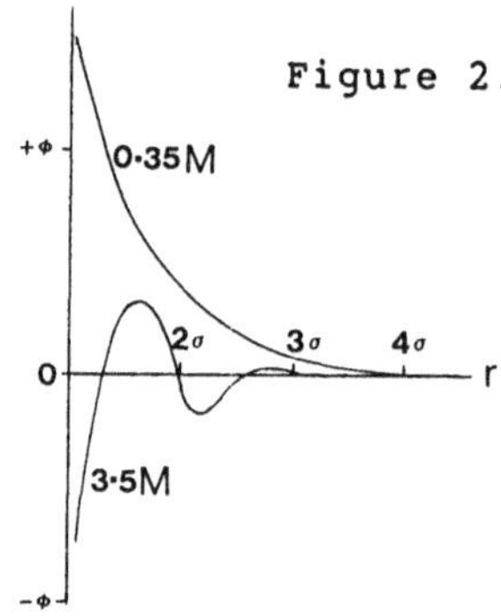

Figure 1. Calculated dispersion curves for the A form homopolymer poly(dG).poly(dC). The phase shift which equals the helix pitch angle (Ψ) is zone center in the lab frame for certain modes (this is a consequence of the use of helical coordinates). Zone center acoustic modes are (1) longitudinal compressional, (2) torsional and (3) bending. Low-frequency Raman bands which have been observed in experiments are shown as dots.

Figure 2. Effective anion-anion interaction potential for low (upper curve) and high (lower curve) salt concentrations. σ is the distance of closest approach of a hydrated ion pair. Note the appearance of new <u>attractive</u> regions as a consequence of ion correlations. Note that the origin is at 1σ.

is everywhere repulsive, falling off in the familiar Debye-Hückel manner expected for screened Coulomb interactions. At high salt concentrations, however, the interaction is changed dramatically. It is not at all monotonic, and <u>attractive</u> in places. This structure reflects the details of the ion-ion correlation functions. Fig. 2 shows this effective anion-anion interaction as calculated by Soumpasis[4] for a 1:1 electrolyte of 0.35 and 3.5 M concentration. Note, in particular, the complete reversal near the origin! The length scale is parameterized in terms of the distance of closest approach of the hydrated ions, σ. This parameter may vary between about 3Å and 8Å, but fits to experimental data are obtained with σ close to 5Å[3,4]. I should emphasize that the theory assumes an electrolyte <u>solution</u>, so that features in the high salt effective potential do not correspond to <u>fixed</u> charge structures. In discussing transitions that occur as <u>solid</u> DNA is diluted, I refer to possible charge structures with the understanding that they may be static, or represent some time average over fluctuating structures [6].

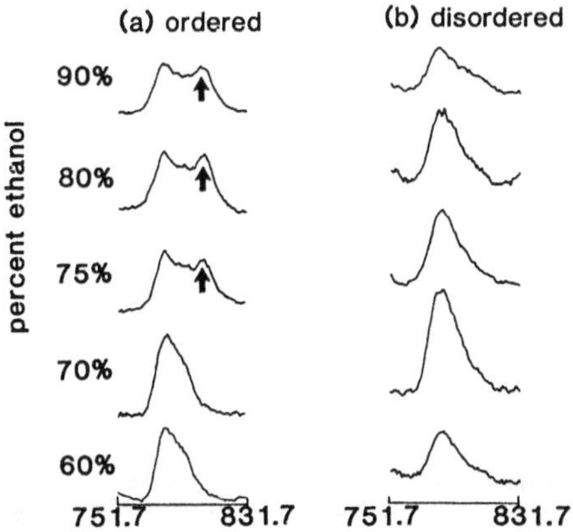

Figure 3. Raman spectra showing (a) the appearance of the 807 cm^{-1} A form marker band (indicated by an arrow) in crystalline samples as ethanol displaces water (percentages marked next to the spectra) and (b) the lack of the A form marker band as the amorphous samples are dehydrated (abscissa are Raman shifts in cm^{-1}).

4. Interhelical interactions and conformation transitions.

The theory of Soumpasis has proved remarkably accurate in describing the salt dependence of the B-Z transition [3]. It is less satisfactory in describing other transitions [4]. However, the A to B transition appears to be driven by solid state interactions [5,6]. (This does not appear to be the case for the B-Z transition [12]). We provide a further demonstration of the effect of crystallinity on the A-B transition in Fig. 3 which shows Raman spectra obtained from crystalline and amorphous films as water is displaced by ethanol [13]. The 807 cm^{-7} 'A-form marker band' appears only in the crystalline sample, demonstrating that aligned strands are needed to form A-DNA. Brillouin scattering experiments show that there are strong interactions between adjacent helices in the solid state[5,14]. Indeed, so strong are these bonds, that highly crystalline films can be nearly insoluble [15]. A-DNA is more closely packed than B-DNA, so the structural fluctuations due to sequence heterogeneity are smaller[16], and thus A-DNA should be capable of forming a more regular pattern of interhelical contacts in the solid.

The nature of the interhelical contacts in the crystalline regions can be obtained from a study of the unit cell geometry. In A-DNA the double helices pack into a monoclinic cell with (at 54% r.h.) a=21.1Å, b=38.9Å, c=27.3Å, β=97° and an additional double helix at (1/2, 1/2, 0) [5]. Although the closest approach (22.1Å center to center) brings the double helices almost into contact (at their O-2 radius), the three-dimensional packing minimizes O-2 clashes, with the backbones of one double helix lying in the major grooves of its neighbors on the whole. However, geometry does not permit the three-dimensional interlocking of screws of the same handedness and clashes do occur. These are illustrated (for the central molecule only) in

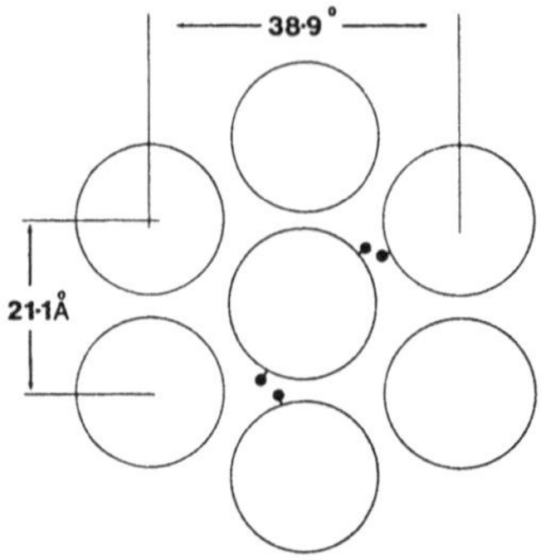

Figure 4. View of the interhelical contacts in A-DNA looking down into the ab plane of the monoclinic unit cell (dimensions are for 54% r.h.). The full circles correspond to the P-P diameter of the double helices and the black dots show the location of the O-2 atoms that come within 3.5Å of each other (for the central molecule only). The breaking of the apparent 6-fold symmetry is a consequence of the incommensurate nature of the 11-fold helix with the nearly 6-fold packing geometry.

Fig. 4 which shows a view down onto the unit cell. The circles correspond to the P-P diameter (nearly 18Å), and the black dots show the location (in the ab plane) of the O-2 atoms in one turn of the central molecule that come within 3.5Å of O-2 atoms on adjacent double helices. There are two such shared clashes per turn (i.e. one per double helix per turn), and it is interesting to note that the 'magic' amount of excess NaCl (1% by weight) required to give good crystalline x-ray patterns gives almost exactly one extra Na^+ per double helical turn.

The O-2 interactions fall into two categories: the close approaches at backbone clashes, and the more common interactions as the backbone of one molecule packs into the major groove of another. It is quite possible that many of the O-2 interactions fall into the attractive regions at 1σ and 2σ in the high salt potential (Fig. 2), and some possible charge 'structures' are given in ref. 6. The strength of the 'electrolyte' is indeed in the molar region for fairly dry (A-DNA) fibers, but of course the applicability of a liquid electrolyte model must be somewhat questionable. Nonetheless, we can explain the mode softening observed as DNA films are hydrated in terms of the measured separation of the charge centers and an assumed attractive Coulomb interaction due to anion-cation clusters [6,14]. (Furthermore we can eliminate mode softening and relaxation as major sources of mode softening [6,14].)

5. Collective excitations and conformation transitions.

In the above picture of the A-B transition, the mechanism might be understood in terms of a crossover to simple screened Coulomb repulsion as the cation cloud is diluted by water addition. Much more must be known about the structure of the hydration shell before a dynamical pathway for the transition can be constructed. However, it is clear that such a pathway must involve collective excitations of DNA and its hydration shell. The frequencies of collective motion involving parts of the double helix as big as a

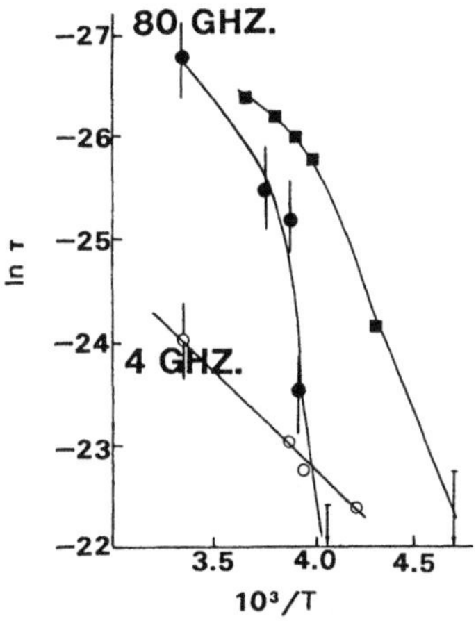

Figure 5. Plot of the natural logarithm of relaxation time with reciprocal relaxation time for a sample with primary hydration (○) and a sample with secondary hydration (●) as measured by Brillouin scattering [7]. The remaining data (■) are for a sample with secondary hydration and are extracted from the Raman measurements of Tominaga et al. [19]. Relaxation frequencies at physiological temperatures are marked in GHz.

basepair lies in the gigahertz region [8] where the coupled dynamics of the DNA-hydration shell system is important [6,7]. We summarize our measurements of the hydration shell relaxation in Fig.5 which shows an Arrhenius plot of relaxation times (in seconds per radian) for the so-called 'primary' and 'secondary' hydration. In the region between the viscoelastic transitions of the primary and secondary shells (4 to 80 GHz.) there are <u>extra degrees of freedom</u> associated with coupled modes of the DNA and <u>the primary shell</u>. This point may be particularly important for the B-Z transition, because the mechanisms that have been proposed (based on 'dry' helix models) involve substantial energy barriers [17,18]. It is possible that a low-energy pathway can be constructed with one or more <u>collective</u> modes of the polymer and its hydration/electrolyte shell.

6. The Hydrogen Bond Melting of the Double Helix.

The bridging hydrogen bonds probably constitute the most nonlinear 'springs' first encountered in the strand separation process. The renormalization of eigenfrequencies due to nonlinearities is straightforward, however the standard selfconsistent phonon theories cannot describe melting which is a consequence of the uneven terms in the potential. Gao and Prohofsky [20] have modified the standard theory to include thermal expansion through the imposition of an additional constraint on the R_T, the thermally expanded mean position. The usual set of equations linking the displacement correlation function, D_{ij}, the eigenvectors and eigenfrequencies, ω_k and the effective spring 'constants, ϕ_{ij},

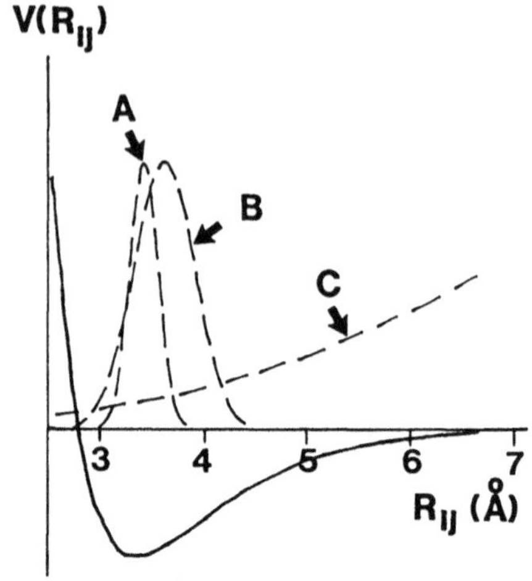

Figure 6. An effective N-N hydrogen bond potential (solid line) together with displacement correlation functions for a) 1K, b) 250K and c) 500K.

$$D_{ij} = D_{ij}(\omega_k, T),$$

$$\omega_k = \omega_k(\phi_{ij}),$$

$$\phi_{ij} = \phi_{ij}(D_{ij}, R_T),$$

are solved self-consistently with an extra equation

$$R_T = R_T(D_{ij})$$

so as to fix R_{ij} halfway between the classical turnaround points. That is

$$V(R_T + u_0) = V(R_T - u_0)$$

determines R_T (u_0 is the classical oscillator amplitude). At high temperatures, this system of equations becomes unstable - no self-consistent solution can be found. This is illustrated in Fig. 6 where the solid line shows an effective N-N potential due to the hydrogen bond, and the dashed lines show the correlation function for N-N distances for a) T= 1K, b) T=250K and c) T= 500K (from ref 20). Calculations for a DNA homopolymer [21,22] predict an instability near the observed melting temperature with no fitting parameters (the instability occurs a little above the measured melting temperature, but this calculation, which is not a free energy calculation, can only set an upper bound). The theory has been reformulated using Green's functions, an approach that allows melting to be studied at local (nucleation) sites. An interesting conclusion is that only certain phonons in very narrow frequency regions contribute to melting. In the perfect polymer this band lies around 85 cm^{-1} [25], close to where modes assigned to hydrogen bond stretch are observed [11]. At a defect, such as an open basepair, different phonon bands are associated with directional melting [26]. That is, low-frequency modes contribute fluctuation amplitude to the hydrogen bonds near the minor groove, while modes near 80 cm^{-1} lead to unwinding in the major groove [26].

Stimulated by this theoretical effort we have begun an experimental study of non-equilibrium phonons in DNA in the low-frequency region (10-200 cm^{-1}).

7. Conclusions.

Although some very elegant theories have given us some feel for the physics of DNA and stimulated many experiments, it is clear that our picture of the polymer is still far too simple. The 'structure' of the surrounding electrolyte is important in bringing about interactions which stabilize conformations. Coupled modes of the DNA and its hydration shell are also of importance. We have discussed the onset of melting as dominated by the bridging hydrogen bonds, but here too we suspect that coupled (non-linear?) modes of the DNA and electrolyte will be important. I end by mentioning the very non-linear optical polarizability of DNA, probably associated with the phosphate groups [27]. The non-linearity of these groups is familiar to the condensed matter community because of the remarkable behavior of such materials as KDP. We expect that DNA will prove at least as interesting.

Acknowledgments

The experimental work referred to was carried out with S.A. Lee, T. Weidlich, N.J. Tao, G. Lewen, J. Powell and C. DeMarco. Support was received from the NSF (PCM8215433), ONR (N00014-84-C-0487)) and EPA (68-02-4105 - this chapter was not subject to EPA review and does not necessarily reflect the agency's views).

References.

1. J.M. Eyster and E.W. Prohofsky, Biopolymers 16, 965-982 (1977).

2. J.A. Krumhansl, G.M. Wysin, D.M. Alexander, A. Garcia, P.S. Lomdahl and S.P. Layne in Structure and Motion: Membranes, Nucleic Acids and Proteins (Eds E. Clementi, G. Corongiu, M.H. Sarma and R.H. Sarma) Adenine, N.Y., 1985, pp 407-416.

3. D.M. Soumpasis, Proc. Nat. Acad. Sci. (USA) 81, 5116-5120 (1984).

4. D.M. Soumpasis, J. Wiechen and T.M. Jovin, "Relative Stabilities and Transitions of DNA Conformations in 1:1 Electrolytes: A Theoretical Study" preprint (1985).

5. S.M. Lindsay in Progress and Challenges in Biological and Synthetic Polymer Research (Eds C. Kawabata and A.R. Bishop) Ohmska, Tokyo, 1986.

6. S.M. Lindsay in Structure and Dynamics in Biomolecules (Ed. E. Clementi) Plenum, NY (1987).

7. N.J. Tao, S.M. Lindsay and A. Rupprecht, Biopolymers, in press (1986).

8. S.M. Lindsay, J. Powell, E.W. Prohofsky and K.V. Devi-Prasad in <u>Structure and Motion: Membranes, Nucleic Acids and Proteins</u> (Eds E. Clementi, G. Corongiu, M.H. Sarma and R.H. Sarma) Adenine, N.Y., 1985, pp 531-551.

9. C. DeMarco, S.M. Lindsay, M. Pokorny, J. Powell and A. Rupprecht, Biopolymers <u>24</u>, 2035-2040 (1985).

10. S.M. Lindsay, J. Powell and A. Rupprecht, Phys. Rev. Lett. <u>53</u>, 1853-1855 (1984).

11. H. Urabe and Y. Tominaga, J. Phys. Soc. Japan <u>50</u>, 3543-3544 (1981).

12. The equilibration of Z-DNA is independent of concentration over six orders of magnitude - F.M. Pohl, A. Ranade and M. Stockburger, Biochim. Biophys. Acta <u>335</u>, 85-92 (1973).

13. This is an extension of an experiment first performed by Herbeck et. al. - R. Herbeck, T.J. Yu and W.L. Peticolas, Biochemistry <u>15</u>, 2656-2660 (1976).

14. S.A. Lee, J. Powell, N.J. Tao, G. Lewen, S.M. Lindsay and A. Rupprecht, to be published.

15. G. Lewen, S.M. Lindsay, N.J. Tao, T. Weidlich, R.J. Graham and A. Rupprecht, Biopolymers <u>25</u>, 765-770 (1986).

16. B.N. Conner, C. Yoon, J.L. Dickerson and R.E. Dickerson, J. Mol. Biol. <u>174</u>, 663-695 (1984).

17. S.C. Harvey, Nucleic Acids Res. <u>11</u>, 4867-4878 (1983).

18. W.K. Olson, A.R. Srivasan, N.L. Marky and V.N. Balaji, Cold Spring Harbor Symp. Quant. Biol. <u>47</u>, 229-242 (1983).

19. Y. Tominaga, M. Shida, K. Kubota, H. Urabe, Y. Nishimura and M. Tsuboi, J. Chem. Phys. <u>83</u>, 5972-5975 (1985).

20. Y. Gao and E.W. Prohofsky, J. Chem. Phys. <u>80</u>, 2242-2243(1984).

21. Y. Gao, K.V. Devi-Prasad and E.W. Prohofsky, J. Chem. Phys. <u>80</u>, 6291-6298 (1984).

22. Y. Kim, K.V. Devi-Prasad and E.W. Prohofsky, Phys. Rev. <u>B32</u>, 5185-5189 (1985).

23. Y. Kim and E.W. Prohofsky, Phys. Rev. <u>B33</u>, 5676-5681 (1985).

24. Y. Kim and E.W. Prohofsky, "Vibrational Fluctuations around a Nucleation Defect of Melting in Poly(dG).Poly(dC)" preprint (1986).

25. E.W. Prohofsky, Comments Mol. Cell. Biophys. <u>2</u>, 65-86 (1983).

26. Y. Kim and E.W Prohofsky, " Predicted Role of Particular Phonons in the Strand Separation Melting of the DNA Double Helix" Preprint (1986).

27. T. Weidlich, S.M. Lindsay and A. Rupprecht, "The Optical Properties of Li- and Na-DNA Films" preprint (1986).

Phonon Anomalies, Structural Transformations, and Precursor Structure in the Martensitic Alloys TiNi(Fe) and $Al_{1-x}Ni_x$

J.A. Krumhansl

Department of Physics, Cornell University, Ithaca, NY 14853, USA

1. ABSTRACT

Recent x-ray and electron diffraction, electron microscope, and neutron studies of the pretransformation and transformation structure of these alloys show that they cannot be attributed to mode softening, charge density waves, or Fermi surface effects, and that tweed structure can have several origins. Rather, it is proposed that these systems exhibit the features expected of highly anisotropic, anharmonically bistable, structural changes involving planes normal to [110] and sliding transversely between alternative metastable positions. Thereby a continuous transition of the nucleation type proposed by deGennes can take place.

2. INTRODUCTION

Structural phase transformations in metallic alloys have been the subject of extensive study [1]. These transformations, taking place as they do at essentially constant composition, and therefore, including thermoelastic martensite transformations, are diffusionless and one can hope to understand their origin through studies of the lattice displacements at the atomic scale. Several experimental techniques are thereby indicated: precision x-ray, electron, and neutron diffraction; inelastic neutron scattering; diffuse scattering and Mössbauer studies; various forms of electron microscopy. It has taken some years for a reasonably complete set of complementary experiments to be done on a given alloy, but this has been accomplished for the martensitic alloys TiNi(Fe) [2] and most recently several Ni-Al alloys [3].

In the most recent work on Ni-Al alloys Shapiro, Noda, Moss and Tanner report experimental results on: (i) TEM of $Ni_{63}Al_{37}$ showing a tweed "precursor" pattern; (ii) elastic scattering neutron diffraction; (iii) phonon dispersion curves for Ni = 63.58, and 50 percent, respectively, show anomalously low [110] TA_2 branches with a dip at $q \neq 0$ which varies with composition. Taken together these provide extremely important information on which to base explanations of the underlying microscopic origins of the structural transformations. Such studies have been interpreted by many authors as follows:

1. That these are soft mode transitions.
2. That the experiments show the existence of a CDW (i.e. Peierls) instability.
3. That Kohn anomalies or other singular Fermi surface effects play a central role in the transformation and precursors.
4. That the highly anisotropic diffuse scattering associated with tweed precursors is due either to special electronic effects or defects. (i.e. embryos). (i.e. Huang scattering).

We proceed to address each of these points.

First, with respect to the soft mode concept as proposed and developed, there is supposed to be a specific lattice displacement mode whose frequency varies strongly with temperature and at some temperature goes to zero, at which point the displacement pattern in this mode provides the structural distortion. While this has been an appealing idea, widely claimed to have been shown, in point of fact there is no experimental example of temperature-dependent softening adequate for transformations among alloys studied to date. Once again this is found to be the case by Shapiro et al. in AlNi(.63). A good discussion of this topic for a series of CuAlNi alloys has been given by Morii and Iizumi [4] (1985); indeed a quote from Ref. 3 (Hoshimo, Shirane, Suezawa, Kajitanii) of that article: "... the temperature-dependent soft phonon was found for $AuCuZn_2$, in which maximum of about 25% reduction of the frequency in the TA_2 phonon branch was observed in the wide temperature range of 270°. However, no critical behavior was observed in this case either. From these experimental results, therefore, the attempt to explain the thermoelastic martensitic transition by the softening of phonon alone is not justified and there should exist some other microscopic mechanism besides the lattice dynamical instability to cause the transition (at Tm) in these cases." We believe that this statement generally stands today, and that the recent studies provide one more case in point. To a large extent the use of the terminology "mode softening" is misleading; we will return to this point below. It is far more descriptive of the true state of affairs to describe the lattice dynamics as anomalously anisotropic (Morii and Iizumi [4]).

One more point in passing: it is almost universally assumed among physicists that continuous phase transformations must be second order, accompanied by a divergent susceptibility. This is an oversimplification; deGennes [5] (1973) provides a clear discussion of the difference between that and a <u>continuous</u> transition to a modulated phase, which he calls a nucleation transition, in which the susceptibility does not diverge (i.e. no mode softens). More attention should be paid to this classification; indeed we believe that it is the latter class which applies to many alloys, including those under discussion here.

With respect to the next points, 2 and 3 above, regarding the evidence for Kohn anomalies or other Fermi surface effects, there are several additional physical facts. First, as pointed out previously for TiNi(Fe) [2] and found again in NiAl the anomalously low phonon branch (TA_2 on [110]) is one for which displacements are transverse to $\underline{q}$, and therefore weakly coupled at best to the electron gas. Second, in both TiNi(Fe) and AlNi(.63) at the temperatures (232K and 295°, respectively) where strong elastic scattering sets in, there are no observed electronic anomalies. Third, an analysis by Varma and Weber [6], and Weber [7] and one by Bruinsma [8] for TiNi(Fe) all lead to the conclusion that Fermi surface singularities at q-phonon = $2K_f$ make little contribution to the anomalous phonon dips at $2K_f$. To quote Weber: "... the anomalies are not caused by structure in the energy denominator (the bare "susceptibility") of Eq. 2; i.e. they are not Fermi-surface anomalies in the Kohn-Overhauser sense. ... very anisotropic electronic band topologies ... are favorable for the occurrence of the phonon anomalies. In Nb this is the "jungle gym" [Fermi surface] ..." To summarize, both circumstantial experimental evidence and realistic theories strongly contradict the Kohn-Overhauser anomaly suggestion as a cause of the anomalous [110] phonon spectrum.

Of course, this line of discussion leads immediately to ask what then determines the phonon spectrum. For example, since the phonon spectrum changes with composition of metal alloy it has been argued that it is changes in the Fermi surface which cause that variation. Any examination in depth of the microscopic factors which determine lattice dynamics reveals the oversimplification, and inadequacy, of such a qualitative argument. Specifically, in references 4, 6, 7, 8 one finds in the transition metals and their compounds that short-range (i.e. core-core) pair, triplet, etc. interactions make the major contributions to the phonon energies; indeed, in insulators [9], where there are manifestly no Fermi surface effects, completely analogous changes in phonon spectra to those seen with changing composition in metallic alloys are found. Thus, many of the phonon anomalies (perhaps most) attributed to the Kohn effect are due, rather, to competing simple interactions between nearest, next nearest, etc. neighbors (e.g. the analyses of Morii and Iizumi [4]) and their change with composition. The coherent potential approximation [10] provides a theoretical framework for such calculations. We wish to point out specifically that if the electronic band functions are made up from tight-binding hydridized s-p-d bonds (which characterize the transition metals) then the "jungle gym" Fermi surfaces referred to by Weber can be obtained; thus, bond bending energies (intrinsically 3-center, next nearest neighbors) fall naturally into Iizumi's scheme for computing phonon spectra. Summarizing these considerations: phonon anomalies may arise from several sources, and in the alloys under discussion it is as likely that they arise from competition between pair, bond-bending, and twisting forces of a localized atomic nature rather than from any Fermi surface effects. What is needed is independent monitoring of the Fermi surface during transformation, such as by EXAFS or photoemission.

What then might be the nature of the dynamical excitations underlying the observed transitions? Following the above arguments we are led to examine transformations in highly anisotropic materials generally, taking the point of view that the experimental features of the phonon spectrum are the truly significant feature, however they arise. Thus we look for transitions in other materials, metals or non-metals, with similar phonon anomalies. In another paper we will discuss the alkali metals in this context. Fortunately, an extensive experimental program has been carried out over several years on the uniaxial ferroelectrics, reviewed recently by Axe, Iizumi, and Shirane [9]. One member of this class (K_2SeO_4) did indeed show reasonably complete mode softening, but in spite of heroic attempts to fit the others to the party line <u>none</u> exhibited significant softening--nonetheless, showing similar structural transformations and precursors. They conclude that an entirely different dynamics, particles fluctuating back and forth between variants (clusters) of the lower temperature state was more representative of the approach to the transformation. This state of affairs can be envisaged in terms of the dynamics of a set of coupled double-well oscillators [11].

This model, entirely different physically from the mode picture, has important consequences. Both the thermal diffuse scattering and the defect-induced Huang scattering are ordinarily computed in the harmonic approximation. De facto, as one approaches a structural phase transition the harmonic approximation is inapplicable. Here, the point we wish to emphasize as strongly as possible is that under those circumstances neither Huang diffuse scattering nor Debye-Waller formula from harmonic theory are of use. Such commonly used general rules as "DW factor can measure mean square displacement" or "defect-induced diffuse scattering falls off as $1/q^2$" are now invalid. Recognizing this, Dash, Johnson, and Visscher [12], and Mair et al. [13] in a series of studies addressed the problem of properly interpreting diffuse scattering under these circumstances. Yet another, and distinct, physical consideration is that these effects become more important the greater the anisotropy [14]. In view of the great phonon (elastic) anisotropy of the alloys under discussion there is then even further reason to be concerned about the breakdown of harmonic formula for diffuse scattering. For these reasons we question the applicability of Huang scattering theory, and the implied "defects" or "embryos", as being required by the data.

On the other hand, whatever the origin, if there is strongly anisotropic diffuse scattering the discussion of transmission electron microscope (TEM) imaging given by Robertson and Wayman [15] demonstrates that the imaging process in itself leads to a tweed structure, where the apparent period in real space depends on the direction of observation. The importance of Tanner's work on these and many other systems showing premartensitic "tweed" is to demonstrate that highly anisotropic diffuse scattering well above the transformation temperature is a ubiquitous phenomenon, which can give important clues to the microscopic processes in the transformation. As a separate matter one may examine possible origins of the diffuse scattering -- thermal diffuse vs. defect Huang -- both in the traditional harmonic approximation, and conclude that the latter fits the observations better, being more "ridge-like" in reciprocal space. On the other hand, intrinsic bistable slow thermal fluctuations of planes transverse to [110] directions, due to the extreme phonon anisotropy plus anharmonicity, can equally well provide suitable diffuse scattering. Undoubtedly, defects could pin such incipient transformation fluctuations, but this would be a secondary rather than primary effect. Both Müller [11] and Iizumi [4,9] have discussed some experimental manifestations of these ideas. To summarize our viewpoint: tweed structure may sometimes be only an imaging artifact of highly anisotropic diffuse scattering, or it may be due to intrinsic modulations, as for example due to arrays of structural discommensurations, but beyond that, to propose that it is specifically electronic in origin or that it demonstrates Huang scattering is inconclusive under the conditions leading up to the structural transformations under discussion in NiTi(Fe) or AlNi.

Perhaps, in closing, it is worth providing an overall summary. First, we reiterate our beginning statement: the experimental programs on TiNi(Fe) and AlNi are of great value in developing new insight into structural transformations in alloys. Second, in contrast to the views of the authors of the recent papers cited we believe that these experiments provide further strong evidence that: (i) soft mode theories are inapplicable to these alloy transformations; (ii) that charge density wave or Fermi surface effects do not likely play a significant role; (iii) that imaging and other studies of transformation precursors show them to be a consequence of the great phonon anisotropy, and do not necessarily imply embryos or defect (Huang) scattering. Third, instead of the traditional interpretations (which we feel have been in question for some time) it is becoming clear that we are dealing with continuous transitions which deGennes [5] refers to as the nucleation type, spatially inhomogeneous and without divergent critical scattering. The kind of theory which should be applied is that discussed by Bruce [14], and now being explored by Kerr and Bishop [15] using computer simulation. We will present the appropriate theory elsewhere in detail.

I am indebted to S. M. Shapiro, J. Z. Larese, S. C. Moss, and L. E. Tanner for many discussions of their results and interpretation prior to publication; the present ideas, dissenting as they do, are put forward in the spirit of widening our views on some very interesting phenomena. I am also indebted to G. R. Barsch and G. B. Olson for ongoing general discussions of martensitic transformations. I wish also to acknowledge the hospitality of Brookhaven National Laboratory, and the continued interest of colleagues at Los Alamos.

References

1. Modulated Structures - 1979 (Kailua Kona, Hawaii) Ed. J. M. Cowley, J. B. Cohen, M. B. Salamon, B. J. Wuensch (AIP, New York, 1979); Proceedings of ASM-MSD Symposium, New Orleans (1986), to be published.
2. S. M. Shapiro, Y. Hada, Y. Fujii, and Y. Yamada, Phys. Rev. B 30, 4314 (1984); M. B. Salamon, M. E. Meichle, and C. M. Wayman, Phys. Rev. B 31, 7306 (1985).
3. S. M. Shapiro, J. Z. Larese, Y. Hada, S. C. Moss and L. E. Tanner (private communication, to be published).
4. Y. Morii and M. Iizumi, J. Phys. Soc. Japan 54, 2948 (1985).
5. P. G. deGennes, in "Fluctuations, Instabilities, and Phase Transitions," Geilo, Norway, 1973, Ed. T. Riste (Plenum, NY, 1973).
6. C. M. Varma and W. Weber, Phys. Rev. Lett. 17, 1094 (1977).
7. W. Weber, p. 131 in "Superconductivity in d- and f-band Metals." Ed. (Academic Press, NY, 1980).
8. R. Bruinsma, Phys. Rev. B 25, 2951 (1982).
9. J. D. Axe, M. Iizumi, and G. Shirane, p. 1 in "Incommensurate Phases in Dielectrics 2" Ed. R. Blinc and A. P. Levanyuk. (Elsevier, B. V. 1986). Also see H. Bestgen, Solid State Commun. 58, 197 (1986), and R. Blinc, J. Dolinsek, P. Prelovsek and K. Hamono, Phys. Rev. Lett. 56, 2387 (1986).
10. R. J. Elliott, J. A. Krumhansl, P. L. Leath, Rev. Mod. Phys. 46, 465 (1974). cf. p. 527.
11. K. A. Müller, p. 68 in "Statics and Dynamics of Nonlinear Systems" Ed. G. Benedek, H. Bilz, and R. Zeyher (Springer, NY, 1983); also p. 1 in "Nonlinear Phenomena at Phase Transitions and Instabilities" (Plenum, NY, 1982); also this conference.
12. J. G. Dash, D. P. Johnson, and W. M. Visscher, Phys. Rev. 168, 1087 (1968).
13. S. L. Mair, Phys. Rev. B 30, 6560 (1984), and references therein.
14. A. D. Bruce, Adv. Phys. 29, 111 (1980).
15. I. M. Robertson and C. M. Wayman, Phil. Mag. A 48, 421 (1983); A 48, 443 (1983); A 48, 629 (1983).
16. W. C. Kerr and A. R. Bishop, Phys. Rev. B (in press), and these proceedings. See also work of T. R. Koehler described in the article of A. R. Bishop in "Proceedings of the International Conference on Lattice Dynamics, Paris 1977, Ed. M. Balkanski (Flamarion, Paris, 1978).

Temperature Dependence of the Soft Mode for Structural Phase Transitions in Highly Anisotropic Systems

A.R. Bishop[1] *and W.C. Kerr*[2]

[1]Theoretical Division, Los Alamos National Laboratory, Los Alamos, NM 87545, USA

[2]Department of Physics, Wake Forest University, Winston-Salem, NC 27109, USA

We describe here one result from an extensive molecular dynamics (MD) study of structural phase transitions [1]. The models which have been used for structural phase transition studies have a one-component displacement variable associated with each lattice site moving in a local double-well potential. This potential gives rise to two different excitations: phonons and domain walls. The frequency of the particular phonon with displacement pattern matching the distortion occurring at the transition decreases toward zero as the critical temperature T_c is approached (the "soft mode"). The domain walls separate correlated regions possessing short-range order. Nonlinear theory indicates that the phonon softening should not be complete and should reach its minimum at a temperature T_s greater than T_c. However for isotropic systems T_s and T_c are very close and phonon softening is almost complete. Therefore these major features expected from nonlinear theory can be probed more precisely by considering weakly coupled chains of double-well local potentials. In the limit of uncoupled chains, $T_c = 0$, whereas T_s is non-zero and depends on the energy to create a 1-d domain wall or soliton.

The potential energy function describing this situation is

$$\Phi = \sum_{i,n} \left\{ \tfrac{1}{2}Au_{i,n}^2 + \tfrac{1}{4}Bu_{i,n}^4 + \tfrac{1}{2}C(u_{i+1,n} - u_{i,n})^2 + \tfrac{1}{2}D(u_{i,n+1} - u_{i,n})^2 \right\}. \tag{1}$$

Here $u_{i,n}$ is the displacement of the particle associated with the i-th position along the n-th chain, the parameters $A < 0$ and $B > 0$ describe the double-well potential at each lattice site, and C and D are the intra- and inter-chain coupling constants, respectively. For the simulations described here $C \gg D$.

The potential function (1) implies a certain nonlinear equation of motion. In the 1-d limit ($D = 0$) and the continuum limit, this equation becomes the 1-d ϕ^4 partial differential equation, which has the soliton solution

$$u(x,t) = u_0 \tanh[(x - Vt)/\sqrt{2}\,\xi] \tag{2}$$

(u_0 and ξ are certain combinations of the parameters in (1)). This solution describes a moving boundary between regions where particles are in the left or right half of the double-well potential. Since our 2-d system ($D \neq 0$) is weakly-coupled, we can expect nearly one-dimensional domain walls to appear in our simulations.

We have carried out computer solutions of the equations of motion implied by (1) for a system of 10 chains, each consisting of 150 particles. The values used for the coupling constants were $C = 4.0$, $D = 0.01$. Thermodynamic properties (specific heat, susceptibility, etc.) have been calculated and indicate that the system has a continuous structural phase transition at the temperature $k_BT_c \approx 0.51(|A|^2/B)$. We have also carried out extensive calculations of dynamic properties, specifically frequency and wavevector-dependent spectral functions for time-dependent correlation functions. Further details about the integration scheme and these results are contained in [1]. Here we concentrate on one particular aspect of the dynamics of this highly anisotropic system, namely the temperature dependence of the soft mode frequency. We will do this with a pictorial device that we call a "soliton detector" along with the power spectrum of the soft mode.

Since each particle is moving in a double-well potential, a coarse-grained way of viewing its motion is to record whether it is in the right- or left-hand well. To do this, we have made pictures which detect the sign of the displacement: as a function of time the two possibilities are distinguished by a black or white mark. Figure 1 shows one of the results from

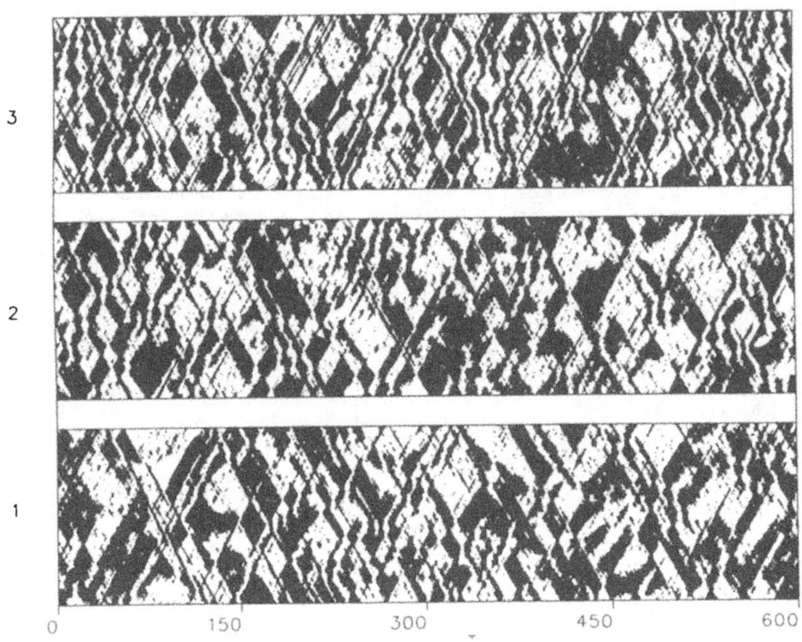

Fig. 1. Soliton detector at $k_BT = 0.90\ (|A|^2/B)$

this "soliton detector". In this picture time is plotted along the horizontal axis [in units of $\sqrt{(M/|A|)}$]. The vertical axis shows three consecutive chains of the system; the chain number n is shown along the left side. Within each chain there are 150 positions corresponding to each of the sites along a given chain, labelled by i; these are also plotted vertically. If the displacement $u_{i,n}(t)$ is positive, then a black dot is put at the (t,i)-th position on the n-th chain. It the displacement is negative, then a white mark is plotted at that position. With this scheme a soliton or domain wall is the boundary between large black and white regions (cf. (2)). This particular picture is at the temperature $k_BT = 0.90\ (|A|^2/B)$. At higher temperatures the soliton detector shows rapid incoherent change back and forth between black and white. At about this temperature the black and white regions begin to last for sufficient time intervals that it becomes reasonable to talk about the propagation of solitons. As T is lowered further, the extent of the black and white regions becomes larger both in the space and time directions, showing the growth of the correlated regions and their longer lifetimes. Below T_c the pictures become almost completely of one color, indicating the long-range order in the low-temperature phase.

Further indications about the atomic motions can be obtained from the spectral function for the order parameter. The order parameter is proportional to the $\vec{q} = 0$ spatial Fourier component of the displacement field. That variable, the correlation function, and the corresponding spectral function are defined by

$$u(\vec{q},t) = \frac{1}{\sqrt{N}}\sum_{\vec{R}} e^{-i\vec{q}.\vec{R}} u_{i,n}(t); \qquad D(\vec{q},t) = \langle u(\vec{q},t)u(-\vec{q},0)\rangle;$$

$$D(\vec{q},\omega) = \int dt\ e^{i\omega t} D(\vec{q},t). \tag{3}$$

[$\vec{R}$ denotes the lattice sites indexed by (i,n)]. Figure 2 shows the order parameter ($\vec{q} = 0$) spectral function at $k_BT = 0.99\ (|A|^2/B)$. A central peak and a phonon peak are clearly

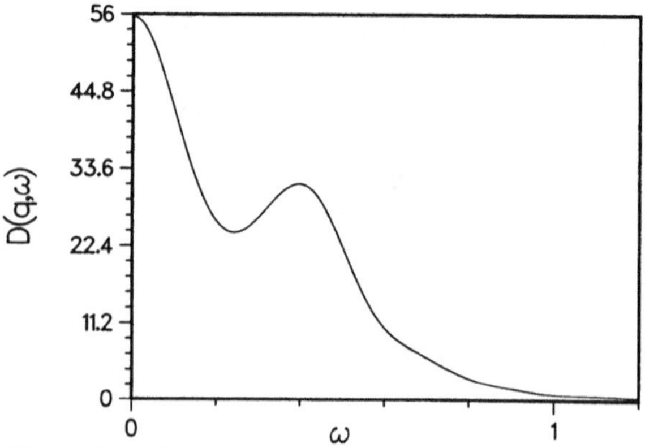

Fig. 2. Order parameter spectral function at $k_BT = 0.99\ (|A|^2/B)$

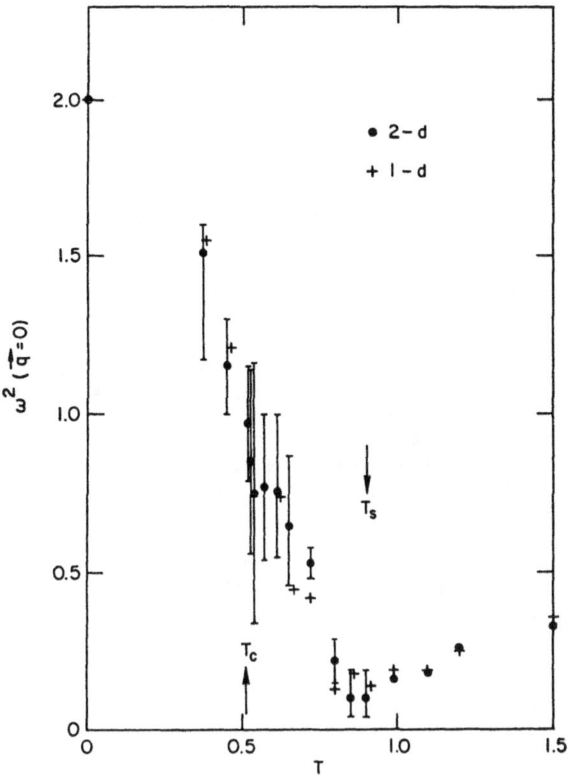

Fig. 3. Square of the soft-mode frequency versus dimensionless temperature $k_BT/(|A|^2/B)$

evident in this spectrum. At higher temperatures the central peak has less weight, and at lower temperatures it grows to become the overwhelmingly dominant feature of the spectrum near T_c.

Figure 3 shows the square of the soft mode frequency ω_s versus temperature both for the 2-d system and for the 1-d system obtained by setting the interchain coupling constant D to zero. The important point to note here is that the minimum in ω_s does not occur at the transition temperature T_c, but is at a so-called "soft mode" temperature T_s, which is near 0.90 $(|A|^2/B)$. The softening occurs with the same temperature dependence for both the 2-d and 1-d systems, indicating that the mechanism is the same in the two cases.

For temperatures much above T_s the soliton detector pictures indicate that the particles oscillate completely across the double-well potential. For T near T_s, where solitons start to become evident in the soliton detector, the particles oscillate within one well and are flipped back and forth between the wells as the solitons pass by. Near T_s a separation of time scales takes place between the oscillation period within one well and the time between flips from one well to the other. Above T_s there is no such distinction. When the particles start to oscillate within one well rather than across both wells, then the soft-mode frequency starts to increase. The correlations producing this temperature dependence are essentially one-dimensional.

Two-dimensional correlations set in within a narrow temperature interval around T_c. These are shown by a strong narrowing and intensity increase of the central peak part of the spectral functions in the 2-d system relative to the central peak in the 1-d system, which is due to soliton propagation. The details of this argument are given in [1].

In conclusion we have shown that describing the dynamics of structural phase transitions in highly anisotropic systems requires the introduction of another characteristic temperature in addition to the transition temperature. This soft-mode temperature is identified with the minimum of the soft-mode frequency and results from the appearance of 1-d soliton propagation along the chains.

This work has been supported at LANL by the U.S. Department of Energy and partially supported at Wake Forest University by a grant from the Research Corporation.

References

1. W. C. Kerr and A. R. Bishop: Phys. Rev. B, to be published.

A Model for the B→Z Transition of DNA Involving Solitary Excitations

Z. Zang[1] *and W. Olson*[2]

[1]Department of Physics, Rutgers University, Piscataway, NJ 08854, USA
[2]Department of Chemistry, Rutgers University, Piscataway, NJ 08854, USA

1. Introduction

Since the discovery of the left-handed Z-form deoxyribonucleic acid (DNA) in 1979 [1], much work has been done both experimentally [2-4] and theoretically [5-7] in trying to understand the detailed mechanism of the transition from the standard right-handed B form to this peculiar left-handed Z form. Among the questions being asked frequently is whether the hydrogen bonds connecting the two bases of a base pair are broken during the transition.

In this report, based on previous work on the existence of open states in DNA and the descriptions of those open states as solitary excitations [8- 11], we propose a simplified three-dimensional plane base rotator model for the B → Z transition. In contrast to HARVEY's model [6] which requires base pair flipping with hydrogen bonds left intact, our model involves breaking and rejoining of hydrogen bonds with the two bases of a pair flipping over separately before the rejoining processes. A detailed presentation of our model is given in Sec.2. We derive in Sec.3 the equations of motion, and show that in a continuum approximation those equations become a set of mutually coupled sine-Gordon (SG) and double sine-Gordon (DSG) equations. Soliton-like solutions are expected from these equations, which is further supported by preliminary numerical calculations. We also predict the possible existence of the Hoogsteen base pairs within our model, which, if experimentally verified, may supply the first unambiguous evidence for a H-bond breaking mechanism of the B → Z transition.

2. Model Hamiltonian

The B and Z conformations of DNA are different in many aspects. Yet the most distinct differences between them are that the B form is a right-handed double helix and that all of its base pairs face up if viewed by some convention, while the Z form is left-handed, with all of the base pairs facing down.

In modelling this dramatic transition, HARVEY [6] proposed that a cavity is allowed to form as a result of unwinding (untwisting) of the duplex. The cavity reaches its maximum value when the twisting angle is zero, with a space of 7 Å on either side of a base pair. Once the cavity is formed a base pair flips over as a unit with the standard Watson-Crick hydrogen bonding left intact. The duplex then twists in the opposite direction, completing the transition from the right-handed to the left-handed double helix. This model, however, violates the nearest neighbor exclusion principle found in planar drug intercalation experiments [12]. In addition, it may require much higher activation energy for a base pair to flip over within the helix than observed experimentally.

With these considerations, we propose the following alternative picture for the transition. The hydrogen bonds of a base pair in B-DNA are first broken, then each of the now separated bases of a pair can rotate out of the helix around an axis parallel to the helical axis (chosen to be the z axis). As suggested by ENGLANDER et al.[9], those open states may be described as solitary excitations, which were also discussed by YOMOSA [11] and by TAKENO and HOMMA [10] using the simplified two-dimensional plane base-rotator model. What is added here to each base is the freedom of rotation around axes perpendicular to the helical axis in a way leading to the flipping of the base, thus changing the plane-base rotator to three-dimensional in character. This flipping process begins when the base is out of the helix, and is completed (i.e., from facing up to facing down) before the base rotates back into the helix and the hydrogen bonds rejoin, ending in a form consistent with Z-DNA. We also assume that accompanying these rotations the sugar and the phosphate of the nucleotide to which the base belongs move in such a way as to have the corresponding conformation consistent with the B or Z form. The whole picture travels down along the B-DNA chain in the form of topological solitary excitations, leaving the DNA in Z form behind.

The picture adopted here is very similar to that suggested by WANG et al. [1] in their first report on Z-DNA, except that we explicitly require the flipping-over processes to occur at the outside of the helix.

To give a mathematical description of the transition stated above, we directly generalize the scheme used by TAKENO and HOMMA [10]. For a given base pair n, four conformationally significant variables are chosen: two (ϕ_n and ϕ_n') of them describe the rotations perpendicular to the z-axis, the other two (χ_n and χ_n') describe the flipping processes. The on-site and intra-strand base-base coupling potentials in ϕ_n and ϕ_n' are given respectively as [10]:

$$\sum_n v_1(\phi_n,\phi_n')=\sum_n\{h_{1n}[(1-\cos\phi_n)+(1-\cos\phi_n')]-\lambda_n[1-\cos(\phi_n-\phi_n')]\} \tag{2.1}$$

$$\sum_n u_1(\phi_n,\phi_n')=\sum_n\{J_n[1-\cos(\phi_{n+1}-\phi_n)]+J_n'[1-\cos(\phi_{n+1}'-\phi_n')]\} \tag{2.2}$$

where (2.1) is the on-site potential, (2.2) the intra-stand potential, h_{1n} and λ_n the interaction constants characterizing the local-field energy and the hydrogen-bonding energy of the nth base pair, J_n and J_n' the interaction constants associated with the intra-strand stacking of nth bases in complementary strands. ϕ_n and ϕ_n' are chosen in such a way that zero or 2π corresponds to the B or/and Z form. For details, see [10].

Acceptable phenomenological interaction energies in χ_n and χ_n' should guarantee the right order of motion in time, i.e., the flipping motion of a base begins after the base is rotated more or less out of the double helix. This suggests that the on-site potential in χ_n and χ_n' should also be dependent on ϕ_n and ϕ_n'. We choose it in the following form:

$$\sum_n v_2(\phi_n,\phi_n';\chi_n,\chi_n')=\sum_n h_{2n}[(1+\cos\phi_n)(1-\cos 2\chi_n)+(1+\cos\phi_n')(1-\cos 2\chi_n')], \tag{2.3}$$

where $h_{2n} >> h_{1n} > 0$, so that the barrier between zero (corresponding to B form) and π (corresponding to Z form) in χ_n or χ_n' is very high unless ϕ_n or ϕ_n' is not too far away from the value of π. At $\phi_n, \phi_n' = \pi$, the on-site potential is flat, which is easily feasible and is consistent with empirical energy estimates [13]. We did not include the correlation term $\cos(\chi_n - \chi_n')$ in our on-site potential, since the flippings of the two bases of a pair are assumed to occur fairly independently.

The intra-strand base-base coupling in χ_n and χ_n' can be written down accordingly:

$$\sum_n u_2(\chi_n, \chi_n') = \sum_n \{K_n[1-\cos(\chi_{n+1}-\chi_n)] + K_n'[1-\cos(\chi_{n+1}'-\chi_n')]\}, \tag{2.4}$$

where K_n and K_n' are coupling constants.

The Hamiltonian can then be readily written as:

$$H = \sum_n [(I_{1n}/2)\dot{\phi}_n^2 + (I_{1n}'/2)\dot{\phi}_n'^2 + (I_{2n}/2)\dot{\chi}_n^2 + (I_{2n}'/2)\dot{\chi}_n'^2 + v_1(\phi_n, \phi_n') + u_1(\phi_n, \phi_n') + v_2(\phi_n, \phi_n'; \chi_n, \chi_n') + u_2(\chi_n, \chi_n')], \tag{2.5}$$

where all I's are moments of inertia.

3. Equations of Motion and Discussion

Using the Hamiltonian in (2.5), assuming n independence for all the coupling constants, and letting J=J', K=K', we can derive straightforwardly equations of motion for the system, which in the continuum approximation can further be simplified as

$$\frac{\partial^2 \phi}{\partial z^2} - \frac{1}{c_1^2}\frac{\partial^2 \phi}{\partial t^2} = \frac{1}{l_1^2}[\sin\phi - \eta\sin(\phi - \phi')] - \delta\sin\phi(1 - \cos 2\chi), \tag{3.1}$$

$$\frac{\partial^2 \phi'}{\partial z^2} - \frac{1}{c_1^2}\frac{\partial^2 \phi'}{\partial t^2} = \frac{1}{l_1^2}[\sin\phi' - \eta\sin(\phi' - \phi)] - \delta\sin\phi'(1 - \cos 2\chi'), \tag{3.2}$$

$$\frac{\partial^2 \chi}{\partial z^2} - \frac{1}{c_2^2}\frac{\partial^2 \chi}{\partial t^2} = \frac{1}{l_2^2}\sin 2\chi\,(1 + \cos\phi), \tag{3.3}$$

$$\frac{\partial^2 \chi'}{\partial z^2} - \frac{1}{c_2^2}\frac{\partial^2 \chi'}{\partial t^2} = \frac{1}{l_2^2}\sin 2\chi'(1 + \cos\phi'). \tag{3.4}$$

In (3.1)-(3.4), $c_1^2 = Ja^2/I_1$, $c_2^2 = Ka^2/I_2$, $l_1^2 = Ja^2/h_1$, $l_2^2 = Ka^2/2h_2$, $\eta = \lambda/h_1$, and $\delta = h_2/Ja^2$, where a is defined as $a = z_{n+1} - z_n$.

Analytical solutions to (3.1)-(3.4) are still lacking. Despite this, for the following reasons we believe in the existence of soliton-like solutions to them which are responsible for the B → Z transition. First, without the couplings between the variables (ϕ,ϕ') on the one hand and (χ,χ') on the other hand, the coupled SG and DSG equations (3.1)-(3.4) are simplified to four uncoupled SG and DSG equations, with explicitly-given analytical 2π kink solutions for ϕ and ϕ' and π kink solutions for χ and χ' [10,14]. When the conformation is close to B form DNA, these coupling terms only give small perturbational contributions, thus the original picture of solitary excitations should not change much. Second, equations (3.1)-(3.4) are invariant in replacing ($\phi,\phi';\chi,\chi'$) by ($\phi+2m\pi,\phi'+2m'\pi;\chi+n\pi,\chi'+n'\pi$). In particular, as shown in Fig.1, if all the coupling constants are properly chosen, the configurations B = (0,0;0,0) and Z = ($2\pi,2\pi;\pi,\pi$) of a base pair correspond to two degenerate ground states of the total on-site potential, indicating that there exists a path connecting them which is solitary in character.

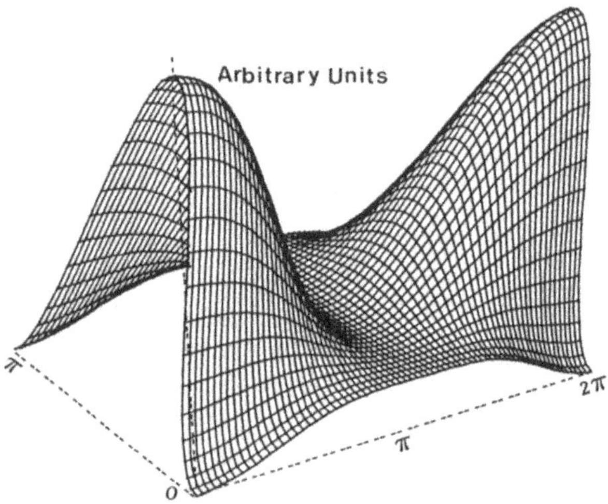

Fig.1. Total On-site Potential (with $h_1 = 1$, $\lambda = -1.5$, $h_2 = 6$)

In Fig.1, we choose $\eta = -1.5$ so that the on-site potential has a relative minimum position at ϕ or $\phi' = \pi$. This relative minimum favors the flipping of the base. In addition, the influences of environmental conditions on the transition could possibly be incorporated as the cross-over from the region $\eta > -0.5$ where there is no relative minimum at π to the region $\eta < -0.5$ where there is one.

Figure 1 also indicates that among the various degrees of freedom of motion of a base pair, there is an effective path along which the energy scale is relatively low. The corresponding effective coordinate may be a good candidate for the hidden degree of freedom suggested by LINDSAY [15]. From Fig.1 we estimate the activation energy of the transition to be around 30 kcal/mole, which is in agreement with the experimental value of 23 kcal /mole [16].

There are two more degenerate ground states available for a given base: (ϕ,χ) = (2π,0) and (0,π). We predict that proper combinations of the four degenerate states for one base and the four for its complement may give us a Hoogsteen base pair as a result of only one base having flipped over. If existing, these kinds of " mutated " base pairs are most likely to occur within the B-Z junction region. WANG et al. [17] and Quigley et al. [18] have observed the occurrence of Hoogsteen base pair in their experiments on drug intercalations to DNA, yet to our knowledge until now there is no direct evidence for the appearance of Hoogsteen base pairs in B → Z transition of pure DNA segments. Verification of the prediction by experiment would be a strong enough support for a H-bond breaking mechanism of the transition.

Another point worth mentioning is the so-called " conjugated " motion of the bases in which $\phi' = -\phi$. Equations (3.1)-(3.4) are then reduced to only two coupled ones. These two equations represent a new two-component field problem, which itself is interesting mathematically.

4. Numerical solutions

To gain more confidence on the existence of topological soliton solutions, we have made an extensive numerical calculation on the difference equations without invoking the continuum approximation. These equations are obtained by the extrema of the total potential with respect to ϕ, ϕ', χ and χ', which in the case of " conjugated " motion take the form:

$$\sin(\phi_{n+1}-\phi_n)-\sin(\phi_n-\phi_{n-1}) = \frac{h_{1n}}{J_{1n}}\{\sin\phi_n[1-\frac{h_{2n}}{h_{1n}}(1-\cos 2\chi_n)]-\frac{\lambda_n}{h_{1n}}\sin 2\phi_n\}, \tag{4.1}$$

$$\sin(\chi_{n+1}-\chi_n)-\sin(\chi_n-\chi_{n-1}) = \frac{2h_{2n}}{J_{2n}}\sin 2\chi_n(1+\cos\phi_n). \tag{4.2}$$

Using the method of standard mapping as used by TAKENO and HOMMA [10], we do find among various kinds of solutions those ones which correspond to the soliton solutions discussed above in the continuum approximation. The spatial length of the solutions which are responsible for the B → Z transition depends critically on the values of coupling constants, and are generally ten to twenty base pairs long if those constants are in the accepted ranges of magnitudes. Details will be published elsewhere.

5. Conclusion

To conclude, we have proposed a model for the B → Z transition of DNA which involves hydrogen bond breaking and solitary excitations. The model may be verified experimentally by looking for the occurrence of Hoogsteen base pairs during the transition.

Acknowledgments

The authors wish to thank M.H. Hao for help on numerical calculations and to acknowledge support from USPHS Gtant GM20861. Z.Y. Zhang thanks S. Ji, X.G. Wen, J. Sak, A. C. Scott and S.E. Truilinger for suggestions and discussions.

References

1. A.H.J. Wang, G.J. Quigley, F.J. Kolpak, J.L. Crawford, J.H. van Boom, G.A. van der Marel and A. Rich: Nature 282, 680 (1979)

2. J. Klysik, S.M. Stirdivant, J.E. Larson, P.A. Hart and R.D. Wells: Nature 290, 672 (1981)
3. M.H. Sarma, G. Gupta, M.M. Dhinga and R.H. Sarma: J. Biomol. Struct. and Dyn. 1, 59 (1983)
4. J. Feigon, A.H.J. Wang, G. van der Marel, J.H. van Boom and A. Rich: Science 230, 82 (1985)
5. W.K. Olson, A.R. Srinivasan, N.L. Marky and V.N. Balaji: Cold Spring Harbor Symp. Quant. Biol. 47, 229 (1983)
6. S.C. Harvey, Nucleic Acids Res. 11, 4867 (1983)
7. A. Sarai and R.L. Jernigan: J. Biomol. Struct. and Dyn. 2, 767 (1985)
8. C. Mandal, N.R. Kallenbach and S.W. Englander: J. Mol. Biol. 135, 391 (1979)
9. S.W. Englander, N.R. kallenbach, A.J. Heeger, J.A. Krumhansl and S. Litwin: Proc. Natl. Acad. Sci. USA 77, 7222 (1980)
10. S. Takeno and S. Homma: Prog. Theor. Phys. 70, 308 (1983); S. Homma and S. Takeno: ibid. 72, 679 (1984)
11. S. Yomosa: Phys. Rev. A 27, 2120 (1983); 30, 474 (1984)
12. P.J. Bond, R. Landgridge, K.W. Jennette and S.J. Lippard: Proc. Natl. Acad. Sci. USA 72, 4825 (1975)
13. W.K. Olson and A.R. Srinivasan: unpublished data
14. C.A. Condat, R.A. Guyer and M.D. Miller: Phys. Rev. B 27, 474 (1983)
15. S.M. Lindsay: This Volume
16. F.M. Pohl and T.M. Jovin: J. Molec. Biol. 67, 375 (1972)
17. A.H.-J. Wang, G. Ughetto, G.J. Quigley, T. Hakoshima, G.A. van der Marel, J.H. van Boom and A. Rich: Science 225, 1115 (1984)
18. G.J. Quigley, Ugheeto, G.A. van der N Marel, J.H. van Boom, A.H.-J. Wang and A. Rich: Science 232, 1255 (1986)

Part VII

Spin-Glasses and Random Field Systems

Recent Experimental Results in Spin-Glass Dynamics

N. Bontemps

Laboratoire d'Optique Physics (ER 5 du CNRS), 10 rue Vauquelin, F-75231 Paris Cedex 05, France

1. Introduction

There seems to be now a number of experimental evidences that dynamics in spin-glasses may provide informations which are relevant to the equilibrium properties : several groups have successfully performed dynamic scaling in a variety of spin-glasses /1-5/; the dynamic exponent is experimentally found to scale between roughly 7 and 8 which compares very well to the exponent determined in the latest numerical simulations /6/; this can be understood through the occurrence of a phase transition at finite temperature.

When extending the dynamic scaling analysis to finite fields, the characteristic relaxation time may be shown to diverge along a line whose equation involves static critical exponents /1/. The divergence of the relaxation time might be a unique feature to characterize such a line. Whether it is an actual transition line is still a matter of controversy /7/.

Though remanent properties of spin-glasses have been known for a very long time /8/, their actual understanding is not yet fully achieved. I shall put emphasis on thermoremanent magnetization (TRM) measurements and I will discuss two of the latest interpretations. The first one relies on the fit of the decay of the TRM to a stretched exponential form and is said to map onto a " simple " theoretical calculation performed in the framework of the random free energy model /9,10/. In the second one, which refers to measurements performed over a larger range of time, the whole decay can be accounted for by using a power law multiplied by a stretched exponential /11,12/. Recent noise measurements /13/ appear to support this analysis. The situation, as will be discussed below, is still controversial.

2. Dynamic Scaling : the Search for the Spin-glass Transition

More and more accurate work has been devoted to the characterization of the spin-glass transition by means of scaling ; the non-linear susceptibility has been convincingly shown to diverge at Tg /14,15,16/ ; the β and γ exponents have been determined by many groups /16/ and the scatter which is exhibited might arise, as pointed out by H.BOUCHIAT /16/ from the temperature and field range investigated, which in turn may be relevant to the Heisenberg regime (high field) or the Ising regime (low field).

In contrast, there have been until recently very few systematic investigations of the dynamic properties of spin-glasses above Tg from the point of view of scaling. Yet dynamic scaling could be considered as a natural consequence of the occurrence of a phase transition in spin-glasses. I shall review the latest work where the analysis of dynamical susceptibility data in a variety of spin-glasses, in presence of a zero or a finite - applied magnetic field, leads to a satisfactory scaling for a characteristic relaxation time and yields a well-identified transition temperature.

One of the most reliable way of defining a characteristic time in spin-glasses is to measure the phase lag of the ac susceptibility with respect to the modulated field /1,17/. One determines the set of the (Ω , T) points which give rise to the same (small) value of the phase lag. Provided this value is small enough compared to 1, the phase lag can be shown to be equal to $\Omega\bar{\tau}$ where $\bar{\tau}$ is an "average" relaxation time /17,1/. This procedure establishes therefore the temperature dependence of the "average" time .

Dynamic scaling in the vicinity of a finite transition temperature Tc is then assumed for all relaxation times, according to the same scaling relation , which states that the time is proportional to a power z of the correlation length

$$\tau \propto \xi^{z} \qquad (1)$$

hence

$$\tau = \tau_0 \{ (T-Tc)/Tc \}^{-z\nu} \qquad (2)$$

since the correlation length diverges at Tc (exponent ν).

An alternative model which assumes a zero transition temperature has been suggested by BINDER et al /18/,giving rise to the so-called "logarithmic scaling" :

$$\mathrm{Ln}(\tau/\tau_0) \propto T^{-z\nu} \qquad (3)$$

Table 1 shows the results which have been obtained by different groups testing (2) ; table 2 shows the results if fitting the data to (3). We point out that not all the groups have used the criterion described above. Some of them /3,4/ have defined the temperatures associated to a given experimental frequency from the inflexion point exhibited by the imaginary part of the susceptibility ; such a choice could alter slightly the values of the dynamical exponent $z\nu$ and of the transition temperature Tc (the latter is in most cases the adjustable parameter in the fitting procedure). BEAUVILLAIN et al /5/ have suggested another criterion, namely the study of the breakaway point between the ac and the dc susceptibility ; they argue that, from an experimental point of view, this point is less difficult to locate that the onset of the phase lag but that both criteria are very similar.

There is quite a remarkable agreement between the various exponents $z\nu$, which range between 7 and 8, the computed value being 7.2 ; also the transition temperatures compare fairly well with the estimates that one can get from static properties (maximum of the dc susceptibility /1/, static scaling /2,3,4/. Finally, the microscopic times derived from

Table 1 : Values of the dynamic exponents $z\nu$, the transition temperature Tc and the microscopic time τ_o from dynamic scaling (D) compared to static (S) determinations in various spin-glasses

Compound	Ref.	$z\nu$	T_c D	T_c S	$\beta+\gamma$ D	$\beta+\gamma$ S	τ_0	h/kT_c
$Eu_{.4}Sr_{.6}S$	/1/	8.2	1.50	1.51	4		$3\ 10^{-12}$	$5\ 10^{-12}$
$CdCr_{.85x2}In_{.15x2}S_4$	/2/	6.9	16.7	16.75	3.5		$5\ 10^{-14}$	$5\ 10^{-13}$
$Al_2Mn_3Si_3O_{12}$	/3/	7.0	2.98	2.95	3.3	4.5		
$(Fe_{.15}Ni_{.85})_{75}P_{16}B_6Al_3$	/4/	8.2	22.4	22.6	3.8	3.78	$2\ 10^{-13}$	$3\ 10^{-13}$
$(MnF_2)_{.65}(BaF_2)_{.15}(NaPO_3)_{.2}$	/5/	7.7	2.79	2.79		4.45		
SIMULATIONS	/6/	7.2						

Table 2 : values of the dynamic exponent $z\nu$ and of the microscopic time from "logarithmic" scaling

Compound	Ref.	$z\nu$	τ_0
$Eu_{0.4}Sr_{0.6}S$	/1/	9 ± 1	10^{-8}
$CdCr_{2x0.85}In_{2x0.15}S_4$	/2/	10.6 ± 1	$8\ 10^{-10}$
$Al_2Mn_3Si_3O_{12}$	/3/	8.5 ± 1	
$(Fe_{0.15}Ni_{0.85})_{75}P_{16}B_6Al_3$	/4/	14.3 +	$2\ 10^{-7}$

the scaling are very close to the natural time scale of the various systems, which is expected to be the inverse of the exchange frequency.

Such an overall agreement is impressive and suggests strongly the occurrence of a phase transition at a finite temperature.

In contrast, in case of the so-called "logarithmic scaling" (3), the exponents are significantly more scattered and the microscopic times are unusually long ; in particular, it was pointed out by SVEDLINDH et al /4/ that shorter times had been measured by neutron experiments in the metallic spin-glasses. Therefore this latter model yields less acceptable results than the former one.

Next, one may wonder whether a similar analysis could be worked out to look for a transition line in the field-temperature plane /19/. Indeed, there is no expected signature of such a critical line through the equilibrium properties. On the contrary, if there is a divergence of the energy barriers between valleys in presence of a finite field, the relaxation times would diverge at the transition ; in order to investigate such a possibility, one has to look for the asymptotic behaviour of the characteristic time which we have defined as being $\bar{\tau}$. To do so, we first assume that dynamic scaling holds in a finite field, hence that the time remains proportional to a power of the correlation length ; we write down the field dependence of the correlation length ; one gets then /17/ :

$$\tau = \tau_o t^{-z\nu} f(H^2 / t^{\beta+\gamma}) \qquad (4)$$

where $t=(T-Tc)/Tc$.

Keeping this time constant defines a set of H and T values which correspond to a given relaxation time contour. Along this contour, since all relaxation times are assumed to behave in the same way, the response of the magnetization after a small step in field does not change. Alternatively, when measuring the phase lag of the ac susceptibility with respect to the excitation frequency, one should observe the same phase lag along this line.

Using such considerations, we have established a set of constant time contours over a large range of frequencies and times (Fig. 1) /1/.

If the scaling relation describes properly these lines, then by plotting all the points of fig.1 in the following reduced units :

$$t_f/t = (T_f-Tc)/(T-Tc) \qquad \text{and} \qquad H^2/t^{\beta+\gamma}$$

(where T_f is the temperature for H=0 on fig.1)

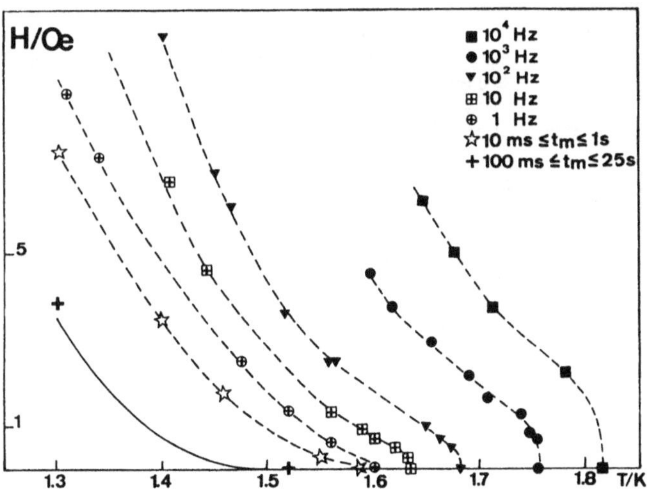

Fig.1 Constant relaxation time contours determined in EuSrS (40% Eu). Dashed lines are a guide for the eye. The solid line is defined by (8) with b=0.17 /1/

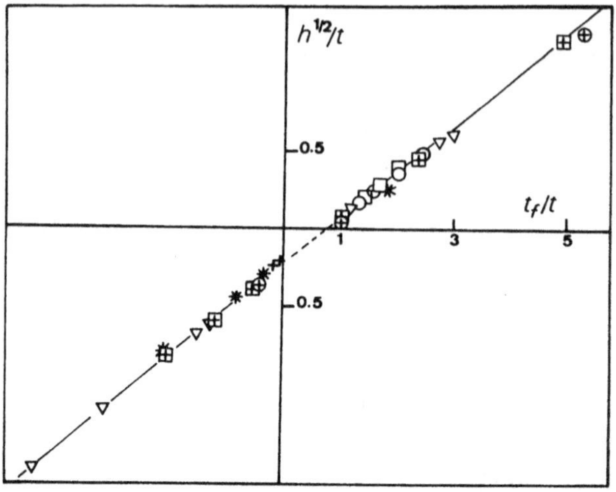

Fig 2 : Scaling of the contours exhibited on fig (1) /1/

one should get a single universal curve if adjusting the exponent $\beta+\gamma$. This is exhibited on fig. 2 which indeed shows that a good scaling can be achieved for $\beta+\gamma = 4$.

A comparable analysis has been performed by SVEDLINDH et al /4/ however using the inflexion point rather that a fixed value of the phase lag. They obtain very similar results : what is quite convincing in their case is that they were able to compare on the same sample the exponents extracted independently from static and dynamic scaling. As shown in table 1, the agreement is excellent and can be taken as an evidence for the consistency of the analysis.

These scalings raise several comments. Firstly, all points fall on a well-defined straight line, except for those which are close to $t_f/t = 1$; this point comprises all the points at H=0. Right at Tc, one expects to find the critical isotherm, which, from elementary considerations is indeed a straight line in this plot :

$$\tau \propto H^{-z\nu/2} \tag{5}$$

More surprising is the fact that even below Tc, the points are still on the same straight line :

$$H^{1/2}/t = a\, t_f/t + b \tag{6}$$

From (2) and (6), we write :

$$\tau \propto (H^{1/2} - Hc^{1/2})^{-z\nu} \tag{7}$$

where $$Hc^{1/2} = b\, t \tag{8}$$

Hc represents the field where t_f is 0 hence when the time is diverging ; this means explicitly that there exists a line defined by (8) where the relaxation times are diverging therefore which can be considered as a transition line.

At T=Tc, (7) yields exactly the critical isotherm ; when H is no longer zero, then (7) shows that the relaxation regime remains critical, but with repect to a finite field value Hc.

It has been very recently argued by MOORE /20/ and FISHER et al /7/ that the mean field picture could not describe realistic short-range Ising spin-glasses. They suggest that the De Almeida Thouless (AT) line does not exist but that the characteristic time should increase extremely rapidly in the vicinity of a "dynamical AT line". If in this framework one can understand that the scaling properties arise from the "cross-over" near the transition temperature in zero field, it is however not quite clear why below Tc, such a relation as (7) seems to hold experimentally if the field Hc is irrelevant.

3. Time Decay of the Thermoremanent Magnetization

Dynamic susceptibility measurements if supporting the evidence for a phase transition have not been extensively used to investigate the low-temperature phase ; below Tc, the experimental work comprises essentially the analysis of the TRM, since the very long times involved in the decay are not easily accessible to a conventional ac susceptibility experiment.

It is now well established that the previously agreed logarithmic /21/ or algebraic /22/ decays are not suitable to account for the TRM in a large temperature range below Tc if measured over several (>2) decades of time. It was first suggested by CHAMBERLIN et al /23/ that the decay of the TRM could be better characterized by a stretched exponential:

$$\sigma = \sigma_o \exp\{-(t/tp)^{1-n}\} \qquad (9)$$

The problem is further complicated by the fact that the system, either field-cooled or zero field-cooled, is not in equilibrium /24,25,11/ : while kept in an external field at a given temperature T (<Tc), the system is continuously evolving ("aging"). This is exemplified by the change in the time decay as a function of the time tw spent in the field-cooled (or zero-field cooled) state /24,25,26,11,12/.

Lately, HOOGERBEETS et al /26/ have focused on the temperature and field variation of the decay of the TRM over a time range smaller than the waiting time tw (tw = 600 s, tmax = 500 s). They have shown that, under these experimental conditions, the fit to the stretched exponential is very good. They compare their data to a dynamical mean-field calculation performed by DE DOMINICIS et al /10/ : he showed that the time decay of the quantity x(t) (which represents the sum of the squares of the spin-glass state occupancies) is of the stretched exponential form. This is the correct approximation

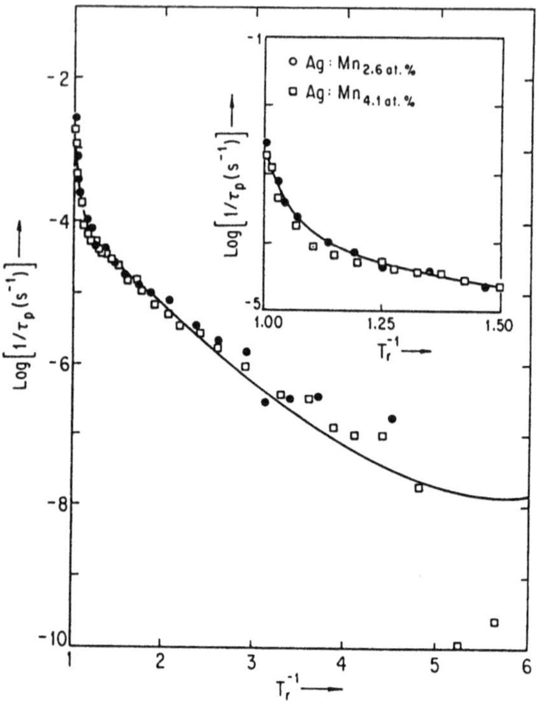

Fig 3 : Response time versus temperature in two metallic spin-glasses /26/ ;the solid line shows the best fit to the model /10/

in the case where t>>tp or t<<tp and if the initial conditions are such that all states are equally occupied ; it is reasonable to assume that all states are mixed when the field is on and thus these initial conditions are appropriate /27/.

Though it is not clear why there should be a relation between x(t) and the TRM, the agreement between the model and the results is actually impressive : the temperature variation of the response time tp measured experimentally reproduces well the theoretical expectation (fig . 3) ; the exponent n should be related to $\bar{y}$ which is a measure of the length of the plateau along which the order parameter q(x) remains constant /10/ ; as a matter of fact, the variation of n versus temperature is in qualitative agreement with what would be expected for $\bar{y}$. When one induces a simultaneous change of n and tp by changing the initial field (close to Tc), the variations of n and of tp are related, as expected from the model /28/. Though there remain some inconsistencies, in particular on a number of states M which are involved in the fitting procedure and which are experimentally found unphysically small, the authors conclude that there are features in the mean-field theory of spin-glasses which may well account for the dynamics in the low-temperature phase.

The recent data that we have collected on EuSrS (40 % Eu) seem to be consistent with this analysis : over 3 decades of time (and even sometimes 4), in a range where t remains smaller than tw (according to the definition of tw by HOOGERBEETS et al /26/, namely the time spent below the glass temperature before removing the field), the TRM in EuSrS is well characterized by a stretched exponential law. Since the field which is switched off is not infinitely small compared to the critical field near the transition temperature /1,29/, we expect to cross from the paramagnetic phase to the spin glass phase at a field-dependent temperature Tc(H) that we know from our previous work /1/ on the same sample. From the model, n should go to 1 at Tc and we argue that this ought to hold at Tc(H). Indeed n goes through a maximum (0.88) at the expected temperature /29/ and decreases above Tc(H).

An alternative interpretation of the time decay of the TRM has been very recently developed relying on a systematic study of the aging effects.

NORDBLAD et al /30/, OCIO et al /11/ and ALBA et al /12/ have found, when measuring towards longer times (t>>tw), that they could not fit their data with a simple stretched exponential form, but that they had to use a different functional form. I will only mention a simplified relation derived from /11,12/ in the limit t<<tw :

$$\sigma = \sigma_o \, (t/tw)^{-x} \exp\{-(t/tp)^{1-n}\} \qquad (10)$$

The recent noise experiments by OCIO et al /13/ provide the strongest support to (10). The noise power spectrum was measured on the insulating spin-glass $CsNiFeF_6$ over 5 decades of frequency in various aging conditions (Fig.4).

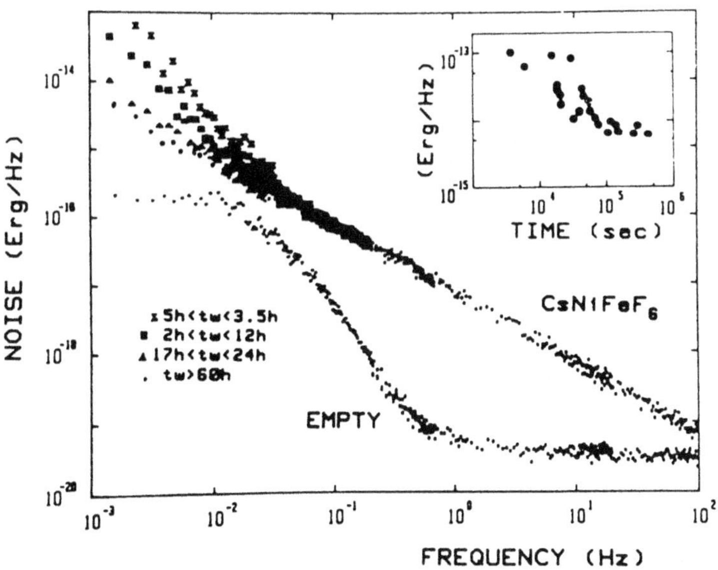

Fig. 4 Noise power spectrum measured for various waiting times in $CsNiFeF_6$ /13/

It was shown that, after waiting long enough, the noise power spectrum is stationary and exhibits a 1/f variation. Moreover, when assuming linear response in order to relate the noise power to the response of the system to a step in magnetic field, the authors have been able to compute the TRM from the noise power spectrum, and they have shown that the computed and the actually measured TRM compare quite well /13/. Therefore it seems justified to argue that, when the system has roughly reached equilibrium e.g. when one has waited long enough, then the equilibrium response is the power law which appears as a prefactor in (10), and that the stretched exponential is due to the aging of the system /24,25/. Because the exponent of the power law is very small, such an equilibrium response would give rise to a noise spectrum which varies as a power of f very close to -1, hence experimentally hardly distinguishable, considering the experimental accuracy, from actual 1/f noise.

If linear response holds, this exponent should be directly comparable to the one which characterizes the decay of the dynamic correlation function computed in the numerical simulations /31/. Figure (5) shows the computed temperature variation of the exponent x , from /6/, and the experimental values found for this exponent in various systems. There is a qualitative agreement between the experimental data and the computed exponent.

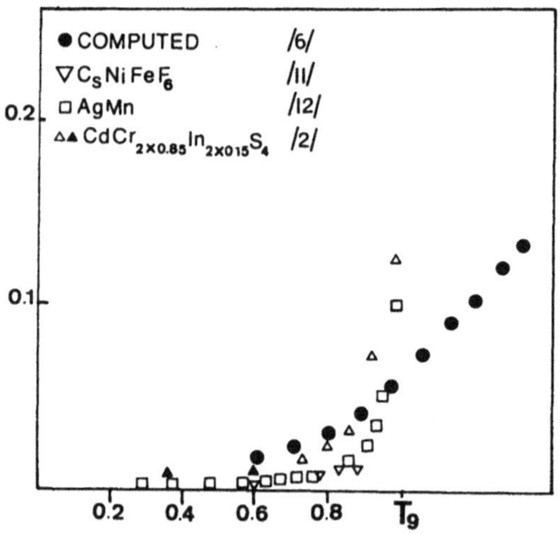

Fig 5 : Comparison of the computed value of the exponent x which characterizes the decay of the auto-correlation function q(t) /6/ and the experimental values of the exponent of the power law in (10)

There remain however also in this framework some inconsistencies : the set of frequencies which belong to the "aging" spectrum in the noise experiment are off by 2 orders of magnitude when compared to the time scales where aging is supposed to interfere with the decay of the TRM experiment. Also, in EuSrS, we have noticed that the stretched exponential form fits better and better as one increases the initially applied field /29/; this seems inconsistent with the statement that the stretched exponential is due to aging, since it has been observed that spin-glasses "age faster" in larger fields /31,32/ and therefore should be closer to equilibrium. Of course one could argue that all times are becoming shorter and shorter, so that the time regime which would be thought to be relevant to the equilibrium response is shifted towards inaccessible short times and that the experimental time scale, which does not change, is set on the long time tail of the response, namely the stretched exponential.

As a matter of fact, the aging means that the system is looking for a lower energy state, while the TRM is decaying whereas the so-called equilibrium response presumably occurs when the system is trapped momentarily in a local minimum. Therefore the stretched exponential decay could be quite essential for our understanding of the low-temperature phase.

4. Conclusion

The occurrence of a phase transition in Ising spin-glasses does not seem to be nowadays a matter of further controversy; a problem arises, as in dynamic scaling, when one determines exponents which are more likely relevant to the Heisenberg regime : for instance, γ should be 2.2 in the Ising regime /33/ but $\beta + \gamma$ is found equal to 4 in EuSrS. Since β is always roughly 1, this yields $\gamma = 3$; 3 is much closer to the Heisenberg exponent /33/ ; this is inconsistent with the Ising behaviour which is expected in low fields when taking into account the dipole-dipole anisotropic coupling /34/.

Mean field theory is strongly disputed in the case of short-range spin-glasses /7,20/ but it has always been a guide for experimentalists. The issue when looking for a transition line or when studying the TRM is essentially a better understanding of the low-temperature phase and its possible relation to the mean field picture.

Acknowledgments

I am grateful to H. Bouchiat, C. De Dominicis, J. Hammann, R. Hoogerbeets and especially R. Orbach for enlightening discussions.

5. References

1. N.Bontemps, J.Rajchenbach, R.Chamberlin, R.Orbach J.Mag.Mag.Mat. 54-57, 1 (1986)
2. E.Vincent, J.Hammann, M.Alba Solid State Commun. 58, 57 (1986) J.Hammann (private communication)
3. J.A Hamida, S.J.Williamson (preprint)
4. P.Svedlindh, L.Lundgren, P.Nordblad, H.S.Chen (preprint)
5. P.Beauvillain, M.Matecki, J.J.Préjean, J.P.Renard to be published in Europhysics Letters

6. A.Ogielski Phys.Rev.B 32, 7384 (1985)
7. D.S.Fisher, D.A.Huse Phys.Rev.Lett. 56, 1601 (1986)
8. R.Tournier Thesis Université de Grenoble, France (1965)
9. R.Hoogerbeets, Wei-Li Luo, R.Orbach
Phys.Rev.Lett. 55, 111 (1985)
10. C.De Dominicis, H.Orland, F.Lainée
J.Physique-Lettres (Paris) 46, L463 (1985)
11. M.Ocio, M.Alba, J.Hammann
J.Physique-Lettres (Paris) 46, L1101 (1985)
12. M.Alba, M.Ocio, J.Hammann (to be published)
13. M.Ocio, H.Bouchiat, P.Monod
J.Mag.Mag.Mat. 54-57, 11 (1986)
14. R.Omari, J.J.Préjean, J.Souletie
J.Physique 44, 1069 (1983)
15. P.Beauvillain, C.Chappert,J.P.Renard
J.Physique-Lettres 45, L665 (1984)
16. H.Bouchiat J.Physique 47, 71 (1986)
This paper presents a good review of the experimental exponents
17. N.Bontemps, J.Rajchenbach, R.Chamberlin, R.Orbach
Phys.Rev.B 30, 6514 (1984)
18. K.Binder, A.P.Young Phys.Rev.B 29, 2864 (1984)
19. J.R.L. de Almeida, D.J.Thouless J.Phys. A 11, 983 (1978)
20. M.A.Moore, A.J.Bray
J.Phys.C Solid State Physics 18, L699 (1985)
21 .See for instance H.V.Löhneysen, J.L.Tholence
Phys.Rev.B 19, 5858 (1979)
22. J.Ferré, J.Rajchenbach, H.Maletta
J.Appl.Phys. 53 (3), 1697 (1981)
23. R.Chamberlin, G.Mozurkewich, R.Orbach
Phys.Rev.Lett. 52, 867 (1984)
24. L.Lundgren, P.Svedlindh, P.Nordblad, O.Beckman
Phys.Rev.lett. 51, 911 (1983)
25. R.Chamberlin Phys.Rev.B 30, 539 (1984)
26. R.Hoogerbeets, Wei-Li Luo, R.Orbach (preprint)
27. R.Orbach (private communication)
28. R.Hoogerbeets, Wei-Li Luo, R.Orbach, D.Fiorani (preprint)
29. N.Bontemps (to be published)
30. P.Nordblad, P.Svedlindh, L.Lundgren, L.Sandlund
Phys.Rev.B 33, 645 (1986)
31. H.Bouchiat Thesis Université Paris-Sud (1986)
32. J.Ferré (private communication)
33. N. de Courtenay, H.Bouchiat, H.Hurdequin, A.Fert
(preprint)
34. G.Kotliar, H.Sompolinsky Phys.Rev.Lett. 53, 1751 (1984)

Spin Glasses: Recent Theoretical Developments

A.P. Young

Department of Physics, University of California, Santa Cruz, CA 95064, USA

The current status of spin glass theory is reviewed. Attention is given to the mean field theory and to numerical results on more realistic models. A brief comparison between theory and experiment is made.

1. Introduction

Spin glasses are magnetic systems where the interactions between the magnetic movements are "in conflict" with each other, due to some frozen-in structural disorder. No conventional long-range magnetic order can be established, but these systems exhibit a "freezing transition" to a state with a new kind of "order" where the spins are aligned in random directions. The nature of this new kind of order has been a matter of much discussion. In particular, is the freezing transition a new type of phase transition [1] or a failure to establish thermal equilibrium during the observation time?

The characteristic phenomena observed in spin glasses, such as the rather sharp "cusp" in the frequency-dependent susceptibility in low fields first found by CANELLA and MYDOSH [2], are a fairly universal feature. They have been seen in dilute metallic alloys, such as CuMn, concentrated insulators, e.g. $Eu_xSr_{1-x}S$, amorphous materials like $CoO{\cdot}Al_2O_3{\cdot}SiO_2$, electric dipolar glasses such as $KTaO_3$ diluted with Li, and in "quadrupolar" glasses like $K(CN)_xBr_{1-x}$. All these materials have two ingredients: randomness and frustration. The latter refers to competition between different terms in the Hamiltonian so no spin configuration simultaneously mimimizes each term. The observed universality suggests that one can study a simple model with these two ingredients. Much of the theory has therefore centered on understanding the model proposed by EDWARDS and ANDERSON [1]. The Hamiltonian is

$$\mathcal{H} = -\sum_{\langle i,j \rangle} J_{ij}\, \vec{S}_i \cdot \vec{S}_j \tag{1}$$

where the $\vec{S}_i$ are m-component classical vectors of unit length on the sites of a regular lattice. The case of m = 1 is an Ising model and m = 3 is a classical Heisenberg system. The interactions J_{ij} are independent random variables with a probability distribution $P(J_{ij})$, often taken to be symmetric.

If the observed freezing is due to a sharp second order transition then the spins become correlated over larger and larger distances as the temperature T approaches the freezing temperature T_c from above. However, some pairs of spins will be correlated antiferromagnetically and others ferromagnetically, so that the ferromagnetic correlation lengths ξ_F and susceptibility χ_F, defined by

$$[\langle S_i S_j \rangle_T]_{av} \sim \exp[-|\vec{R}_i - \vec{R}_j|/\xi_F] \tag{2}$$

$$T\chi_F = \frac{1}{N} \sum_{i,j} [\langle S_i S_j \rangle_T]_{av} \quad , \tag{3}$$

stay small. Here $\langle ... \rangle_T$ denotes a statistical mechanics average and $[...]_{av}$ an average over the random interactions. For simplicity we have used Ising spins here, m = 1, and throughout we set Boltzmann's constant equal to unity. However, if we square the thermal averages in (2) and (3) the cancellations between terms of different signs will not occur. Hence we expect the spin glass correlation length ξ_{SG} and spin glass susceptibility χ_{SG}, defined by

$$[\langle S_i S_j \rangle_T^2]_{av} \sim \exp[-|\vec{R}_i - \vec{R}_j|/\xi_{SG}] \tag{4}$$

$$\chi_{SG} = \frac{1}{N} \sum_{i,j} [\langle S_i S_j \rangle_T^2]_{av} \quad , \tag{5}$$

to diverge at the transition, ie.

$$\xi_{SG} \sim (T - T_c)^{-\nu} \tag{6}$$

$$\chi_{SG} \sim (T - T_c)^{-\gamma} \quad . \tag{7}$$

Relaxation times will also diverge at T_c. If T_c is finite one expects the characteristic time τ to follow conventional critical slowing down [3]

$$\tau \sim (T - T_c)^{-z\nu} \tag{8}$$

where z is the dynamical exponent. Below T_c the spins freeze in random orientations which can be characterized by an order parameter

$$q = \frac{1}{N} \sum_{i} [\langle S_i \rangle_T^2]_{av} \tag{9}$$

which vanishes as $T \to T_c^-$ like $q \sim (T_c - T)^{\beta}$.

Fortunately, one can measure χ_{SG}, or something very close to it experimentally. If, for $T > T_c$, one expands the magnetization in powers of the magnetic field h, ie.

$$m = \chi_F h - \chi_{n\ell} h^3 + \ldots \quad , \tag{10}$$

then the non-linear susceptibility $\chi_{n\ell}$, which is a four spin cumulant average, reduces to essentially χ_{SG} for an Edwards Anderson model with a symmetric-distribution. Even for a more realistic model the difference between χ_{SG} and $\chi_{n\ell}$ comes from ferromagnetic correlations which are short range. Hence χ_{SG} and $\chi_{n\ell}$ are expected to have the same divergence at T_c. By now there is substantial evidence that $\chi_{n\ell}$ becomes extremely large close to T_c and probably diverges there [4] with an exponent

$$\gamma \simeq 2.2 - 3.4 \quad . \tag{11}$$

The spread of values may well be due to different groups using data in different temperature ranges and from different methods of analysis.

Recent careful studies of dynamics [5] find a corresponding dynamical exponent

$$z\nu \simeq 7\text{-}8 \quad . \tag{12}$$

In the rest of this article we discuss recent theoretical developments which attempt to understand these observations. We will discuss the Mean Field theory in section 2 and refer to studies of more realistic models in section 3. A recent review [6] gives further details.

2. Mean Field Theory

The current Mean Field theory of spin glasses is due to EDWARDS and ANDERSON [1]. Subsequently SHERRINGTON and KIRKPATRICK [7], SK, realized that this should be exact for an infinite range model where the bond distribution $P(J_{ij})$ is the same for all pairs of spins, no matter how far they are apart. A lot of the subsequent work has therefore focused on the SK model. The high-temperature region is easily treated [1,7] and one finds there is indeed a transition where $\chi_{n\ell}$ diverges. The critical exponents are

$$\gamma = 1 \quad ; \quad \beta = 1/2 \tag{13}$$

$$z\nu = 2 \quad .$$

It is very instructive to also include a magnetic field. One finds [8] a line of transitions, known as the Almeida Thouless, AT, line, terminating at $T = T_c$, $h = 0$. Along this line the relaxation time diverges but, except for $h = 0$, $\chi_{n\ell}$ does not. At temperatures below the AT line the solution, due to PARISI [9], is very complicated. There are many thermodynamic states [10] separated from each other by barriers which diverge in the thermodymanic limit. The order parameter is a probability distribution [11] $P(q)$ (rather than a single number as is usually the case) and the "landscape" of valleys and barriers has interesting structure [12].

Above the AT line there is only a single thermodynamic state and the original solution of EDWARDS and ANDERSON [1] and SK [7] is correct.

It is interesting to ask when Mean Field theory should work. In conventional systems such as ferromagnets, Mean Field theory predicts correct exponents when the dimension d is greater than the upper critical dimension, d_u. Additionally, fluctuations destroy the transition for $d < d_\ell$, the lower critical dimension, so Mean Field theory is very inaccurate in this region. The scenario for spin glasses may be a little more complicated. The upper critical dimension is $d_u = 6$, but some quantities, such as the shape of the AT line (if it exists, see below) change below d = 8 [27]. For d_ℓ, it may be necessary to distinguish the cases of h = 0 and h ≠ 0 (AT line). In particular, it has been suggested [13] that the AT line is an artifact of the infinite range interactions, in which case d_ℓ (h≠0) = ∞. We will discuss this and recent results of d_ℓ (h=0) in the next section.

3. Numerical Results on More Realistic Models.

The infinite range SK model was proposed because analytic results can be obtained for it. However, one would like to know whether the predictions for the SK model go over to more realistic systems where the interactions fall off with distance. In this case, analytic results are hard to get and one does not have a good intuition with which to formulate reasonable approximations, so numerical studies have played an important role.

Because of their relative simplicity for numerical work Ising models have been extensively studied. The simplest such spin glass system is a nearest neighbor Edwards-Anderson model, as in (1) with the spins on a square (d = 2) or simple cubic (d = 3) lattice and a symmetric distribution for the interactions. There is considerable evidence [14,15] that in two dimensions a transition only occurs at T = 0. However, a three-dimensional Ising spin glass does [15-18] appear to have a finite T_c in zero field. Furthermore, the critical exponents obtained, $\gamma \simeq 3.2$ [17] and $z\nu = 7.2 \pm 1$ [18] agree well with the experimental values noted in (11) and (12). Further discussion of these numerical results and comparison with experiment is given in Reference 19.

While it seems encouraging that Ising model results agree fairly well with experiment, most experimental systems have relatively small anisotropy and so should be better described by a Heisenberg model. However it now appears [20-22] that $T_c = 0$ in a three-dimensional short-range Heisenberg model and probably the lower critical dimension, d_ℓ, is around four. In particular, while the non-linear susceptibility of the d = 3 Ising simulations and experiment behave in a similar fashion, the d = 3 Heisenberg model shows qualitatively different behavior [21].

Probably two factors have to be included to resolve this paradox. First of all anisotropy, though small in magnitude, is crucial [22] in getting a finite T_c. However, for weak anisotropy, one would expect to observe a wide range of temperatures where Heisenberg behavior occurs, followed by a crossover to Ising behavior close to T_c. As far as we are aware this crossover is not seen. One may well, therefore, also need a second factor, namely the RKKY interaction, which falls off like $(\text{distance})^{-3}$ with an essentially random sign, and which occurs in the dilute metallic alloys such as CuMn. It has been suggested [23] that RKKY Heisenberg systems are at, rather than below their lower critical dimension. Though there are recent numerical results which come to a different conclusion [24], the sizes used were rather small and the question remains open. If it is true that $d_\ell = 3$, then probably $\chi_{n\ell}$ diverges exponentially as $T \to 0$, which is not very easy to distinguish from the observed behavior [21].

So far in this section we have discussed the transition in zero field. Less is known about whether an AT line occurs in non-zero field. This would separate a high-temperature region of the h - T plane where there is a single thermodynamic state from the low-temperature region where there are many degenerate states separated by infinite barriers. Experimentally the AT line would most easily be seen by a divergence of relaxation times and this has apparently been observed [25]. However there are theoretical arguments [13,26] which suggest that there should be no AT line in d = 3, and it is possible that the experiments [25] see a crossover region where relaxation times become large but not actually infinite. Further experimental work would be most valuable.

4. Conclusions.

The mean field theory of spin glasses is now rather well understood. Numerical results show that T_c is finite for a three-dimensional Ising model but $T_c = 0$ for a short-range Heisenberg system. The role of anisotropy and RKKY interactions needs to be better understood in order to make a precise comparison with experiment. It is not completely clear whether or not real systems have an AT transition in a magnetic field. If not, one might question whether the SK model, which does have an AT line, is the appropriate Mean Field theory. Unfortunately, there does not seem to be another viable candidate at present.

Acknowledgements.

I have benefited from stimulating collaborations with R. N. Bhatt and J. A. Olive. I would also like to thank D. S. Fisher, D. A. Huse, A. T. Ogielski and D. Sherrington for interesting discussions. This work is supported by the NSF under grant DMR 8419536.

References

1. S. F Edwards and P. W. Anderson, J. Phys. F 5, 965 (1975).

2. V. Canella and J. A. Mydosh, Phys. Rev. B 6, 4220 (1972).

3. P. C. Hohenberg and B. I. Halperin, Rev. Mod. Phys. 49, 435 (1977).

4. See e.g. A. P. Malozemoff, Y. Imry and B. Barbara, J. Appl. Phys. 53, 7622 (1982); R. Omari, J.J. Préjean and J. Soulelie, J. de Phys. 44, 1069 (1983); P. Beauvillain, C. Chappert and J. P. Renard, J. de Phys. Lett. 45, L665 (1985); other references are given in J. L. Tholence, Physica 126B, 157 (1984).

5. N. Bontemps, J. Rajchenbach, R. V. Chamberlin and R. Orbach, Phys. Rev. B 30, 6514 (1984).

6. K. Binder and A. P. Young, Rev. Mod. Phys. (to be published).

7. D. Sherrington and S. Kirkpatrick, Phys. Rev. Lett. 32, 1792 (1975).

8. J. R. L. de Almeida and D. J. Thouless, J. Phys. A 11, 983 (1978).

9. G. Parisi, Phys. Rev. Lett. 43 1974 (1979); J. Phys. A 13, L115, 1001, 1887 (1980); Phil. Mag. 41, 677 (1980).

10. A. J. Bray and M. A. Moore, J. Phys. C13, L469 (1980); C. de Dominicis, M. Gabay, T. Garel, and H. Orland, J. de Phys. 42, L523 (1980); F. Tanaka and S. R. Edwards, J. Phys. F 10, 2471 (1980).

11. G. Parisi, Phys. Rev. Lett. 50, 1946 (1980); A. Houghton, S. Jain and A. P. Young, J. Phys. C 16, L375 (1980).

12. M. Mézard, G. Parisi, N. Sourlas, G. Toulouse and M. Virasoro, Phys. Rev. Lett. 52, 843 (1984); J. de Phys. 45, 843 (1984).

13. D. S. Fisher and D. A. Huse, Phys. Rev. Lett. 56, 1601 (1986).

14. I. Morgenstern and K. Binder, Phys. Rev. Lett. B 22, 288 (1980); W. L. McMillan, Phys. Rev. B 28, 5216 (1983); A. P. Young, Phys. Rev. Lett. 50, 917 (1983); J. D. Reger and A. Zippelius, Phys. Rev. B 31, 5900 (1985).

15. R. Singh and S. Chakravarty (to be published).

16. B. W. Southern and A. P. Young, J. Phys. C 10, 2179 (1977); S. Kirkpatrick, Phys. Rev. B 31, 631 (1985).

17. R. N. Bhatt and A. P. Young, Phys. Rev. Lett. 54, 924 (1985).

18. A. T. Ogielski, Phys. Rev B 32, 7384 (1985).

19. A. P. Young and R. N. Bhatt, J. Mag. Mat. 54-57, 6 (1986).

20. J. R. Banavar and M. Cieplak, Phys. Rev. Lett. 48, 832 (1982); W. L. McMillan, Phys. Rev. B 31, 342 (1985).

21. J. A. Olive, A. P. Young and D. Sherrington (to be published).

22. B. W. Morris, S. G. Colborne, M. A. Moore, A. J. Bray and J. Canisus, J. Phys. C 19, 1157 (1986).

23. A. J. Bray, M. A. Moore and A. P. Young, Phys. Rev. Lett. 56, 2641 (1986).

24. A. Chakrabarti and C. Dasgupta, Phys. Rev. Lett. 56, 1404 (1986).

25. N. Bontemps, J. Rajchenbach, R. V. Chamberlin and R. Orbach, J. Mag. Mat. 54-57, 1 (1986).

26. M. A. Moore and A. J. Bray, J. Phys. C 18, L699 (1985); W. L. McMillan, J. Phys. C 17, 3179 (1984).

27. D. S. Fisher and H. Sompolinsky, Phys. Rev. Lett. 54, 1063 (1985).

The Random Field Problem – Facts and Fiction

V. Jaccarino

Department of Physics, University of California,
Santa Barbara, CA 93106, USA

Site random fields have profound effects on the ground state properties and critical behavior of magnetically ordered systems. For the random field Ising model (RFIM), it has been proven theoretically that the lower critcal dimension $d_\ell = 2$. A variety of experimental studies by the author and his colleagues on RFIM systems have demonstrated the phase transition to be destroyed at d = 2 and a sharp, continuous phase transition at $T_c(H)$ to exist at d = 3 with new critical exponents corresponding to an effective dimensionality $\bar{d} \simeq 2$; suggestive of a dimensionality reduction $\bar{d} \simeq d-1$. Furthermore, extreme critical slowing down has been observed as $T \to T_c(H)$. This explains why field cooling through $T_c(H)$ always traps the d=3 system into a nonequilibrium domain state on what are laboratory time scales. The predicted crossover scaling has been observed for the field dependence of all thermodynamic functions and "phase boundaries" in the distinctly different d=2 and d=3 RFIM systems. Contrary to current theory and the results described above, Birgeneau, Cowley and Shirane - from neutron scattering studies alone - have 1) assumed the low-temperature, field-cooled configuration is always the ground state of the RFIM, 2) "unambiguously" determined $d_\ell \geq 3$ and 3) discovered a "temporal" and/or discontinuous phase transition occurs at d=3. I suggest the errors in all of their conclusions represent a confluence of the limitations of the use of a single experimental technique, not doing the key experiments with that technique, incorrect interpretations and a concomitant use of poor and/or badly characterized materials.

Full paper is to appear in the Festschrift Book: Condensed Matter Physics: The Theodore D. Holstein Symposium (Springer-Verlag)

Statics and Dynamics of the Random Field Ising Model (Theory)

R. Bruinsma

Physics Department and Solid State Science Center,
University of California, Los Angeles, CA 90024, USA

I will give in this lecture an introductory overview of our present understanding of the Random Field Ising Model (RFIM). As this is a subject of a large literature and a long, turbulent history, I will restrict myself to a small number of topics, whose choice is very subjective. In particular, I will neither discuss experimental nor numerical work, except when it cannot be avoided. It is, however, important to realize why such a disproportionately large amount of attention has been devoted to this model. The first reason is that the RFIM is the simplest model which incorporates the dramatic effect of quenched-in disorder coupled linearly to an order-parameter. This effect has been seen in a steadily growing list of experiments (1). It was extensively studied for the first time in diluted antiferromagnets in a magnetic field (2) but the RFIM is now used to explain how impurities affect displacive transitions, how phase separation of binary fluids proceeds in porous media or gels, and how hydrogen dissolves in metallic alloys. Random fields also play an important role in charge-density wave systems.

The second reason is that the physics of a phase transition in the presence of a random field appears to be unlike anything we are familiar with from non-random systems. The hyperscaling relation is violated, below the "critical" temperature the RFIM does not want to equilibrate, and above the critical temperature puzzling Griffiths singularities lead to problems with (dynamic) scaling.

A. *Statics: Mean-Field Theory and Linear Response*

The RFIM is defined as a collection of Ising spins on a lattice with an associated Hamiltonian

$$\mathcal{H} = \sum_{R,R'} J S(\vec{R}) S(\vec{R}') - \sum_{R} h(\vec{R}) S(\vec{R}) \quad , \tag{1}$$

where $S(\vec{R}) = \pm 1$, J is the exchange constant between nearest neighbors, and $h(\vec{R})$ is the random field. Whereas S can change with time due to thermal fluctuations, $h(\vec{R})$ is frozen in time. The random fields obey a statistical distribution P(h) such that their average is zero and their square deviation h^2:

$$\overline{h(\vec{R})} = 0 \tag{2a}$$

$$\overline{h(\vec{R})\ h(\vec{R}')} = h^2 \delta_{\vec{R},\vec{R}'} \quad . \tag{2b}$$

Randomness in the value of $S(\vec{R})$ is due both to thermal fluctuations as in the pure Ising model, and quenched fluctuations which cause the time-average $\langle S(\vec{R})\rangle$ to vary from site to site over the lattice. The simplest way to deal with fluctuations is, of course, to ignore them altogether, but depending on whether we ignore thermal or quenched fluctuations, we get very different answers. First we ignore both and we replace in Eq. (1) the neighbors of $S(\vec{R})$ by the lattice average $\overline{\langle S\rangle}$ of $\langle S(\vec{R})\rangle$ so

$$\mathcal{H} \simeq - Jz\overline{\langle S\rangle} \sum_{R} S(\vec{R}) - \sum_{R} h(\vec{R}) S(\vec{R}) \tag{3}$$

where

$$\overline{\langle S\rangle} = \sum_{\vec{R}} \langle S(\vec{R})\rangle \Big/ \sum_{\vec{R}} \tag{4}$$

and z is the number of nearest neighbors. If we now compute from Eq. (3) the average of $S(\vec{R})$ and identify this with $\overline{\langle S\rangle}$ we find a self-consistency condition:

$$\overline{\langle S\rangle} = \int_{-\infty}^{\infty} dh\, P(h)\, \tanh([Jz\overline{\langle S\rangle} + h]/k_B T) \quad . \tag{5}$$

The tanh results from the thermal average of $S(\vec{R})$ which is then averaged over the distribution of random fields at $\vec{R}$. The resulting phase diagram (3) looks as follows:

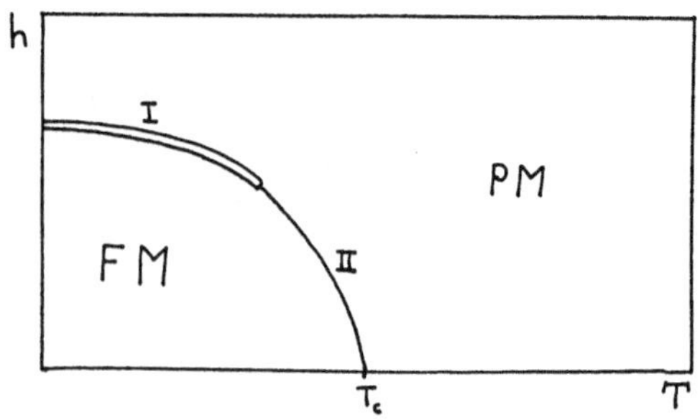

Figure 1. Mean-Field phase diagram.

where we assumed a binary distribution ($h(\vec{R}) = \pm h$). There is a normal paramagnetic ($\overline{\langle S\rangle} = 0$) to ferromagnetic ($\overline{\langle S\rangle} \neq 0$) phase transition which can be either first-order (double line) or second-order (single line). We will consider this to be our mean-field phase diagram, which is hoped to be reasonably accurate for higher dimensions. Weak first-order transitions have been reported in some experiments (4) and numerical simulations (5), but critical behavior observed in the specific heat and the correlation length (6,7) does not agree with mean-field exponents. To improve on mean-field theory we will next include the effect of thermal fluctuations above the critical temperature T_C, including the random field in the same way we include the effect of low concentrations of impurities in solids, namely through the use of linear-response theory (8).

Assume that in the absence of any random fields the exact spin-spin correlation function can be computed:

$$\chi(\vec{R}-\vec{R}') = \langle S(\vec{R}) S(\vec{R}')\rangle \ . \tag{6}$$

The thermodynamic potential then has the form

$$\Phi_o \cong \frac{1}{2} \sum_{RR'} \langle S(\vec{R})\rangle\langle S(\vec{R}')\rangle\, \chi^{-1}(\vec{R}-\vec{R}') \tag{7}$$

for $\langle S(\vec{R})\rangle$ small. This can be checked by noting that

$$\frac{\partial^2 \Phi}{\partial\langle S(\vec{R})\rangle \partial\langle S(\vec{R}')\rangle} = \chi^{-1}(\vec{R}-\vec{R}') \tag{8}$$

as it should. Next, add a single random field at the origin: our impurity. So now

$$\Phi = \Phi_o - h\langle S(0)\rangle \tag{9}$$

If we minimize Φ with respect to $\langle S(\vec{R})\rangle$ then the Fourier transform $S(\vec{q})$ of $\langle S(\vec{R})\rangle$ is

$$S(\vec{q}) = h\chi(\vec{q}) \tag{10}$$

where $\chi(\vec{q})$ is the Fourier transform of the correlation function. We can reasonably well approximate $\chi(\vec{q})$ by a Lorentzian:

$$\chi(\vec{q}) = \chi/[1 + q^2\xi^2] \tag{11}$$

where χ is the magnetic susceptibility and ξ the correlation length. The magnetization in the neighborhood of the random field [Eq. (10)] dies off with ξ as the characteristic length.

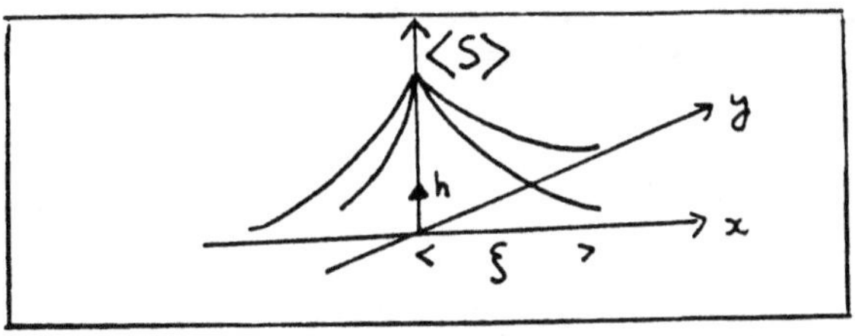

Figure 2. Magnetization induced by a random field impurity.

If we compute the thermodynamic potential, Eq. (9), we find a correction term

$$\Phi = \Phi_o(h=0) - \frac{1}{2}h^2 c \int d^d q\, \chi(\vec{q})$$

$$\simeq \Phi_o(h=0) - \frac{1}{2}h^2 c\, \chi/\xi^d \quad , \tag{12}$$

where we assumed a low concentration c of impurities. We can now compute from Φ the specific heat of the RFIM in the limit $c \to 0$:

$$C \sim \frac{1}{(T-T_c)^\alpha} + \frac{ch^2}{J}\left(\frac{T}{T_c} - 1\right)^{d\nu-\gamma-2} \tag{13}$$

We took for the specific heat, susceptibility and correlation length of the pure system the scaling forms

$$C_o \sim 1/(T-T_c)^\alpha \tag{14a}$$

$$\chi \sim 1/(T-T_c)^\gamma \tag{14b}$$

$$\xi \sim 1/(T-T_c)^{\nu} \tag{14c}$$

This is an interesting result because the correction term to the specific heat diverges more rapidly as T approaches T_c than the specific heat of the pure system. The new specific heat exponent $\bar{\alpha}$

$$\bar{\alpha} = 2 - d\nu + \gamma \tag{15}$$

violates the hyperscaling relation $\alpha = 2-d\nu$ obeyed by the critical exponents of pure systems. The measured exponents ($\alpha \sim 0$, $\nu \sim 1$) in d = 3 indeed violate hyperscaling, but they do not obey Eq. (15) either.

Linear response also allows us to compute the structure factor S_q, which is be measured through neutron scattering:

$$S_q = \overline{\int d^dR <S(0)\ S(\vec{R})> e^{i\vec{q}\cdot\vec{R}}} \tag{16}$$

which can be expanded as

$$S_q = \overline{<S(\vec{q})>^2} + \overline{<|S(\vec{q}) - <S(\vec{q})>|^2>} \quad . \tag{17}$$

The first term, which vanishes for the pure Ising model, reflects the spatial variation of $<S(\vec{R})>$ due to the random field and can be computed using Eq. (10). The second term describes the thermal fluctuations of $S(\vec{R})$ around its mean value and is not much affected by the random field. Using Eq. (11) in the second term gives the "Lorentzian + Lorentzian^2" line-shape

$$S_q \simeq \frac{k_BT\ \chi}{1 + q^2\xi^2} + \frac{h^2\ \chi^2}{(1 + q^2\xi^2)^2} \tag{18}$$

which indeed accounts reasonably well for the lineshape observed by neutron scattering experiments on diluted anti-ferromagnets (9).

How good is linear response? We should expect it to break down when the correction term to Φ_0 in Eq. (12) becomes comparable to Φ_0 so when ($\Phi_0 \sim ch^2J/\xi^d$)

$$ch^2\chi \gtrsim J \tag{19}$$

which will happen as we approach T_C. In terms of the correlation length, ξ roughly must exceed (J/H) times the lattice constant in order for linear response to fail. For diluted anti-ferromagnets in a field, J/h ~ 10-100, so one must approach T_C very closely in order to observe the true critical behavior.

B. *Statics: Scaling and Renormalization Group*

We now approach T_C and try to establish this true critical behavior. This is done using the same method which has proven to be so successful for understanding the critical properties of non-random systems: scaling and renormalization. The scaling ansatz asserts that near T_C we can express the singular part of the free energy as a homogeneous function of ξ and the fields. In our case this means that we can express the free energy F as (2).

$$F \sim \frac{1}{\xi^d} f(h^2 \xi^{\phi/\nu}) \tag{20}$$

where f(x) is an as yet unknown function and ϕ is called the cross-over exponent. In the limit $h \to 0$ we should recover the linear response result Eq. (12). This can only happen if for small x, $f(x) \sim \text{constant}+x$, and if $\phi = \gamma$. Thus, as we approach T_c where ξ diverges, the argument of f becomes larger and larger, and we move away from the Ising limit f(0). Suppose we are now below T_c in the ordered phase and increase h from zero. How large should h be to destroy the ordered phase? The scaling form Eq. (20) shows that the random field becomes noticeable when the argument of f is of order one. If $T_c(h)$ is the critical temperature for finite h, then

$$T_c(h) - T_c(h=0) \sim h^{2/\phi} + \mathcal{O}(h^2) \quad . \tag{21}$$

This relation between T_c and h was confirmed experimentally (10). It must be realized that the scaling ansatz Eq. (20) for the random field is a much more radical assumption than the scaling ansatz for, say, a uniform magnetic field on a pure Ising magnet. The random field is really characterized by the distribution P(h), and Eq. (20) asserts that only the standard deviation h^2 is important for the critical behavior.

The divergence of the random field term in the scaling function means that the random field is a relevant perturbation on the Ising fixed point. The next step is to assume that there must be a finite h fixed point which at sufficiently low dimensions is stable with respect to the Gaussian fixed point below the upper critical dimension $d_u = 4$. The new fixed point is assumed to lie at T = 0 so the fixed-point structure looks as follows:

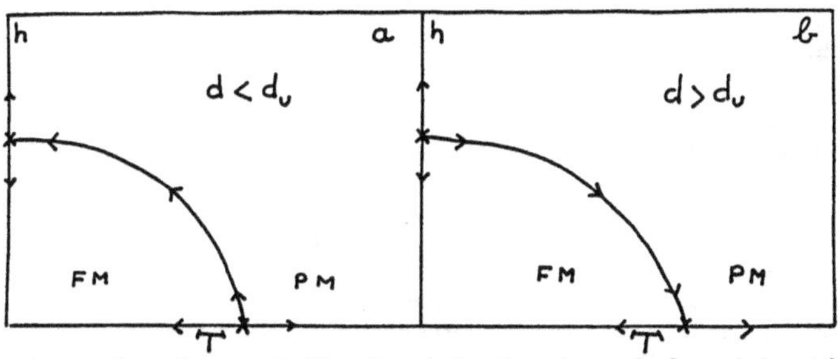

Figure 3. Assumed fixed-point structure below (a) and above (b) the upper critical dimension d_u.

To compute the critical exponents of the new fixed-point, we first go to the continuum limit. That is, we divide the lattice into cells and define $\phi(\vec{r})$ as the cell average of $S(\vec{R})$. The Hamiltonian of this coarse-grained variable is, for small enough ϕ, of the Ginzburg-Landau form

$$\mathcal{H} = \int d^d \vec{r} \left\{ \frac{1}{2} (\vec{\nabla}\phi(\vec{r}))^2 + \frac{1}{2} r\phi^2(\vec{r}) + \frac{U}{4} \phi^4(\vec{r}) + h(\vec{r})\, \phi(\vec{r}) \right\} \tag{22}$$

where r is proportional to $T-T_c$. Since the new fixed point lies at T = 0, we ignore thermal fluctuations and simply minimize $\mathcal{H}$ with respect to $\phi(r)$. This results in a stochastic differential equation

$$\nabla^2\phi(\vec{r}) - r\phi(\vec{r}) - u\phi^3(\vec{r}) = h(\vec{r}) \quad . \tag{23}$$

Because of the randomness of $h(\vec{r})$, it is not easy to solve this equation.

We start, as usual for Ginzburg-Landau equations, with the linearized form u = 0,

$$\nabla^2\phi(\vec{r})-r\phi(\vec{r}) = h(\vec{r}) \tag{24}$$

which can be solved through Fourier transformation. After performing the quenched average we retrieve the Lorentizian[2] form, as expected from linear response:

$$\overline{|\phi(\vec{q})|^2} = \overline{|h(\vec{q})|^2} \Big/ (r+q^2)^2 \quad . \tag{25}$$

where we must identify ξ with $1/\sqrt{r}$ to agree with Eq. (18). The nonlinear term involving u is handled through perturbation theory. To lowest order we treat it in the Hartree approximation and absorb it in a redefinition of r:

$$r \to r + 3u \int d^dq \overline{|\phi(\vec{q})|^2}|_{r=0} \tag{26}$$

or, using diagrammatic language and Eq. (25),

$$r \to r + 3uh^2 \int d^dq \frac{1}{q^4} \tag{27}$$

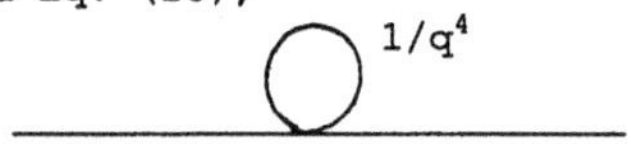

This correction to T_c due to quenched fluctuations must be compared to the effect of thermal fluctuations on the transition temperature of the *pure* Ising model in D dimensions

$$r \to r+3uT \int d^Dq \frac{1}{q^2} \tag{28}$$

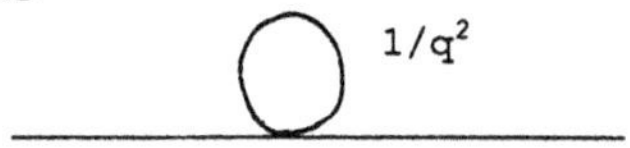

If we choose D = d-2 and T = h^2 then these two terms in the perturbation expansion look very similar! In a series of elegant papers (11) it was shown that this mapping can be carried through term by term for the most divergent terms in the perturbation theory. It looks as if the new fixed-point is the Wilson-Fisher fixed-point but in a different dimension. This led to the idea of the dimensional reduction:

"A d dimensional RFIM is in the same universality class as a d-2 dimensional Ising model."

Dimensional reduction has a number of immediate consequences: since the upper critical dimension of the pure Ising model is 4, the upper critical dimension of the RFIM is 6 and since the lower critical dimension, below which there is no longe-range order, is 1 for the pure Ising model, it must be 3 for the RFIM. How well does dimensional reduction work? Unfortunately, disastrously badly. There is a rigorous result (12) that there is long-range order in 3 dimensions. The best numerical and critical exponents, from numerical simulations in d = 3 (13), are $\nu \sim 1.3$ and $\eta \sim 0.5$, which does not bear out dimensional reduction. It was suggested (14) that below the upper critical dimension the propagator $1/q^2$ really should be replaced by $1/q^{2-\eta}$, which would lead to a dimensional reduction.

$$D = d - 2 + \eta(D) \quad . \tag{29}$$

This gives a lower critical dimension of 2. However, critical exponents computed in a 2+ε expansion (15) do not obey any dimensional reduction.

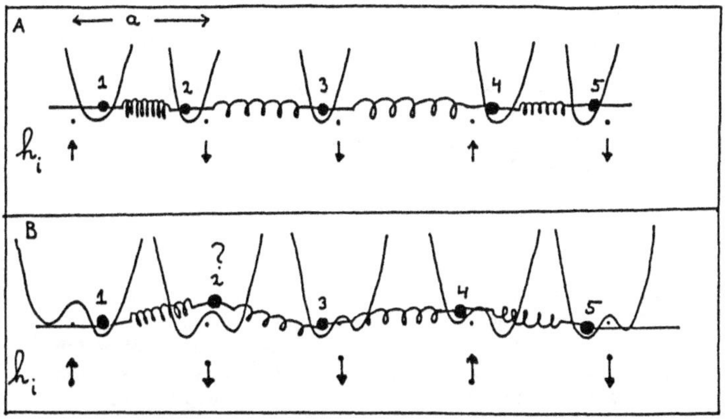

Fig. 4. Mechanical analog of the RFIM. For r > 0 (A) and r < 0 (B). The particles are on a lattice with lattice constant a and are connected by springs. The local random field biases the potential.

Dimensional reduction does not work. This failure is not just a technical problem; it goes to the heart of the question how to deal with quenched randomness in statistical mechanics. We actually know where the problem lies. If we return to our stochastic differential equation, Eq. (23), we can think of this equation as representing a mechanical model with particles i = 1,2,..., connected by harmonic springs. The particles are in an anharmonic potential $r(x(i)-ia)^2 + u(x(i)-ia)^4 + h(i)(x(i)-ia)$ with a the lattice constant (Fig. 4). If r is positive, then there is no potential barrier for the particles to overcome in order to reach the lowest energy configuration, and the solution of Eq. (24) is unique (Fig. 4a). For negative r, the potential is of the double-well form and every particle must make a difficult choice whether to choose the lowest well favored by h(i) or satisfy the preferences of its neighbors (e.g., #2). There are many solutions separated by energy barriers. Thus, below the transition temperature of the pure Ising model, we expect many solutions to Eq. (23) and we have not provided an algorithm which to choose. In order words, the problem was ill posed! If we assign a configurational coordinate to the different competing spin configurations then we should expect the configurational free energy to look somewhat as follows:

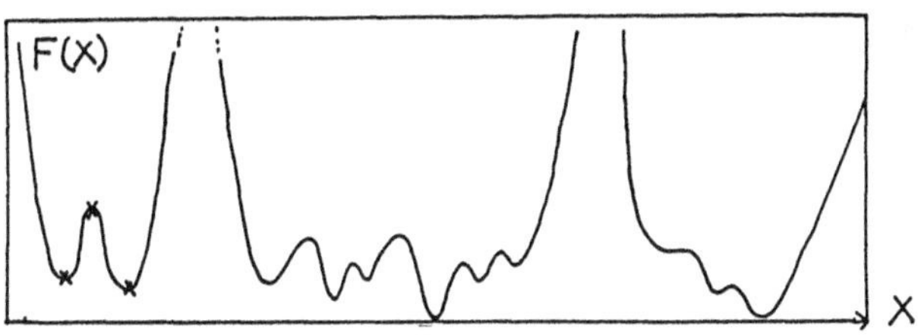

Fig. 5. Configurational free energy F(X) of the RFIM. Crosses are solutions of Eq. (23).

Note that Eq. (23) can define an energy maximum as well as a minimum. The proper procedure would be to find all energy minima whose energy is comparable to the ground state (i.e., the energy difference is of order 1 and not of order N) and assign a Boltzmann distribution. If the distribution of energy minima is finite in the limit of zero energy difference, one can use the usual arguments for two-level glassy systems (16) to show that the specific heat should be linear in T as $T \to 0$.

The multiple minima description of Fig. 5 will play an important role in our discussion of the dynamics, but first we will try to gain more insight into the physical nature of the minima.

C. *Statics: Ground State Configurations*

Already in 1975 it was suggested (17) how insight could be gained into the low-lying minima of Fig. 5 through an explicit real-space construction: the Imry-Ma argument. We now believe this argument to be essentially correct. It goes as follows: Above the critical dimension, there must be long range order in the ground state. However, statistical fluctuations will cause certain regions to have a high concentration of random fields pointing opposite to the magnetization, and the spins in these regions could follow the random field. If such a minority region is of size R, then the

$$\Delta E \sim JR^{d-1} - \sum_{\substack{\text{minority}\\ \text{region}}} h(\vec{r}) \quad . \tag{30}$$

The first term is the interface energy between majority and minority spins, the second term is the gain in random field energy. To estimate the second term we assume R to be large so we can use the central limit theorem, which then states that the second term is a Gaussian random variable of RMS deviation $hR^{d/2}$ so

$$\Delta E \sim JR^{d-1} - hR^{d/2} \quad . \tag{31}$$

Minimizing ΔE with respect to R gives

$$R \sim (h/J)^{2/d-2} \quad . \tag{32}$$

For $d > 2$, R is finite so our description makes sense, but as d approaches 2, R becomes larger and larger. Thus we identify $d = 2$ as the lower critical dimension. The resulting ground state looks as follows ($d > 2$)

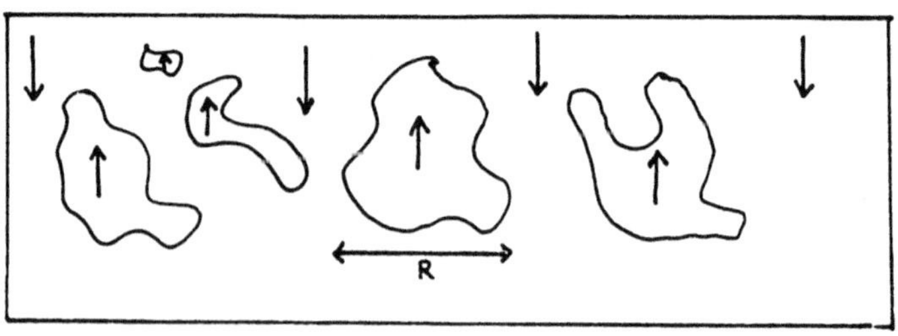

Fig. 6. Ground state configuration for $d > 2$ according to the Imry-Ma argument.

There is of course still a great deal of arbitrariness in our choice of the location of the minority "droplets," and we identify these different choices for the interfaces with the minima of Fig. 5. One could argue that these interfaces are unlikely to be smooth, as is implicit in the Imry-Ma argument, and this was a basis for early criticisms. To check

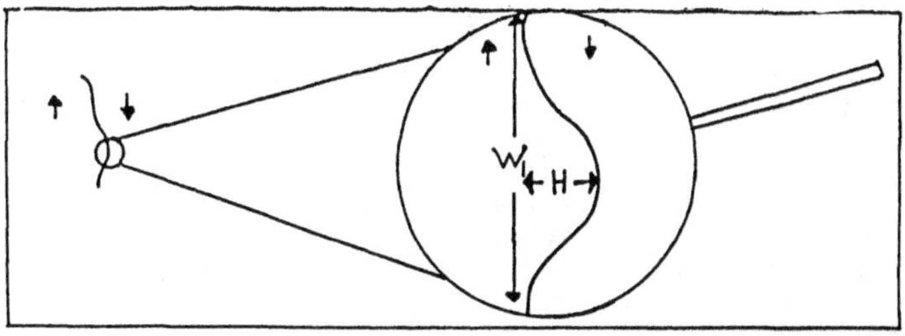

Fig. 7. Magnified portion of a section of interface from Fig. 6.

this we magnify a section of the interface in Fig. 6 and re-apply Imry-Ma (18).

Let H be the roughness on a scale W. The energy cost of the bump in Fig. 7 is then, according to Imry Ma:

$$\Delta E = JHW^{d-2} - h(HW^{d-1})^{1/2} \tag{33}$$

because HW^{d-2} is the surface area increase of the droplet and HW^{d-1} the increase in volume. Minimizing ΔE with respect to H gives

$$\Delta E \simeq \left(\frac{h^2}{J} W^{2-d}\right) W^{d-1} \quad . \tag{34}$$

ΔE is the contribution to the interfacial energy of the surface roughness induced by the random field on a length scale W. In the limit $W \to \infty$, it is much smaller than the interface energy of Eq. (30) if $d > 2$ and we can safely neglect it. More elaborate calculations (19) have shown that interface "overhangs" do not affect the Imry-Ma argument either, although numerical simulations have yet to confirm Eq. (32).

We can summarize our results for the statics as follows:

1) There is an ordered ground state in $d = 3$ and the Imry-ma argument is almost surely correct, indicating a lower critical dimension of 2.
2) Linear response gives a good description of the high-temperature phase for small enough random fields.
3) We have no theory for the critical behavior of the RFIM. In fact, as I will argue later, we have not even a viable mean-field theory with a correct description of the multiple minima for the low-temperature phase in higher dimensions.
4) The $6-\varepsilon$ expansion cannot be used below the critical temperature of the pure Ising model, as it fails to identify the true free energy minimum. Scaling theory works as long as we are near the Ising fixed point, and correctly predicts the dependence of T_c on h. We have no evidence that the fixed point structure of Fig. 3 is valid for finite h. In section F, I will argue why, in fact, it is invalidated by Griffiths singularities.

D. *Dynamics*

Irreversibility has long been known to play an important role in the physics of impure magnets (20). In the case of realizations of the RFIM, hysteresis is due to the difficulty of identifying the true ground state. Diluted antiferromagnets cooled in the presence of the field do not show magnetic order, but if they are cooled in zero field, long-range order is seen (21). If we identify a critical temperature from the maximum in the specific heat T_c, then the freezing temperature T_f, below which irreversibility sets in, is slightly above T_c in $d = 3$ (22) and clearly below T_c in $d = 2$ (23). In the latter case no true phase transition takes place. We will restrict ourselves to $d > 2$ and assume $T_c = T_f$. As a function of temperature, we will distinguish three different regions. Above T_c relaxation to equilibrium is seen but the relaxation time τ diverges rapidly on approaching T_c (region A). Right at the freezing temperature, very slow relaxation (proportional to the logarithm of the time) is observed in the specific heat (region B) (23), while further below T_c, hardly any relaxation is seen in the neutron-scattering line shape (region C) (24).

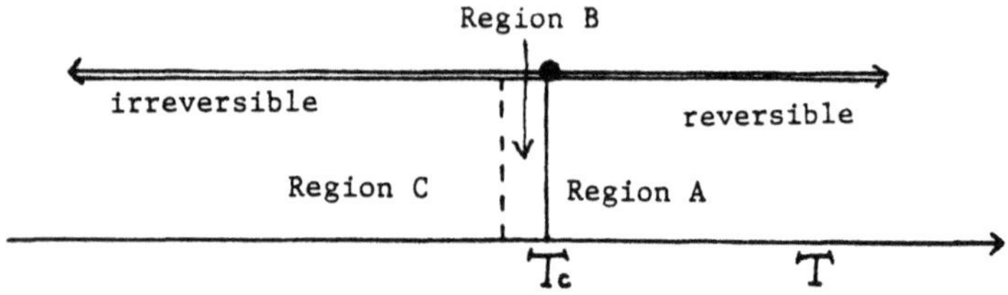

Region A (25)

Close to T_c we will assume that we still can reach equilibrium, and we try to use dynamical scaling. If we are below the transition temperature of the pure Ising model, then the multiple minima description of Fig. 6 should already be relevant. Thus, we must expect relaxation to take place through thermal activation over potential barriers. If we assign a *single* relaxation time to this process, then relaxation of the magnetization will depend on time as $\exp(-t/\tau)$, with

$$\tau = \tau_o \exp \Delta/k_B T \quad , \tag{35}$$

where τ_o is the spin-flip time and Δ is the mean barrier heignt near the minimum of the configurational free energy. In section E we will discuss the effect of fluctuations in the barrier height. In the spirit of scaling we should expect that near T_c there is only a single relevant energy scale determining the specific heat, free energy as well as Δ. This implies that

$$\Delta \simeq \xi^d \, F(\xi) \quad , \tag{36}$$

where F is the free energy density. For normal critical behavior, the hyperscaling relation $F \sim 1/\xi^d$ would imply that Δ is approaching a constant value at T_c. However, we saw before that hyperscaling fails for the RFIM. If we define θ as a measure of the breakdown of hyperscaling:

$$\theta = d + (\alpha-2)/\nu \tag{37}$$

then we can write Δ as

$$\Delta \sim \xi^\theta \quad . \tag{38}$$

From measured exponents, it follows that $\theta \sim 1$ so Δ diverges near T_c and

$$\tau \sim \tau_o \exp(1/(T-T_c)^{\nu\theta}) \quad . \tag{39}$$

This exponential divergence was indeed observed numerically (13). The breakdown of hyperscaling is intimately related with the appearance of very long relaxation times near T_c. This exponential divergence should be contrasted with conventional critical slowing down where τ diverges as $1/(T-T_c)^{\nu z}$, a power law divergence. We now are in a position to predict the frequency dependence of the susceptibility if we use dynamical scaling. For instance, the frequency-dependence susceptibility $\chi(\omega)$ is

$$\chi(\omega) = \chi g(\omega\tau) , \tag{40}$$

with g(x) to be determined. By demanding that $\chi(\omega)$ saturates for $\omega\tau > 1$ and using Eq. (39), we find for the susceptibility at $T = T_c$:

$$\chi(\omega) \sim (\ell n(\omega))^{\gamma/\nu\theta} . \tag{41}$$

Unfortunately, we will see in section E that this is only the *typical* relaxation process. The true asymptotic relaxation in the limit $\omega \to 0$ is dominated by atypical quenched fluctuations, so Eqs. (39) and (41) are at best valid over an intermediate range of frequencies.

Region C (26)

At temperature in this region, the spin-configuration is history dependent and, according to experiment, essentially frozen in time. What does such a spin arrangement look like? The simplest way is to do a numerical experiment and the result is the "domain-wall state" (27):

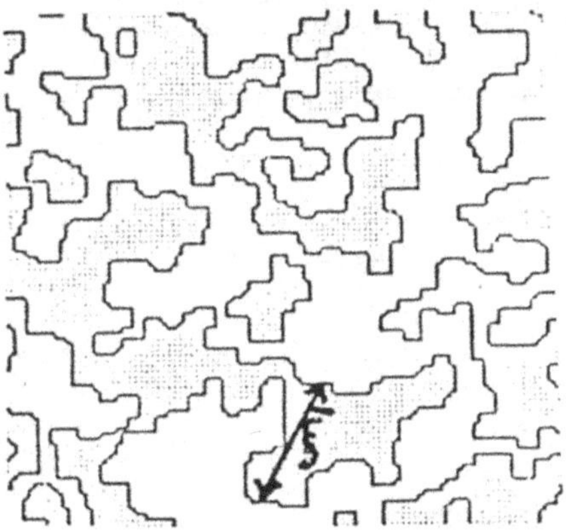

Fig. 8. Domain-wall state for field-cooling (h/J = 0.1).

Let us characterize this non-equilibrium pattern with a typical length scale $\bar{\xi}$: the mean radius of interface curvature. If we compute the structure factor of such a pattern, then we encounter a surprise. A smooth interface contributes for large q a factor $1/q^4$ to S_q (28). On the other hand, for $q \ll \bar{\xi}$, S_q should be independent of q so a reasonable approximation is

$$S_q \sim 1/(q^2 + \bar{\xi}^2)^2 . \tag{42}$$

This is just the Lorentzian^2 lineshape from linear response [Eq. (28)]. An unpleasant coincidence which makes it hard to distinguish the frozen domain-wall state below T_c from the paramagnetic phase using neutron scattering. Our problem is now to compute $\bar{\xi}$. From Fig. 8 it is clear that for relaxation we should look at the motion of the domain walls. The surface energy of a domain wall is roughly (29)

$$E_s \simeq \int d^{d-1}r \left\{\sigma(\vec{\nabla}f)^2 + \int^{f(\vec{r})} h(z,\vec{r})dz\right\} \tag{43}$$

where σ is the surface tension ($\sim J/a^{d-1}$) and where $f(\vec{r})$ is the interface height at $\vec{r}$. At first we will neglect thermal fluctuations. The interface equation of motion is

$$\Gamma \frac{\partial f}{\partial t} = \delta E_s/\delta f \quad , \tag{44}$$

with Γ a friction constant. So

$$\Gamma \frac{\partial f}{\partial t} = \sigma\nabla^2 f + h(f,\vec{r}) \quad , \tag{45}$$

an interface moving in a random medium. We now return to the domain wall state and compute whether a droplet of size ξ will collapse or not. Let $\bar{f}$ be the deviation from a spherical droplet of radius $\bar{\xi}$. The equation of motion of $\bar{f}$ is then

$$\Gamma \frac{\partial \bar{f}}{\partial t} = -\frac{\sigma}{\bar{\xi}} + \sigma\nabla^2\bar{f} + h(\bar{f},\vec{r}) \ . \tag{46}$$

The first term attempts to contract the droplet while the random-field term leads to interface pinning and tries to prevent contraction. We can estimate the competition between these two terms by comparing the gain in surface energy $(\sigma\bar{\xi}^{d-2})a$ on contracting the radius by one lattice constant with a typical energy barrier $h(\bar{\xi}^{d-1}a)^{1/2}$, using the Imry-Ma argument. This means that the condition for the survival of droplets of radius $\bar{\xi}$ at low temperatures is

$$\bar{\xi} > a(J/h)^{2/(3-d)} \ .$$

Similarly, we expect that interfaces with mean radius of curvature $\bar{\xi}$ satisfying this condition are frozen while interfaces with smaller radii of curvature either erode or collapse until they do satisfy the condition. As before in the Imry-Ma argument, we must allow surface roughening. This leads to an improved requirement (26) on the mean radius of curvature:

$$\bar{\xi} > a(J/h)^{4/(5-d)} \ . \tag{47}$$

This inequality has a number of important consequences:

1) Field-cooling vs. zero-field cooling:
During zero-field cooling, long-range order is established first so we start with $\xi = \infty$ which satisfies the inequality. Increasing h from zero only increases the pinning of an interface so the domains, if any, are frozen. During field-cooling we expect to go out of equilibrium near T_c because of the divergence of τ. Starting with an interface pattern of mean radius of curvature ξ, the equilibrium correlation length at the temperature where we left equilibrium, the domains will coarsen and collapse until Eq. (47) is satisfied so no long-range order is established.
2) Mean domain size:
The average domain size for field cooling in d = 3 is thus of order $\bar{\xi} \sim a(J/h)^2$. This is roughly 1000 Å for diluted anti-ferromagnets. Both the order of magnitude as well as the h dependence of $\bar{\xi}$ are in good agreement with the neutron scattering results (30).
3) There is a critical dimension, d = 5, above which there is no interface pinning.

Region B (31)

As we increase the temperature and approach T_C, thermal activation of the interfaces must lead to further relaxation beyond the inequality, Eq. (47). This should be done by adding a thermal noise term to Eq. (45). We can, however, estimate the temperature-dependent relaxation time using a simple argument. Assume a droplet of radius R is slowly collapsing under the combined effect of surface tension, thermal fluctuations, and the random field. Allow the radius to shrink by δ. The change in surface energy is

$$\Delta \sim - \sigma \delta R^{d-2} + h(\delta R^{d-1})^{1/2} \quad . \tag{48}$$

Minimizing Δ with respect to δ gives

$$\Delta \sim Rh^2/\sigma \quad . \tag{49}$$

independent of d. We now use Eq. (35) for the relaxation time to estimate the collapse time of a droplet of radius R:

$$\tau(R) \sim \tau_o \exp(Rh^2/\sigma k_B T) \quad . \tag{50}$$

Returning to the domain wall state of Fig. 9, as time proceeds the mean radius of curvature must increase because the smaller droplets collapse in a time of order Eq. (50), so we can use Eq. (50) to estimate $\bar{\xi}(t)$:

$$\bar{\xi}(t) \sim a\left(\frac{\sigma T}{h^2}\right) \ell n(t/\tau_o) \quad . \tag{51}$$

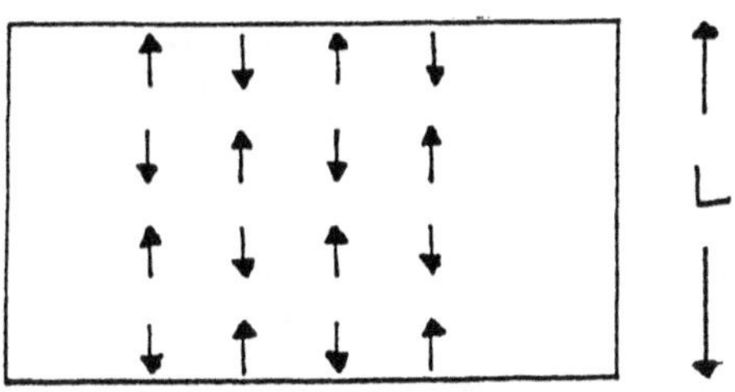

Fig. 9. Special random field arrangements leading to Griffiths singularities.

Assuming other thermodynamic quantities to have a similar time dependence, this provides us with an explanation of the logarithmic relaxation observed in the specific heat of a non-equilibrium state close to T_C (23). On the other hand, neither the linear temperature dependence nor the log time decay are observed further below T_C (24,30). It is not clear why Eq. (51) does not hold over a larger temperature interval. Note that Eq. (51) predicts the observed $1/h^2$ dependence of $\bar{\xi}$ and that it should hold in *any* dimension, not just d = 3.

It should be kept in mind that $\bar{\xi}$ is not the equilibrium correlation length, only the typical length scale of the non-equilibrium domain pattern. If we compute the same length scale for the pure Ising model then the Lifshitz law $\bar{\xi} \sim 1/t^{1/2}$ is found. Clearly, the pure Ising model equilibrates much faster than the RFIM.

In summary, using a combination of scaling arguments and the Imry-Ma argument, there is now some understanding, albeit heuristic, of the dynamics of the RFIM. There is still a painful lack of a single, unified theory which can explain the dynamics over the whole temperature interval.

All our work so far, in fact, made a crucial assumption: the quenched randomness can be characterized by a single quantity: the mean square deviation h^2. The implicit assumption is that the critical properties and the long-time relaxation are "universal" and independent of the detailed form of the random field distribution function. This we will now see is in general not the case.

E. *Griffiths Singularities*

Imagine a special random field arrangement with the fields alternating from site to site in an antiferromagnetic pattern, and assume the size of this region to be L.
The probability of such a region is $(1/2)^{L^d}$, and it contributes a term to the free energy of the form

$$\delta f \sim \int d^d r \left(\frac{1}{2}\right)^{r^d} \frac{1}{\xi^d} f(r/\xi) \ , \tag{52}$$

where f(x) is the finite size scaling function associated with the paramagnetic to ferromagnetic transition in the presence of the staggered field. It is easy to show (32) that this transition occurs between the critical temperature of the pure Ising model and the RFIM. Evaluating the expression by steepest descent gives a contribution of order $\exp(-\xi^d \ell n(2))$ to the free energy. This essential singularity appears thus above the actual phase-transition temperature and is called a "Griffiths Singularity." It depends on the appearance of unlikely random-field configurations. If we simply remove the staggered field configurations from the random field distribution function then it would hardly affect h^2. Now consider the contribution of Griffiths singularities to the dynamics. Borrowing an argument from the theory of spin-glasses (33), assume that we are above T_c but below the critical temperature for the staggered field configuration and that we have a staggered domain of size L. If the moments are aligned in the wrong direction in Fig. 10, then an energy barrier $\Delta(L) \sim JL^{d-1}$ must be overcome to relax the spins. The associated relaxation time is $\tau \sim \tau_o \exp(\Delta(L)/kT)$ and on averaging over L one finds a contribution to the relaxation of the magnetization M(t) of order

$$M(T) \sim \int d^d r \exp(-r^d \ell n\ 2 - t/\tau(r)) \sim \exp -\left(\frac{kT}{J} \ell n \left[\frac{t}{t_o}\frac{J}{kT}\right]\right)^{d/d-1} \tag{53}$$

again using steepest descent. This contribution apparently relaxes *slower* than exponential. Thus, if we wait sufficiently long, it will eventually dominate the exponential relaxation predicted by the scaling argument (22). Non-exponential relaxation above T_c is indeed seen for spin glasses and should also be expected for the RFIM. Long-time relaxation thus appears to depend on the *details* of the random field distribution. We could think of this result as indicating that there is a *distribution* of relaxation times whose maximum is given by Eq. (39) and whose long time tail is governed by Griffiths singularities. If the maximum is sufficiently pronounced, we could neglect the tail, but this remains an open question.

F. *Future Prospects*

Having been requested to comment on possible future developments in the theory of the RFIM, I will restrict myself to the statics. I believe that the mean-field theory of Eq. (5) is not a good starting point for understanding the critical properties of the RFIM and, more importantly, in predicting even the qualitative features of the phase diagram of recent experimental realizations of the RFIM. This claim is based on results which I obtained in a study of the RFIM on the Bethe lattice. The Bethe lattice represents in essence a system with infinite dimensionality. Accordingly we should expect to recover the phase diagram of Fig. 1. A paramagnetic to ferromagnetic transition indeed appears but preceded by an infinite sequence of Griffiths singularities. More importantly, at T = 0 non-trivial critical behavior was obtained (both analytically and numerically) on approaching the phase transition. Equation (5) with Gaussian randomness gives trivial critical behavior. It thus appears as if Eq. (5) is wrong in *any* dimension. Recently, it was shown that the spin-glass on a Bethe lattice exhibits in the presence of a magnetic field a transition to a glassy phase: the de Almeida-Thouless line (34). Interestingly, the spin-glass on a Bethe lattice in a uniform field can be mapped onto the RFIM on the Bethe lattice by a gauge transformation. The de Almeida-Thouless line is then found to lie somewhere below the paramagnetic to ferromagnetic transition line (35). This would confirm our general intuition that the RFIM is similar to a spin-glass in a field. In d = 1 this relationship has been known for some time (36). The resulting phase diagram would look as follows:

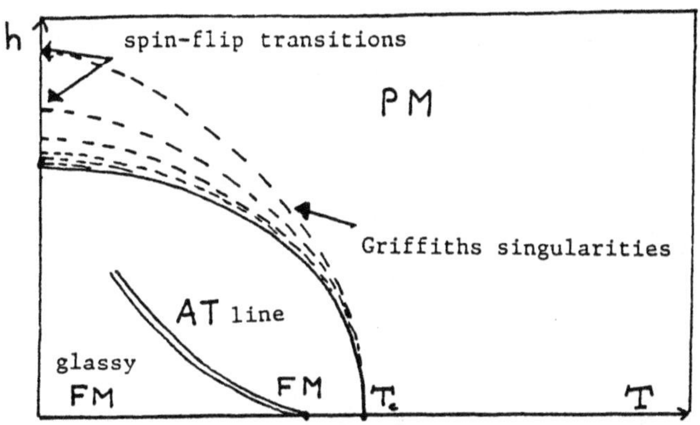

Fig. 10. Phase diagram of the RFIM on a Bethe lattice.

The Griffiths singularities terminate at T = 0 on critical h values where the RFIM undergoes a first-order transition called "spin-flip" transition (37). The low-temperature phase is both ferromagnetic (in the sense that the spins at the center of the Bethe lattice are ordered) and glassy (in the sense that the spin configuration is sensitive to the boundary conditions). However, the Bethe lattice is unusually sensitive to the boundary conditions, and the speculation that the low-temperature phase of the RFIM is glassy also on other lattices lacks further evidence. In fact, even for h = 0, Fig. 10 would predict glassiness, which certainly should not be true in the actual mean-field phase diagram.

Acknowledgments: My own work was done in collaboration with G. Aeppli. I benefited from numerous discussions with colleagues, especially D. Belanger, B. Birgeneau, J. Cardy, R. Cowley, G. Grinstein, V. Jaccarino, A. King, E. Pytte, J. Villain, H. Yoshizawa, P. Young, and P. z. Wong. The section on linear response relied on a review article by J. Villain.

References

The literature on the RFIM is now truly enormous. The references below are only meant as indicating where more material can be found.

1) For various applications of the RFIM see: Y. Imry, J. Stat. Phys. 34, 849 (1984); J. Villain, J. Phys. Lett. (Paris) 43, L551 (1982); B. I. Halperin and C. M. Varma, Phys Rev. B 14, 4030 (1976); F. Brochard and P. G. de Gennes, J. Phys. Lett. (Paris) 44, L785 (1983); H. Fukuyama and P. A. Lee, Phys. Rev. B 17, 535 (1977), on the application of the random field X-Y model to charge-density wave systems.

2) For the relationship between the RFIM and diluted antiferromagnets in a uniform field, see: S. Fishman and A. Aharony, J. Phys. C 12, L729 (1979).

3) A. Aharony, Phys. Rev. B 18, 3318 (1978).

4) R. J. Birgeneau, R. A. Cowley, G. Shirane, and H. Yoshizawa, Phys. Rev. Lett. 54, 2147 (1985).

5) A. P. Young and M. Nauenberg, Phys. Rev. Lett. 55, 2499 (1985).

6) D. P. Belanger, A. R. King, V. Jaccarino, Phys. Rev. Lett. 49, 1050 (1982), J. Appl. Phys. 55, 2383 (1984); D. P. Belanger, A. R. King, V. Jaccarino, and J. Cardy, Phys. Rev. B 29, 2522 (1983).

7) R. A. Cowley, H. Yoshizawa, G. Shirane, and R. J. Birgeneau, Z. Phys. B 58, 15 (1984); R. J. Birgeneau, R. A. Cowley, G. Shirane, and H. Yoshizawa, Phys. Rev. Lett. 54, 2147 (1985).

8) A. P. Levanyuk, V. V. Osipov, A. S. Sigov, and A. A. Sobyanin, Sov. Phys. JETP 49, 176 (1979).

9) For a review on neutron scattering, see: R. J. Birgeneau, R. A. Cowley, G. Shirane, and H. Yoshizawa, J. Stat. Phys. 34, 817 (1984).

10) Both the equilibrium phase boundary as well as the threshold temperature for the onset of irreversibility obey this scaling law. D. P. Belanger, A. R. King, and V. Jaccarino, Phys. Rev. Lett. 54, 577 (1985); A. R. King, V. Jaccarino, D. P. Belanger, and S. M. Rezende 32, 503 (1985).

11) G. Grinstein, Phys. Rev. Lett. 37, 944 (1976); A. Aharony, Y. Imry, and S. k. Ma, *ibid.* 37, 1376 (1976); A. P. Young, J. Phys. C 10, L257 (1977); G. Parisi and N. Sourlas, Phys. Rev. Lett. 43, 744 (1979); J. Cardy, Phys. Lett. 125B, 470 (1983); D. Blankschtein, Y. Shapir, and A. Aharony, Phys. Rev. B 29, 1363 (1984).

12) J. Z. Imbrie, Phys. Rev. Lett. 53, 1747 (1984).

13) A. T. Ogielski and D. A. Huse, Phys. Rev. Lett. 56, 1298 (1986).

14) A. Aharony, Y. Imry, and S. k. Ma, Phys. Rev. Lett. 37, 1364 (1976).

15) A. J. Bray and M. A. Moore, preprint.

16) P. W. Anderson, B. Halperin, and C. Varma, Phil. Mag. 25, 1 (1972).

17) Y. Imry and S. k. Ma, Phys. Rev. Lett. 35, 1399 (1975).

18) G. Grinstein and S. k. Ma, Phys. Rev. B 28, 2588 (1983).

19) D. Fisher, J. Fröhlich, and T. Spencer, J. Stat. Phys. 34, 863 (1983).

20) L. Néel, J. Phys. Rad. 11, 49 (1950).

21) See, for instance: P. z. Wong and J. W. Cable, Phys. Rev. B 28, 536 (1983).

22) D. P. Belanger, et al., Phys. Rev. 29, 2636 (1984).

23) D. P. Belanger, A. R. King, and V. Jaccarino, Phys. Rev. Lett. 54, 577 (1985).

24) R. J. Birgeneau, R. A. Cowley, G. Shirane, and H. Yoshizawa, Phys. Rev. Lett. 54, 2147 (1985).

25) D. S. Fisher, Phys. Rev. Lett. 56, 416 (1986); J. Villain, J. Phys. (Paris), to be published.

26) R. Bruinsma and G. Aeppli, Phys. Rev. Lett. 52, 1547 (1984); G. Aeppli and R. Bruinsma, J. Magn. 54-57, 25 (1986).

27) C. Ro, G. Grest, C. Soukoulis, and K. Levin, Phys. Rev. B 31, 1682 (1985).

28) P. Debye, H. R. Anderson, and H. Brumberger, J. Appl. Phys. 28, 679 (1957); P.z. Wong, J. W. Cable, and P. Dimon, J. Appl. Phys. 55, 2377 (1984).

29) See ref. 18. This expression is valid for Ising-Heisenberg magnets, not for the Ising mode, in which case E is a non-analytic function of f. For most applications, Eqs. (43) should be relevant.

30) M. Hagen, R. A. Cowley, S. K. Satija, H. Yoshizawa, G. Shirane, R. J. Birgeneau, and H. J. Guggenheim, Phys. Rev. B 28, 2602 (1983).

31) J. Villain, Phys. Rev. Lett. 52, 1543 (1984); G. Grinstein and J. Fernandez, Phys. Rev. B 29, 6389 (1984).

32) R. Bruinsma, Phys. Rev. B 30, 289 (1984).

33) M. Randeria, J. Sethna, and R. Palmer, Phys. Rev. Lett. 54, 1321 (1981).

34) D. J. Thouless, Phys. Rev. Lett. 56, 1082 (1986).

35) J. Shapir, Phys. Rev. Lett. 57, 271 (1986); R. Bruinsma, *ibid.*, 272; D. Thouless, *ibid.*, 273.

36) B. Derrida, J. Vanninemus, and Y. Pomeau, J. Phys. C 11, 4749 (1978).

37) I. Morgenstern, K. Binder, and R. M. Hornreich, Phys. Rev. B 23, 287 (1981).

A Driven Random Field System: CDW Dynamics

L. Mihaly and G. Grüner

Department of Physics and Solid State Science Center,
University of California, Los Angeles, CA 90024, USA

Pinned charge density waves can be regarded as a model for the study of the dynamics of random systems. The nonlinear and frequency dependent conductivity of several materials is discussed, with particular emphasis on the nonlinear ac effects and higher harmonic generation, on the remanent polarization and on the critical properties as the threshold field is approached from below.

1. Introduction

A considerable amount of experimental and theoretical effort has recently been directed towards understanding the highly unusual conduction process associated with moving charge density waves (CDW) [1]. It is by now well established that the CDW is pinned to the underlying lattice, but the pinning is weak enough to allow CDW conduction at high enough electric fields. The competition between the pinning forces and the external drive leads to nonlinear and frequency dependent response at small ac fields and low frequencies. Several controversial issues still exist, such as the question whether the dynamics should be described in terms of a classical equation of motion or genuine quantum mechanical processes (like tunneling) are important. On the experimental side much of the remaining disagreement is due to the fact that pinning, in general, may come from impurities (in which case the randomness is essential), or from grain boundaries, surface effects, etc. (in which case we are faced with a boundary problem, where the disorder plays a much reduced role).

In this contribution we give a short summary of the phenomena occurring when random impurity pinning is dominant, and the problem is that of a driven random system with analogies to spin glasses, random field systems, and similar materials. As will be discussed later, the response to an applied ac and dc field is both nonlinear and frequency dependent, and in general the linear response theory is not appropriate. The following topics will be discussed:

a. ac response in the small amplitude limit: frequency dependent conductivity;

b. nonlinear ac response, especially third harmonic generation;

c. time dependent response in the long time limit;

d. approaching the threshold from below: polarization divergence and critical slowing down.

The above questions are different manifestations of the complex problem of charge density wave dynamics. Several experiments have indicated the important role of randomness in these materials [2,3]. In this note we are not going to discuss phenomena related to the current-carrying state, such as the nonlinear dc current-voltage characteristics. Instead we focus our attention to the broad variety of interesting and in many cases unique phenomena occurring in the pinned state, below the threshold field of the nonlinear conduction.

2. Impurity Pinned Charge Density Waves

The most widely used microscopic description of charge density waves interacting with random impurities has been introduced by FUKUYAMA and LEE [4]. The Hamiltonian is

$$H = \frac{1}{2}\int K\,(\nabla\Phi)^2\,dr + \sum_i V_i \cos(2k_F\,r_i + \Phi_i) \tag{1}$$

where the first term represents the elastic energy of the CDW, characterized by an elastic constant K, and the second term describes the interaction of the collective mode with impurities of strength V_i, distributed at random sites i. Here the amplitude fluctuations of the CDW are neglected (which is by no means a trivial assumption; see Ref. 5), and the impurities couple directly to the phase Φ. The discrete version of Hamiltonian Eq. (1) is due to FISHER [6], in the form

$$H = \frac{1}{2}\sum J_{nm}\,(\Phi_n-\Phi_m)^2 + \sum V_n \cos(\beta_n + \Phi_n) \tag{2}$$

where n and m refer to the impurity sites, J_{nm} is the coupling constant, and β_n is the random pinning phase at site n. Both Eq. (1) and (2) represent a competition between two terms: the elastic term prefers a smooth, uniform phase, while the pinning term favors distortions due to the impurities. It was first noted by IMRY and MA, and later by SHAM and PATTON, that in these situations the phase-phase correlation function is finite for any dimension $d < 4$, and thus there is no long range order [7]. Scaling arguments lead to a typical correlation length (referred to in the CDW literature as the size associated to the so called Lee-Rice domains):

$$L \sim \left(\frac{K}{V_i\ c^{1/2}}\right)^{2/(4-\alpha)} \tag{3}$$

where c is the impurity concentration.

Equation (2) is formally identical to the random field X-Y model (RFXY model)

$$H = \sum J'_n\,S_n\,S_{n+1} + \sum h_n\,S_n \tag{4}$$

in the limit where the angle between the neighboring spin directions is small and the interaction J_{nm} is short ranged. Here S represents the classical magnetic moment, and h is the random field. A major difference between the CDW and RFXY model appears when the coupling to the external drive is considered. For charge density waves, the electric field couples directly to the phase,

$$H_E = \sum eE\Phi_n \tag{5}$$

whereas in the RFXY model the magnetic field adds to the local field, $h' = h+H$. Equation (5) corresponds to an applied torque in the RFXY model, leading to the rotation of the spins in the X-Y plane. The electric polarization P of the CDW mode corresponds to a "rotational magnetization" in the RFXY model. Due to this difference in the Hamiltonian, the two systems have a qualitatively different behavior at high fields: in the RFXY model the increase of the magnetic field leads to a smoothly increasing magnetization with a saturation in the infinite field limit, while in the CDW high electric field results in a phase transition from the pinned to the depinned state. In this context, the depinned CDW corresponds to a "phase winding" state in the RFXY model. On the basis of these arguments, we expect that the small amplitude responses of the two systems are identical, but for large amplitude ac signals, or in the presence of large dc bias, the RFXY model and the driven CDW behave differently.

3. Frequency Dependent Response

First we will discuss the overall features of the response to ac electric fields reviewing experimental results obtained mainly in TaS_3 and $K_{0.3}MoO_3$. In both materials the relevant correlation length L [see Eq. (3)] is much smaller than the dimensions of the specimens, and therefore finite size effects are not important. The study of $K_{0.3}MoO_3$ was performed at low temperatures, when long time relaxation effects are dominant. For both cases our discussion focuses at the low-frequency response, where the quantum description is most probably not appropriate.

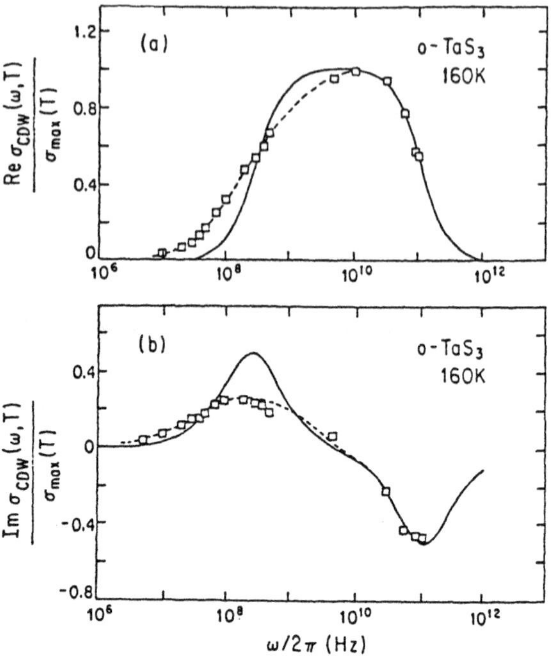

Fig. 1. Frequency dependent conductivity of TaS_3. The full line is a harmonic oscillaotr fit with resonant frequency ω=10 GHz, damping time τ=50 GHz, and effective mass $m^*=100m_{electron}$.

3.1 Small Amplitude Response

The in-phase and out-of-phase component of the ac conductivity measured for low amplitude ac drives (Re σ and Im σ) are displayed in Fig. 1 for TaS_3. The full line is a simple harmonic oscillator response characterized by three parameters: pinning frequency ω_0 , damping constant τ, and effective mass m^*. Although this "classical harmonic oscillator" approach gives a good overall description, a strong low-frequency tail in the experimental data suggests an anomalously enhanced

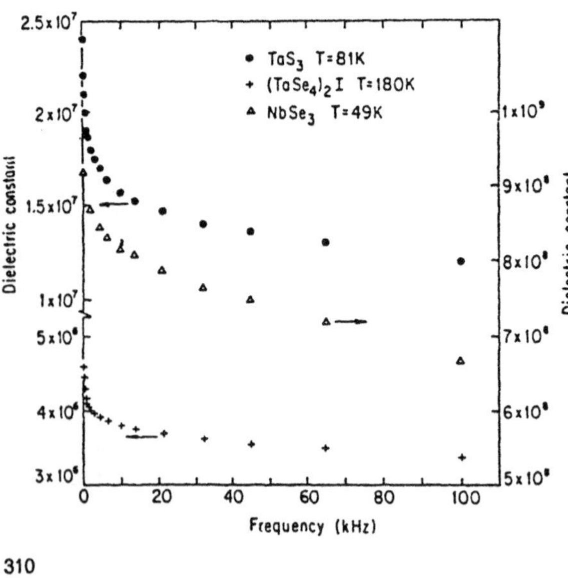

Fig. 2. Frequency dependence of the dielectric function Re ε(ω) for several charge density wave compounds at low frequencies.

spectral weight as $\omega \to 0$ [8]. This feature has been recently accounted for by assuming that tunneling events are important in both the ω and E dependent response [9]. Disorder-induced metastable states, similar to those found in a broad range of glasses and amorphous solids, may also lead to excess low-frequency conduction [10].

The low-frequency tail in Fig. 1 can be well described by the empirical formula [11]

$$\sigma(\omega) = A \left(\frac{i\omega}{\omega_0}\right)^{\alpha}, \qquad \alpha \approx 0.8 \quad . \tag{6}$$

This expression is widely used to account for the ω dependent response of various random systems. Similar behavior has been found in other materials like $NbSe_3$, $K_{0.3}MoO_3$, and $(TaSe_4)_2I$ [6]. The dielectric constant, Re $\varepsilon(\omega) = 4\pi$ Im $\sigma(\omega)/\omega$ diverges as $\omega \to 0$. This behavior is shown for several compounds in Fig. 2. Recent experiments [12] suggest that at low frequencies the ω dependence is better represented by a so-called Cole-Davidson formula [13]:

$$\varepsilon(\omega) = \frac{\varepsilon_0}{[1 + (i\omega\tau_0)^{\alpha}]^{\beta}} \tag{7}$$

with α and β less than one. Eq. (7) is also typical of random conductors.

3.2 Nonlinear Response and Third Harmonic Generation

The ac conductivity, as displayed in Figs. 1 and 2, represents the response of the system to small amplitude excitations. The response, however, strongly depends on the ac amplitude [8,15,16]. We have also found that higher order harmonics at $\Omega = 3\omega$, $\Omega = 5\omega$, etc., appear in the response to a sinusoidal excitation, $E = E_0 \cos \omega t$, even for drive amplitudes far below the threshold field of the nonlinear conduction.

Before discussing the experiments let us extend the single harmonic oscillator approach by including a nonlinear term in the potential, $V(x) = (1/2)kx^2 - (1/4)bx^4$. The equation of motion is

$$1/\tau\, \dot{x} + \omega_0^2\, x + (b/m)\, x^3 = eE_0/m^* \cos \omega t \ , \tag{8}$$

where τ is the damping constant, e is the charge, m^* is the mass of the particle, and $\omega_0^2 = k/m^*$. In the low-frequency limit the linear (b=0) response is given by the complex dielectric function

$$\varepsilon_0^{(1)} = 4\pi \frac{e^2}{m\omega_0^2} \frac{1}{1 + i(\omega/\omega_0^2\tau)} \ . \tag{9}$$

This is the formula used to obtain the full lines in Fig. 1. The first order correction due to the nonlinearity leads to an amplitude-dependent first harmonic and nonzero higher harmonics:

$$\varepsilon^{(1)} = \varepsilon_0^{(1)} + b\zeta(\omega)E_0^2 \ , \quad \varepsilon^{(3)} = b\eta(\omega)E_0^2 \ , \tag{10}$$

where $\zeta(\omega)$ and $\eta(\omega)$ are simple, analytical functions of the frequency. Equation (10) leads to specific predictions concerning the field and frequency dependence of the nonlinear response. At low frequencies, $\omega \ll \omega_0^2\tau$, both the first and third harmonic nonlinear response are independent of frequency and both are proportional to the square of the drive amplitude. The leading term in the frequency dependence of the linear response Re $\varepsilon_0^{(1)} \sim (1-a\omega^2)$, in contrast to Eqs. (6) and (7), where the frequency dependence is singular around $\omega = 0$. Similarly, the leading frequency dependence in the third harmonic is nonsingular, Re $\varepsilon^{(3)} \sim (1-b\omega^2)$.

The first and third harmonic response was measured both in $K_{0.3}MoO_3$ and in TaS_3 [14]. For $K_{0.3}MoO_3$, the experiments were performed at 4.2 K where the conductivity due to the normal electrons is negligible and the dielectric response of the CDW can be detected directly. In TaS_3, the experiments were conducted at higher temperatures, where the conductivity due to the normal

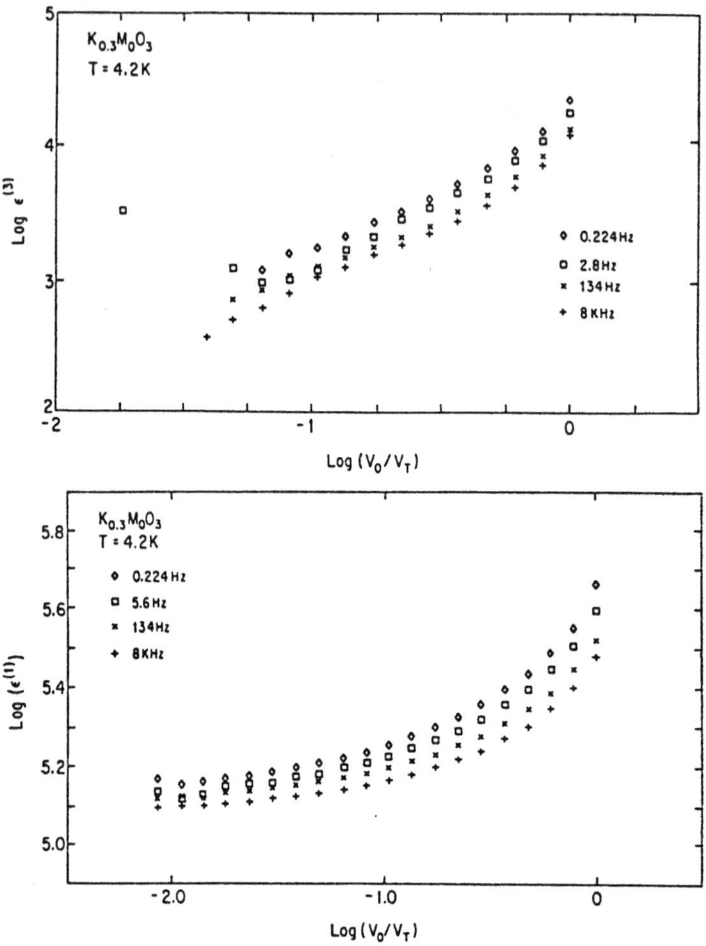

Fig. 3. Nonlinear first harmonic (a) and third harmonic (b) response at various frequencies in $K_{0.3}MoO_3$. The third harmonic is a universal function of the scaling variable V_o/ω^η (c).

electrons has been substracted by a specially designed electronic circuit. In both cases a spectrum analyzer was also employed to record the intensity of the higher harmonics. Here we present the data obtained on $K_{0.3}MoO_3$, but similar results were obtained for the other material.

The excitation amplitude dependence of $\varepsilon^{(1)}$ is displayed in Fig. 3a for various applied V_o voltages, normalized to the threshold voltage V_T for the appearance of nonlinear dc conduction. Although $\varepsilon^{(1)}$ appears to saturate as $V_o/V_T \to 0$, the response is strongly nonlinear and frequency dependent, with $\varepsilon^{(1)}$ slowly increasing with decreasing frequencies. Similar enhancement is observed in the out-of-phase component. The third harmonic response $\varepsilon^{(3)}$ is displayed in Fig. 3b in the same parameter range. Again, $\varepsilon^{(3)}$ is both V_o and ω dependent with increasing response with increasing V_o and decreasing ω. The small amplitude linear response is well described by Eq. (6) in the measured frequency range with $\alpha = 0.9$.

The amplitude-dependent fundamental response has been noted earlier [8,15,16]. The response at any given excitation amplitude V_o can be analyzed by assuming that linear response theory is appropriate, and increasing V_0 leads to changes in the parameters which characterize the response. Such an approach was adopted by CAVA, et al. [16], who argued that the mean relaxation time τ_0 which appears in a Cole-Davidson expression, Eq. (7), increases with increasing ac drive. The interpretation implies that linear response theory is relevant, and that the response to an excitation $V = V_0 \cos \omega t$ is given by a sinusoidal function with no higher harmonics, in contrast to our findings.

An alternative, empirical description of the nonlinear fundamental response may be possible by scaling the amplitude and frequency dependence into a universal function

$$\varepsilon^{(1)} = \varepsilon_0^{(1)}(\omega, V_0{=}0) + g(V_0/\omega^{\eta}) \quad . \tag{11a}$$

This approach follows from the generalization of Eq. (10), and indeed such a description is possible. The uncertainty with this analysis lies in the (perhaps artificial) distinction of $\varepsilon^{(1)}$ into linear and nonlinear part, with $\varepsilon_0^{(1)}$ dominating for small excitation amplitudes.

Such problems do not arise in the third harmonic response which disappears in the $V_0 \to 0$ limit. The full V_0 and ω dependence can be absorbed into a universal function $f(V_0/\omega^{\eta})$ as demonstrated in Fig. 3c. Moreover, for V_0 not close to V_T, $\varepsilon^{(3)}$ can well be represented by

$$\varepsilon^{(3)}(\omega, V_0) = A\, V_0^{\nu}/\omega^{\eta} \tag{11b}$$

with $\nu = 0.9$ and $\eta = 0.05$. Similar results were obtained for TaS_3 at $T = 77$ K.

Just as the frequency-dependent small amplitude response [Eq. (6) or Eq. (7)] is different from that of a simple harmonic oscillator [Eq.(9)], the third harmonic response is also different from that given by Eq. (10). Instead of smooth frequency dependence, predicted by the simple model based on Eq. (8), the third harmonic response diverges as $\omega \to 0$.

4. Time Domain Studies

The amplitude-dependent ac response and the enhancement of the conductivity at low frequencies indicate that the behavior of charge density wave systems at very low frequencies may be better studied by applying step-like excitations and recording the time-dependent relaxation processes. This approach has been succesfully pursued for spin glasses, where the long time relaxation is described by logarithmic or stretched exponential formulas. A typical expression of this class is the Kohlraush formula [17]:

$$P(t) = P_0 \left[1 - \exp\left\{-(t/\tau)^{\alpha}\right\}\right] \tag{12}$$

where P(t) is the time-dependent polarization, τ is the relaxation time, and $\alpha < 1$. For $\alpha \to 0$ the time dependence approaches the $P(t) \sim \log t$, while $\alpha = 1$ corresponds to the simple Debye relaxation. For a linear system the time and the frequency dependence are related to each other by a Fourier transform, and no further information is obtained by studying the relaxation processes in the time domain; however, for a nonlinear system the time and frequency dependence are related only by the microscopic details of the Hamiltonian. It is possible, for example, to describe the observed frequency and field dependence by assuming independent, nonlinear oscillators with a proper distribution of the parameters k, τ, and b in the equation of motion (8). However, the set of oscillators thus selected does not necessarily have the experimentally observed time-dependent response. In fact, we will find that the proper description calls for the appropriate treatment of the collective nature of the response, and the relaxation is best described in terms of strongly interacting nonlinear modes.

4.1 Long Time Relaxation Phenomena

In Fig. 4 typical relaxation curves obtained on $K_{0.3}MoO_3$ at $T = 4.2$ K are displayed [18]. Since the low field resistance of the sample is extremely high, the response was investigated in a configuration most appropriate for the measurement of capacitive signals (see inset). Figure 4a

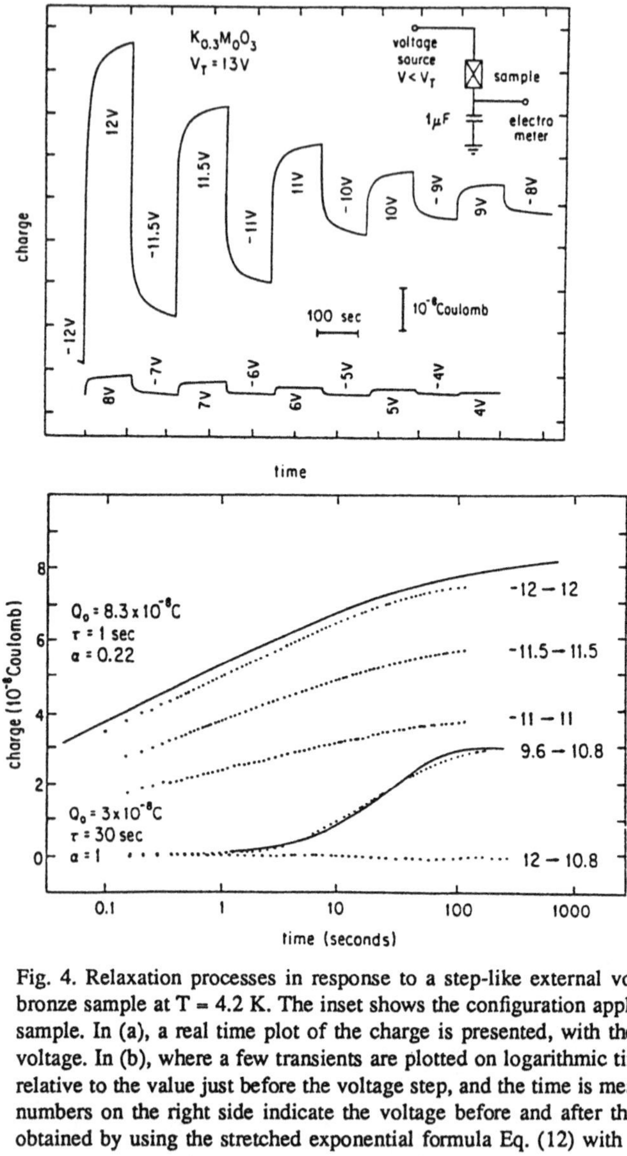

Fig. 4. Relaxation processes in response to a step-like external voltage drive, measured on a blue bronze sample at T = 4.2 K. The inset shows the configuration applied to measure the charge on the sample. In (a), a real time plot of the charge is presented, with the numbers indicating the applied voltage. In (b), where a few transients are plotted on logarithmic time scale, the charge is measured relative to the value just before the voltage step, and the time is measured from the voltage step. The numbers on the right side indicate the voltage before and after the step. The continous lines were obtained by using the stretched exponential formula Eq. (12) with parameters shown on the left side.

shows the charging processes in response to sudden voltage steps. Note the strong nonlinearity and the very long time scale of the response. In general, the time dependence of the charge is not exponential; this is apparent when the response is plotted on a logarithmic time scale (Fig. 4b). The stretched exponential expression, Eq. (12), gives a good fit to the data, with parameters indicated in the Figure. However, the exponent α and the relaxation time τ are not universal, and the response strongly depends on the history of the sample. This feature is illustrated by the lower two curves in Fig. 4b. For comparison we have also plotted a simple exponential relaxation ($\alpha = 1$). It seems that for moderate perturbations (e.g., a voltage step of 9.6 V→10.8 V in Fig. 4b) the response of the system is close to a Debye relaxation, whereas large perturbations activate additional relaxation processes with a broad distribution of time scales. Measurements performed on another sample under the same circumstances resulted in a similar response, but the exponent α fitting the charging process after a large voltage step was smaller, and the time dependence of the charge was very close to logarithmic.

Similar relaxation processes have been observed recently at higher temperatures in TaS_3 [19]. In this study the zero-field relaxation current was recorded after the application of a voltage pulse. Again, the Kohlraush formula, Eq. (12), describes the time dependence of the charge.

We believe that the essential feature in these relaxation experiments is the "local depinning" of the charge density waves. The coupling between CDW segments then leads to a series of relaxations [20], instead of the the independent, parallel relaxation events. Due to the distribution of pinning energies it is possible that for voltages comparable to the threshold voltage of the nonlinear conduction, $V \lesssim V_T$, a considerable part of the charge density wave is temporarily depinned, and it comes to rest due to the interaction with other, more strongly pinned parts. The size of these depinned domains sensitively depends on the voltage, and it approaches zero as the field is decreased. The transient currents associated to the rearrangement of the CDW are responsible to the nonlinear, long time relaxation processes [21].

In order to study the phenomena related to the "partial depinning," we performed computer simulations based on Eq. (3) [22]. The CDW was modeled by a one-dimensional chain of N massless, charged balls, with springs connecting the balls. Impurity pinning was represented by a static friction force. The polarization is given by $P(t) = \Sigma Q_i\xi_i(t)$ where ξ_i is the displacment and Q_i is the charge of the ith ball. Randomness was introduced either by random positions x_i or by random pinning strengths V_i, and both systems led to closely similar results.

The time dependence of the polarization, after an electric field $E_o < E_T$ is applied at $t = 0$ is shown in Fig. 5 for two representative E_o values. The behavior for both applied fields describes a gradual nonexponential increase of P(t) and it can be well described by Eq. (13) with $\alpha = 0.5$. Although the numerical value of the exponent α is different, the computer simulation correctly reproduces the qualitative features of the experiment. This indicates the importance of the local depinning processes in the response to fields below E_T.

4.2 Polarization Effects and Critical Phenomena

The development of the current-carrying state at a critical field E_T has been suggested to arise as the consequence of a dynamical critical phenomenon, with various critical exponents describing the frequency scale and the dielelectric constant below E_T, and the current above the threshold field

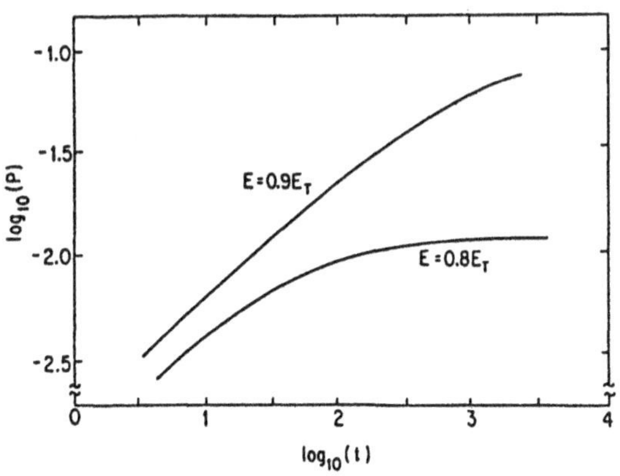

Fig. 5. Computer simulated time-dependent polarization P(t) plotted in logarithmic scales for two representative fields.

[23]. As before, we first discuss the simplest description of the depinning transition, based on the single-particle model [24]. The equation of motion is

$$\ddot{x} + \Gamma \dot{x} + \omega_0^2/2k_F \sin 2k_F x = eE/m^* \tag{13}$$

where it is assumed that the pinning leads to a sinusoidal potential. For small amplitude ac response, Eq. (12) is recovered, and this equation of motion also reproduces Eq. (11) for moderate drives. A threshold field, $E_T = m^*\omega_0^2/2k_Fe$, follows from Eq. (14). Below E_T the low-frequency linear response, measured at low ac amplitudes in the presence of dc bias fields, diverges as the bias approaches E_T, $\varepsilon(\omega \to 0, E) \approx (E_T-E)^{-\mu}$, $\mu = 1/2$. The static polarization, induced by a dc field, shows saturation as $E \to E_T$ from below. Above the threshold the field-dependent current is given by $\langle I_{CDW}\rangle \approx (E_T-E)^{\delta}$ where the exponent is $\delta = 1/2$ and where the bracket indicates time average.

The experiments are in clear contradiction with this single-particle picture. The dielectric constant ε does not show the divergence predicted by Eq. (15) [25]. A closer look at the behavior below E_T indicates that the critical parameter is the static polarization, which is hysteretic and divergent as $E \to E_T$. This feature is entirely absent from the single-particle model.

In Fig. 6a a quantity characteristic of the saturation polarization, $P\alpha$, is plotted for several electric fields. The data were evaluated from relaxation measurements similar to those presented in Fig. 4. The exponent α is not expected to change dramatically, therefore most of the variation is due to the field dependence of the polarization. As the voltage approaches V_T, the polarization seems to diverge, and close to E_T the experimental results can be described with the expression:

$$P_0 \approx (E_T-E)^{-\beta}, \qquad \beta \approx 3. \tag{14}$$

A qualitative understanding of these phenomena can be reached by computer simulations using the model based on Eqs. (2) and (5). As the field is increased, the CDW is locally depinned and the

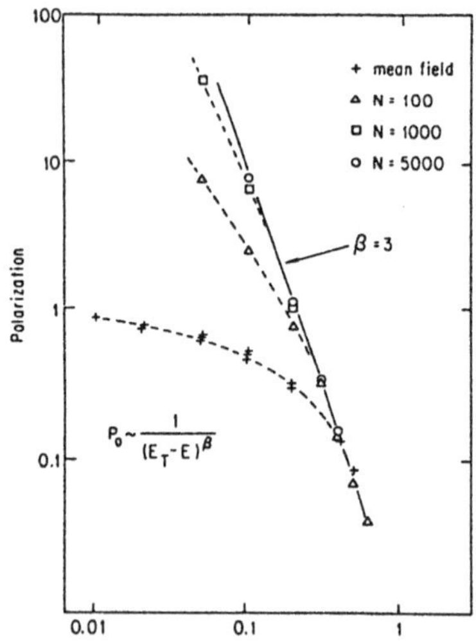

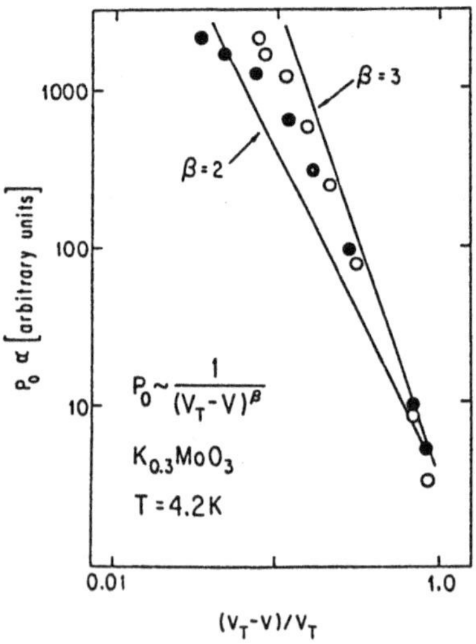

Fig. 6. (a) Critical divergence of the saturation polarization P_0, obtained from relaxation measurements similar to those presented in Fig. 4. Due to the slow (and incomplete) relaxation close to the threshold, only the quantity $P_0\alpha$ was evaluated. The full lines correspond to power law divergence with exponents indicated in the Figure. (b) Computer simulated critical behavior of a one-dimensional CDW model. As the size of the system (N) increases, the saturation polarization approaches power law divergence with exponent $\beta = 3$. The mean field results do not exhibit divergence.

resulting deformations carry a static polarization. When the field is decreased, the deformation is stabilized by the pinning centers, leading to hysteresis. The magnitude of the remanent polarization diverges for $E \to E_T$, and in the one-dimensional model the simulation results in a critical exponent of $\beta = 3$, as indicated in Fig. 6b.

5. Conclusions

We have discussed the various dielectric properties of charge density wave compounds below the threshold field of the nonlinear conduction. The experimental results suggest a qualitative understanding of the depinning transition in random systems: For $E < E_T$ the static polarization diverges as $E \to E_T$, and longer time is required to reach the zero current state. For fields just above E_T there is a nonzero current even in the $t \to \infty$ limit.

There is an overall similarity between the dynamics of pinned charge density waves and other random systems like spin and dipole glasses and various random field systems. The coupling to the external field is different for the random field magnets and CDW. In contrast to the uniform magnetic field acting on a magnetic moment, the electric field acting on the CDW can be represented by a torque, and therefore there is a possibility for a phase transition driven by the external forces.

The initial progress in this area suggests that the driven charge density waves may be regarded as excellent model systems for the study of the dynamics of random systems. Several numerical simulations are in a broad qualitative agreement with the experiments, but further investigations, especially on the area of higher harmonic generation and critical dynamics, are highly desirable.

This work was supported by the National Science Foundation Grant DMR 81-21394.

References

* Permanent address: Central Research Institute for Physics, Budapest 1525, Pf 49, Hungary

1. For a review see G. Grüner and A. Zettl, Phys. Reports 119, 117 (1985); P. Monceau: In Electronic Properties of Inorganic Quasi One-Dimensional Materials, ed. by P. Monceau (Riedel, 1985)
2. R.M. Fleming, R.G. Dunn and L.F. Schneemeyer, Phys. Rev. B 31, 4099 (1985)
3. J. C. Gill, Solid State Commun. 39, 1203 (1981); G. Mihaly and L. Mihaly, Solid State Commun. 48, 449 (1983); R. J. Cava, R. M. Fleming, E. A. Rietman, R. G. Dunn, and L. F. Schneemeyer, Phys. Rev. Lett. 53, 1677 (1984)
4. H. Fukuyama and P.A. Lee, Phys. Rev. B 17, 535 (1977)
5. P. A. Lee, T. M. Rice, and P. W. Anderson, Solid State Commun. 14, 703 (1974); H. Fukuyama, J. Phys. Soc, Japan 41, 513 (1976)
6. D. S. Fisher, Phys. Rev. B 31, 1396 (1985)
7. Y. Imry and S. Ma, Phys. Rev. Letters, 35, 1399 (1975); L. J. Sham and B. R. Patton, Phys. Rev. B 13, 3151 (1976)
8. Wei-yu Wu, L. Mihaly, G. Mozurkevich and G. Grüner, Phys. Rev. B 33, 2444 (1986)
9. J. H. Miller, J. Richard, R. E. Thorne, W. G. Lyons, and J. Bardeen, Phys. Rev. B 29, 2328 (1984); J. H. Miller, J. Richard, J. R. Tucker, and J. Bardeen, Phys. Rev. Letters, 51, 1592 (1983)
10. P. B. Littlewood: In Proceedings of the International Conference on Charge Density Waves in Solids, ed. by Gy. Hutiray and J. Solyom (Springer, 1985)
11. Wei-yu Wu, L. Mihaly, G. Mozurkewich and G. Grüner, Phys. Rev. Lett. 52, 2382 (1984).
12. R. J. Cava, P. Littlewood, R. M. Fleming, R. G. Dunn, and E. A. Rietman, Phys. Rev. B 33, 2439 (1986)
13. R. W. Davidson and R. M. Cole, J. Chem. Phys. 19, 1484 (1951)
14. Ting Chen, L. Mihaly, and G. Grüner, to be bublished
15. J. P. Stokes, M. O. Robbins, and S. Bhattacharya, Phys. Rev. B 32, 6939 (1986)

16. R. J. Cava, R. M. Fleming, P. Littlewood, E. A. Rietman, L. F. Schneemeyer, and R. G. Dunn, Phys. Rev. B 30 3228 (1985)
17. R. Kohlrausch, Ann. Phys. (Leipzig) 12, 393 (1847); G. Williams and D. C. Watts, Trans. Faraday Soc. 66, 80 (1970).
18. R. G. Palmer, D. L. Stein, E. Abrahams, and P. W. Anderson, Phys. Rev. Lett. 53, 958 (1984)
19. L. Mihaly and G. X. Tessema, Phys. Rev. B 33, 5858 (1986)
20. G. Kriza and G. Mihaly, Phys. Rev. Letters, to appear
21. P. B. Littlewood and C. M. Varma, to be published
22. L. Mihaly, M. Crommie, and G. Grüner, to be published
23. L. Sneddon, M. C. Cross, and D. S. Fisher, Phys. Rev. Letters 49, 292 (1982)
24. G. Grüner, A. Zawadowski, and P. Chaikin, Phys. Rev. Letters
25. Wei-yu Wu, L. Mihaly, and G. Grüner, Solid State Commun. 55, 663 (1985)

Part VIII

Frustrated, Incommensurate, and Nonequilibrium Systems

Topography and Diffusion of Kinks in Incommensurate Insulators: Relationship Between Phase Conjugation and Thermal Hysteresis in Incommensurate Phases, and a Physical Analog for Circadian Biorhythms

J.F. Scott

Department of Physics, University of Colorado, Boulder, CO 80309, USA

I. Introduction

In this paper I attempt to describe the role of nonequilibrium defects in incommensurate insulators. I confine my discussion primarily to two systems: $BaMnF_4$ and $Ba_2NaNb_5O_{15}$. These two materials have formed the basis for studies in my own laboratory over the past ten years and are particularly suitable for this review because of the large number of experimental techniques used to elucidate the statics and dynamics of their incommensurate (IC) phases.

Both $BaMnF_4$ and $Ba_2NaNb_5O_{15}$ are known to have mobile defects. In $BaMnF_4$ the magnitude, temperature dependence, and anisotropy of the (ionic) conductivity are compatible [1] with fluorine ion vacancy hopping. And in $Ba_2NaNb_5O_{15}$, which has non-stoichiometry involving all of its ions, especially oxygen, sodium vacancy hopping is well known [2]. Thus we might expect a priori an interesting interplay between mobile defects and soliton-like kinks in the IC phases of both crystals.

$BaMnF_4$ and $Ba_2NaNb_5O_{15}$ share a number of additional characteristics: 1) both have [3,4] four-dimensional order parameters; 2) both are ferroelectric within their IC phases (in most IC insulators the ferroelectric phase is the phase below the IC phase); 3) both exhibit point group symmetry changes [5,6] at the normal-to-incommensurate transition temperatures ($BaMnF_4$ goes from orthorhombic to monoclinic at $T_I \approx 254K$; $Ba_2NaNb_5O_{15}$ goes from tetragonal to orthorhombic at $T_I = 573K$, and then back to tetragonal at $T_{II} = 105K$). Throughout this paper I will term "incommensurate transition temperature" the upper transition of a modulated phase and denote it as T_I (or T_{II} if there are two different modulated phases). The lower transition temperature of these modulated phases will be called T_0 (it is often called T_c in the literature because in the K_2SeO_4 family it is the ferroelectric Curie temperature; however T_C is unrelated to the IC phases in $BaMnF_4$ and $Ba_2NaNb_5O_{15}$).

II. Barium Manganese Fluoride

$BaMnF_4$ has a phase transition [7] at $T_I \sim 254K$ which doubles the size of the primitive cell in the bc-plane [8] and modulates the structure along [9] the a-axis. The critical wavevector lies [9] at $q_0 = (0.389, 1/2, 1/2)$.

The first evidence that defects play a strong role in the dynamics near T_I came from the specific heat data [10] shown in Fig. 1. The transition is now known [11] to lie almost exactly at a tricritical point. In accord with that, the critical exponent α above T_I was found to be 0.5, in agreement with mean field predictions. α' below T_I is 1.1±0.2.

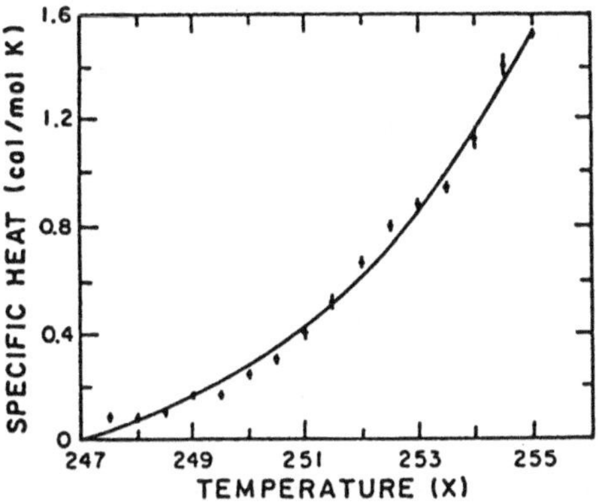

Fig. 1. Specific heat versus temperature in $BaMnF_4$

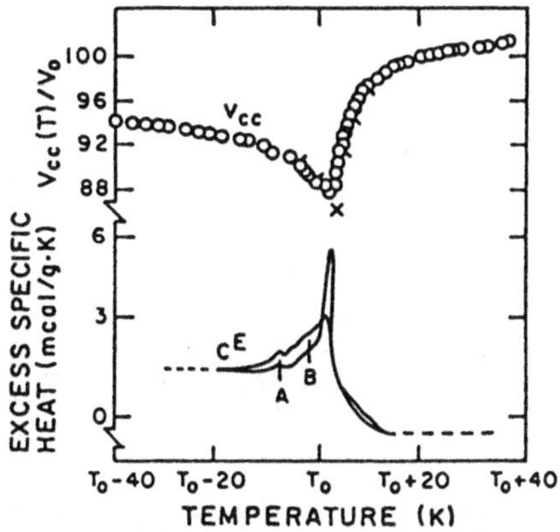

Fig. 2. Comparison of specific heat and logitudinal sound velocity near T_I in $BaMnF_4$. $C_V(T)$ for two samples are shown. Features A and B relate to Fig. 4

Of course no intrinsic statistical mechanical theory predicts α as large as unity, and indeed $\alpha' \gtrsim 1$ predicts a diverging energy at T_I. Nevertheless, we will see that this value of α' is consistent with other measurements as well as with a defect theory.

Figure 2 compares the sound velocity data for longitudinal waves [12] in $BaMnF_4$ with the specific heat data. It is required from the Pippard relationship [13] that $\Delta V(T) \sim t^{-\alpha'}$, i.e., that $\Delta C(T)$ and $-\Delta V(T)$ are proportional. This theory, developed initially for liquid helium, has often been successfully applied to solid state phase transitions (GARLAND [14], JANOVEC [15], and STRUKOV et al. [16]). It is important that it is a thermodynamic equality independent of microscopic details. In TGS (triglycine sulfate) STRUKOV et al. have shown [16]) that this proportionality holds even when both $\Delta C(T)$ and $\Delta V(T)$ arise entirely from nonequilibrium defects which can be annealed away and subsequently re-introduced by irradiation.

Thus the critical exponent $\alpha' \approx 1.1$ is consistently obtained from both specific heat $\Delta C(T)$ and sound velocity data $\Delta V(T)$. Is it reasonable? Using similar defect theories, HOCHLI and BRUCE [17] and LEVANYUK et al. [18] calculate effective values of α in defect-dominated regimes. HOCHLI and BRUCE calculate $\alpha = 1.3$, in excellent agreement with our measured 1.1. The earlier result of LEVANYUK et al. for the ultrasonic attenuation exponent η is

$$\eta = (2\beta - \gamma) + 5\nu \tag{1}$$

which in mean-field theory is 2.5. Since dynamic scaling [19] gives $\alpha = \eta - 1$, in the classical limit $(z\nu = 1)$ this yields $\alpha \approx 1.5$, which is also in reasonable agreement with our 1.1. A slightly different theoretical value of $\alpha = 1.25$ results from REHWALD'S [19] result that $\eta = 2\alpha$. The reason α' can be > 1, stated simply, is that the theory [18] employed is not asymptotic and in fact not valid very near T_I. It assumes a distribution of immobile defects whose contribution to specific heat is dominated by the correlation length $\xi(T)$. $\xi(T)$ diverges as $T \to T_C$ until it is approximately equal to r_d, the average distance between defects. At this point it becomes constant. This produces a "plateau" (~50mK wide). In our data in Fig. 1 the plateau is ~50mK wide and T_I is determined to greater accuracy than that independently from dielectric data on the same specimen.

In summary, we see that the specific heat data and sound velocity data are self-consistent and agree with two defect theories. It is most important to note that whatever these defects are, they disappear above T_I. This immediately suggests that they are soliton-like discommensurations (since $BaMnF_4$ and $Ba_2NaNb_5O_{15}$ are three-dimensional lattices their nonlinear excitations are the kink-like structures discussed by KAWASAKI [20] and not true solitons).

Before leaving the specific heat data in $BaMnF_4$, let us note a second small crimp in C(T) near 247K in Fig. 1. It was suggested [10] that this might signify a lock-in transition. However, inelastic neutron scattering at Brookhaven [21] produced the result that there was no second transition. They found that the IC structure was frozen, astonishingly, with q_o = (0.39, 1/2, 1/2) independent of temperature. Since their work, a variety of independent measurements have shown the Brookhaven sample probably has its IC phase pinned by defects (perhaps fluorine vacancies?). Subsequent GRENOBLE measurements [22] revealed highly sample-dependent q_o, with some specimens showing $q_o(T)$ strongly dependent upon temperature (Fig. 3). The behavior shown in Fig. 3 for one specimen reveals that q_o = (0.399, 1/2, 1/2) below ~60K on cooling and ~160K on heating. The authors comment that within their experimental error, 0.399 = 2/5, and so I regard their "transition" at 60K (cooling) as the lock-in transition, T_o.

Other evidence for a second transition between 0K and 254K comes from piezoelectric resonance data and thermal diffusivity [23], both of which reveal a distinct transition at 226K. It is not certain that this is the same lock-in transition found in Grenoble, since different samples were used.

Another open question is why the piezoelectric resonance data (Fig. 4) display several sharp anomalies at temperatures where no other experiments

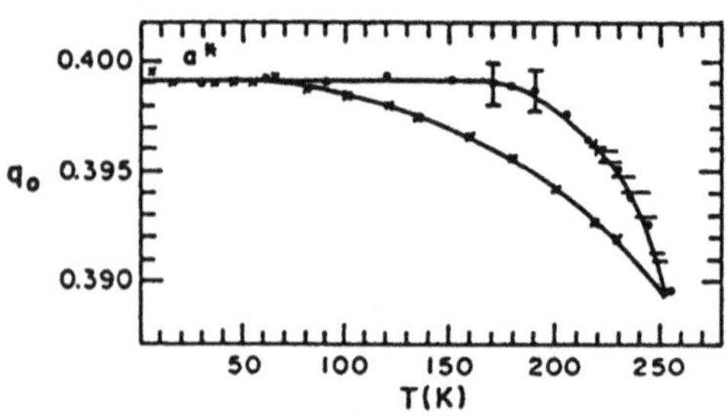

Fig. 3. Incommensurate wavevector $q_o(T)$ in $BaMnF_4$

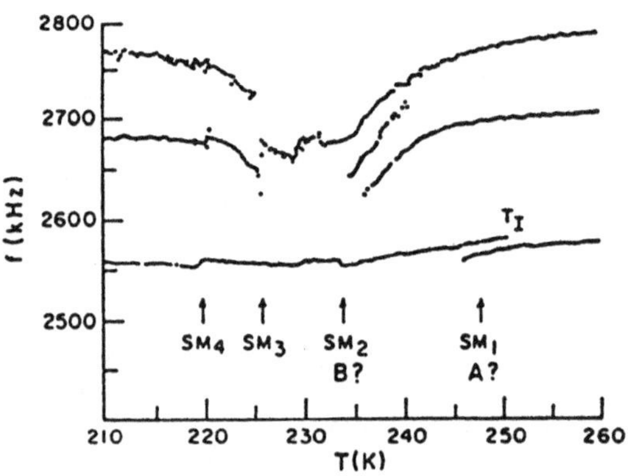

Fig. 4. Piezoelectric resonance data in incommensurate $BaMnF_4$

suggest transitions. The piezoelectric resonance data in Fig. 4 are from absorption of a.c. electric fields by free-standing $BaMnF_4$ crystals. The frequencies plotted are those of natural resonance in the specimen. Although this is a classical physics phenomenon, there seems to be no detailed theory for it. Our results show [23] that it is an extremely sensitive way of detecting IC phase transitions, but it is not clear that each temperature at which an anomaly occurs is a real phase transition. Using the piezoelectric resonance data alone, one might hypothesize that a Devil's staircase of transitions occurs in $BaMnF_4$. The transition temperatures shown approximate $q_o(T)$ of $9a^*/23$, $11a^*/28$, $13a^*/33$, and $15a^*/38$. But such a conclusion requires a better theoretical understanding of the data and/or application of the technique to a better understood system, such as thiourea.

B. "Chiral Anomalies"

I hesitate to use this phrase for theoretician readers, but there are some fascinating and explicit anomalies in IC insulators.

It was first observed in $BaMnF_4$ by IAN FRITZ [24] that the transverse sound velocities in $BaMnF_4$ below I_I were asymmetric with respect to interchange of subscripts: $V_{ji} \neq V_{ij}$. Here ij denote respectively polarization and propagation directions. Fritz made this observation before the phase below T_I was understood to be incommensurate, so he was unable to make any connection between the two properties. His observation went largely ignored, except for a review article I wrote [25], until it was rediscovered in Leningrad by ESAYAN et al. [26] working on $RbH_3(SeO_3)_2$. Their data are shown in Fig. 5.

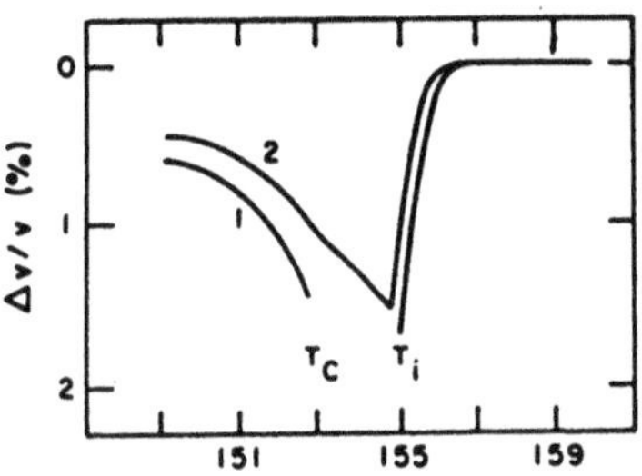

Fig. 5. V_{ij} sound velocities in $RbH_3(SeO_3)_2$ near T_I

These data present a real paradox because sound velocities V_{ij} are simple functions of elastic coefficients C_{ij}, and the C_{ij} are required thermodynamically (e.g., AULD [27]) to be symmetric $C_{ij} = C_{ji}$. Here C_{ij} is reduced notation (e.g., C_{14}) and not to be confused with the V_{ij} subscripts (e.g., $V_{ij} = V_{xy}$).

A possible resolution of this paradox was suggested by DVORAK and ESAYAN [28] and, independently, by SCOTT [25]. Dvorak's theory, simply, is that $C_{ij} = C_{ji}$ at zero frequency whereas V_{ij} and V_{ji} are measured at small but finite frequencies. $|V_{ij} - V_{ji}| \neq 0$ is thus an effect of dispersion. Their theory predicted explicitly that $|V_{ij} - V_{ji}| \propto \omega$. Unfortunately for the theory, we have measured [29] $|V_{ji} - V_{ij}|$ from a few kHz to a few GHz and find it is independent of ω over six orders of magnitude. Thus, it appears that dispersion can be ruled out.

I believe that the problem is simply that $C_{ij} \neq C_{ji}$ in most IC phases. $C_{ij} = C_{ji}$ requires mechanical equilibrium — i.e., no internal torques. Yet many cases are known in physics (e.g., rare earth-containing crystals in magnetic fields) where this condition is not satisfied. It is useful to note that $BaMnF_4$ and $Ba_2NaNb_5O_{15}$ both have screw axes. In such lattices IC discommensurations will have a handedness that produces "local rotations" or internal torques such that $C_{ij} \neq C_{ji}$. (Such an effect could also occur in a commensurate crystal with a screw axis, such as quartz, but in an IC crystal the number density of such defects is vastly greater, because in Serge Aubry's words [30], they are "intrinsic defects".) For example, in $BaMnF_4$ we know from Fritz's [24] that the order parameter at T_I is a screw-like rotation around the incommensurate a-axis. An antiphase boundary in this IC phase will produce an internal torque.

Before producing some independent evidence for this $C_{ij} \neq C_{ji}$ hypothesis, let me comment that experimental measurement of $|V_{ij}(T) - V_{ji}(T)|$ is a new technique which should be exploited in solid state physics. It should be dependent upon the annealing and thermal history of the IC phase, display hysteresis, be sensitive to defect densities and stoichiometries, etc.

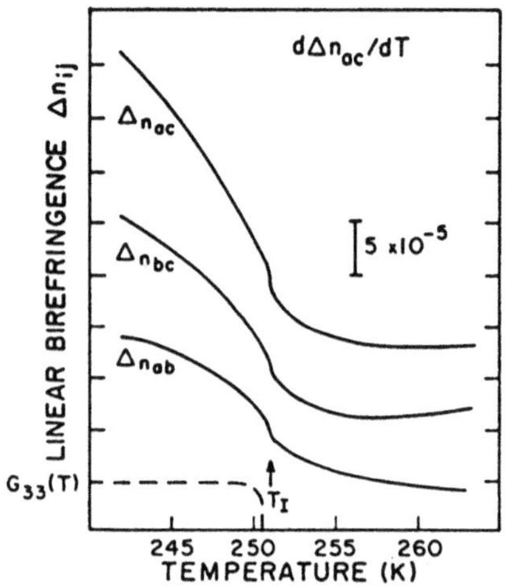

Fig. 6. Birefringence $\Delta n_{ij}(T)$ and optical activity $G_{33}(T)$ in $BaMnF_4$ near T_I

Well, if this chiral hypothesis is correct, how can we test it? Optical activity is the most explicitly chiral phenomenon in a crystal. What does it do at T_I in $BaMnF_4$? The answer, shown in Fig. 6, is that it (G_{33} tensor component) jumps from zero to a large finite value. That was originally unexplained [31] because the linear birefringence (also shown in Fig. 6) goes smoothly to zero at T_I, and both optical activity and birefringence are predicted [32,33] to vary as the square of the order parameter $\phi(T)$. Thus we see that optical activity $G_{33}(T)$ is probably pinned by the chiral strain fields associated with IC discommensurations and prevented from following $\phi(T)$, whereas the birefringence, which is not chiral, follows $\phi(T)$ nicely to zero.

I emphasize that the preceding is not a real theory and I encourage others to develop a true theory of $V_{ij} \neq V_{ji}$ anomalies. There is an analogy with the "local rotation" theory of NELSON and LAX [34] for the fourth order photoelastic tensor: $P_{ijkl} \neq P_{ijlk}$.

C. Thermal Expansion

As an aside, peripheral to the chiral arguments presented above, I note that $BaMnF_4$ and many other IC crystals have zero thermal expansion along their IC axes. Let me suggest a qualitative explanation: In $BaMnF_4$ the MnF_6 octahedra form a helix in the IC phase that resembles a bed-spring. Now you can store energy in a bed-spring by stretching it [thermal expansion a(T)] or by winding (or unwinding) the pitch; you don't have to do both. In $BaMnF_4$ the pitch changes from 2.6a (a is the lattice constant) at T_I = 254K to 2.5a at 4K, eliminating the need for a(T) to contract.

There is no published theory of thermal expansion (or, more precisely, the lack thereof) along incommensurate axes.

D. Small-Angle Brillouin Scattering: Nanosecond Relaxation Times in IC Phases

It is well known in many crystals that the sound velocity $V(\omega)$ is sharply dependent upon frequency ω in the hypersonic regime (GHz). Small-angle Brillouin scattering is an excellent tool to probe this dispersion because frequencies may be varied from ~1GHz to ~50GHz by varying scattering angle θ ($\omega \propto \sin\theta/2$). Fig. 7 shows some sound velocity data [35] for $BaMnF_4$. If we fit these data to the predictions of Landau-Khalatnikov theory

$$C(\omega) = C_\infty + \Delta C[1 + i\omega\tau]^{-1} \qquad (2)$$

$$\rho V^2(T) = C_\infty + \Delta C[1 + \omega^2\tau^2]^{-1} \qquad (3)$$

$$= \mathrm{Re}C \qquad (4)$$

we can extract the relaxation time and its temperature dependence, $\tau(T)$. The result is

$$\tau^{-1}(T) = t\tau_1^{-1}(T) + \tau_2^{-1} \qquad (5)$$

with $\tau_2 \approx$ 1ps. ($t = [T_I - T]/T_I$)

This analysis does not tell us anything about what is relaxing. A more detailed picture emerges from the $Ba_2NaNb_5O_{15}$ data discussed below.

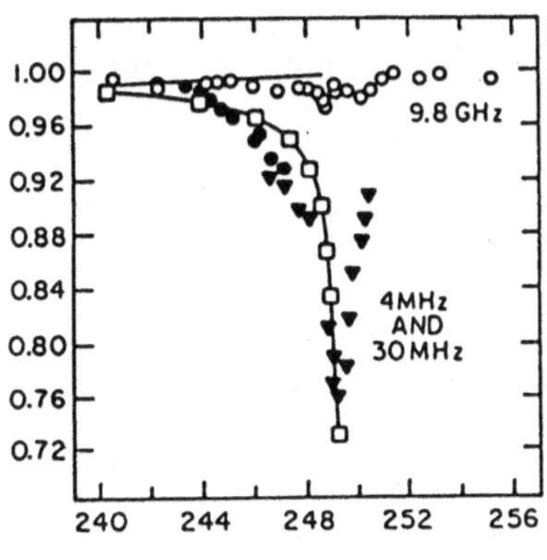

Fig. 7. $V_{jj}(T)$ in $BaMnF_4$ near T_I

II. Barium Sodium Niobate

A. Picosecond IC Relaxation Times

In $Ba_2NaNb_5O_{15}$ we were able to obtain very good linewidth data $\Gamma(T)$, in addition to sound velocity data $V(T)$ near $T_I = 573K$. The $\Gamma(T)$ values satisfy

$$\Gamma(T) = \Gamma_\infty + \tau q^2 \Delta C[2\pi\rho(1 + \omega^2\tau^2)]^{-1} \tag{6}$$

$$= (q/2\pi\rho V)\ \mathrm{Im}C \tag{7}$$

and may be combined with (3) to give a particularly simple expression for $\tau(T)$. Note that if $\Gamma(T)$ is proportional to $\Delta V(T)$ (with opposite sign), $\tau(T)$ is a constant independent of T. Remarkably, as shown in Fig. 8, this is exactly the case for the b-axis in $Ba_2NaNb_5O_{15}$. In contrast, $\Gamma(T)$ and $\Delta V(T)$ are quite different along the a-axis, as shown in Fig. 8. This yields $\tau_0^{-1}(T)$, as in (5) with $\tau_0 = 2\pi \times 1.7\times10^{-13}$s. Combining $\Gamma(T)$ and $\Delta V(T)$ data gives more precise results than were obtainable in $BaMnF_4$ where only $\Delta V(T)$ data existed.

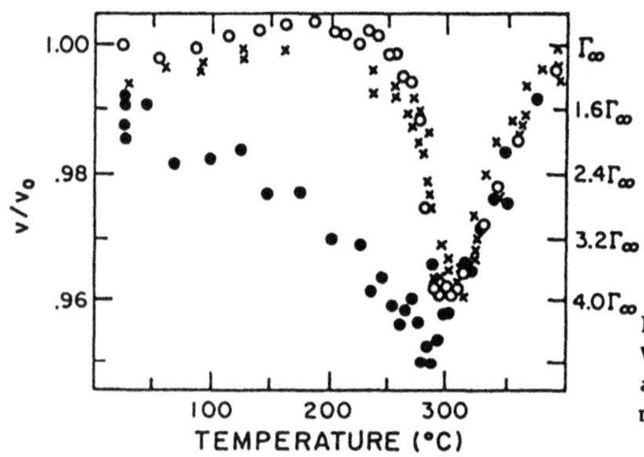

Fig. 8. $V_{aa}(T)$ (solid curves), $V_{bb}(T)$ (open circles), and $\Gamma_{aa}(T)$ and $\Gamma_{bb}(T)$ (crosses) in $Ba_2NaNb_5O_{15}$ near T_I

A surprise afforded by this analysis is that τ^{-1} varies as q^2, where q is the momentum transfer in the laser scattering experiment. A q^2-dependence in inverse lifetime is a signature of diffusion. The diffusion constant inferred from this experiment is $D = 1.3$ cm^2/s. That is extremely fast.

Ionic diffusion in $Ba_2NaNb_5O_{15}$ is $D_I \sim 10^{-4}$ cm^2 and thermal diffusion is $D_T = 2\times10^{-2}$ cm^2/s. Our inferred diffusivity is 65× faster than thermal. Can it be due to kink diffusion D_k of discommensurations? In a 1D model $D_k \ll D_T$ (D_k can even be subdiffusive). In our 2D model [36] $D_k \approx 2D_T$. For steric reasons, D_k/D_T in a 3D model — which no one has calculated — should be bigger. So we think our experimental value $D = 65D_T$ is reasonable for an interpretation as D_{kink}.

What do these kinks look like? In Fig. 9 we show TEM micrographs of IC $Ba_2NaNb_5O_{15}$ from Professor Feng Duan [37]. They consist of an array of defects, each of which has four lines of disclination terminating on it. These four-line vertices are required structures in the quenched four-state clock model of SROLOVITZ and SCOTT [38]. The four states in the clock model correspond to four equivalent tilt directions of thin films of NbO_6 octahedra in the micro-thinned (< 1μm) $Ba_2NaNb_5O_{15}$ TEM specimens.

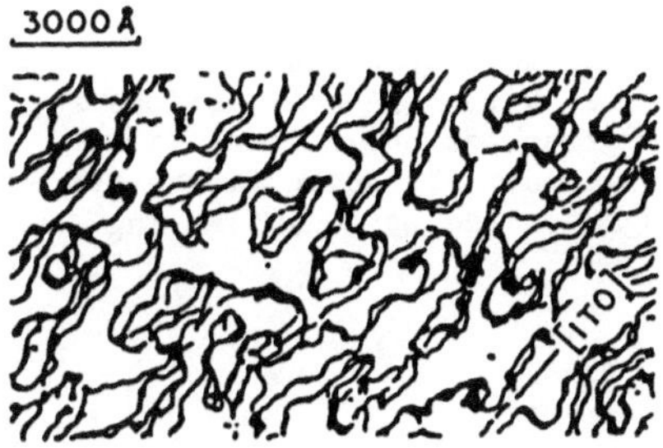

Fig. 9. TEM micrographs of antiphase boundaries in incommensurate $Ba_2NaNb_5O_{15}$

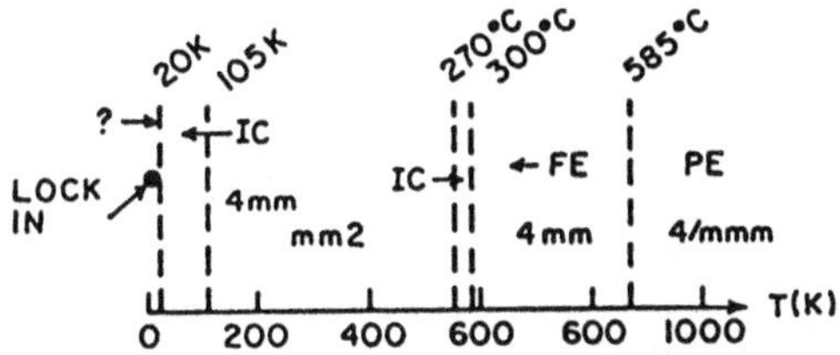

Fig. 10. Phases of $Ba_2NaNb_5O_{15}$

B. The Second IC Phase: Thermal Memory

At 105K = T_{II} $Ba_2NaNb_5O_{15}$ enters a second IC phase in which modulated structure develops perpendicular to the high-temperature modulation, making the structure tetragonal. The phases are diagrammed in Fig. 10. This phase has very similar relaxation times $\tau(T)$ as near T_I = 573K. In addition, it exhibits thermal memory effects similar to those first observed in IC crystals by Professor Hamano's group [39]. In particular, it displays strong hysteresis in various properties (dielectric constant, sound velocity) even when the thermal cycle crosses no discommensurations by defects.

The thermal memory effect is caused by decoration of the modulated structure by charged defects. If the crystal temperature is set at a particular temperature t_i within the IC phase, the modulation vector $q_o(T_i)$ will be precisely determined, but it requires time of order hours for the charged defects to diffuse into the same periodicity. If the sample temperature is changed, even by a fraction of a degree, the defects retain a memory of the original modulation. Only after many hours do they diffuse to a new modulation $q_o(T_f)$.

A similar non-thermal mechanism involving decoration by charged defects of an external optical modulation is the process by which phase conjugation occurs in these materials [40]: A standing wave is produced by shining a laser into the crystal and charged defects (e.g., Fe in $LiNbO_3$) decorate this external modulation. By doing phase conjugation in $Ba_2NaNb_5O_{15}$ we have shown [41] that the charged defects have a negative sign. This suggests that they might be sodium vacancies, as hypothesized by Toledano. Soon we should be able to measure their diffusion constant. Will it be $10^{-4}cm^2s$ for Na^+ or $1cm^2/s$ for kinks? (Janovec has shown [42] that these kinks are in general not neutrally charged.)

It is possible that the two diffusivities are related: The defect sites shown as vertices in the micrograph in Fig. 9 may diffuse slowly ($D \sim 10^{-4}$) whereas the wiggly disclination lines that terminate on them may diffuse very quickly ($D \sim 1$).

Finally, I would like to end by pointing out an analogy: If we have a certain thermal periodicity $q_o(T)$ in an IC and then suddenly shine a coherent laser beam into the crystal with a new periodicity $q_o(L)$, we produce "jet lag". The crystal will take minutes or even hours to redecorate the new modulation $q_0(L)$. This is distinctly analogous to the mammalian circadian biorhythyms, which are known to respond [43] to two independent stimuli: thermal (body temperature, with a 24-hour period) and optical (sunrise/sunset, with a slightly different period). There is no satisfactory mathematical theory of this two-stimuli circadian biorhythym, and perhaps our observations on phase conjugation and optical "jet lag" will help.

References

1. J. F. Scott, Ref. Prog. Phys. 42, 1055 (1979).
2. J. C. Toledano, J. Schneck, and G. Errandonea, Incommensurate Phase in Dielectrics, edited by R. Blinc and A. P. Levanyuk (North-Holland: Amsterdam, 1986, Vol. 2), p.233.
3. D. E. Cox et al., Phys. Rev. B19, 5754 (1979).
4. J. C. Toledano, J. Schneck, and G. Errandonea, Symmetries and Broken Symmetries in Condensed Matter Physics, edited by N. Boccara (IDSET: Paris, 1981), p. 217.
5. R. V. Pisarev et al., Phys. Rev. B28, 2677 (1983).
6. J. Schneck and F. Denover, Phys. Rev. B23, 383 (1981).
7. E. G. Spencer et al., Appl. Phys. Lett. 17, 300 (1970).
8. J. F. Ryan and J. R. Scott, Sol. St. Commun. 14, 5 (1974).
9. D. E. Cox et al., Phys. Rev. B28, 1640 (1983).
10. J. F. Scott, F. Habbal, and M. Hidaka, Phys. Rev. B25, 1805 (1982).
11. F. J. Schafer, W. Kleeman, and T. Tsuboi, J. Phys. C16, 3987 (1983); W. Kleeman, F. J. Schafer, and J. Nouet, J. Phys. C14, 4447 (1981).
12. J. F. Scott, Ferroelectrics 38, 375 (1981).
13. A. B. Pippard, Phil. Mag. 1, 473 (1956).
14. C. W. Garland, J. Chem. Phys. 41, 1005 (1964).
15. V. Janovec, J. Chem. Phys. 45, 1874 (1966).
16. B. A. Strukov et al., Ferroelectrics 25, 399 (1980).
17. U. T. Hochli and A. D. Bruce, J. Phys. C13, 1963 (1980).
18. A. P. Levanyuk et al., JETP 49, 176 (1979).
19. J. O. Fossum, J. Phys. C18, 5531 (1985); W. Rehwald, Ferroelectrics 24, 281 (1980).
20. K. Kawasaki, Ann. Phys. 154, 319 (1984); Physica 124B, 156 (1984).
21. D. E. Cox et al., Bull. Am. Phys. Soc. 26, 303 (1981).
22. M. Barthes-Regis et al., J. Physique Lett. 19, L-829 (1983).
23. M. Hidaka, T. Nakayama, J. F. Scott, and J. S. Storey, Physica 133B, 1 (1985); 123B, 291 (1984).
24. I. J. Fritz, Phys. Lett. 51A, 219 (1975); Phys. Rev. Lett. 35, 1511 (1975).
25. J. F. Scott, Ferroelectrics 47, 33 (1983).
26. S. Kh. Esayan et al., Sov. Phys. Crytallogr. 26, 619 (1981).
27. B. A. Auld, Acoustic Fields and Waves in Solids, (Wiley: New York, 1973), Vol. 1.
28. V. Dvorak and S. Kh. Esayan, Sol. St. Commun. 44, 901 (1982).
29. T. Yagi, S. Sakai, and J. F. Scott, Proc. Ann. Meeting Phys. Soc. Japan, Chiba, Oct. 1985 (Bussei-ken Dayori 24, Oct. 1985).
30. S. Aubry and G. Andre, Ann. Isr. Phys. Soc. 3, 133 (1980); Ferroelectrics 24, 53 (1980).
31. See Reference 5.
32. G. Gehring, J. Phys. C10, 531 (1977).
33. V. Dvorak and J. Fousek, Phys. Stat. Sol. 61A, 99 (1980).
34. D. F. Nelson and M. Lax, Phys. Rev. Lett. 24, 379 (1970); D. F. Nelson and P. D. Lazay, Phys. Rev. Lett. 25, 1187 (1970).
35. D. W. Bechtle, J. F. Scott, and D. J. Lockwood, Phys. Rev. B18, 6213 (1978).

36. G. N. Hassold et al., Phys. Rev. B33, 3581 (1986).
37. Pan Xiao-qing et al., Physica B (1986, in press).
38. D. Srolovitz and J. F. Scott, Phys. Rev. B (in press, August 1, 1986).
39. K. Hamano et al., J. Phys. Soc. Japan 49, 2278 (1980).
40. See, for example, J. Feinberg, Proc. Spie 388, 106 (1983).
41. W. F. Oliver, M. Erie, D. Z. Anderson, and J. F. Scott (unpublished).
42. V. Janovec, Phys. Lett. 99A, 384 (1983).
43. R. A. Wever, The Circadian System of Man, (Springer-Verlag: Berlin, 1979).

Model of Icosahedral Order

M. Widom

Department of Physics, Carnegie-Mellon University,
Pittsburgh, PA 15217, USA

1. Abstract

Metallic glasses, quasicrystals, and crystals may share identical local icosahedral order. This type of ordering extends to fill a three-dimensional curved space, producing an icosahedral "polytope" with perfect short and long-range icosahedral order. In this paper I demonstrate how to flatten the polytope and fill space with structures possessing the short-range order of the polytope but various types of long-range order. Both the rhombohedral packing units required to construct a three-dimensional Penrose pattern and long-range orientational order arise from rolling the polytope along special paths in three-dimensional flat space.

2. Introduction

The discovery of sharp diffraction patterns with icosahedral symmetry[1] emphasizes the importance of icosahedral order in solid metals. In 1952 F. C. Frank suggested local icosahedral order as an explanation for super-coding of liquid metals.[2] Others have extended his ideas in recent years to form a theory of metallic glass.[3] Although this theory succeeds qualitatively,[4] direct experimental confirmation has not been possible. Perhaps understanding the structure of quasicrystals will lead to a deeper understanding of metallic glass. This paper reviews one approach to this problem by presenting a unified model of icosahedral order in glassy, quasicrystalline, and crystalline materials.

This model utilizes a perfect icosahedral crystal, known as a "polytope", which exists in the three-dimensional curved surface of a sphere embedded in four-dimensional space.[5] Because curved and flat space are locally indistinguishable, structures in the curved space map into flat space with little distortion. Over larger distances defects arise in flat space because of the curvature mismatch. I demonstrate how to incorporate defects to model: glass with only short order; quasicrystals with long-range orientational order and either no long-range translational order or quasi-periodicity; and Frank-Kasper crystalline phases.

To map from curved to flat space, roll the four-dimensional polytope along three-dimensional flat space.[6] Everywhere an atom of the polytope touches three-dimensional space, place a three-dimensional atom. I show that long-ranged orientational and translational order arise from rolling the polytope in a straight line. Rolling around a closed loop places defects within the loop.[7] For certain types of loops, these defects maintain orientational order. These special loops bound the faces of the oblate and prolate rhombohedra of the three-dimensional Penrose pattern.[8]

3. Local Icosahedral Order

Frank proposed that icosahedra would be present in supercooled liquid metals as a result of the tendency to form tetrahedral clusters of four atoms. The tetrahedron provides an extremely dense, rigid, energetically favorable configuration. If it could, a metal would consist entirely of tetrahedra. Tetrahedra, however, do not fill space. An interplay of energetic preferences and geometrical constraints determines the real structure of a metal.

Try to fill a small region of space with tetrahedra. Note that any two faces of a tetrahedron meet at an angle of $\cos^{-1}(1/3) = 70.53°$. Thus a total of $360/70.53 = 5.104$ tetrahedra may share a common edge. Of course, tetrahedra only exist in integral units, so only five will fit without distortion. The small gap left over prevents tetrahedra from filling space.

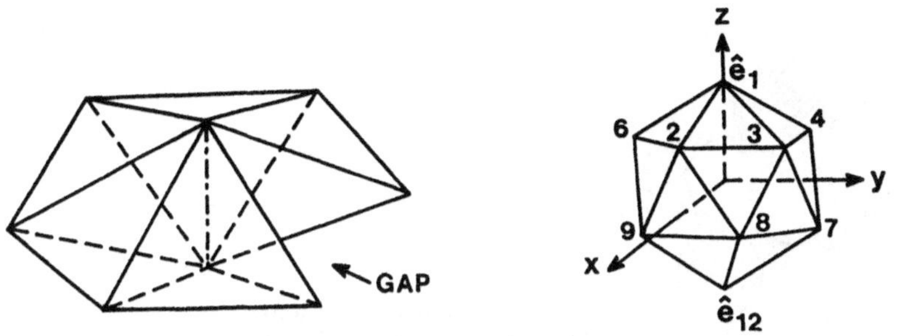

Figure 1a shows five tetrahedra sharing a bond, Fig. 1b shows an icosahedron in the standard orientation

If we distort the five tetrahedra slightly to fill in the gap, the resulting structure forms a fragment of an icosahedron. The distortion of the tetrahedra results in the edge length of the icosahedron being 5% greater than the center to vertex length. Despite the distortion, the icosahedron inherits the traits of high density and low energy. Thus icosahedral clusters should be prevalent in liquid metals near or below their freezing points.

Ideally no distinction exists between the atom at the center of a cluster and the atoms on the surface. Each atom has an equal right to form the center of a cluster. Larger clusters form in such a way that atoms previously at the surface find themselves at the centers of new icosahedra.

Icosahedra themselves cannot pack to fill space. The gap in the cluster of tetrahedra reappears as gaps between atoms on the surfaces of icosahedral clusters. As we enlarge the clusters, the gaps grow in size causing icosahedral frustration.[3] Curving space eliminates the frustration, as I show in the next section.

4. Icosahedral Polytope

Gaps between atoms may be filled by choosing slightly larger atoms to place on the surface of the icosahedral clusters. As more and more atoms are added to the cluster, the atoms on the outer coordination shells must become increasingly large to fill the gaps. Equivalently, one could keep the atoms at a given size but shrink the metric so that the distance between the centers of atoms is always equal to their diameter. The metric found on the curved three-dimensional curved surface of S^3, a sphere in four dimensions, accomplishes this.

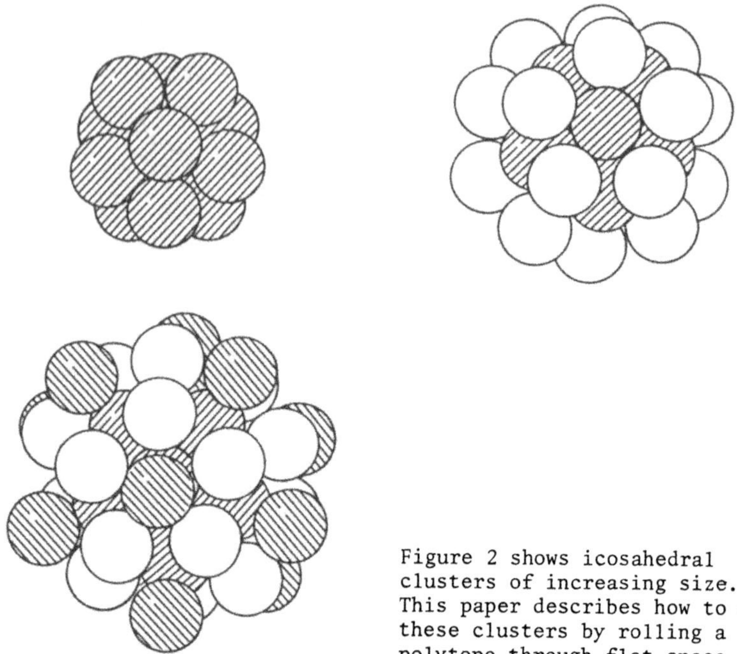

Figure 2 shows icosahedral clusters of increasing size. This paper describes how to create these clusters by rolling a polytope through flat space

Polytope {3,3,5} places 120 atoms in S^3 so that each atom sits at the center of an icosahedron formed by its twelve neighbors. All 600 cells form perfect tetrahedra. All frustration disappears because there are no gaps between atoms. The polytope forms a non-Euclidean crystal with rotational and translational symmetries.

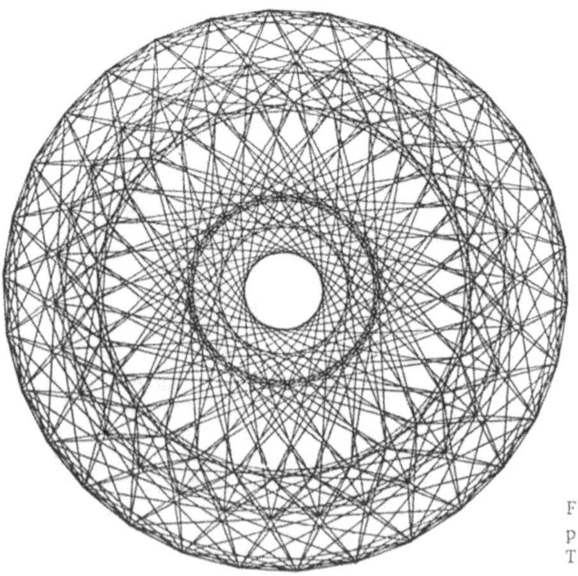

Figure 3. Poyltope {3,3,5} projected from S^3 onto R^2. Taken from reference 5

The structure and symmetries of the polytope have been described thoroughly elsewhere.[5,7,9] I will mention just two facts necessary for the remainder of this paper. Denote the separation between neighboring atoms by d and the radius of S^3 by R. A remarkable relationship between the structure of the polytope and the symmetry group of an icosahedron[10] guarantees that the ratio d/R is one half the smallest nonzero angle of rotation which leaves an icosahedron invariant. Thus

$$d = \frac{\pi}{5}R. \quad (1)$$

There is a special, intrinsic coordinate system for S^3 in which the structure of the polytope is especially simple. In this special coordinate system the polytope looks like a crystal with a dodecahedral unit cell translated rigidly along geodesics connecting neighboring atoms. Curiously, when viewed from a parallel transported coordinate system the intrinsic coordinates twist and turn.[11] The dodecahedral unit cell thus rotates around the geodesic by an angle equal to the geodesic distance travelled divided by the radius of S^3.

We fill flat space with atoms by rolling the polytope on R^3. As the polytope rolls in a straight line the point of contact between the sphere S^3 and flat space R^3 traces out a geodesic curve. Thus, if we construct an icosahedron at some point $\vec{R} \in R^3$, roll the polytope to $\vec{R} + \vec{V}$ and construct a new icosahedron, the new icosahedron will be identical to the old icosahedron rotated by

$$\theta = |\vec{V}|/R \quad (2)$$

around the axis $\vec{V}$. Equations (1) and (2) provide the key to creating long-range order.

5. Filling Space With Atoms

How can we use this non-Euclidean crystal to describe noncrystallographic structures in ordinary space? We must flatten the polytope. Strains introduced by flattening create the icosahedral frustration and open up gaps between atoms. Introducing defects into the polytope structure relieves this frustration. Recalling that 5.104 tetrahedra fit around a single bond, we note that placing a sixth tetrahedron on a bond will be a common type of defect. Such a sixfold bond may be viewed as a -72° disclination line in the polytope.

Rolling the polytope on R^3 provides a convenient scheme for placing atoms in flat space. This method automatically incorporates disclination line defects, because rolling the polytope in a straight line is equivalent to a screw transformation in which an icosahedron is simultaneously translated and rotated.[7] Equation 2) makes this notion precise. Because the rotation group So(3) is nonabelian, rolling around a closed loop generally results in a net rotation of the icosahedron. Thus the loop encloses a disclination line.

Imagine the sphere S^3 placed in four dimensions so that the atom at its northpole touches three-dimensional space at the origin, and the twelve neighbors of the northpole sit in the standard orientation shown in Fig. 1. We will place an atom in flat space wherever an atom of the polytope touches R^3. Thus we begin by placing an atom at the origin.

Now roll the polytope a distance d in the direction $\hat{e}_k$ where d is the geodesic separation between atoms in the polytope and $\hat{e}_k$ labels one of the vertices of the icosahedron in Fig. 1. Now one of the atoms neighboring the northpole of the polytope touches R^3 at the point $d\hat{e}_k$. Repeat this procedure for all twelve directions $\hat{e}_k$. The result is a thirteen atom icosahedral cluster such as is shown in Fig. 2a.

Let's try to extend this construction along the z axis. Figure 4 shows five intersecting icosahedra generated by rolling along the z axis and constructing the local icosahedron at each point.[12] Note how the orientation of the icosahedron centered at (o,o,d) has been rotated by $\pi/5$ around the z axis from the standard orientation at the origin. This is a direct consequence of (1) and (2). The original orientation is restored at (o,o,2d) because rotations by $2\pi/5$ belong to the icosahedral symmetry group.

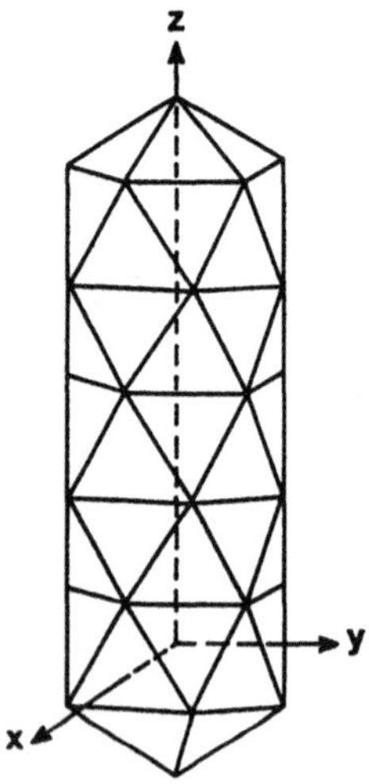

Figure 4 rolls the polytope along the z axis. Constructing the local icosahedron around each point (o,o,nd) where n is an integer. Note how the orientations of the icosahedra at odd values of n differ by $\pi/5$ from the standard orientation at even values of n.

Things get more complicated when the path we roll along turns a corner. Consider a path formed inductively by constructing the icosahedron surrounding the point of contact between the polytope and R^3, rolling the polytope to one of the vertices of this icosahedra, then repeating these two steps over and over. Figure 4 shows the special case in which the chosen vertex is always $\hat{e}_1$. In general the path will zig-zag through R^3.

Every time we roll a distance d we rotate the icosahedron through an angle of $\pi/5$. This rotation is not in the symmetry group of an icosahedron. Therefore the new orientation is always distinct from the previous one. Rolling off in some other direction will result in another rotation not in the icosahedral symmetry group. Rolling twice multiplies the individual rotations producing a net rotation which is not, in general, in the icosahedral symmetry group. In fact, we can create any rotation in So(3) with arbitrary precision by rolling a finite distance.

Angular correlations along the path will decay exponentially. The result of filling space with atoms in this manner is a compelling model of metallic glass. One can formulate a Landau theory for rapidly quenched metals based on this picture.[7] The calculated structure function agrees well with experiments.[4] In the final section of this paper I describe how to construct quasicrystalline and crystalline long-range order.

6. Long-Ranged Order

To maintain orientational order over large distances simply roll the polytope an even number of steps in each direction. Because the icosahedron gets rotated by $\pi/5$ in each step, rolling two steps rotates the icosahedron by $2\pi/5$. This larger rotation belongs to the icosahedral symmetry group, so the orientation after rolling two steps is equivalent to the starting orientation.

The simplest closed loop we can roll around in this manner forms a rhombus with edge length 2d. Roll up the z axis along $\hat{e}_1$ to (o,o,2d). Then turn and roll in the direction $\hat{e}_2$ (which is the direction the old $\hat{e}_6$ vertex has been rotated into) to (1.79d,o,2.89d). Then head back down along $\hat{e}_{12} = -\hat{e}_1$ to (1.79d,o,0.89d). Finally roll back to the origin along $\hat{e}_{11} = -\hat{e}_2$. This loop forms a rhombus with angles 63.43° and 116.57°. The net result of rolling around this loop is to rotate the icosahedron by $-2\pi/5$ around $\hat{e}_8$. Thus a -72° disclination line threads the loop.

Such loops may be joined along their edges to form the faces of solid bodies. Six loops combine in two ways to form rhombahedra. If three 116.57° angles meet at a point (as $\hat{e}_2$, $\hat{e}_4$, and $\hat{e}_{12}$ meet at the origin), the three associated rhombuses bound three of the six faces of the oblate rhombus shown in Fig. 5a. If three 63.43° angles meet at a point (as $\hat{e}_1$, e_2, and e_3 meet at the origin), the three associated rhombuses bound three of the six faces of the prolate rhombus shown in Fig. 5b.

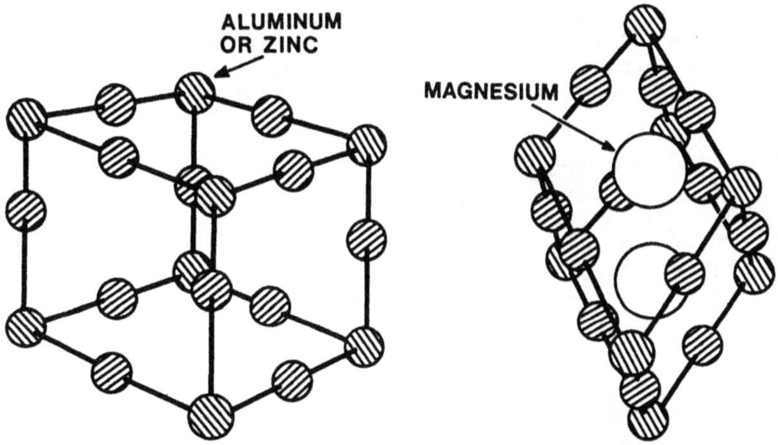

Figure 5a) and b) show oblate and prolate rhombahedra with decoration by atoms resulting from rolling polytope along edges of rhombic faces.

The clusters of Fig. 2 are both[9] polytope fragments rolled into flat space and components of the unit cell of the Frank-Kasper phase[13] of $Mg_{32}(Al,Zn)_{49}$. We identify the 20 atoms forming a dodecahedral shell in Fig. 2b as Magnesium atoms. The remainder are aluminum or zinc. Thos motivates Henley and Elser's[8] identification of the atoms decorating the rhombahedra in Fig. 5. Each face of a rhombahedron surrounds a -72° disclination line defect. Henley and Elser[8] note that +72° disclinations occur within the oblate rhombahedron.

This pair of rhombahedra is sufficient to create a three-dimensional Penrose pattern. Just join up the rhombahedra along their faces while imposing Penrose pattern-matching rules and the resulting structure's diffraction pattern will have delta function peaks and icosahedral symmetry.[14] In addition the structure will possess inflation and deflation symmetries whose physical significance is questionable. Note that the decorated rhombahedra in Fig. 5 possess no matching rules. Elser has shown that packing the rhombahedra without regard to matching rules can eliminate the inflation and deflation symmetries without significantly altering the diffraction pattern.[15]

The oblate rhombahedron in Fig. 5a) is unlikely to occur with great frequency because the +72° disclinations correspond to an unfavorable arrangement of four tetrahedra around a single bond. It is more likely that pairs of oblate rhombahedra will combine with pairs of prolate rhombahedra to form a rhombic dodecahadron as described by Henley and Elser.[8] This combination is common in the Penrose pattern and also in the $Mg_{32}(Al,Zn)_{49}$ Frank-Kasper phase. We can produce it also by rolling the polytope along its edges, which are parallel to the $\hat{e}_k$ of an icosahedra. The resulting decoration of the rhombic dodecahedron with atoms differs from the decorations of the component rhombahedra because all +72° disclinations have been removed.

The rhombic dodecahedron may pack together with the prolate rhombahedron to create the structure of the Frank-Kasper crystalline phase of $Mg_{32}(Al,Zn)_{49}$. This structure possesses long-ranged icosahedral orientational order, translational periodicity, and cubic symmetry in its disclination-like network. Alternatively, the rhombic dodecahedron may pack together with prolate and oblate rhombahedra to create a three-dimensional Penrose pattern with matching rules observed or not as we choose. These structures may model the quasicrystalline phase[16] of $Mg_{32}(Al,Zn)_{49}$.

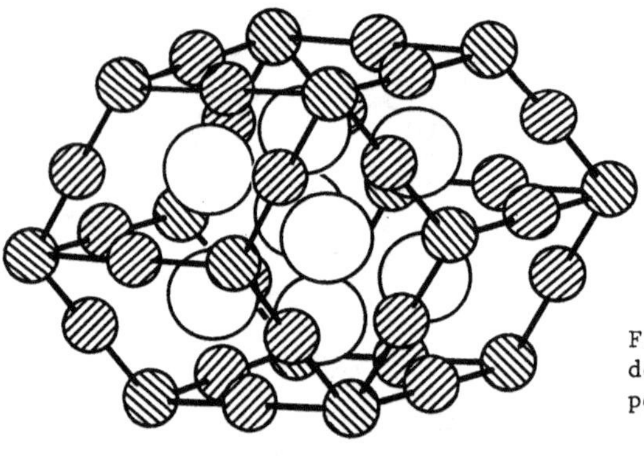

Figure 6. Rhombic dodecahedron decorated with atoms by rolling polytope along each edge

In conclusion, rolling out the polytope fills space with atoms in structures which model metallic glass when the path we roll on turns as often as one atomic diameter. If the paths always continue an even number of atomic diametrics in a given direction, long-range orientational order emerges. Rolling around rhombic paths of edge length 2d leads to the prolate rhombus, the oblate rhombus, and the rhombic dodecahedron, each decorated with atoms and enclosing disclination line defects. The prolate rhombus and rhombic dodecahedron pack to form the Frank-Kasper phase of $Mg_{32}(Al,Zn)_{49}$. One must include the oblate rhombus as well to obtain a three-dimensional Penrose pattern which may model this material's icosahedral phase.

7. References

1. D. Schechtman, I. Blech, G. Gratias, and J.W. Cahn, Phys. Rev. Lett. 53, 1951 (1984)
2. F.C. Frank, Proc. Roy. Soc. 215, 43 (1952)
3. M. Kléman and J.F. Sadoc, J. Phys. (Paris) Lett. 40, L569 (1979); D.R. Nelson, Phys. Rev. B28, 5515 (1983)
4. S. Sachdev and D.R. Nelson, Phys. Rev. Lett 53, 1947 (1984)
5. H.S.M. Coxeter, "Regular Polytopes" (Dover, N.Y., 1973)
6. J.P. Sethna, Phys. Rev. Lett. 51, 2198 (1983)
7. D.R. Nelson and M. Widom, Nucl. Phys. B240[FS12], 113 (1984)
8. C.L. Henley and V. Elser, Phil. Mag. B53, L59 (1986)
9. M. Widom, Phys. Rev. B31, 6456 (1985)
10. P. DuVal, "Homographics, Quaternions, and Rotations" (Oxford, London, 1964)
11. M. Widom, Phys. Rev. B33, July 15 (1986)
12. Actually an imperceptible adjustment must be made because the vertices below the equator of the icosahedron centered at (o,o,d) do not match precisely the locations of the vertices above the equator of the icosahedron at the origin.
13. G. Bergman, J.L.T. Waugh, and L. Pauling, Acta Cryst. 10, 254 (1957)
14. D. Levine and P.J. Steinhardt, Phys. Rev. Lett. 53, 2477 (1984)
15. V. Elser, Phys. Rev. Lett., 54, 1730 (1985); See also article by P. Horn in these proceedings.
16. P. Ramachandrarao and G.V.S. Sastry, Pramana 25, L225 (1985)

Systematics of Disorder in Quasiperiodic Material

P.M. Horn, W. Malzfeldt, D.P. Divincenzo, J. Toner, and R. Gambino

IBM Thomas J. Watson Research Center, P.O. Box 218,
Yorktown Heights, NY 10598, USA

Abstract

X-ray diffraction and transmission electron microscope (TEM) measurements were performed on icosahedral AlMn and AlMnSi samples. Quasicrystalline order was obtained via different sample treatments: quenching, laser annealing, and furnace annealing of amorphous samples. In quenched bulk AlMn samples, the icosahedral grains are embedded in a crystalline Al matrix. The Al matrix is strongly reduced or rendered noncrystalline by the addition of Si. Samples prepared by laser annealing and quenching show a high degree of disorder. This shows up, in x-ray diffraction, as a large average peak line width. The furnace annealed samples show apparent homogeneous grains by TEM darkfield methods, and the smallest peak widths in x-ray diffraction. However, the grain sizes are much larger than the x-ray correlation lengths. These samples have peak widths which correlate strongly with $G_{\perp}$ (phason momentum). We show that the data are consistent with either frozen-in phason strain or random icosahedral packing. These two models predict surprisingly similar but potentially distinguishable scattering widths, differing in their detailed dependence on $G_{\perp}$.

Friction and Forced Flow: Collective Transport in Disordered Media

D.S. Fisher

AT&T Bell Laboratories, Murray Hill, NJ 07974, USA

Various problems of non-linear collective transport through disordered media are discussed, using sliding charge density waves as a paradigm. These systems all exhibit features of nonlinear dynamics, statistical mechanics and critical phenomena.

1. Introduction

Transport phenomena can be loosely divided into two classes, one involving approximately independent motion of a large number of particles and the other involving *collective* motion in which there is a high degree of coherence and the forces between the particles are as important as the applied driving force. Although there have been dramatic advances recently in the understanding of independent transport in disordered systems (e.g. localization [1], the general area of collective transport in disordered systems has, with a few exceptions, received rather little attention. Examples of this kind of phenomena are charge density wave transport [2-11], flux flow in type II superconductors [12-13], motion of an interface between two fluids in a porous medium [14], and motion of domain walls in disordered magnets [15-17].

In this paper we will raise some simple questions about collective transport in disordered systems and attempt to draw parallels between a variety of problems in this field, using charge density wave transport as a paradigm. The phenomena which we will discuss are all nonlinear in an essential way and involve interactions between large numbers of degrees of freedom. Thus they exhibit aspects of both dynamical systems and statistical mechanics. In particular, in the limit that thermal fluctuations are negligible, these systems should exhibit a whole new class of nonlinear dynamic critical phenomena.

This paper is organized as follows: We first introduce a class of models and pose a series of questions, then in Section 3 the current understanding of a particular example which has been relatively well studied, sliding charge density waves [2-11], is discussed in some detail. Section 4 lists several other examples and some of their associated problems and finally in Section 5 some tentative conclusions are drawn.

2. Models and Questions

The kind of collective transport phenomena we are interested in can typically be modeled by a large collection of objects with positions $\{\vec{R}_j\}$ in d-dimensional space interacting via a potential $V_I\{\vec{R}_j\}$ and moving in a random static potential $\Phi(\vec{r})$ under the influence of uniform applied forces $\vec{F}(t)$. Without applied forces the hamiltonian is simply

$$\mathcal{H} = \sum_j \Phi(\vec{R}_j) + V_I\{\vec{R}_j\} \ . \qquad (1)$$

For most purposes, we will neglect inertial effects and take classical dissipative equations of motion

$$\frac{\partial \vec{R}_j}{\partial t} = -\frac{\delta \mathcal{H}}{\delta \vec{R}_j} + \vec{F} \ . \qquad (2)$$

For some problems, it is important to consider the effects of a small thermal noise term added to Eq. (2), although we will primarily be concerned with the zero temperature deterministic dynamics. Collective aspects of the transport arise if the interaction forces, $\frac{\partial V_I}{\partial R}$, are as important as the random and applied forces. The kind of questions we would like to address are:

(1) Is there a *threshold force*, F_T, below which the system is stationary in steady state and above which it moves? If so, is the threshold unique (i.e. non-hysteretic)?

(2) What is the steady state spatially averaged *velocity*

$$v \equiv \left< \frac{\partial \vec{R}_j}{\partial t} \right>_{j,t} \qquad (3)$$

as a function of the applied force?

(3) What is the *response* of the system to a small additional spatially dependent *ac force* $\epsilon(q,\omega)$, both above and below threshold? Below threshold this should probe aspects of the *metastability* and *hysteresis* which will occur generically.[3,4]

Finally, (4) what is the *critical behavior* near threshold? In particular are there diverging length scales as well as diverging time scales?

In some cases, the limit that the randomness is weak compared to the interparticle forces will be especially relevant; one would like to know whether the neglect of *thermal fluctuations* can be justified in this or other limits.

Most of the systems we will be interested in can be approximated by an elastic medium so that the $\vec{R}_j$ can be labeled in terms of a $D(\leqslant d)$ dimensional internal coordinate, s, of the elastic medium: $\vec{R}_j \rightarrow \vec{R}(s)$. The interaction potential can then be approximated by a discrete or continuum elastic interaction for the deformations $\vec{u}(s) \equiv \vec{R}(s) - \vec{R}_0(s)$ of the elastic medium away from its undistorted state, $\vec{R}_0(s)$.

3. Sliding Charge Density Waves

We now turn to a particular system which has been extensively studied in recent years both experimentally and theoretically [2-11], although many problems, some associated with general issues of collective transport, still remain.

A large class of materials in which charge density waves (CDW's) form that are incommensurate in one of the directions exhibit non-linear electrical transport associated with the CDW [2]. At low electric fields in the incommensurate $\hat{z}$

direction the current is ohmic, but above a small (0.01 - 1 V/cm) sharp threshold field there is an additional non-linear contribution to the current [2] believed to be caused by overall motion of the CDW which contributes a current density proportional to the CDW velocity, v. Above threshold the velocity is a unique function of the field, while below threshold hysteretic phenomena are observed [4]. At scales large compared to the CDW coherence length and times long compared to the order parameter relaxation time, the system can be well modeled by effective time dependent Ginzburg-Landau -like *classical* equations of motion for phase deformations of the order parameter [3,5,6]. A simplified version of the hamiltonian introduced by Fukuyama, Lee and Rice [6] takes into account the *strains* in the $\hat{z}$ direction of the CDW [3,5]:

$$\mathcal{H} = \sum_{s} \tfrac{1}{2} K (\nabla_s u)^2 + \Phi[s + \hat{z} u(s)] \tag{4}$$

where s is the lattice coordinate of the undistorted charge density maxima of the CDW spaced by the wavelength a, K is the CDW elastic constant, and Φ a random potential due to impurities (smeared by the structure factor of the CDW) with typical strength $\tilde{\Phi}$ and correlations over distances of order a. Neglecting inertia, which is well justified experimentally, the purely dissipative equation of motion is [3,6]

$$\frac{\partial u(s)}{\partial t} = \frac{-\partial \mathcal{H}}{\partial u} + F \tag{5}$$

where F is proportional to the $\hat{z}$ component of the applied electric field and we have scaled time to get rid of a damping coefficient. We have neglected dislocations and other defects in the CDW, and also thermal fluctuations. These can both be justified, as discussed below, in the limit that the dimensionless randomness,

$$\delta = \tilde{\Phi}/K \tag{6}$$

is small.

The model Eqs. (4,5) has been studied by perturbation theory in δ, by numerical simulations [7] by mean field theory [3] which is valid in the limit of infinite range stiffness of the CDW, and by renormalization group and scaling arguments [3]. The following picture, which is at least qualitatively consistent with most experiments, emerges from this body of work. There exists a unique threshold field and v(F) is non-hysteretic, although there is no general understanding of this. Well above threshold, motional averaging makes the effects of random pinning negligible and $v \approx F$. Corrections to this can be analyzed perturbatively [5], however perturbation theory fails when the velocity is small or zero, and may even have problems in other regimes [3].

A simple argument, originally due to Lee and Rice [6], yields the form of the threshold field for weak pinning, i.e. $\delta \ll 1$. In this limit, there is a characteristic length scale, ξ_0, at which the cumulative effect of the random pinning force, $(\tilde{\Phi}/a)\ (\xi_0/a)^{d/2}$, is the same magnitude as the elastic force with a one-wavelength deformation, $(K/a)\ (\xi_0/a)^{d-2}$, yielding a coherence length

$$\xi_0 \sim a\delta^{\frac{2}{4-d}} \tag{7}$$

(for d >4, ξ_0 is exponentially small in δ). This length scale, which is known as the Lee-Rice length [6,12], plays a role analogous to the bare BCS coherence length in superconductivity. Volumes of size ξ_0^d effectively act like single degrees of freedom and on scales longer than ξ_0 non-linear collective effects come in and perturbation theory breaks down. The threshold force, which is non-zero for arbitrarily weak disorder [3,6], is proportional to the average random force in a region of size ξ_0^d, thus

$$F_T \sim \frac{K}{a}(\xi_0/a)^{-2} \quad . \tag{8}$$

A. Above Threshold

Above threshold, there are fluctuations in the local instantaneous velocities driven by the randomness [5]. These are correlated over distances $\sim\xi$, the velocity correlation length, defined as the decay length of the instantaneous velocity correlation function

$$C(s) \equiv <(\frac{dR(s,t)}{dt}-v)\,(\frac{dR(s+s',t)}{dt}-v)>_{s'} \quad . \tag{9}$$

At about twice threshold, ξ is of order ξ_0, however as the velocity decreases, ξ grows. Just above threshold, the motion consists of regions of size $\sim\xi_0$ moving jerkily with correlations in this jerky motion over lengths $\sim\xi$ [3]. As the threshold is approached, v goes to zero as

$$v \sim f^{\zeta} \tag{10}$$

with the reduced force defined as

$$f \equiv \frac{F-F_T}{F_T} \quad . \tag{11}$$

Concomitantly, ξ diverges as [3]

$$\xi/\xi_0 \sim f^{-\nu} \quad . \tag{12}$$

There are several natural time scales in the system. The simplest is associated with the periodicity of the CDW motion in the lab frame: in a time a/v, the inverse of the "washboard" frequency, the CDW moves one wavelength. Apparently coherent ac noise at the washboard frequency is observed in many experiments [8]. However recent measurements have confirmed the expectation that, since the *instantaneous* velocity correlations decay [3], the noise will decrease as the volume grows, although the detailed interpretation is still controversial [8]. Note however that the *time averaged* velocity will be the same in all regions of the sample, at least in the absence of dislocations.

In addition to the washboard frequency, there is a more important time scale, which has not yet been measured, that is associated with the jerky motion [3]. This can be probed by measuring the differential linear response to a finite frequency additional force $\epsilon(\omega)e^{-i\omega t}$. The polarizability is defined by

$$\chi(\omega) = \frac{d\bar{u}(\omega)}{d\epsilon(\omega)} \tag{13}$$

where $\bar{u} \equiv \langle u(s) \rangle_s$ In the limit of low frequency, this is related to the differential conductivity by

$$-i\omega\chi(\omega) \approx \frac{dv(F)}{dF} \quad . \tag{14}$$

A characteristic frequency Ω can be defined as the scale at which $-i\omega\chi(\omega)$ deviates from this limit. This is dominated by the response of regions which are almost unstable and just about to jump forward [3]. The frequency Ω goes to zero as [3]

$$\Omega \sim f^{\mu} \quad . \tag{15}$$

Because of the diverging correlation length ξ and time Ω^{-1}, there are non-trivial dynamic critical phenomena near threshold [3]. The exponents have been calculated in mean field theory which presumably is valid above some currently unknown upper critical dimension, d_c. The results are ζ=3/2, μ=½ and υ=½ [3]. In addition to the dc motion and ac linear response, there are many fascinating non-linear phenomena above threshold which involve combinations of ac and dc forces [9,10], etc, but we do not have the space to discuss these here. Many of these phenomena are infinite number of degrees of freedom versions of known phenomena, such as mode locking, which exist in simple non-linear systems.

B. Below Threshold

We turn next to the behavior below threshold, where there is no motion in steady state. One can, however, probe the system by, for example, measuring the ac polarizability [11] or by adiabatically changing the dc force and looking for history dependence [4]. In principle it may be possible to prepare the system in a locally stable configuration (e.g. the ground state at F=0) whose total energy $\mathcal{H} - F\sum u(s)$ cannot be lowered by moving any finite region over a small barrier. However for most histories of the system, there will be regions which can be moved over *arbitrarily small barriers* to lower the energy [3]. These regions are of size $\sim\xi_0^d$, well below threshold. A schematic picture of the response due to these almost unstable regions can be obtained from a simple model where each sits in the upper well of a local assymetric double well potential which has a small curvature in the upper well [3]. This leads to a singular low frequency response arising from the distribution of such curvatures. Generically, this has the form [3]

$$\text{Re}\chi(\omega) \approx \chi(0) - C|\omega| \quad . \tag{16}$$

Singular response consistent with this form has been observed [11]. Associated with this singular linear response is a divergent non-linear response arising from the jumps from upper to lower wells. [3]

Well below threshold, the volume of CDW that moves appreciably during a local jump is of order ξ_0^d. However as threshold is approached, the motion of one region makes other nearby regions unstable and the overall effect becomes larger. Exactly at threshold, one region will push another which will push another and the system will start to creep forward. Thus there is a natural correlation length ξ_- which roughly measures the size of the moved region in response to a local disturbance [3]; ξ_- will diverge as $F \rightarrow F_T$, presumably as a power law

$$\xi_-/\xi_0 \sim |f|^{-\nu'} \quad , \tag{17}$$

and the natural expectation is that $\nu' = \nu$. The correlation length ξ_- also measures the density of metastable configurations: in a volume V, one expects roughly $\exp(V/\xi_-^d)$ such configurations [3].

The polarizability, $\chi_0 \equiv d\bar{u}/dF$, will have a singularity, or possibly a divergence, at threshold due to the jumping regions, but note that generally $\chi_0 \neq \chi(\omega=0)$ [3].

C. Problems

We now briefly mention several relatively open questions. Although there have been many qualitative experiments on hysteretic phenomena below threshold [4], some of which exhibit similar behavior to that found in the simple mean field model [3], no attempts have been made at quantitative analysis. In general, the critical behavior near the threshold has also not yet received much experimental attention. On the theoretical side, there are various assumptions made in writing down Eqs. (4,5) which we now discuss.

The neglect of defects in the CDW is justified for weak pinning [3], although if defects are present, then many new questions arise (see below). Addition of *thermal effects* will generally destroy the threshold and there will be creep at any value of the applied force, however a simple estimate [3] shows that the effects are negligible for weak pinning in $d>2$: Consider a coherence volume ξ_0^d (also known as a Lee-Rice domain) which typically has to surmount a barrier

$$B \sim F_T a \xi_0^d \sim \xi_0^{d-2} \tag{18}$$

to move. This is large for $d>2$, and thus the rounding of the threshold will be small. Even larger factors suppress quantum tunneling of coherence volumes.

As mentioned above, the uniqueness of the threshold is not well understood. The presence of a large inertial term in the equations of motion will certainly cause hysteresis, but it appears that a small mass does not do so [3].

Even far above threshold, where one might expect perturbative methods to work, there are unanswered questions; for example: what is the motionally narrowed x-ray (or better neutron) structure factor of the CDW due to the disorder, and is there diffusive response to a small local force, as there will be in the absence of disorder?

4. Other Systems

We now briefly discuss various related problems in collective transport.

A. Interfaces in disordered solids

Interfaces between two coexisting phases in solids with randomly placed defects provide examples of $D=d-1$ dimensional elastic media which can move in a higher dimensional random environment. One example is a domain wall between the two broken symmetry phases in diluted Ising magnets [18]. There has been considerable study of the effects of the slow thermal motion of such domain walls and their effect on the equilibration of disordered magnets [15-18]. Here, however, we are interested in the behavior of *driven* interfaces at very low temperatures. Thus we consider applying a small uniform symmetry breaking field which prefers one phase. The applied force on the interface, F, will then be proportional to this applied field and must compete, with the aid of the surface tension σ, with the quenched random forces. For simplicity, we assume that both the typical length scale and typical separation of the impurities are of order a. The randomness can be of two kinds [18]: *random exchange* disorder which does not distinguish between the two phases but couples directly to the interface position (this is possible only if the phases are symmetry related), and *random field* disorder which prefers one or the other phase and acts like a random *force*, h, on the interface. We will restrict our discussion to the more generic random field case. By analogy with the CDW, the characteristic scale ξ_0 at which the disorder becomes important is

$$\xi_0/a \sim (h/\sigma a^{d-2})^{\frac{-2}{5-d}} \tag{19}$$

(for $d<5$) [18] and the threshold force per unit area is $F_T \sim \sigma a \xi_0^{-2}$. The barriers for motion at $F \sim \frac{1}{2} F_T$ are $B \sim \sigma a^2 \xi_0^{d-3}$ and thus in $d=3$ weak disorder does not by itself supress thermal creep, although large scale size, a, of the disorder can. Thus for magnetic interfaces [15,18], it is necessary to work at low temperatures in order to observe the deterministic dynamics. However the crossover from thermal creep [16] to driven motion [15] is certainly of interest here.

The first question we must ask is whether the interface stays close to flat on long length scales. This is likely to be the case for weak disorder in $d>2$, where the roughness due to random field disorder does not cause large angle deviations [18]. Thus a "solid on solid", single-valued interface model in $D=d-1$ dimensions should be appropriate [18]. All the questions posed for the CDW system are of potential interest here. Note only that, due to the lack of periodicity in the direction of motion, there is no sharp washboard frequency of the motion.

Systems with *structural transitions*, in which stress can be used to preferentially select one phase may be better to study than disordered magnets, since optical and other techniques can directly follow the interface motion and measure its correlations.

B. Fluid Interfaces in Porous Media

Another driven interface problem is the interface between two fluids in a porous medium [14,19]. If the more viscous fluid pushes out the less viscous one, then there is no macroscopic hydrodynamic instability [19]. The contact angle between the fluid interface and the wall plays an important role; it is a function of the asymmetry of the surface-fluid interactions. For certain ranges of contact angle, it seems likely that the interface between the two fluids remains well defined on scales large compared to the pore size, a.

For this system, it is not clear what the dimensionless disorder parameter, δ, is, however it is likely to be a complicated function of the contact angle and the geometry. It is possible that there is a regime in which the characteristic scale ξ_0 above which the disorder dominates is somewhat larger than a. In this regime, it is natural to conjecture that there will be a well-defined coarse grained interface and an effective surface tension on long length scales. Breakoff of droplets from the moving interface should then only occur (if at all) on small scales. Futhermore, thermal creep effects will be negligible if the pore size is large. Thus, except for the dynamics, the system is similar to the random field interface with intermediate strength pinning and we expect flow above a threshold value of the pressure drop across the interface. The dynamics are very different, however, due to mass conservation which implies that flow associated with a deformation on scale λ will penetrate a distance of order λ into the two phases. This will presumably change the universality class of the dynamic critical phenomena, although the statics below threshold will be similar. Measuring the analog of the velocity, v, as a function of the force is complicated by the contribution to the pressure drop from flow in the bulk fluids, so it may be easiest to drive at constant flow rate, i.e. constant v.

For small velocities, the motion should be jerky with regions of size ξ_0^{d-1} jumping distances which are probably of order a. The velocity correlations should be measurable by light scattering. Below threshold, responses to ac pressure pulses could perhaps be used to study the metastable regime. A singular response analogous to Eq. (15), although not necessarily of the same form, should occur.

Note added: Experiments in progress [14] indicate that the interface can stay well defined on long scales if the invading fluid partially wets. This behavior is found in the opposite regime to various "fingering" phenomena [19] which occur when the invading fluid does not preferentially wet, or when it is less viscous than the displaced fluid [14].

C. Flux Flow in Type II Superconductors

In type II superconductors in a magnetic field, a *flux lattice* forms in the absence of disorder and resistance is caused by transverse motion of the flux lattice under the Lorentz force from an applied current. In the presence of *weak disorder*, the flux lattice still exists (at least out to some length scale) but is pinned by disorder [12]. Both three dimensional (bulk) and two dimensional (film) systems can be studied [13]. If dislocations in the flux lattice are ignored, the motion, which is strongly damped, is directly analogous to CDW's, except that deformations transverse to the average direction of motion will occur; presumably this changes the critical behavior.

Because superconductors can be well characterized and many parameters can be varied at will (including the effects of thermal fluctuations), they provide an excellent system for studying collective transport phenomena. Surprisingly, there have not been detailed studies of the behavior near the critical current threshold in the weak pinning flux lattice limit.

In the *strong pinning* limit when the random forces are comparable to the intervortex-line forces, new phenomena come into play. The flux lattice is destroyed and the time averaged local velocity, in contrast to the systems we have considered thus far, can vary from one part of the sample to another. In the absence of thermal creep, this behavior is more analogous to a fluid moving in a random potential than to the elastic media we have been discussing so far. Wholly new questions arise in this limit. For example, what is the spatial distribution of the flux line velocities near threshold and what is the nature of the spatial and temporal response above and below threshold? Note that because of dislocations and tearing of the flux lattice, on large scales and very close to the critical current threshold, fluid-like effects will generally appear even for weak pinning. However, for very weak pinning there should be a large regime in which these effects are negligible.

D. Fluid Flow Down a Bumpy Surface

We have argued above that in the strong pinning limit flux flow should be analogous to motion of a fluid in a random potential. Another, two dimensional, realization of such a system is a thin fluid film on a randomly bumpy plane which is below the percolation threshold for formation of an "ocean". If the plane is tilted, there will be presumably be a critical angle at which the fluid will flow at a non-zero rate. Is this angle history dependent (ignoring fluid inertia)? What is the behavior for the flow rate as a function of the angle? This problem has some aspects of percolation, but it is not obviously related to known percolation problems. A diverging correlation length as the flow rate is decreased by decreasing the angle can be defined as the scale above which the coarse grained flow is approximately uniform.

E. Other Problems

Several other closely related problems have recently been discussed by deGennes [20,21]: Motion of the interface between the fluids in a random Hele-Shaw cell [20], which is a simpler, two dimensional version of the two fluid interface in porous media; and motion of the front of a wetting fluid which is film spreading on a dirty substate [21]. In both of these problems lengths analogous to ξ_0 have been found [20,21], but little is known otherwise.

5. Conclusions

We have discussed various problems in collective transport all of which represent non-linear dynamical systems in which large numbers of degrees of freedom play an essential role. Because of this, their behavior must be analyzed statistically, and new kinds of critical phenomena appear. Since they can exhibit both driven and thermally activated dynamics, advances in understanding them may yield useful clues to help unravel the complex interplay between equilibrium and non-equilibrium phenomena in disordered thermodynamic systems such as glasses and spin glasses.

At this stage it is not clear which theoretical methods other than numerical simulations are likely to yield progress in understanding these collective transport phenomena. Mean field theory [3] can be used for all the problems involving elastic media, although what physics it leaves out and how to improve upon it are uncertain. For the fluid problems, mean field theory is less instructive, and perhaps ideas from percolation theory will be more useful.

Finally, it is interesting to consider problems in other fields which are in the same general category of problems. Natural examples are crack propagation in disordered alloys and glasses, stickslip behavior of geological faults, and the old problem of solid-on-solid friction.

Acknowledgements

I would like to thank Peter Littlewood, Susan Coppersmith, and Michael Cross for contributing to my understanding of charge density waves, and Jerry Gollub and Pierre-Giles deGennes for stimulating discussions of their work on flow in porous media and related problems.

REFERENCES

1. P. A. Lee and T. V. Ramakrishnan: Rev. Mod. Phys. *57*, 287 (1985).
2. For a recent collection of papers on CDW transport see *Charge Density Waves in Solids*, Vol. 217 of *Lecture Notes in Physics*, ed. by G. Hutiray and J. Solyom (Springer-Verlag, Berlin, 1985).
3. D. S. Fisher: Phys. Rev. Lett. *50*, 1486 (1983); Phys. Rev. B*31*, 1396 (1985); for a summary of results see D. S. Fisher: Physica 126B, 409 (1984).
4. N. P. Ong, D. D. Duggan, C. B. Kalem, T. W. Jing and P .A. Lee: in *Charge Density Waves in Solids*, Ref. 2, p. 387.
5. L. Sneddon, M. C. Cross and D. S. Fisher: Phys. Rev. Lett. *49*, 292 (1982); L. Sneddon: Phys. Rev. B*29*, 719 (1984); and *29*, 725 (1984).
6. H. Fukuyama and P. A. Lee: Phys. Rev. B*17*, 535 (1977); P. A. Lee and T. M. Rice: *19*, 3970 (1979).
7. N. Teranishi and R. Kubo: J. Phys. So. Jpn. *47*, 720 (1979); J. B. Sokoloff: Phys. Rev. B*23*, 1992 (1981); L. Pietronero and S. Strassler: *28*, 5863 (1983); P. B. Littlewood: *33*, 6694 (1986).
8. R. M. Fleming and C. C. Grimes: Phys. Rev. Lett. *42*, 1423 (1979) G. Mozurkewich and G. Gruner: *51*, 2206 (1983); G. Verma and N. P. Ong: Phys. Rev. B*30*, 2928 (1984).
9. S. N. Coppersmith and P. B. Littlewood: Phys. Rev. Lett, Oct. 1986.
10. S. E. Brown, G. Mozurkewich and G. Gruner: Phys. Rev. Lett. *52*, 2277 (1984).
11. R. J. Cava, R. M. Fleming, P. B. Littlewood, E. A. Rietman L. F. Schneemeyer and R. G. Dunn: Phys. Rev. B*30*, 3228 (1984).
12. A. I. Larkin and Yu N. Ovchinikov: Sov. Phys.-JETP *38*, 854 (1974); and *41*, 960 (1975).
13. See for example, A. T. Fiory: Phys. Rev. B*7*, 1881 (1973) and *8*, 5039 (1973). Recently developed a-In films may be a good system in which to study weakly pinned flux flow; see A. T. Fiory, A. F. Hebard and W. I. Glaberson: Phys. Rev. B*28*, 5075 (1983).
14. J. B. Stokes, D. A. Weitz, J. P. Gollub, A. Dougherty, M. O. Robbins, P. M. Chaikin and H. M. Lindsay: Phys. Rev. Lett. *57*, 1718 (1986); and J. P. Gollub: private communication.

15. G. Bruinsma and G. Aeppli: Phys. Rev. Lett. *52*, 1547 (1984).

16. G. Grinstein and J. Fernandez: Phys. Rev. B*29*, 6389 (1984); J. Villain: Phys. Rev. Lett *52*, 1543 (1984).

17. R. J. Birgeneau, Y. Shapira, G. Shirane, R. A. Cowley and H. Yoshizawa: Physica, to be published.

18. G. Grinstein and S -K. Ma: Phys. Rev. B*28*, 2588 (1983); D. A. Huse and C. L. Henley: Phys. Rev. Lett. *54*, 2708 (1985); D. S. Fisher *56*, 1964 (1986).

19. L. P. Kadanoff, D. Bensimon, S. Liang, B. Shraiman and S. Tang: Rev. Mod. Phys., to be published,.

20. P. G. deGennes: to be published.

21. P. G. deGennes: Rev. Mod. Phys. *57*, 827 (1985); D. A. Huse: unpublished.

Non-Equilibrium Behavior, Hysteresis and Condensate Quakes in the Pinned Charge Density Wave

N.P. Ong and Z.Z. Wang

Department of Physics, Princeton University, Princeton, NJ 08544, USA

An experimental investigation of the thermal hysteresis in the pinned charge-density-wave (CDW) in TaS_3 shows that the R vs. T features are independent of the thermal velocity $\dot{T}$ and acceleration $\ddot{T}$. The relaxation observed when T is held fixed is either suppressed or dominated by charges induced by a changing T. A technique analogous to ac demagnetisation is used to prepare a state which is observed to be free of relaxation for 8 hours. The disequilibration of the pinned CDW by slight ΔT is then demonstrated. We propose that thermal changes induce stress on the CDW condensate. Release of the stress usually proceeds by spontaneous condensate quakes which occur throughout the length of the sample. Brief discussions of the repinned state, Preisach dynamics and the nature of the quakes are given.

1. INTRODUCTION

The quasi-one-dimensional compounds $NbSe_3$, TaS_3, $K_{0.3}MoO_3$, $(TaSe_4)_2I$, etc. are examples of a distinctive class of solids called charge-density-wave (CDW) conductors [1,2]. The CDW condensate when depinned displays a remarkable variety of "anomalous" properties which distinguish CDW transport from more well-known conductivity mechanisms. The sliding regime (accessed when the field E exceeds a threshold E_T) has been intensively investigated, and much has been learned about the behavior of the moving condensate. In recent years a number of studies [3-6] have shown that the transport properties of the CDW system in the pinned state are also of great intrinsic interest. Broadly speaking these unusual properties reflect the enormous number of degrees of freedom of the condensate, and the large number of configurations with closely similar energies accessible when the CDW is pinned [7,8]. Behavior analogous to that of glassy systems such as disordered polymers, spin-glasses, and random field magnets is observed even though the host lattice is quite clean.

In this paper we examine in some detail the hysteretic behavior [3,4] of the pinned CDW when the temperature T is cycled independently. The configurations accessed by the pinned CDW are monitored by measuring the small-E resistance R. (It turns out that because of the electrical hysteresis [3, 6] the system's electrical conductivity is never "Ohmic" even when $E \ll E_T$. R depends on the history of the sample as well as the time scale of observation.) We take R to be the value observed at time t with a small dc current (1-10 [μA]). Many of the properties to be described here are also seen in the dielectric constant $\varepsilon(\omega)$ [9,10].

2. THERMAL HYSTERESIS

The "thermal hysteresis" (hysteretic behavior of R in the R-T plane) was first reported in TaS_3 by HIGS and GILL [3], and independently by HUTIRAY et al.[4]. As shown in Fig. 1 the R vs. T trajectory when cooling is substantially lower than the curve obtained when warming. Plotted in Arrherius fashion the two curves $R_{warm}(t)$ and $R_{cool}(t)$ are found to be activated (for T>100 [K]) with different gaps <u>and</u> prefactors. In Sample 1 the difference $\Delta R \equiv R_{warm} - R_{cool}$ is as large as 37% at 120 [K]. (Below 120 [K] the slope of ℓn R vs 1/T decreases, possibly because of the ionisation of impurity states. The slight break in slope near 160 [K] arises from a slight admixture of the monoclinic polytype.)

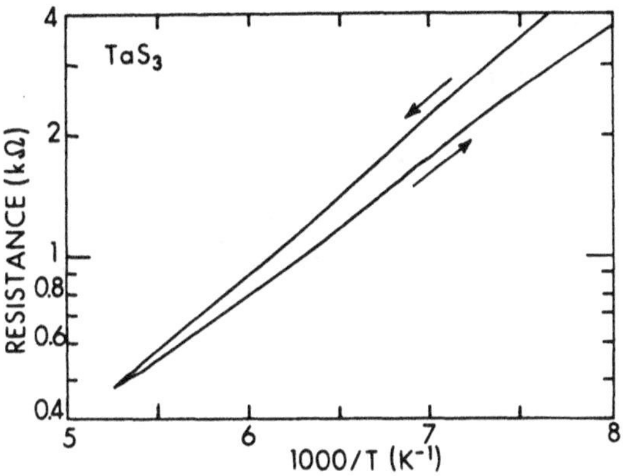

Fig.1 Plot of log R vs 1/T during warming and cooling in orthorhombic $TaSe_3$. In this sample the difference is ~38% at 120 [K]. The small break in slope near 160 [K] is due to a slight admixture of the monoclinic polytype. Note that the effective gap and prefactor are different in the two curves.

The first question we address is how intrinsic is the thermal hysteresis and how universal is it in the class of CDW conductors? In all samples of TaS_3 examined by us (~10) the thermal hysteresis is observed with maximum ΔR of the order of 20 to 40%. Furthermore, in $NbSe_3$ which is a metallic host thermal hysteresis of the same sign ($\Delta R>0$ for T<59 [K]) has been demonstrated [11] at the 1-3% level. At the upper CDW in $NbSe_3$ thermal hysteresis (also with $\Delta R>0$) has been demonstrated [11] indirectly in the R-E plane. In blue bronze $K_{0.3}MoO_3$ thermal hysteresis with $\Delta R>0$ has been reported [12] down to 50 [K]. Thus this anomaly is a rather general property of the pinned CDW, observable over a wide range of T in both semiconducting and metallic hosts. It is also worth noting that, whereas the CDW wave vector Q is T-dependent [13] between ~110 [K] and 200 [K] in TaS_3, Q is independent [14] of T in $NbSe_3$, and in $K_{0.3}MoO_3$ below the "lock-in" transition [15] at 100 [K]. Thus the thermal hysteresis need not be associated with the motion of Q alone. Other mechanisms may be significant.

3. STRUCTURE of HYSTERESIS

To explore further the hysteretic behavior we have used a computer to control T with the sample in a bath dewar. T can be swept at predetermined rates from 0.3 [mK/s] to 10 [mK/s]. During a sweep $\dot{T}$ (= dT/dt) is computed and compared with the present value. Via feedback to the temperature controller (Lake Shore 81C) $\dot{T}$ can be accurately controlled for arbitrarily long periods. For runs when T is held constant it is necessary to decouple the computer from the T-controller to avoid perturbations caused by computer interrogation. Fluctuations in T are kept below ±30 [mK] for 8-12 hrs. By comparing the R vs T hysteresis loop generated at constant speed with that generated with some arbitrary T vs t (time) profile we found that the hysteresis loops are insensitive to either $\dot{T}$ or the magnitude of the acceleration $\ddot{T}$. In Fig. 2a we show a particular T vs t profile in which $\dot{T}$ is changed in step-wise manner. The R-T loop generated by this profile (data points in Fig. 2b) is compared with a loop generated by changing T at the constant rate 10 [mK/s] (solid lines). For clarity the latter is displaced downwards by 50 [Ω]. Thus we conclude that the R vs T curves are in some sense dynamically stable and independent of the T vs t profile except for the extremal T values. The curves traversed with $\dot{T}$ of one sign are called [16] cooling (or warming) saturation curves. Paths which bridge the region between the saturation curves (described after $\dot{T}$ changes sign) are called transit curves. Closed loops which merge with the appropriate saturation curve at the two extremal temperatures are called major loops while all other loops are minor. Similarities with the B-H hysteresis in feromagnets are apparent. The analogy is in fact quite close.

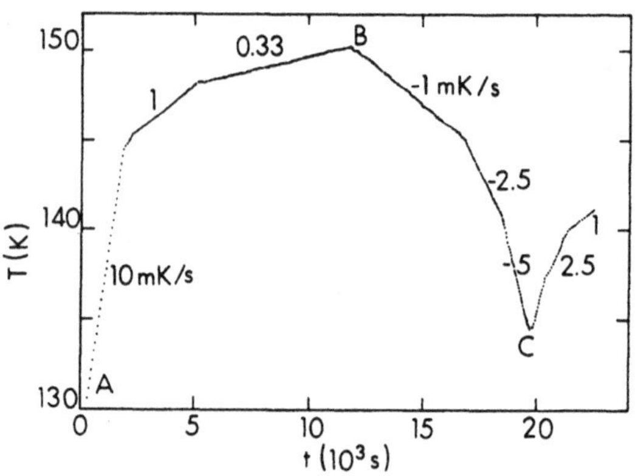

Fig. 2a Temperature vs time profile used to investigate sensitivity of R vs T hysteresis to $\dot{T}$ and $\ddot{T}$. The numbers indicate the rate $\dot{T}$ during each leg.

When $\dot{T}$ is changed in sign, say, from negative to positive, the system abruptly leaves the cooling saturation curve and moves towards the warming saturation line along a transit curve. It is tempting to regard this as a slow "relaxation" from the cooling to the warming saturation line, in response to the abrupt change in sign of $\dot{T}$. However, this view is incorrect. In Fig. 3 we show this transition from cooling to warming executed at two rates, 1 [mK/s] and 10 [mK/s]. (R is referenced to the warming saturation value at each T to emphasize the relaxation viewpoint.) Again, the transit

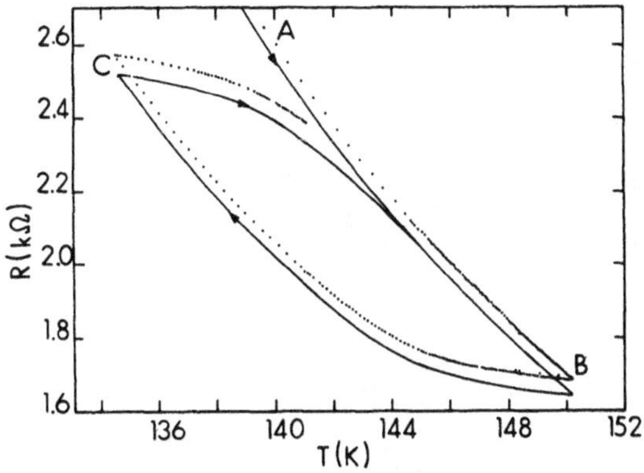

Figure 2b Variation of R vs T executed with the T vs t profile in Fig. 2a (data points) compared with curve obtained when $\dot{T}$=10 [mK/s] (solid lines). The latter is displaced downwards by 50 [Ω] for clarity. Note that the two hysteresis loops are identical.

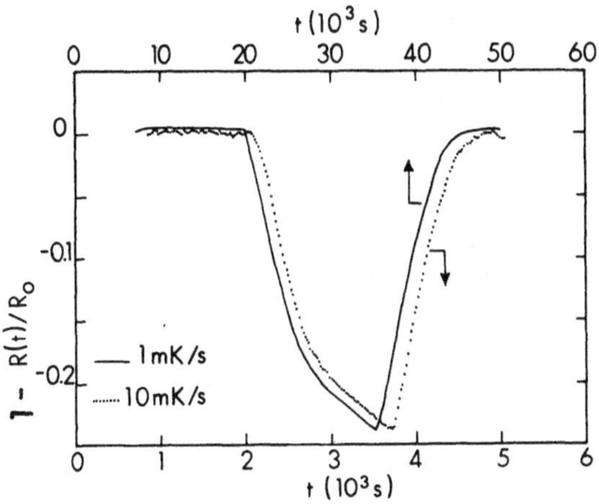

Fig. 3 Plot of $\Delta R = R_{warm}(T) - R$ vs time for two runs executed at 10 [mK/s] (solid lines) and 1 [mK/s] (data points). By rescaling the time axis we show that the variation of R vs T is identical for the two speeds.

curve is insensitive to $\dot{T}$. The two data sets coincide if the time axis is rescaled by 10, i.e., there is no time scale (or scales) intrinsic to the system during the "transition". Where the system is on the transit curve at time t depends only on the value of T(t), and not on $\dot{T}$.

4. NON-EQUILIBRIUM AND RELAXATION

MIHALY and MIHALY [5] found that when T is held constant slow relaxation of R (~ℓn t) occurs towards a value roughly midway between R_{warm} and R_{cool} (Fig. 4). The magnitude of this creep at 121 [K] is ~1% after 10 hrs. (If $R(t) = R(0)\,(1-\alpha \log_{10} t)$, α equals 3.3×10^{-3}.) We have found that this decay (which is <u>not</u> associated with polarisation changes of the sample) is

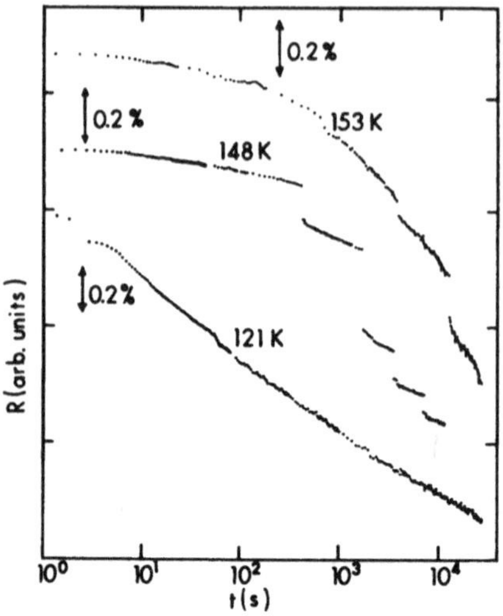

Fig. 4 Examples of relaxation of R when T is held constant. The lowest curve (R ~ log t) is the process reported by MIHALY and MIHALY [5]. Some samples show a stretched exponential behavior (upper curve). However, most samples relax by a series of spontaneous jumps in R (middle curve).

sensitive to the history of the sample, displaying significant variations from run to run under closely similar conditions. In addition, clean samples that have not been repeatedly cycled in T relax primarily by a series of jumps in R. These jumps, called condensate quakes, are quasi-periodic and spontaneous. Typically the cumulative ΔR due to jumps is 3 to 5% in 10 hrs. Nonetheless, relaxation either by creep or jumps proceeds to move the system closer to equilibrium when T is held constant. Interestingly, these relaxation processes are either totally suppressed or dominated by changes in R described in Sec. 3 when $\dot{T}$ is non-zero. For instance, the quakes are rarely observed when $\dot{T}$ is non-zero whereas at fixed T they occur on a time scale of a few 1000 [s]. Apparently, relaxation is effectively suppressed by changes in T as slow as 0.33 [mK/s]. This point is reinforced by studying the dynamics in the R-T plane. Note that the relaxation processes at fixed T are invariably inwards in direction (from the saturation curves towards the mid-point). In contrast we find that changes in T of either sign invariably drive the system towards saturation. Thus the two processes act in opposition in a very uneven contest.

The flow pattern in the R-T plane was previously described by DUGGAN and ONG [16] (Fig. 5). Following an earlier report [4] that a pulse "conditions" TaS_3 DUGGAN et al.[6] have found that when a field ($E > E_T$) is applied (and subsequently removed) the CDW in the repinned state is moved much closer to equilibrium. (R changes by 10 to 15% towards the equilibrium line.) This large change (vertical arrows in Fig. 4) occurs uniformly over the bulk and is to be distinguished from polarisation effects [17] which are at the 1-2% level. (As shown by MIHALY et al.[17] the sample is left in an asymmetric state due to remanent polarisation. Although the polarisation implies that the repinned state is not in equilibrum it cannot account for the uniform change in R of 10%.)

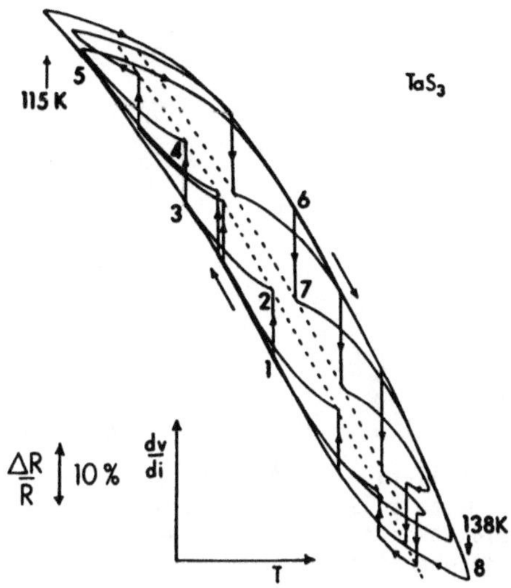

Fig. 5 Structure of the R vs T plane explored by applying a short E-field pulse to depin the CDW. After repinning, R changes by 10%, and the system moves closer to equilibrium. The broken lines indicate the locus of the repinned state. The vertical arrows indicate the change in R induced by a pulse. (From Ref. [16].)

In Ref. [16] it was found that R in the repinned state does not coincide with the equilibrium line (midway between R_{warm} and R_{cool}). If the sample is warming (cooling) at the moment the pulse is applied it ends up in a repinned state with R slightly higher (lower) than the equilibrium value. Thus, there are two repinned states at each T. They are shown as two broken lines in Fig. 5. The qualitative difference between the repinned configurations and equilibrium is probably the remanent polarisation induced by the applied field. (A further creep in R towards equilibrium is discernible in the repinned state.)

Nonetheless it is clear from Fig. 5 that the system is driven back to the saturation lines as T continues to change, i.e., when $\dot{T}$ is non-zero the flow is outwards. One might then infer that this effect of a non-zero $\dot{T}$ is quite general. In particular, if the system is prepared in the equilibrium state one would expect that a slight change in T drives it to the saturation lines. This is indeed the case, as shown in the next section.

5. DISEQUILIBRATION by ΔT

We have used a technique analogous to the ac demagnetisation of a feromagnet to arrive at a state which is free of long-term creep. By slowly oscillating T between successively smaller symmetric minor loops we obtain a state with R very close to midway between R_{warm} and R_{cool} (Fig. 6). This procedure involves the execution of 10 to 20 loops over a period of 10 hrs. Subsequently, T is held constant for 8 hrs. and R is found to be free of relaxation ($\frac{\Delta R}{R} < 3 \times 10^{-4}$) during this period (Fig. 7). Although this is

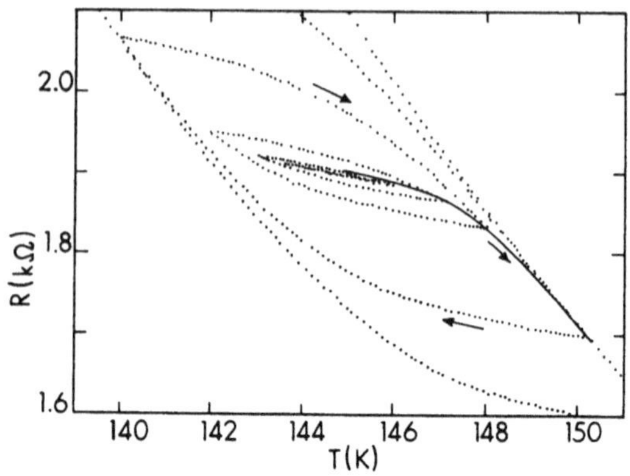

Fig. 6 Approach to stable state by slow oscillation of the temperature about 145 [K]. The procedure takes approximately 12 hours and results in a state which is observed to be free of relaxation for 8 hours (Fig. 7). The solid line (virgin curve) indicates the return of the system to the warming saturation curve when T is warmed at the rate 0.33 [mK/s].

an operational demonstration of equilibrium, we obviously cannot prove that it is in fact true equilibrium. We will simply refer to this state as the unstrained state. Certainly, the stability of the unstrained state shown in Fig. 7 is superior to any other in the CDW literature.

The significant point is the disequilibration of the system when T is subsequently changed. The solid line ("virgin curve") in Fig. 7 shows that R is forced back to the warming saturation line as T is increased frm 145 [K]. The shape of the curve is again independent of $\dot{T}$. (The solid line in Fig. 7 is for $\dot{T}$ = 0.33 [mK/s].) The sensitivity of the system to ΔT is emphasized by examining the slope dT/dt on the virgin curve for small ΔT. The system departs from the equilibrium line at a large angle; a ΔT as small as 10 [mK] is sufficient to disequilibrate the pinned CDW.

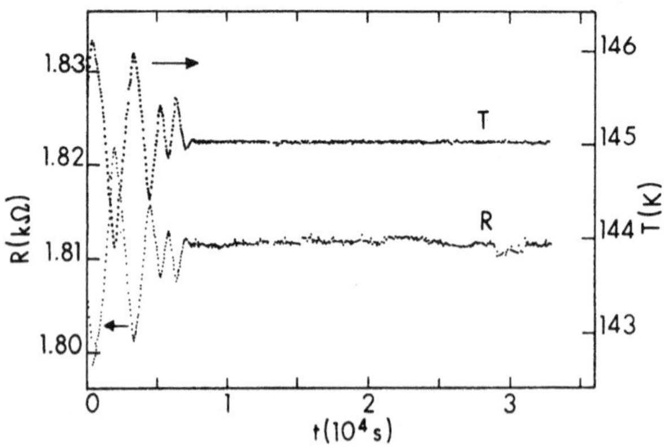

Fig. 7 The stability of the state prepared in the technique in Fig. 6. The temperature (upper curve) is held constant for 8 hours while R is monitored. The last few oscillations of T during the approach to the state are shown on the extreme left.

The results in this section and the last imply that if the system is in either the unstrained or the repinned state, a non-zero $\dot{T}$ drives it back to the saturation curves. We have also examined metastable states reached by spontaneous jumps in R. When T is held fixed these jumps can move R towards equilibrium by as much as 3 to 5%. Again, a subsequent non-zero $\dot{T}$ drives R back to the saturation values.

Quite generally we conclude that changes in T executed as slowly as 0.3 [mK/s] drive the system from the interior of the hysteresis loop outwards towards saturation. Before we discuss an interpretation of these results we will compare a particular feature of the present data with the familiar B-H hysteresis in feromagnets.

6. THE PREISACH MODEL

An interesting point about the virgin curve in Fig. 7 (solid line) is that it passes close to the warm cusp of each of the previously executed minor loops. This behavior is quite generic to hysteretic systems. When the H field is increased after a feromagnet sample has been demagnetised the remagnetisation curve passes through the cusps of the minor loops.

Such behavior is often called Preisach [18-20] dynamics. In this model of hysteresis the magnetisation M is expressed as an integral over a domain [20] in the 2-d space defined by (α,β) where $\alpha(\beta)$ is the extremal value of H when dH/dt is changed from positive to negative (negative to positive)

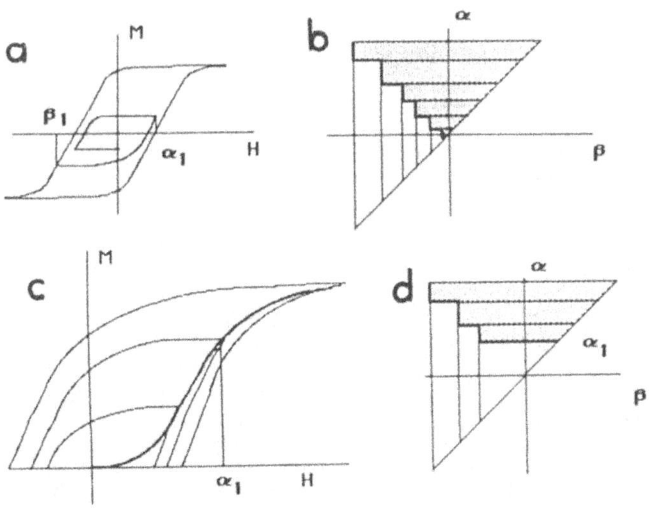

Fig. 8 The Preisach [18] representation for magnetic hysteresis. a) M vs H hysteresis showing execution of a minor loop during ac demagnetisation procedure. The maximum and minimum of the first minor loop is denoted by α_1 and β_1. b) The domain used [20] to compute M. The contribution from the shaded (unshaded) region is negative (positive). The boundary L depends on the extremal values (α_n,β_n) of each minor loop. The state shown is the fully ac demagnetised state (M=0). c) the virgin cuve executed after demagnitisation. The curve passes through the cusp of each minor loop previously executed. d) Preisach representation during remagnetisation (virgin curve). The boundary L is identical to the shape obtained when H equals the nth maximum during the process of demagnetisation. (See text).

during the execution of a loop (Fig. 8). The contribution from the shaded region in the domain is negative while the unshaded region is positive. The boundary L(t) between these regions is a staircase whose corners are determined by the extremal values of a particular loop. The most recently created segment of L (the one touching the line $\alpha=\beta$) is horizontal when $\dot{H}$ is positive and vertical when $\dot{H}$ is negative. If H is increased to $H=\alpha_n$ the last segment moves to the line $\alpha=\alpha_n$, erasing all segments of L below this line. Next, if H is decreased to β_n ($<\alpha_n$) the last segment (now vertical) moves to the line $\beta=\beta_n$, erasing all segments of L with $\beta>\beta_n$. Following this construction it is clear that ac demagnetisation creates a staircase which splits the domain evenly, i.e., the integral over the whole domain vanishes and M=0. The details of how M got to zero are encoded [20] in the individual corners of L since the extremal values of each minor loop are the coordinates of some corner (α_m, β_m).

Starting from the M=0 state let us increase H to α_n. The last branch of L moves to the line $\alpha=\alpha_n$, erasing the lower portion of the staircase in the demagnetised state. However, this is precisely the shape of L during the previous execution of the nth minor loop when H was last increased to the local maximum value α_n. Thus, M in the virgin curve coincides with M at the cusp of the nth loop. Since this argument is quite general we conclude that the virgin curve passes through the cusp of each previously executed loop.

7. STRESS and STRAIN

The dramatic effect that ΔT has on the equilibrium state of the pinned CDW suggests that the role of ΔT (in generating ΔR) is not associated with the usual Boltzmann factor. The results described above suggest that ΔT is changing a parameter in the problem which alters a thermodynamic variable not associated with the electronic entropy. The most reasonable hypothesis is that thermal expansion of the host lattice is playing a central role. We suggest that changes in the host lattice parameter exert a significant stress on the CDW condensate because the CDW wave vector cannot follow the lattice parameter. If ΔT is positive (negative) the condensate is left in a compressively (dilatively) stressed state relative to the host. The response of the host lattice to slight changes in T is determined by the velocity of sound and therefore the stress on the condensate follows changes in T on time scales of microseconds, certainly "instantaneous" on the time scales used here. When the condensate is left in a state of compressive stress the single-particle gap may be slightly larger than the unstrained value. Similarly in the dilatively stressed state the gap is smaller. This would account for the positive sign of ΔR in the three systems studied. There is transport evidence from the slope of the two lines in Fig. 1 that the gap is larger when $\dot{T}$ is positive. However, a direct measurement of the gap is clearly desirable.

In this model the thermal hysteresis is properly viewed as stress vs strain hysteresis. The quantity $R(t) - R_{equil}$ is a measure of the strain of the condensate in response to the stress exerted by the expanding (or contracting) host lattice. When T is held fixed after a change at a slow rate the condensate attempts to move to the equilibrium value either by "creep" or drastic rearrangement. We discuss the latter in the next section.

8. CONDENSATE QUAKES

As discussed above relaxation at constant T proceeds primarily by spontaneous, abrupt jumps in R towards the equilibrium line (Fig. 4). Insofar as these jumps appear to relieve stress in the condensate, and may be quasiperiodic, they resemble quakes. In one particular sample we observed 9 quakes spaced 4400±50 [s] apart. In general, though, the spacing fluctuates by 30% around a mean of the order of 2000 to 5000 [s] for T between 120 and 150 [K].

By examining different regions of the sample we determined that when the quake magnitude exceeds 0.2% of the total R the whole sample participates in the jump. However, because our t resolution is 10 [s] we are unable to say if the quake occurs instantaneously over the length of the sample or is triggered by a propagating wave which runs down the length. The origin of these quakes is rather interesting. We have not found a way to trigger them in a controlled fashion. Because the magnitude is generally large at the start of the relaxation and no quakes are observed in the unstrained state they are most likely stress-relief events that move the system towards equilibrium. In rather general terms we may discuss the quakes as an abrupt response of the condensate to a slow build-up of stress. The quasiperiodic mechanism is then analogous to what occurs in a relaxation oscillator (e.g., snow piling up on a roof). However, in our experiments $\dot{T}$ which is presumably the source of stress is held at zero during the relaxation. Therefore the above analogy is not entirely satisfactory.

The microscopic nature of these quakes is also quite unclear at present. By studying the relative magnitudes of the quakes at different regions of the sample we obtained some evidence that they are propagating waves. Apparently after a critical waiting period the whole sample is poised for a violent rearrangement which lowers the free energy considerably. A wave which spontaneously originates at some point in the sample triggers this rearrangement as it propagates down the sample. Experiments to test these ideas are underway.

9. CONCLUSION

Although the experimental investigation of the thermal hysteresis is continuing several properties are already apparent, and clues to the correct picture are emerging. The speed with which the hysteretic loops are executed is irrelevant to the shapes in the R-T plane. The deviation of R from the equilibrium value at each T apparently depends only on T and the extremal values of the thermal cycle. By studying how the system disequilibrates from the "unstrained" state we have shown that very slight ΔT's executed as slowly as 0.33 [mK/s] are sufficient to force the system out of equilibrium. Return to equilibrium is not observed unless T is held fixed; and then it proceeds by large jumps in R (which reflect a spontaneous rearrangement of the condensate) interspersed with slow logarithmic creep. We have argued that the thermal hysteresis shares many dynamic features with the well-studied B-H hysteresis in ferromagnets. On this basis we proposed that the condensate is placed under stress wherever T changes.

Clearly many questions raised by these studies remain unanswered. For example, why are there two stable limiting trajectories (the saturation lines)? What is the physical nature of the quakes? What prevents the condensate from rapidly adjusting to the conditions of the host lattice after a slight ΔT?

Discussions with E. Abrahams, P.W. Anderson, D. Stein and S. Trugman are grateful acknowledged.

REFERENCES

1. For recent developments see Charge Density Waves in Solids, Lecture Notes in Physics, Vol. 217 (Springer-Verlag, Berlin, 1985)
2. For a review see G. Gruner and A. Zettl: Phys. Reports 119, 117 (1985); P. Monceau: in Electronic Properties of Inorganic Quasi-One-Dimensional Materials, ed. by P. Monceau, Vol. 2 (Reidel, Dordecht, 1985) p.139
3. A.W. Higgs and J.C. Gill: Solid State Commun. 47, 737 (1983)
4. Gy. Hutiray, G. Mihály and L. Mihály: Solid State Commun. 47, 121 (1983)
5. G. Mihály and L. Mihály: Phys. Rev. Lett. 52, 149 (1984)
6. D.M. Duggan, T.W. Jing, N.P. Ong and P.A. Lee: Phys. Rev. B 32, 1397 (1985); N.P. Ong et al., in Ref. 1, p.385
7. D.S. Fisher: Phys. Rev. B 31, 1396 (1985)
8. P. B. Littlewood: preprint (1986)
9. R.J. Cava: private communication
10. C.B. Kalem, J.C. Eckert, and N.P. Ong: to be published; C.B. Kalem: Ph.D. thesis, University of Southern California (1986), unpublished
11. D.M. Duggan: Ph.D. thesis, University of Southern California (1986), unpublished
12. Reported by K. Tsutsumi at International Conference on Charge Density Waves in Solids, Budapest (1984)
13. C. Roucau: Jnl. de Phys. 44, C3-1725 (1983); Z.Z. Wang, H. Salva, P. Monceau, M. Renard, C. Roucau, R. Ayrolles, R. Levy, L. Guemas and A. Meerschaut: Jnl. de Phys. Lett. 44, L-311 (1983)
14. R.M. Fleming, C.H. Chen and D.E. Moncton: Jnl. de Phys. 44, C3-1651 (1983); R.M. Fleming, D.E. Moncton, J.D. Axe and G.S. Brown: Phys. Rev. B 30, 1877 (1984)
15. R.M. Fleming, L.F. Schneemeyer and D.E. Moncton: Phys. Rev. B 31, 899 (1985)
16. D.M. Duggan and N.P. Ong: Phy. Rev. B 34, xxx (1986)
17. L. Mihaly and A. Janossy: Phys. Rev. B 30, 3530 (1984)
18. F. Preisach: Z. Phys. 94, 277 (1935)
19. J.A. Barker, D.E. Schreiber, B.G. Huth, and D.H. Everett: Proc. Roy. Soc. London, Ser. A 386 251 (1983)
20. I.D. Mayergoyz: Phys. Rev. Lett. 56 1518 (1986)

Vortex Dynamics in Two-Dimensional Arrays of Superconducting Weak Links

P. Martinoli, Ch. Leemann, and Ph. Lerch

Institut de Physique, Université de Neuchâtel, CH-2000 Neuchâtel, Switzerland

1. Introduction

Ordering in a physical system strongly depends on its dimensionality d and on the number of degrees of freedom, n, of the quantity used to describe the order [1]. For instance, in a magnetic system n is the number of components of the spin (n = 1 for the Ising model, n = 2 for the XY model, n = 3 for the Heisenberg model), while the description of the ordered phase of a superfluid in terms of a complex order parameter implies n = 2. In this context, two-dimensional systems (d = 2) with n ≥ 2 hold, on account of the delicate problems they raise in connection with the possibility of a phase transition and with the nature of the ordered state that may occur at low temperatures [2-5], a very special position. Since the pioneering work of KOSTERLITZ and THOULESS [6] in the early 1970s, there has been considerable progress in the theoretical understanding of the critical behaviour of two-dimensional (2D) systems. It is now well established that topological excitations, such as vortices in 2D superfluids or dislocations in 2D solids, play a central role in critical phenomena occurring in these systems. For n = 2 there is a sharp transition, triggered by the unbinding of pairs of topological excitations, from a low-temperature phase showing quasi long-range (or topological) order to a disordered phase at high temperatures, while for n > 2 no phase transition at all occurs in two dimensions. Advances in material science, experimental techniques and numerical simulations have contributed to assess, in a variety of systems, the validity of the various theoretical ideas underlying the physics of phase transitions in two dimensions.

Recently, experiments on 2D regular arrays of Josephson junctions [7-14] and on 2D periodic superconducting networks [15,16] have stimulated a great deal of theoretical interest [17-28] in the critical behaviour of 2D lattice spin models. In zero magnetic field 2D periodic arrays of superconducting weak links are a physical realization of the XY model, a 2D lattice of planar (n = 2) classical spins. For this system the Kosterlitz-Thouless (KT) theory [6] predicts, at a critical temperature T_c, a transition from a topologically ordered phase ($T < T_c$) to a disordered phase ($T > T_c$) driven by the unbinding of thermally excited vortex-antivortex pairs. The physics of 2D arrays of weak links exposed to a perpendicular magnetic field $\vec{B}$ is more complex and only partially understood at present. The interaction of the vortices induced by the field with the pinning potential provided by the periodic structure of the array creates new and interesting phenomena which are most simply described by a uniformly frustrated XY model [17-25]. The degree of frustration is governed by the ratio $f = \Phi/\Phi_0$, where Φ is the magnetic flux threading a unit cell of the array and $\Phi_0 = hc/2e$ is the superconducting flux quantum. The frustration parameter f determines the ground-state (T = 0) configuration of the vortex lattice and, in addition, has a profound effect on the nature of the phase transition at $T_c(f)$. For instance, symmetry considerations for the fully frustrated ($f = 1/2$) XY model [17,20-22] show that two kinds of topological excitations, vortices (points defects) and domain walls (line defects), are relevant in determining the critical behaviour of the array. As a consequence, one is faced with the possibility of two competing phase transitions, one driven by vortex unbinding (KT-like transition) the other by the proliferation of domain walls (Ising-like transition). The question of whether these two transitions are distinct or merge into a single transition belonging to a new universality class is presently a subject of theoretical speculations [17-25].

Studies of the resistive transition and of the current-voltage characteristics of 2D arrays in zero field ($f = 0$) were found consistent with the vortex-unbinding idea [7,8,10-12]. Moreover, resistance and critical current measurements [9-12, 16, 17] show a complex periodic dependence on on f, reflecting the fact that, by varying f, the array is driven through a periodic sequence of pinned commensurate vortex phases and "floating" incommensurate vortex phases. A somewhat similar behaviour was observed earlier in other periodic 2D superconducting structures [29, 30].

In a recent Letter [13] we reported a study of vortex dynamics in a 2D square array of proximity - effect coupled Josephson junctions. In those experiments the complex ac response of vortex excitations to a driving oscillating field was inferred from measurements of the screening properties of the array. For an integer number of flux quanta per unit cell a peak in dissipation and a drop in superfluid density were observed in the critical region. The shape of these structures, which are reminiscent of those observed at the superfluid transition of uniform 2D superconductors [31] and of 2D superfluid helium films [32], as well as their frequency dependence were found to agree with predictions of the KT theory [6] and of its extensions [33,34] to finite frequencies.

In this paper we give a fairly detailed account of vortex dynamics in 2D arrays of Josephson junctions. We shall mainly focus on the critical behaviour of the unfrustrated system, deferring the discussion of the effect of frustration on the nature of the phase transition at $T_c(f)$ to a subsequent publication. Some qualitative aspects of the dynamics of field-induced vortices in commensurate and incommensurate vortex phases at low temperatures are also discussed.

2. Arrays in Zero Magnetic Field

Consider a 2D square lattice of superconducting islands connected by Josephson weak links. Only nearest-neighbour islands are assumed to interact with each other. The interaction energy E_{ij} of a pair of islands is the energy necessary to create a difference $(\phi_j - \phi_i)$ in the phase of the superconducting order parameter between the nearest-neighbour lattice sites i and j. Using Josephson's fundamental equations, E_{ij} turns out to be given by :

$$E_{ij} = [\hbar i_c(T)/2e]\,[1 - \cos(\phi_j - \phi_i)] \quad , \tag{1}$$

where $i_c(T)$ is the critical current of the (isolated) junction <ij> in the absence of thermal fluctuations [35]. Summing over all nearest-neighbour pairs <ij>, the Hamiltonian of the system will be :

$$H = [\hbar i_c(T)/2e] \sum_{\langle ij \rangle} [1 - \cos(\phi_j - \phi_i)] \quad . \tag{2}$$

If one identifies the phase variable ϕ_i with the angle a planar (n = 2) spin at the site i makes with a fixed arbitrary direction in the lattice plane, then Eq. (2) shows quite clearly that a 2D array of Josephson junctions is isomorphic to an XY model with a temperature-dependent coupling energy $J = \hbar i_c(T)/2e$. As a consequence, the statistical mechanics of the of the system is conveniently described in terms of a dimensionless temperature parameter $\tilde{T} \equiv k_B T/J(T) = 2ek_B T/\hbar i_c(T)$ [35].

2.1 The Kosterlitz - Thouless Phase Transition

According to an important theorem due to MERMIN and WAGNER [4] and to HOHENBERG [5], a 2D array of weak links is precluded from developing, at finite temperatures, a state showing <u>conventional</u> long-range order in the phases ϕ_i by spontaneously breaking a <u>continuous</u> symmetry, in this case a uniform global rotation of the ϕ_i's. However, calculations based on "spin-wave" excitations of the phase show that, at sufficiently low temperatures, the "spin-spin" correlation function decays algebraically with distance, a behaviour requiring some kind of phase transition to the high-temperature régime where the correlation function is expected to decay exponentially, as in a liquid. The nature of this novel phase transition was explained by KOSTERLITZ and THOULESS [6] with the introduction of vortices and antivortices as additional topological excitations resulting from 2D fluctuations in the phase of the order parameter.

According to the KT theory, for $T < T_c$ the system is populated by bound pairs of vortices with opposite circulation. In an array the potential energy of a vortex-antivortex pair with cores separated by a distance r is given by [36] :

$$U(r) = [hi_c(T)/2e] \ln(r/a) \quad , \tag{3}$$

where a is the lattice spacing. The logarithmic dependence of U(r) on r is valid as long as r is less than the effective penetration depth $\Lambda(T) = (c\Phi_0/4\pi^2)/i_c(T)$ of the array [36]. It turns out, however, that, at $T = T_c$, $\Lambda(T_c)$ is approximately given by the relation $\Lambda(T_c)\, T_c \approx 1.96$ cmK. It follows that for typical values of T_c (a few Kelvin) $\Lambda(T_c)$ is a macroscopic length. Thus, in the interesting critical region the condition $r < \Lambda$ is nearly satisfied even by the largest vortex pairs (for which r is of the order of the linear dimension, Na, of the array) excited in a typical array (Na $\lesssim$ 1 cm).

The above expression for U(r) shows that vortex excitations in a 2D array interact as electric charges in a 2D Coulomb gas. This analogy allows one to define a "vortex charge" q_v as follows :

$$2q_v^2/\ell = hi_c(T)/2e \quad , \tag{4}$$

where ℓ is the length of the vortex charge.

Below T_c , the array is in a superconducting state with <u>quasi</u> long-range (or <u>topological</u>) order in the phase of the order parameter. Thermally excited vortex-antivortex pairs, together with "spin wave" excitations, lead to an algebraically decaying correlation function of the form [6] :

$$\langle \exp\{i[\phi(\underline{R}) - \phi(\underline{0})]\} \rangle \sim R^{-\eta_R(T)} \quad , \tag{5}$$

where the exponent $\eta_R(T)$ is related to J_R, the coupling energy "renormalized" by the presence of bound vortex pairs (see below), by $\eta_R(T)=k_BT/2\pi J_R$.

The phase transition of an infinite 2D array at $T = T_c$ is triggered by the undinding of the vortex-antivortex pairs of largest separation ($r \to \infty$). As the temperature increases above T_c, the unbinding mechanism progressively extends to pairs of smaller and smaller size, a process creating an increasing number of free vortices which destroy the topological order existing below T_c. In the disordered state above T_c short-range correlations in the phases ϕ_i are described by [37] :

$$\langle \exp\{i[\phi(\underline{R}) - \phi(\underline{0})]\} \rangle \sim e^{-R/\xi_+(T)} \quad , \tag{6}$$

where the correlation length $\xi_+(T)$ has an unusual temperature dependence reflecting the very peculiar nature of the KT transition :

$$\xi_+(T) \approx a \exp[b(\tilde{T}-\tilde{T}_c)^{-1/2}] \quad , \tag{7}$$

where b is a non-universal constant of order unity. Physically, $\xi_+(T)$ is a measure of the average distance between free vortices. Accordingly, the free-vortex areal density $n_{f_\square}$ is approximately given by $n_{f_\square} \approx \xi_+^{-2}(T)$.

The potential energy (3) of a vortex-antivortex pair with separation r is reduced (or "renormalized") by the presence of intervening pairs of smaller size. Referring to the 2D Coulomb gas analogue, the resulting screened interaction between the members of a pair of size r is conveniently described in terms of a scale-dependent dielectric constant $\varepsilon(r)$ which increases with r and is such that $\varepsilon(a) = 1$. The calculation of $\varepsilon(r)$ is based on the Kosterlitz scaling equations [36, 37], a set of coupled recursion relations for the reduced "spin-wave" stiffness constant $K(r) = (J/k_BT)\,\varepsilon^{-1}(r) = [\hbar i_c(T)/2ek_BT]\,\varepsilon^{-1}(r)$ and the vortex thermal activity $y(r)$. An important prediction of the renormalized KT theory is that at T_c and for infinite scale ($r \to \infty$) $K(\infty) \equiv K_R$ has the universal value $2/\pi$. For a 2D array of weak links the universal relation can be written in the form [35, 36] :

$$i_c(T_c)/\varepsilon_c T_c = 8ek_B/h \approx 27 \text{ nA/K} \quad , \tag{8}$$

where ε_c is the value of $\varepsilon(r)$ at T_c and for $r \to \infty$. Since, at T_c, $J_R = J/\varepsilon_c$, an alternative formulation of the KT universal result is that the exponent $\eta_R(T)$ in Eq. (5) has the value 1/4 at the transition. The value of ε_c depends on the parameters which determine the detailed form of the core potential, $\mu_c(T)$, of a vortex excitation, a quantity related to the vortex activity at the shortest scale $r = a$ by $y(a) \sim \exp[-\mu_c(T)/k_BT]$ [36]. For the XY model on a square lattice, KOSTERLITZ and THOULESS found $\varepsilon_c \approx 1.175$, while from Monte Carlo calculations [38] one infers $\varepsilon_c \approx 1.75$.

2.2. Vortex Dynamics

The dynamics of vortex excitations near the phase transition of 2D systems has unique features which, when seen in an appropriately conceived experiment, allow a crucial and detailed test of the KT theory [13, 31, 32].

The dynamical properties of the KT transition were studied by AMBEGAOKAR et al. [33] for 2D superfluids and, more recently, by SHENOY [34] for 2D arrays of Josephson junctions. In these models the response, near T_c, of bound vortex pairs and of free vortices to a driving oscillating field of angular frequency ω is described in terms of a temperature and frequency-dependent complex dielectric constant $\varepsilon(T,\omega)$. Carrying on the analogy with the 2D Coulomb gas, $\varepsilon(T,\omega)$ can be written as the sum of a contribution $[\varepsilon_b(T,\omega)-1]$ due to the polarization of vortex-antivortex dipoles and of a "Drude term" $\varepsilon_f(T,\omega) = \{1 + [4\pi\sigma_v(T)/i\omega]\}$, where $\sigma_v(T)$ is the vortex conductivity, describing dynamical screening of the free-vortex plasma, i.e.:

$$\varepsilon(T,\omega) = \varepsilon_b(T,\omega) + [4\pi\sigma_v(T)/i\omega] \quad . \tag{9}$$

The motion of a bound vortex pair with separation r in response to the driving ac field is characterized by the time $\tau(r) \sim r^2/D_v$ which governs its diffusive relaxation towards the equilibrium orientation (D_v is the vortex diffusivity). In the dynamical theories [33, 34] the bound-pair contribution $\varepsilon_b(T,\omega)$ in Eq. (9) is obtained from the scale-dependent dielectric constant $\varepsilon(r)$ introduced in Sec. 2.1 by integrating over a continuous distribution of relaxation times $\tau(r)$. This procedure singles out a characteristic scale, the vortex diffusion length

$$r_\omega \approx (14D_v/\omega)^{1/2} \quad , \tag{10}$$

which determines the separation of those vortex pairs which dominate the response. Then, an approximate calculation shows that $\varepsilon_b(T,\omega)$ is related to $\varepsilon(r_\omega)$ in the following way :

$$\begin{aligned} \text{Re}\ \varepsilon_b(T,\omega) &\approx \varepsilon(r_\omega) \\ \text{Im}\ \varepsilon_b(T,\omega) &\approx (\pi/4)\ [r d\varepsilon/dr]_{r\,=\,r_\omega} \quad . \end{aligned} \tag{11}$$

Recalling that $\xi_+(T)$ is a measure of the average distance between free vortices, the existence of a characteristic length scale r_ω at finite frequencies indicates that there is a crossover in response due to the unbinding of bound vortex pairs with separation r_ω into free vortices at a temperature $T_\omega > T_c$ such that :

$$r_\omega \approx \xi_+(T_\omega) \quad . \tag{12}$$

This relation, together with Eqs.(7) and (10), shows that, by probing the system with increasing frequency, the vortex-unbinding transition should be seen to shift to higher temperatures.

Above T_c, the contribution of free vortices to $\varepsilon(T,\omega)$ can be easily estimated by writing the free-vortex conductivity $\sigma_v(T)$, in the spirit of the 2D Coulomb gas analogue, as $\sigma_v(T) = q_v^2\, n_f(T)\, \mu_v = (q_v^2/\ell)\, n_{f_\square}(T)\, \mu_v$, where the vortex charge q_v is given by Eq.(4) and $n_{f_\square} \approx \xi_+^{-2}(T)$ (see Sec. 2.1). The vortex mobility μ_v can be easily expressed in terms of array parameters following the analysis of the flux-flow régime in arrays by LOBB et al.[35]. It turns out that, in the critical region, μ_v can be written, to a very good approximation, as :

$$\mu_v \approx (c/\Phi_0)^2\ a^2 r_n \quad , \tag{13}$$

where r_n is the normal-state resistance of an individual junction. Finally, using Eqs. (4) and (10), the vortex conductivity becomes :

$$\sigma_v(T) = (e/h)\,[a/\xi_+(T)]^2\, i_c(T) r_n \quad , \tag{14}$$

an expression showing that σ_v is proportional to the fraction of area occupied by the vortex cores (of radius ~ a) and to the $i_c r_n$ - product of a single weak link. Notice that, using Eq. (13) and Einstein's relation $D_v = \mu_v k_B T$, the vortex diffusivity entering Eq. (10) can be expressed as $D_v = (c/\Phi_0)^2\, a^2 r_n\, k_B T$.

In Sec. 4 it will be shown that, near the phase transition, the signal voltage $\delta V_\omega(T)$ which measures the ac response of the array in our experiments is inversely proportional to $\varepsilon(T,\omega)$. Calculations of $\varepsilon^{-1}(T,\omega)$ based on the previous discussion and on the Kosterlitz scaling equations show that, as T crosses T_ω from below, there is a drop in $\mathrm{Re}(\varepsilon^{-1})$, while a peak emerges in $\mathrm{Im}(\varepsilon^{-1})$. The fall-off in $\mathrm{Re}(\varepsilon^{-1})$ reflects the smooth drop in superfluid density one expects when vortex unbinding is probed at a finite scale r_ω (a discontinuous jump occurs at infinite scale only). The peak in $\mathrm{Im}(\varepsilon^{-1})$, on the other hand, is a manifestation of the crossover from a bound-pair dominated dissipation $[r_\omega < \xi_+(T)]$ to a free-vortex - dominated dissipation $[r_\omega > \xi_+(T)]$.

3. Arrays in a Magnetic Field

Two-dimensional arrays of Josephson junctions exposed to a perpendicular magnetic field $\vec{B}$ are isomorphic to a frustrated XY model with the Hamiltonian [17-20, 39] :

$$H = [\hbar i_c(T)/2e] \sum_{\langle ij\rangle} [1-\cos(\phi_j-\phi_i-A_{ij})] \quad , \tag{15}$$

where A_{ij} is proportional to the line integral of the vector potential $\vec{A}$ from the site i to the site j : $A_{ij} = (2\pi/\Phi_0)\int_i^j \vec{A}.d\vec{s}$. In a square lattice each plaquette contains four weak links with a phase difference$(\phi_j-\phi_i-A_{ij})$ across the link <ij>. Then, the requirement that the phase of the superconducting wave function must change by an integral multiple of 2π in going around a closed path implies [9, 17-20, 39] :

$$\sum_p (\phi_j-\phi_i-A_{ij}) = -2\pi f \pmod{2\pi} \quad , \tag{16}$$

where the sum is over the four links in the plaquette. If the transverse penetration depth is large compared to the sample size ($\Lambda > Na$), screening currents are negligible and, as a consequence, the magnetic field penetrates the array uniformly. In this case $f\Phi_0$ is constant throughout the array and equal to the magnetic flux, $\Phi = Ba^2$, threading a unit cell.

The Hamiltonian (15) is a periodic even function of f with period 1. As a consequence, we can restrict our attention to the interval $0 < f < 1/2$. While for an isolated plaquette the ground-state phase configuration is such that Eq. (16) is satisfied by taking $(\phi_j-\phi_i-A_{ij}) = -(\pi/2)f$, in an array frustration effects resulting from competing interactions with neighbouring cells make the determination of the ground state a difficult problem. For $1/3 < f < 1/2$ an interesting quasi one-dimensional solution has been proposed by HALSEY [20]. For a square lattice the supercurrent distribution in the ground state consists of a sequence of "staircase" supercurrents generating a striped quasi 1D structure along a diagonal of the lattice. Staircase supercurrent patterns for f = 1/2 and f = 1/3 are shown in Fig. 1(a) along with the corresponding phase configurations. We refer to this description in terms of phases and currents as the "phase-current picture". The gauge-invariant phase differences $(\phi_j-\phi_i-A_{ij})$ are identical for all links located on a given staircase. Therefore, magnitude and sign of the supercurrent are constant along the same staircase. As required by Kirchhoff's law, the supercurrent is conserved at each lattice site. Numerical calculations agree with the staircase model for $1/3 < f < 1/2$. For other values of f, no analytical solution exists so far and the ground state has been constructed by Monte Carlo and mean field methods [17, 18].

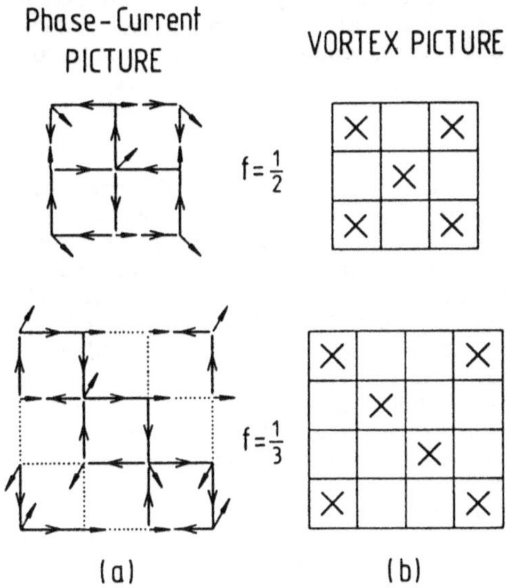

Fig. 1 Ground states of a 2D array for f = 1/2 and f = 1/3 (a) Staircase current patterns and phase configurations in the phase-current model. (b) Commensurate vortex phases in the vortex model.

An equivalent and, in several circumstances, useful way of describing ground states of a frustrated 2D array of weak links is provided by what we call the "vortex picture". In this model the ground state is determined by considering the interaction of the vortex lattice induced by the magnetic field with the pinning potential created by the periodic structure of the array. Since there are two competing periodicities, as f changes the array is driven through a sequence of commensurate (C) (f = p/q rational, where p and q are integers such that p/q is an irreducible fraction) and incommensurate (I) (f irrational) vortex phases, whose detailed structure can be found by mapping the Hamiltonian (15), via a Villain duality transformation [40], onto a lattice 2D Coulomb gas [17, 20]. C-vortex configurations for f = 1/2 and f = 1/3 are shown in Fig. 1 (b). The ground state of a C-phase is characterized by the formation of a superlattice with a (qa x qa) unit cell.

At low temperatures $[T \ll T_c(f)]$, C-vortex phases are pinned by the periodic potential provided by the array. As a function of f = p/q, the pinning strength, as measured by the critical current, $i_c(f=p/q)$, of the array, has a strongly discontinuous upper bound $\pi(e/\hbar)\ \varepsilon(f=p/q)\ q^{-1}$, where

$\varepsilon(f=p/q)$ is the (non-monotonic) ground-state energy per junction [17]. In an I-phase, on the other hand, the vortex lattice can slide freely [$\varepsilon(f=p/q)/q=0$ for $q \to \infty$]. Thus, in experiments probing vortex dynamics as a function of f, marked structures should appear in the complex ac response of an array in correspondence to low-order (small q) C-phases. Such experiments will be discussed in some detail in Sec. 4.

Concerning the nature of the phase transition at $T_c(f)$, in this paper we restrict our discussion to the unfrustrated ($f = p$) case. Since the Hamiltonian (15) is periodic in f with period 1, the superconducting transition of a 2D array of weak links is expected to be KT-like also for $f = p \neq 0$. In this case the thermally activated point defects are positive and negative vacancies associated, respectively, with $(p - 1)$ and $(p + 1)$ quantized vortices per unit cell. These vacancies of opposite sign can be viewed as free-moving vortex-antivortex excitations immersed in a pinned commensurate background of field-induced vortices with p flux quanta per unit cell [13].

4. Experimental Results and Discussion

We have studied [13] the dynamics of vortices in 2D arrays of proximity-effect Pb/Cu/Pb junctions using a modified version of the two-coil technique devised by FIORY and HEBARD [41]. The experiments reported here were performed on two arrays, A1 and A2, consisting of $N \times N \approx 5 \times 10^5$ square Pb islands forming a square lattice with $a = 8$ µm on a Cu layer, the length L of the Cu bridges connecting adjacent Pb islands being of the order 1.7 µm. A1 and A2 differ in the thicknesses of the superconducting islands and of the normal film, thereby having slightly different coupling energies. The zero-field dc resistance of both arrays shows, with decreasing temperature, the two distinct transitions observed in similar systems by other groups [7, 8, 10]. For the ac measurements the arrays were positioned directly under a system of coaxial cylindrical coils consisting of an external driving coil and an inner astatic-pair receiving coil [42]. An ac current of amplitude $I_{D\omega}$ and angular frequency ω was applied to the driving coil and the signal voltage, δV_ω, at the receiving coil due merely to the screening currents flowing in the array was phase-sensitively detected. It turns out that, for our sample-coil geometrical configuration, δV_ω is related to the complex sheet impedance, $Z_\square$, of the array by [41] :

$$\delta V_\omega = -\omega^2 I_{D\omega} Z_\square^{-1} \int_0^\infty e^{-x} f(x) [x + (H/\Lambda)\ (i\omega L_{K_\square}/Z_\square)]^{-1} dx \quad , \tag{17}$$

where $L_{K_\square}(T) = (2\pi/c^2)\Lambda(T) = \hbar/2ei_c(T)$ is the bare sheet kinetic inductance of the array, H the distance of the first (lowest) winding of the driving coil from the array ($H \approx 0.3$mm) and f(x) an oscillating function of x, such that $f(0) = 0$, decaying as $1/x^2$ at large x and depending only on the geometrical parameters of the coils.

If one assumes that normal currents do not appreciably contribute to the total current flowing in the array, i.e. if $\omega \ll r_n/L_{K_\square}$, then for $f = 0$ $Z_\square(T,\omega)$ can be written in the form [13, 31] :

$$Z_\square(T,\omega) = i\omega L_{K_\square}(T)\ \varepsilon(T,\omega) \quad , \tag{18}$$

an expression which explicitly shows that vortex excitations described by $\varepsilon(T,\omega)$ [Eq. (9)] renormalize the bare areal superfluid density

$n^*_\square(T) = m^*/4e^2 L_{K_\square}(T)$ downwards. In our experiments H is much smaller than Λ in the transition region. Accordingly, near T_c the term proportional to (H/Λ) in Eq. (17) can be ignored and δV_ω becomes, using Eq. (18) :

$$\delta V_\omega \approx C\ i\omega I_{D\omega} i_c(T)/\varepsilon(T,\omega) \quad , \tag{19}$$

where C is a purely geometrical constant. Since the monotonic temperature dependence of $i_c(T)$ does not significantly affect the response of the array near T_c, the behaviour of $\delta V_\omega(T)$ in the transition region should reflect the characteristic features of $\varepsilon^{-1}(T,\omega)$ predicted by the dynamical extensions [33, 34] of the KT theory (see Sec. 2.2). This is quite clearly demonstrated by the peak in Re $[\delta V_\omega(T)]$ and the drop in Im $[\delta V_\omega(T)]$ shown in Fig.2 for the array A2. From these data, using an iterative procedure to solve Eq.(17), it is possible to extract $\varepsilon^{-1}(T,\omega)$. The $i_c(T)$ curves needed to calculate $L_{K_\square}(T)$ were obtained by fitting low-temperature i_c measurements of A2 to the expression [43] :

$$i_c(T) = i_0[1-(T/T_{cs})]^2 \exp[-L/\xi_N(T)] \quad , \tag{20}$$

[using T_{cs} = 6.8 K for the BCS transition temperature of the Pb islands, $\xi_N(T_{cs})$ = 1060 Å for the Cu coherence length and i_0 = 0.38 A] and by extrapolating the theoretical curves in the critical region. Results exhibiting the expected temperature dependence of $\mathrm{Re}(\varepsilon^{-1})$ and Im (ε^{-1}) near the transition are shown in Fig. 3. A detailed comparison of these data with the theoretical predictions of Sec. 2.2 should shed further light on the dynamics of vortex excitations in 2D arrays.

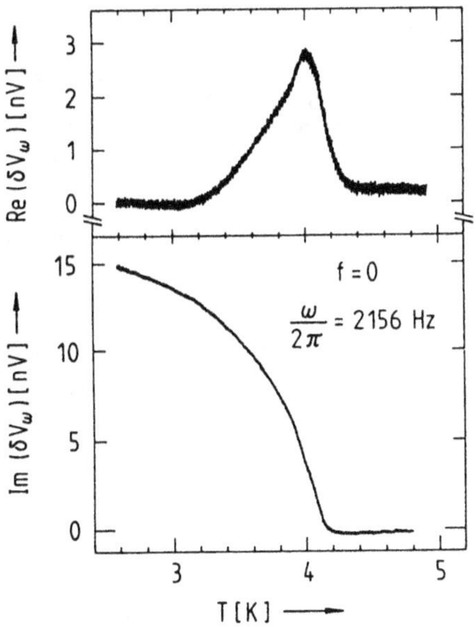

Fig.2 Temperature dependence of the complex ac response of array A2 in zero magnetic field (f = 0)

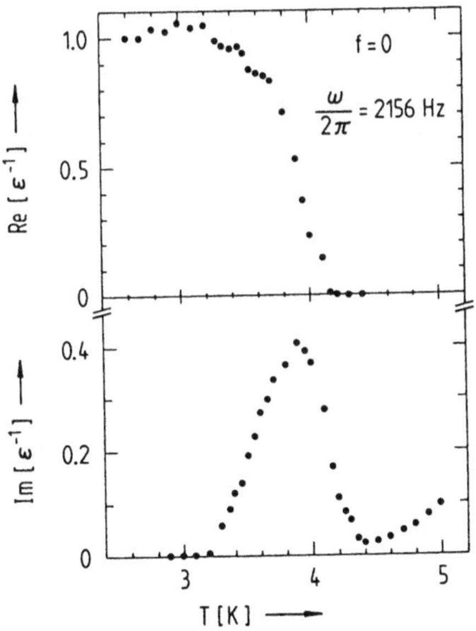

Fig.3 Temperature dependence of the inverse complex vortex dielectric constant of array A2 as deduced from the ac response shown in Fig.2 and from Eq. (17)

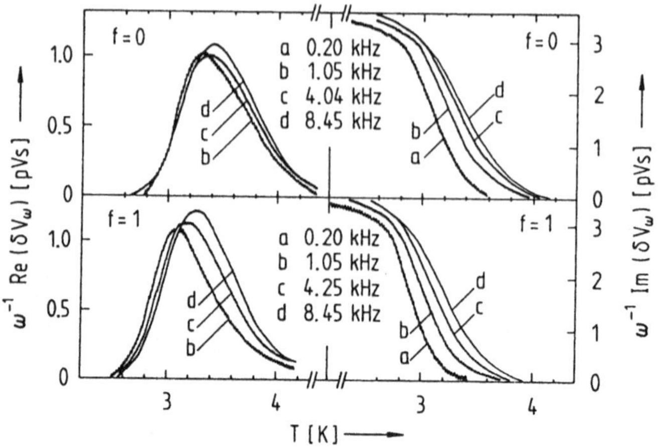

Fig.4 Temperature dependence of the normalized complex ac response of array A1 measured at different frequencies for f = 0 and f = 1. From Ref. 13

As predicted by Eq.(12) and demonstrated by the f = 0 ac response of A1 shown in Fig. 4, the peak in Re $[\delta V_\omega(T)]$ shifts to higher temperatures and the fall-off of $\mathrm{Im}[\delta V_\omega(T)]$ broadens with increasing ω. To analyze these results, we define T_ω by extrapolating to zero the portions of the $\mathrm{Im}(\delta V_\omega)$ vs T curves [31] and introduce a scale parameter $\ell_\omega \equiv \ln(r_\omega/a)$. Then, Eq. (12) can be written in the form :

$$\ell_\omega^{-2} = b^{-2}(\tilde{T}_\omega - \tilde{T}_c) \quad . \tag{21}$$

The scale ℓ_ω and the dimensionless temperature $\tilde{T}_\omega = 2ek_BT_\omega/\hbar i_c(T_\omega)$ were calculated using, for A1, $r_n = 2.2$ mΩ to determine D_v and taking $T_{cs} = 7$ K, $\xi_N(T_{cs}) = 850$ Å and $i_o = 0.78$ A to deduce $i_c(T_\omega)$ from Eq. (20). A plot for f = 0 of ℓ_ω^{-2} vs $\tilde{T}_\omega$ is shown in Fig. 5. A good fit to Eq. (21) is obtained for b = 1.29. Extrapolation of the linear plot to infinite scale ($\omega \to 0$) leads to $i_c(T_c)/T_c = 49$ nA/K. Inserting this result in the universal prediction (8) gives $\varepsilon_c = 1.75$, a value in good agreement with an estimate ($\varepsilon_c = 1.81$) based on Monte Carlo simulations [38].

According to the KT picture discussed in Sec. 3, the response $\delta V_\omega(T)$ for $f = p \neq 0$ should be similar to that for f = 0, a conjecture confirmed by the experimental results for f = 1 of A1 shown in Fig. 4. When compared with the case f = 0, however, the transition is seen to occur at a lower temperature. This is easily understood if one realizes that the finite size of the junction makes i_c field dependent. More precisely, the critical current of an <u>isolated</u> junction of the array is a quadratically decreasing function of the magnetic field as long as $SB/\Phi_0 \ll 1$, where S is an effective area of the junction. From Eq. (8), it then follows that the KT transition is pushed to lower temperatures by increasing f=p. The analysis of the frequency dependence of the response for f = 1 shows, as demonstrated by the data of Fig. 5, that the vortex-unbinding transition at T_ω shifts to higher temperatures with increasing ω as predicted by Eq.(21). From the linear fit of Fig. 5, which was obtained using $i_0(f{=}1) = 0.26$ A in Eq. (20), one deduces b (f=1) = 0.95 and $i_c(T_c,f{=}1)/T_c(f{=}1) = 143$ nA/K, a value leading to $\varepsilon_c(f{=}1) = 5.3$ [Eq.(8)]. Because of additional dielectric screening provided by the commensurate background of field-induced

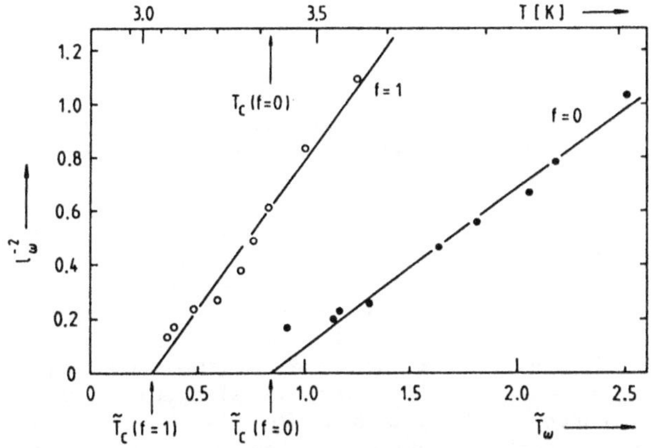

Fig. 5 Dependence of the scale parameter $\ell_\omega \equiv \ln(r_\omega/a)$ on the dimensionless vortex-unbinding temperature $\tilde{T}_\omega$. Solid lines are fits according to Eq.(21). The upper axis is a real temperature scale for f=0. From Ref.13

vortices, renormalization effects should be more important for f = 1 than for f = 0, a conjecture consistent with the experimental observation $\varepsilon_C(f=1) > \varepsilon_C(f=0)$. Additional screening alone, however, seems to be insufficient to account for the large value of ε_C found for f = 1.

The magnetic field dependence of the ac response δV_ω has also been investigated [13]. In Fig. 6 we show the real and imaginary parts of δV_ω measured at 4.033 kHz as a function of f for different temperatures. Prominent structures emerge in both Re $[\delta V_\omega(f)]$ and Im$[\delta V_\omega(f)]$ in correspondence to C-vortex phases defined by f = p and f = p/2. Structures at f = p/3 are also clearly resolved in most of the data of Fig. 6. A detailed interpretation of these curves requires a theory describing the dynamics

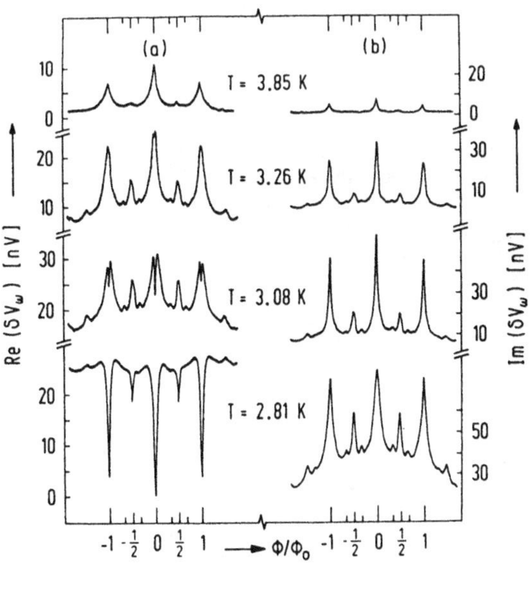

Fig. 6 (a) Real and (b) imaginary parts of the complex ac response at 4.033 kHz of array A1 as a function of the frustration parameter $f = \Phi/\Phi_0$. From Ref. 13

of field-induced vortices and of thermally generated topological excitations (vortices, domain walls, dislocations) in a periodic force field. Such a theory is not available so far. However, some of the essential features can be understood in terms of the "vortex picture" discussed in Sec. 3 and, for f = p, of the dynamics of thermally excited vortices. At low temperatures ($T \ll T_c$), where the influence of topological excitations is negligible, structures in $\delta V_\omega(f)$ at f = p/q reflect the drastic change in pinning occuring at a CI transition. As discussed in Sec. 3, in a low-order (small q) C-phase the mobility of the field-induced vortices is considerably reduced by the periodic pinning potential provided by the array, while the vortex lattice can slide freely in an I-phase. As a consequence, there is a marked reduction of dissipation in a low-order C-phase, a process resulting in the periodic sequence of dips one observes in the $\mathrm{Re}[\delta V_\omega(f)]$ signals of Fig. 6 (a) at low temperatures. On the other hand, since pinning is important in low-order C-phases, considerable lag in response is expected for such vortex configurations. This is the mechanism responsible for the commensurate peaks occurring in the inductive component of $\delta V_\omega(f)$ shown in Fig. 6 (b). As the temperature rises and approaches $T_c(f)$, the low-temperature commensurate dips in $\mathrm{Re}[\delta V_\omega(f)]$ gradually transform into peaks which finally vanish above $T_c(f)$ [Fig. 6 (a)]. Simultaneously, a rapid degradation of the commensurate peaks in $\mathrm{Im}[\delta V_\omega(f)]$ is observed in the critical region [Fig. 6 (b)].

The evolution of the commensurate structures in $\delta V_\omega(f)$ suggests the presence of thermally activated defects which, on account of their high mobility, generate additional dissipation and reduce the lag in response in a C-phase. We have shown that, for f=p, the dynamics of these defects has features allowing their identification as the vortex excitations of the KT theory. For a frustrated array (f=p/q) further work is needed to ascertain the nature of the defects and their possible role in triggering the phase transition.

Acknowledgements

We thank G.-A. Racine for clarifying several aspects of the detection technique. This work was supported by the Swiss National Science Foundation.

References

1. J.M. Kosterlitz and D.J. Thouless : in Progress in Low-Temperature Physics, ed. by D.F. Brewer, Vol VII B (North-Holland, Amsterdam 1978) p. 371
2. R. Peierls : Helv. Phys. Acta 7 (Suppl. II), 81 (1934)
3. L.D. Landau : Zh. Eksp. Teor. Fiz. 7, 627 (1937)
4. N. Mermin and H. Wagner : Phys. Rev. Lett. 17, 1133 (1966)
5. P.C. Hohenberg : Phys. Rev. 158, 383 (1967)
6. J.M. Kosterlitz and D.J. Thouless : J. Phys. C 6, 1181 (1973)
7. D.J. Resnick, J.C. Garland, J.T. Boyd, S. Schoemaker, and R.S. Newrock : Phys. Rev. Lett. 47, 1542 (1981)
8. D.W. Abraham, C.J. Lobb, M. Thinkham, and T.M. Klapwijk : Phys. Rev. B 26, 5268 (1982)
9. M. Tinkham, D.W. Abraham, and C.J. Lobb : Phys. Rev. B 28, 6578 (1983)
10. D. Kimhi, F. Leyvraz, and D. Ariosa : Phys. Rev. B 29, 1487 (1984)
11. R.F. Voss and R.A. Webb : Phys. Rev. B 25, 3446 (1982)
12. R.A. Webb, R.F. Voss, G. Grinstein, and P.M. Horn : Phys. Rev. Lett. 51, 690 (1983)

13. Ch. Leemann, Ph. Lerch, G.A. Racine, and P. Martinoli : Phys. Rev. Lett. 56, 1291 (1986)
14. R.K. Brown and J.C. Garland : Phys. Rev. B 33, 7827 (1986)
15. B. Pannetier, J. Chaussy, and R. Rammal : J. Physique Lett. 44, L-853 (1983)
16. B. Pannetier, J. Chaussy, R. Rammal, and J.C. Villegier : Phys. Rev. Lett. 53, 1845 (1984)
17. S. Teitel and C. Jayaprakash : Phys. Rev. B 27, 598 (1983), and Phys. Rev. Lett. 51, 1999 (1983), and J. Physique Lett. 46, L-33 (1985)
18. W.Y. Shih and D. Stroud : Phys. Rev. B 28, 6575 (1983), and 30, 6774 (1984), and 32, 158 (1985)
19. M.Y. Choi and S. Doniach : Phys. Rev. B 31, 4516 (1985)
20. T.C. Hasley : Phys. Rev. B 31, 5728 (1985), and J. Phys. C 18, 2437, (1985), and Phys. Rev. Lett. 55, 1018 (1985)
21. D.H. Lee, J.D. Joannopoulos, J.W. Negele, and D.P. Landau : Phys. Rev. Lett 52, 433 (1984), and Phys. Rev. B 33, 450 (1986)
22. V.S. Dotsenko and G.V. Uimin : J. Phys. C 18, 5019 (1985)
23. M. Yosefin and E. Domany : Phys. Rev. B 32, 1778 (1985)
24. E. Granato and J.M. Kosterlitz : Phys. Rev. B 33, 4767 (1986)
25. S. Miyashita and H. Shiba : J. Phys. Soc. Jpn. 53, 1145 (1984)
26. S. Alexander : Phys. Rev. B 27, 1541 (1983), and J. Physique 44, 805 (1983)
27. R. Rammal, T.C. Lubensky, and G. Toulouse : Phys. Rev. B 27, 2820 (1983)
28. R. Young : Phys. Rev. B 31, 4294 (1985)
29. O. Daldini, P. Martinoli, J.L. Olsen, and G. Berner : Phys. Rev. Lett. 32, 218 (1974)
30. A.T. Fiory, A.F. Hebard, and S. Somekh : Appl. Phys. Lett. 32, 73 (1978)
31. A.F. Hebard and A.T. Fiory : Phys. Rev. Lett. 44, 291 (1980), and Physica (Amsterdam) 109 & 110 B+C, 1637 (1982)
32. D.J. Bishop and J.D. Reppy : Phys. Rev. Lett. 40, 1727 (1978), and Phys. Rev. B 22, 5171 (1980)
33. V. Ambegaokar, B.I. Halperin, D.R. Nelson, and E. D. Siggia : Phys. Rev. Lett. 40, 783 (1978), and Phys. Rev. B 21, 1806 (1980)
34. S.R. Shenoy : J. Phys. C 18, 5163 (1985)
35. C.J. Lobb, D.W. Abraham, and M. Tinkham : Phys. Rev. B 27, 150 (1983)
36. J.E. Mooij : In Percolation, Localization, and Superconductivity, ed. by A.M. Goldman and S.A. Wolf, NATO ASI Series B, Phys. Vol. 109 (Plenum, New-York 1984) p. 325
37. J.M. Kosterlitz : J. Phys. C 7, 1046 (1974)
38. J. Tobochnik and G.V. Chester: Phys. Rev. B 20, 3761 (1979)
39. J.C. Lobb : Physica (Amsterdam) 126B, 319 (1984)
40. J. Villain : J. Phys. C 10, 1717 and 4793 (1977)
41. A.T. Fiory and A.F. Hebard : in Inhomogeneous Supreconductors, ed. by D.U. Gubser, T.L. Francavilla, S.A. Wolf, and J.R. Leibowitz, AIP Conference Proceedings No 58 (American Institute of Physics, New York 1980) p. 293
42. Ch. Leemann, Ph. Lerch, G.A. Racine, A. Strupler, and P. Martinoli : in SQUID 85, ed. by H.D. Hahlbohm and H. Lübbig (Walter de Gruyter, Berlin 1985) p. 1065
43. P.G. de Gennes : Rev. Mod. Phys. 36, 225 (1964)

Interfacial Pattern Formation: A Progress Report

H. Levine

Schlumberger-Doll Research, Old Quarry Road, Ridgefield, CT 06877, USA

Interfacial patterns such as those which arise during dendritic solidification, viscous fingering and electrochemical deposition have long been of scientific and technological interest. Until recently, however, even the simplest steady-state structures have eluded theoretical treatment. In this talk, I review the progress made over the past several years towards remedying this situation. Specifically, I explain how the paradigm of *microscopic solvability* can be used to explain unique steady-state patterns and why secondary instabilities such as tip-splitting or sidebranching are inevitable. I focus on the Saffman-Taylor finger in a channel to illustrate the general framework.

1. Introduction

Nature abounds in patterns which are generated by the motion of an interface between differing phases. Examples range from the snowflake and the distribution of oil in an underground reservoir all the way to the shape of unicellular algae. Understanding these patterns remains a major scientific challenge as well as an ingredient for technological progress in a wide variety of fields.

It is usually fairly straightforward to determine *why* any particular system gives rise to patterns. For instance, in solidification there is the well-known Mullins-Sekerka instability which prevents simple crystal shapes from continuing to grow stably. This instability arises because the rate of crystal growth is limited by the ability of the surrounding fluid to transport energy and/or impurities away from the interface. Sharply curved outward protrusions are more effective at initiating diffusive transport and hence grow more quickly. Similar instabilities lie at the core of all systems of interfacial pattern formation.

The hard problem, then, is not understanding why patterns form. Rather, we must focus on the question of *what* becomes of the initial instability far beyond the validity of any linear approximation. We are then faced with an at first bewildering array of possible patterns depending on the material being considered, boundary conditions, initial conditions, etc. In order to make progress, we must focus our attention on simple systems and on the understanding of experimentally reproducible features of the pattern. Remarkably, this approach has been an extremely successful way to proceed, and our understanding of patterns has increased quite rapidly over the past several years.

This talk will concentrate on one such pattern forming system, the motion of an air-water interface in a Hele-Shaw cell. Many of the characteristic features of diffusion-limited interface evolution occur in this simple example. We will trace the behavior of the

(in collaboration with D. Kessler and J. Koplik)

interface from its initial instability, through the steady-state *finger* regime, to the occurrence of secondary instabilities. The most important idea that will be presented is that of *microscopic solvability*, the idea that the microscopic cutoff (here embodied by surface tension) plays the crucial role in determining the macroscopic pattern. This approach will enable us to calculate the unique steady-state behavior and understand its inevitable breakdown into a more disordered structure. At the end, we will briefly review the progress to date in applying microscopic solvability to other problems of pattern formation.

2. The Hele–Shaw Cell

The problem of multiphase flow in porous media is one of obvious technological importance for the oil industry. Unfortunately, it is rather difficult to do controlled experiments on the scale of a geological reservoir. To circumvent this problem, one can use a simple laboratory device, invented by HELE-SHAW [1], to obtain a flow which approximately mimics the flow which arises in a porous media. Specifically, a Hele-Shaw cell is a pair of parallel plates (usually glass) which force the flow to occur in the narrow gap between them. A typical cell [2] might have width 5 cm with a gap of .1 cm. The problem of multiphase flow in such a cell has been with us ever since 1898, coupled to the additional issue of the relationship between flow in the device and real flow in the reservoir.

The usual treatment [3] of fluid flow in this device starts by recognizing that the no-slip condition along the top and bottom plates force an approximately parabolic profile in the gap direction. Specifically, we assume that

$$\vec{v} = \frac{3}{2}\left(1 - \frac{4z^2}{b^2}\right)\vec{v}_\perp(x,y) \tag{1}$$

where $\vec{v}$ vanishes at $z = \pm b/2$ for gap thickness b and the mean velocity $v_\perp$ only has components in the x, y plane. Under the conditions of creeping flow, the fluid velocity obeys the Stokes law

$$\mu \nabla^2 \vec{v} = \vec{\nabla} p \tag{2}$$

Substituting (1) into (2) and keeping only the derivative of the rapid variation with z, we arrive at the flow equation (Darcy's law)

$$\vec{v} = \frac{-b^2}{12\mu} \vec{\nabla} p \tag{3}$$

The fact that the velocity is proportional to pressure gradient is the reason why this device is useful for modelling flow in porous media [4]. Actually, this derivation is only approximate and better treatments exist in the literature [5]. We will comment later on which details of the pattern might require a more sophisticated approach.

We now turn to the problem of multiphase flow. We assume that we are displacing the fluid in the cell by air (assumed to have zero viscosity). In the fluid, incompressibility yields the macroscopic equations

$$\nabla^2 p = 0 \quad \text{(inside the fluid)} \tag{4a}$$

$$-\hat{n} \cdot \vec{\nabla} p = \frac{12\mu}{b^2} v_n \tag{4b}$$

v_n is the normal velocity of the air-fluid boundary. Finally, we must specify the pressure on the boundary. In accordance with simple force balance, $p = p_{air} - \gamma \varkappa$, where $\varkappa$ is interface curvature and γ is the surface tension. We can define dimensionless variables by using the half-channel width a as the unit of length and the asymptotic flow rate v_∞ as a unit of velocity. The final set of equations is then

$$\nabla^2 \phi = 0$$

$$-\hat{n} \cdot \nabla \phi \mid_{int} = v_n$$

$$\phi(x_{int}) = -\tilde{\gamma} \varkappa_\perp \tag{5}$$

for in-plane curvature $\varkappa_\perp$ and $\tilde{\gamma} = \frac{\gamma}{12\mu v_\infty}(\frac{b}{a})^2$. The boundary conditions are $\frac{\partial \phi}{\partial y} \mid_{y=\pm 1} = 0$ and $\phi \sim -x$ at large distances downstream.

Why does this system generate patterns? Consider a small deviation away from a planar interface,

$$x(y) = x_0 + \delta_k e^{iky}$$

$$\phi(x,y) \sim -x + \phi_k e^{iky-kx}$$

The pressure boundary condition requires $-\delta_k + \phi_k = -\gamma k^2 \delta_k$ or equivalently

$$\phi_k = \delta_k(1 - \tilde{\gamma} k^2)$$

and the velocity condition yields

$$\dot{\delta}_k = k\phi_k = k\delta_k(1 - \tilde{\gamma} k^2) \tag{6}$$

For small wavevector k, perturbations grow in time. At much larger k, surface tension restabilizes the interface. This dispersion is completely analogous to the one found for a crystal-fluid interface as discussed by MULLINS and SEKERKA [6].

So, a simple planar interface cannot occur. The pattern that does occur must represent a compromise between the destabilizing macroscopic instability and the stabilizing effect due to surface tension. We now turn to a discussion of the simplest possible structure, a steady-state propagating finger.

3. The Saffman–Taylor Finger

For values of $\tilde{\gamma}$ that are not too small ($\cong 10^{-3}$), a single finger pattern is observed. This finger asymptotes rapidly from its tip to a region of constant width λ. By fluid conservation, the finger moves with velocity $1/\lambda$. λ is a *unique* function of γ, independent of the initial conditions and of the individual components that make up γ. How can we understand this behavior?

The first attempt to predict the steady-state pattern was the work of SAFFMAN and TAYLOR [7]. They recognized that $\tilde{\gamma}$ was relatively small (10^{-2} – 10^{-3}) and guessed that to first approximation it could be safely set to zero. Then, the interface becomes an isobar and the steady-state problem can be solved exactly via conformal mapping. They found a solution

$$x(y) = \frac{2(1-\lambda)}{\pi} \log \cos \frac{\pi y}{2\lambda}$$

for arbitrary λ! That is, steady-state fingers existed at all λ and no selection mechanism could be discerned. The problem was compounded by the work of McLEAN and SAFFMAN [8] which showed that perturbative corrections due to finite $\tilde{\gamma}$ did not resolve the selection dilemma. Finally, it was shown by AREF and TRYGVASSON [9] that selection of a unique width did indeed occur in direct numerical simulations of the equations of motion, and therefore could not be attributed to effects not taken into account in the theoretical model.

Faced with a lack of analytic understanding, McLEAN and SAFFMAN [8] and later VANDEN-BROECK [10] turned to a numerical solution to the steady-state problem. Let us describe a variation of Vanden-Broeck's approach derived by us in a recent paper [11]. Equation (5) for the interface can be rewritten as an integro-differential equation by eliminating the field ϕ. This results in the steady-state equation

$$\frac{1}{\lambda} \int_{-\lambda}^{\lambda} G(y,x(y);\, y',x(y'))\, dy' = \int_{-\lambda}^{\lambda} \hat{n}' \cdot \vec{\nabla}' G(y,x(y);\, y',x(y'))\tilde{\gamma}\varkappa_{\perp}(y')ds' \tag{7}$$

with the Green's function relevant for the channel

$$G(x,y;\, x',y') = -\frac{1}{2\pi}\log(1+e^{-2\pi\,|y-y'|} - 2\cos\pi(x-x')e^{-\pi\,|y-y'|}) - \frac{1}{2}\,|x-x'|\,\theta(x-x')$$

and ds' is the unit of arclength. It is easy to show that solutions to (7) must approach λ exponentially; $y(x) \sim \lambda + 0(e^{+\alpha x})$ as $x \to -\infty$ for some α. We then discretize these equations at equidistant points y_i and find the corresponding $x(y_i)$ via Newton's algorithm. At $\tilde{\gamma} = 0$, this can be done for all λ yielding the Saffman-Taylor solution described above.

At finite $\tilde{\gamma}$, smooth solutions do not exist at arbitrary λ. Following Vanden-Broeck, we look for solutions with a cusp at the tip and plot the cusp magnitude vs λ at fixed $\tilde{\gamma}$. A

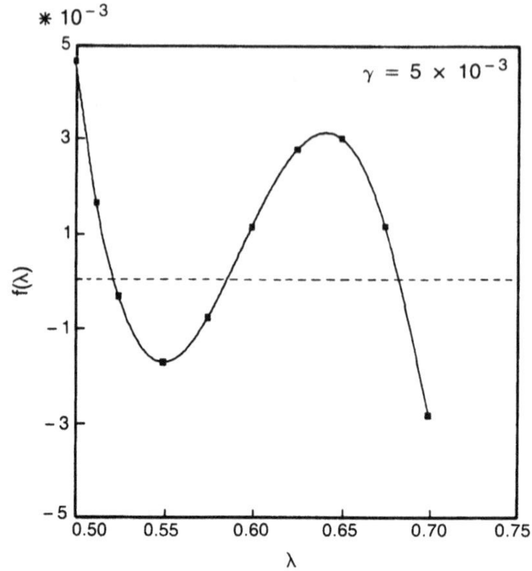

Fig. 1. Cusp magnitude f vs λ at $\tilde{\gamma} = .005$

typical plot is shown in Fig. 1. Physical solutions correspond to vanishing cusp magnitudes. Clearly, this occurs only for a discrete set of λ values. Finite $\tilde{\gamma}$ has indeed given rise to a selection of possible shapes.

We can better understand the effect of finite $\tilde{\gamma}$, by plotting the cusp magnitude vs $\tilde{\gamma}$ at $\lambda = ½$. It was shown in [8,10] that ½ is the asymptotic width approached from above as $\tilde{\gamma} \to 0$. This means that at any finite $\tilde{\gamma}$ there would be a non-zero cusp magnitude which would go to zero as $\tilde{\gamma} \to 0$. The graph in Fig. 2 shows that this dependence of cusp magnitude is $f \sim \exp - c/\sqrt{\tilde{\gamma}}$. This result, first demonstrated in [12], explains the

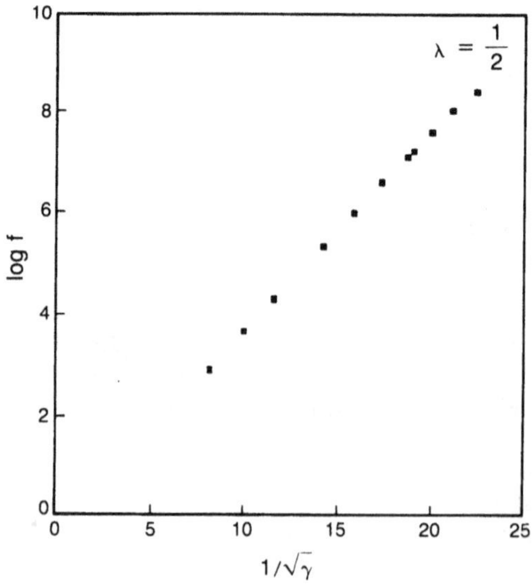

Fig. 2. Cusp magnitude vs $\tilde{\gamma}$

failure of standard asymptotic methods since this behavior is beyond all orders in an asymptotic series in $\tilde{\gamma}$.

We refer to the requirement that the cusp vanish at finite $\tilde{\gamma}$ as the condition of *microscopic solvability*. The problem at $\tilde{\gamma} = 0$ incorporates only *macroscopic* physics with no microscopic cutoff. As we derived above, the spectrum of oscillations in this limit is unbounded above. This suggests that the problem is ill-defined in much the same way that quantum field theories make no sense without some cutoff on small scale fluctuations. In fact, it has been shown [13] that an initially smooth interface will invariably generate singularities at finite time. Therefore, finite $\tilde{\gamma}$ no matter how small, is an essential ingredient of the evolution process. Qualitatively, we may think of the restriction to the discrete set as the requirement that the pattern be consistent not only with the macroscopic conservation law but also with the microscopic dynamics underlying the actual interface motion.

The fact that selection takes place via essentially singular terms suggests that some type of WKB approach might be fruitful. This idea has been applied by several groups [14] who have derived that all allowed widths scale as $\lambda - ½ \sim \tilde{\gamma}^{2/3}$. Recently [15], a more exact asymptotic method based on the work of KRUSKAL and SEGUR [16] has verified the above relation. This scaling is in good agreement with the numerical calculations. These calculations verify the essential point that surface tension controls the possible steady-state structures via non-perturbative effects missed in all earlier treatments.

4. Linear Stability and Unique Structure

Up to now, we have seen how a discrete set of possible widths are permitted by the finite $\tilde{\gamma}$ steady-state equation. Experimentally, only the narrowest branch is observed. In this section we would like to explain this phenomena and compare the theory to experiment.

The basic idea we need to employ is that steady-state solutions can exist in nature only if they are linearly stable. In fact, the lack of linear stability for a planar interface was the reason for patterns in the first place. The only results known analytically are that at $\gamma = 0$, all fingers are unstable. This result is neither surprising nor useful.

Recently, two different numerical stability calculations have been performed, one by us [12,17] and one by BENSIMON [18]. Both of these calculations are approximate in the sense that the analysis is carried out around the analytically determined Saffman-Taylor solution (plus possibly perturbative corrections) rather than around the exact steady-state solution which is known only numerically. Because of this, the stability operator exists not only at the selected set of widths, but in fact, at all λ. This will turn out to be somewhat useful in what follows.

The details of the analyses are too lengthy to be repeated here, so we just give the results. The lowest width solution at all $\tilde{\gamma}$ is linearly stable. That is, the entire spectrum which consists of a real continuum, a complex continuum and both symmetric and antisymmetric discrete modes all have Re $\omega \leq 0$. At very small $\tilde{\gamma}$, the numerics become increasingly more difficult and hence, strictly speaking, these results have only been established for $\tilde{\gamma} \geq .001$. Nevertheless, there are at least some indications that this will persist to much lower γ. This particular issue of whether a linear instability occurs at some non-zero $\tilde{\gamma}$ is being addressed by a new set of computations in progress.

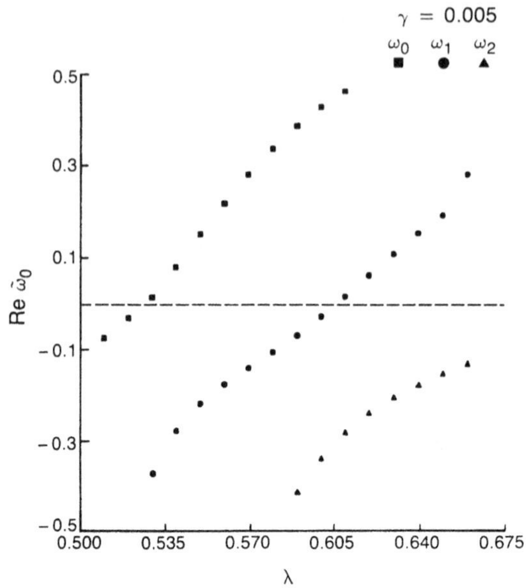

Fig. 3. Lowest three symmetric discrete eigenvalues

In Fig. 3, we have plotted Re ω for several of the lowest discrete modes. At the lowest allowed value of λ, the spectrum consists of one mode at $\omega = 0$ (translation zero mode) and other modes at negative real ω; hence the finger of this width is stable. As we increase λ and approach the next allowed width, one mode goes *unstable* and, further allowed values of λ are even more unstable. This analysis (albeit with the caveat that it must at some point be checked by an exact stability analysis) shows that there is only *unique* steady-state structure permitted.

Before comparing the theoretical results to experiments, it is worthwhile to try to understand why small surface tension can nonetheless ensure linearly stability. The basic idea is that perturbations near the tip are coupled to perturbations near the flat tail region. This latter region is stable because the sidewall suppresses the macroscopic *diffusive* instability. This can be seen directly by computing the dispersion relation for waves in this region in the lab frame,

$$\omega = -\tilde{\gamma} k^3 \tan k(1-\lambda) \tag{8}$$

The result of the stability calculations can be interpreted as the fact that would-be instabilities near the tip are stabilized by being coupled to the tail region. From a dynamic point of view, a bump which starts at the tip will be advected far from the tip and will eventually decay in amplitude. This way of looking at the stability was discussed by deGREGORIA and SCHWARTZ [19] and by the Chicago Group [20] following similar ideas by ZELDOVICH [21].

In Fig. 4, we show a recent comparison [2] between theory and experiment at intermediate values of $\tilde{\gamma}$. Actually, $\tilde{\gamma}$ has been corrected for the fact that the water forms a wetting layer on the plates [22] which effectively reduces the gap distance b. After this

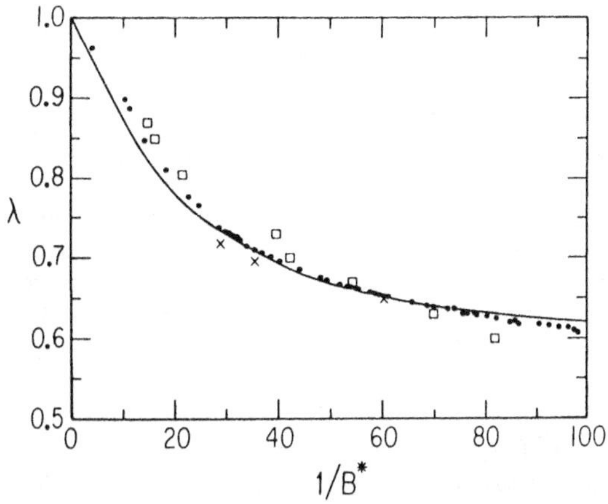

Fig. 4. Comparison between experiments and theory (solid line) from ref. 2. ($B^* = \tilde{\gamma}/4$)

correction, the steady-state shape predicted by the solvability condition coupled with the stability analysis agrees well with the experimental results.

At smaller $\tilde{\gamma}$, the experimental results diverge from the limit of the theoretical results and the width goes below ½, the limit of the theoretical curve at $\tilde{\gamma} = 0$. However, it is quite clear experimentally that this discrepency reflects not a problem with the theoretical method but the breakdown of the simple two-dimensional model. And, there is evidence that very small changes in the equation can alter quantitatively the selected λ. For example, the imposition of a finite flow anisotropy [11] allows steady-state solution widths all the way down to zero. We expect that more sophisticated models which include three-dimensional effects will be able to explain the pattern selection without any changes in the principles elucidated up to now.

5. Non−Linear Instabilities

At small $\tilde{\gamma}$, instabilities begin to be seen which interfere with steady-state finger propagation. In the channel geometry, work by TABELING and LIBCHABER [22] and by PARK and HOMSY [23] have demonstrated this phenomena. (In radial Hele-Shaw cells, many groups following PATTERSON [24,25] have shown the breakdown of fingers at all $\tilde{\gamma}$ as they grow radially outward.) Given the assumption that further investigation will continue to predict linear stability for all $\tilde{\gamma}$, there must be a different explanation for what has been observed.

Actually, there is a simple physical reason why we might expect there to be a finite amplitude instability. Recall that the steady-state solution was determined by a solvability *signal* proportional to $\exp -c/\sqrt{\tilde{\gamma}}$. If there is noise in the system due either to experimental imperfections, thermal fluctuations or applied external perturbation which is greater than the signal, we might expect the system to be unable to sustain steady-state motion. This idea can be studied more quantitatively by performing a stability analysis around a finger modified by *random* noise. A calculation of this sort by Bensimon demonstrated the above exponential form of the noise threshhold. Experimental results as to the value of $\tilde{\gamma}$ at onset are qualitatively in agreement with this prediction.

The idea of an inevitable noise-induced destabilization of steady-state solution is an attractive feature of the solvability concept. In real flow in porous media [26], and in other noisy systems such as flow in non-Newtonian fluids [27], a disordered fractal-like interface is observed. These systems have been modeled by the DLA (diffusion-limited aggregation) algorithm of WITTEN and SANDER [28]. It has been shown [29] that DLA is sort of a noisy version of the Saffman-Taylor problem with the noise due to the finite arrival of particles. The fact that noise can eventually induce a fractal growth regime by causing many competing instabilities may eventually lead to a better understanding of disordered structure.

Let us return to the breakdown of the finger solution. Experimentally, the first instability that occurs is an antisymmetric *thumbing* mode; later on a symmetric tip-splitting mode also appears. What causes this particular sequence of modes? The answer awaits a better method for treating the finite amplitude instabilities, but one possible clue arises from considering the lowest lying modes of the linear stability operator. Specifically, we might expect that the modes closest to Re $\omega = 0$ are the first to be excited by the finite noise. This approach [12] does indeed predict that the first excitation should be antisymmetric.

Finally, changing various aspects of the growth problem can not only change the steady-state structure, but also the nature of the secondary instabilities which occur. The most striking occurrence of this to date is the transition from tip-splitting to dendritic sidebranching which occurs when we mimic the effects of crystal anisotropy by using grooved plates [30]. An example of the resulting pattern is shown in Fig. 5. This type of experiment goes a long way toward showing that all of the different manifestations of diffusion-limited growth that have been observed in *differing* physical systems can arise from the same fundamental theory in different regimes of parameter space.

Fig. 5. Dendritic sidebranches in a grooved cell

6. Lessons and Directions

We have outlined the approach of *microscopic solvability* to the Saffman-Taylor problem. This approach posits that a unique steady-state structure arises during interface growth via the competition between microscopic stability and macroscopic instability. Using this idea, we have been able to explain finger width selection, the onset of secondary instabilities and the relationship of this problem to other examples of interfacial pattern formation.

To date, this set of ideas has been applied to varying degrees in other classic pattern forming systems. The case of the free space dendrite [31] closely parallels the Saffman-Taylor finger. Specifically, the introduction of anisotropic surface tension gives rise via a solvability condition to a discrete set of possible velocities [32]. Presumably, stability considerations then choose the fastest of the set. The WKB approach [33] can be used to derive the scaling $\frac{vd_0}{D} \sim \epsilon^{7/4}$ with d_0 the capillary length, D the thermal diffusion constant. Finally, there is some evidence using the Zeldovich approach that sidebranching can be understood via a finite amplitude instability [34]. Other solidification problems for which pattern selection has been verified include solidification in a channel [11] and directional solidification [35].

Other pattern-forming systems currently under investigation include electrochemical deposition [36] and the Taylor bubble [37]. The transitions between different growth regimes during electrochemical deposition are at least in qualitative agreement with the ideas presented here; a more quantitative comparison has not yet been performed. For the Taylor bubble, it has been shown that finite surface tension selects a discrete set from the continuum, although it has not yet been verified that this occurs via the essentially singular terms characteristic of microscopic *solvability*. In the future, these methods will be applied to flame propagation in channels, to flow through real porous media and, possibly, to examples from cellular biology.

There are several lessons that can be drawn from the history of research on interfacial pattern formation. First, one should never trust an asymptotic expansion for a highly non-linear problem. For many years, progress in this field was stifled by the mistaken assumption that what was true as an asymptotic series in the microscopic parameters was actually true for the full solution of the equations. The second thing this work has demonstrated is that not all non-linear systems evolve chaotically. Over the past several years, spurred by the ideas of Feigenbaum, the study of non-linear dynamics has focussed primarily on the transition to chaos. It is certainly true that most dynamical systems will eventually lead to chaos, represented here by fractal, disorderly growth. Interfacial pattern forming systems clearly show, however, that highly complex non-linear equations can also exhibit non-trivial *ordered* structures which are stably reproduced upon variations in initial conditions. This type of dynamical behavior might be crucial for predicting biological structures. Finally, we have learned that microscopic information such as crystal structure, finite particle size and thermal fluctuations can play an important role in controlling *macroscopic* patterns on a much larger length scale. In this sense, *microscopic solvability* is a sort of information amplifier for at least certain aspects of the microscopic dynamics of interface growth. The occurrence of this type of behavior was not foreseen by many researchers who tacitly assumed that macroscopic conservation laws were all that was necessary to understand macroscopic systems.

References

1. H.J.S. Hele-Shaw: Nature (London) 58, 34 (1898)
2. P. Tabeling, G. Zocchi, A. Libchaber: "An Experimental Study of the Saffman-Taylor Instability", Chicago preprint (1986)
3. P.G. Saffman, G.I. Taylor: Proc. Roy. Soc. A245, 312 (1958)
4. See for example A. Scheidigger: Physics of Flow Through Porous Media, (University of Toronto Press, Toronto, 1974)
5. C.W. Park, G.M. Homsy: J. Fluid. Mech. 139, 291 (1984)
6. W.W. Mullins, R.F. Sekerka: J. Appl. Phys. 34, 323 (1968), ibid. 35, 444 (1964)
7. P.G. Saffman, G.I. Taylor: ref. 3. For later developments see E. Pitts: J. Fluid Mech. 97, 53 (1980)
8. J.W. McLean, P.G. Saffman: J. Fluid Mech. 102, 455 (1981)
9. G. Trygvasson, H. Aref: J. Fluid. Mech. 136, 1 (1983)
10. J.M. Vanden-Broeck: Phys. Fluids 26, 2033 (1983)
11. D. Kessler, J. Koplik, H. Levine: "Dendritic Growth in a Channel", Schlumberger preprint (1986)
12. D. Kessler, H. Levine: Phys. Rev. A33, 2621, 2634 (1986)
13. B.I. Shraiman, D. Bensimon: Phys. Rev. A 30, 2840 (1984), S.D. Howison: to appear
14. B.I. Shraiman: Phys. Rev. Lett. 56, 2028 (1986); T. Pombre, V. Hakim, Y. Pomeau: C. R. Acad. Sci. 302, 803 (1986); D.C. Hong, J.S. Langer: Phys. Rev. Lett. 56, 2032 (1986)
15. R. Combescot, T. Dombre, V. Hakim, Y. Pomeau, A. Pumir: Phys. Rev. Lett. 56, 2036 (1986)
16. M. Kruskal, H. Segur: "Asymptotics Beyond All Orders in a Model of Dendritic Crystal Growth", A.R.A.P. Tech. Memo 85-25; For a different approach, see R. Dashen, D. Kessler, H. Levine, R. Savit: "The Geometric Model of Dendritic Growth; the Small Velocity Limit", Physica D to appear
17. D. Kessler, H. Levine: Phys. Rev. A32, 1930 (1985)
18. D. Bensimon: Phys. Rev. A33, 1302 (1986)
19. A.J. DeGregoria, L.W. Schwartz: "A Boundary Integral Method for Two-Phase Displacement in Hele-shaw Cells", Exxon preprints (1985)
20. D. Bensimon, L. Kadanoff, S. Liang, B. Shraiman, C. Tang: "Viscous Flows in Two Dimensions", Chicago Preprint 1985
21. Ya.B. Zeldovich, A.G. Istratov, N.I. Kidin, V.B. Librovitch: Comb. Sci. and Tech. 24, 1 (1980)
22. See ref. 2 and references therein
23. C.W. Park and G.M. Homsy, Phys. Fluids 28, 1583 (1985)

24. L. Patterson: J. Fluid Mech. 113, 513 (1981)

25. D. Kessler, J. Koplik, H. Levine: Phys. Rev. A30, 2820 (1984); also see L. Sander, P. Ramanlal, E. Ben-Jacob: Phys. Rev. A32, 3160 (1985)

26. J.D. Chen, D. Wilkinson: Phys. Rev. Lett. 55, 1982 (1985); K.J. Maloy, J. Feder, J. Jossang: Phys. Rev. Lett. 55, 2681 (1985)

27. J. Nittman, G. Daccord, H.E. Stanley: Nature 314, 141 (1985)

28. T.A. Witten, L.M. Sander: Phys. Rev. Lett. 47, 1400 (1981)

29. L. Patterson: Phys. Rev. Lett. 52, 1621 (1984); C. Tang: Phys. Rev. A31, 1977 (1985); L. Kadanoff: J. Stat. Phys. 39, 267 (1985)

30. E. Ben-Jacob, R. Godbey, N. Goldenfeld, J. Koplik, H. Levine, T. Mueller, L. Sander: Phys. Rev. Lett. 55, 1315 (1985)

31. For a recent review, see D. Kessler, J. Koplik, H. Levine: "Pattern Formation Far From Equilibrium - The Free Space Dendrite", NATO A.R.W. on "Patterns, Defects and Microstructures", to appear

32. D. Kessler, H. Levine: Phys. Rev. B, to appear; D. Meiron: Phys. Rev. A33, 2704 (1986); D. Kessler, J. Koplik, H. Levine: Phys. Rev. A33, 3352 (1986)

33. See ref. 31 and also, A. Barbieri, D. Hong, J. Langer: "Velocity Selection in the Symmetric Model of Dendritic Crystal Growth", ITP 86-65 preprint; M. Ben-Amar, Y. Pomeau: "Theory of Dendritic Growth in a Weakly Undercooled Melt", Saclay preprint (1986)

34. P. Pelce: "Dynamique des Fronts Courbes", Thesis (May 1986)

35. A. Karma: Caltech preprints (1986)

36. Y. Sawada, A. Dougherty, J. Gollub: Phys. Rev. Lett. 56, 1260 (1986); D. Grier, E. Ben-Jacob, R. Clarke, L. Sander: Phys. Rev. Lett. 56, 1264 (1986)

37. J.M. Vanden-Broeck: Phys. Fluids 27, 2604 (1984)

Intrinsic Defect States in Protein Assemblages[1]

C.J. Ritz-Gold[2]

Department of Chemistry, San Francisco State University,
San Francisco, CA 94132, USA

1. Introduction

Certain protein assemblies are known in which the bonding specificity is not conserved. This is thought to arise from a <u>localized switching</u> of the subunit conformation or intersubunit bonds into an inequivalent state. Such assemblies also commonly possess alternative <u>polymorphic states</u> - structural forms representing different spatial arrangements of subunit conformational or bonding states [1,2]. In the supercoiled helical flagellum, a strand of subunits is switched into an inequivalent "long" conformational state. The "normal" and "curly" polymorphic states of the flagellum are associated with different numbers of these switched-conformation strands [2].

Evidence also exists for localized switching and polymorphic-like states in organelle-level structures. Under certain conditions, segmental regions of myofibrils or myofibril bundles appeared to exist in an "active" oscillatory state. Such regions were thought to have both short (contracted) and long (relaxed) structural forms [3]. It is evident that these alternative forms of the region may appear as polymorphic-like states of the myofibril. In addition, the dynamic behavior of spin-labeled rigor-state fibers is thought to be consistent with the switching of localized regions into a state in which crossbridges exist in a detached "melted" form [4]. Fig. 1 is representative of the observed changes in the distribution of crossbridges between mobile and immobile states as a function of time.

We use the term <u>assemblage</u> to denote both assembly and organelle types of structure. Since the origin of localized switching and polymorphism in these structures is not entirely clear, we seek a better understanding by applying

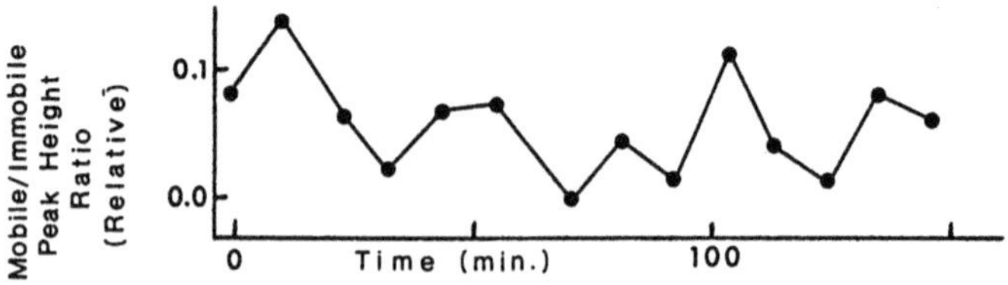

<u>Fig. 1</u> Changes in the distribution of crossbridges between mobile and immobile states as a function of time. EPR spectra were taken on rigor-state psoas fibers labeled with a maleimide spin probe rigidly and specifically attached to myosin "crossbridge" domains

---

[1] Supported by NSF RII-8310749

[2] Present address: Biomaterial Sciences, 38451 Timpanogas, Fremont CA, USA

certain concepts from condensed matter physics. In this preliminary report, we give a brief sketch of these concepts and focus on their application to the helical flagellum and the rigor-state myofibril.

2. Protein Assemblages as Condensed Matter Systems

The approach taken here rests on a view of protein assemblages as condensed matter systems. Here we take this view of two other properties known to characterize assemblages in which localized switching and polymorphism arise.

2.1 Organized Protein Structures as Extended Systems

Protein assemblages exist as organized supramolecular structures made up of a large number of fundamental structural units. The helical flagellum is a tubular array of protofilament-like strands; each strand in turn contains many flagellin subunits [2]. The myofibril is a linear array of half-sarcomere structural units; each half-sarcomere in turn contains many sliding filament actomyosin assemblies [5]. We propose that organized protein structures like these may be viewed as polymeric materials with certain physical properties in common with extended condensed matter systems. This view is similar to that taken in recent theoretical studies of structural transitions in biopolymers like double-helical DNA [6,7].

2.2 Quasiequivalence as Structural Modulation

In certain protein assemblies, geometric constraints intrinsic to the structture as a whole lead to a <u>built-in</u> departure from regular packing order. (Chemically identical) subunits or bonds are found to suffer varying degrees of distortion. This geometry-induced distortion is referred to as <u>quasi-</u> equivalence [2,8]. In the flagellum, a certain misfitting must be overcome to close a 2D subunit sheet into a tube. Closure give rise to a helical surface lattice in which the different subunit strands are found to suffer varying degrees of distortion [2]. Geometric constraints in the actomyosin assembly (the "Vernier Effect" [5]) lead to a similar type of intrinsic distortion in the rigor-state myofibril. The regular array of rigor crossbridges protruding from the myosin filament triad (dM3 = 42.9 nm) will suffer distortion when strongly attached to the periodic array of actin binding sites on the actin filament (dA = 37.0 nm) (Fig. 2).

We propose that quasiequivalence in certain protein assemblages may be considered as a type of structural modulation similar in principle to that found in certain classes of physical systems where the presence of competing

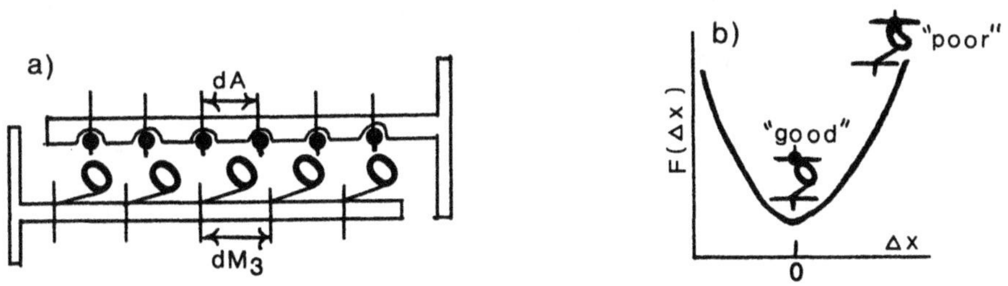

<u>Fig. 2</u> (a) Actomyosin "Vernier Effect"; (b) Rigor complex free energy (F) as a function of crossbridge alignment with an actin binding site (Δx). Alignment is "good" when $\Delta x = 0$ (no distortion) and "poor" when $\Delta x >> 0$ (severe distortion)

periodicities gives rise to incommensurate frustration [9,10]. In the flagellum, a helical surface lattice is present in which subunit strands having different rest lengths are found to coexist. This mixture of length scales appears to represent a type of incommensurate frustration similar to that found in certain Zr-Nb alloys [11] and binary liquid crystal systems [12] where structural elements with conflicting lattice constants coexist. In the myofibril, the incommensurate periodicities of the interacting lattices within the sliding filament actomyosin assembly appears likewise to represent a type of incommensurate frustration similar to that found in certain strained-layer epitaxial materials [13].

3. Intrinsic Defect States in Condensed Matter Systems

In many types of discrete physical systems subjected to conflicting periodicities, localized cluster-like defect regions ("discommensurations") arise when the degree of conflict is above a certain threshold value. These defect states represent intrinsic (i.e., ground-state) structural inhomogeneities, and their creation corresponds to a commensurate-incommensurate phase transition [7,9-10]. In composite structures such as Zr-Nb alloys above about 17-wt. % Nb, discommensuration defects arise in the mass density and take the form of stacking faults [11]. In strained-layer epitaxial systems, similar incommensurate ("conventional" [13]) phase defects are expected to arise at the interface for lattice discrepancies of sufficient magnitude.

When the degree of conflict (length scale discrepancy) is far enough above threshold, a regularly-spaced (periodic) array of discommensuration defects will appear. As the degree of conflict is increased further, a series of different periodic arrays may arise ("Devil's staircase") [9]. In composite structures, different periodic arrays of stacking fault defects would arise in the form of polytype-like states; in the binary liquid crystal system, e.g., these states take the form of different ribbon lattices [12].

Close to the commensurate-incommensurate transition point, however, the density of defects will be low. Large domains of the commensurate phase extending over most of the system will be separated by small cluster-like defect regions. These defects will tend to be pinned by the discreteness of the lattice and occur at randomly-spaced intervals (spatial chaos). Many such inhomogeneous defect states (different cluster-like distributions) will be available to the system and will represent alternative near-degenerate metastable "ground state" configurations of the system [7,9].

4. Predicted Defect States in Protein Assemblages

It is proposed that the presence of structural modulation in protein assemblages such as the helical flagellum and rigor-state myofibril will lead to the creation of intrinsic defect states. Defects are predicted to arise as a consequence of incommensurate frustration according to the same principle whereby discommensurations are thought to arise in certain sufficiently frustrated physical systems (above).

In the helical flagellum, the switching of a subunit strand into an inequivalent conformational state has been described as an energy-minimizing maneuver whereby the deformed polymer achieves relief of strain energy [2]. We propose that this maneuver appears to correspond to the creation of an intrinsic defect state similar in principle to the creation of stacking faults in composite structures (above). The mixed-conformation state produced by this maneuver may similarly correspond to an inhomogeneous "ground-state" configuration of the system.

It follows that the "normal" and "curly" polymorphic states of the flagellum may correspond to alternative polytype-like states of the system.

In the rigor-state myofibril, we propose that intrinsic cluster-like defect regions may arise as a consequence of the incommensurate periodicities of the actomyosin assembly. Large commensurate-phase domains would extend over most of the myofibril; in these domains, crossbridges would exist in an immobile rigor complex state (a type of "strained layer" epitaxial phase) These domains would be separated by small incommensurate-phase regions in which crossbridges would exist in a "pseudorelaxed" mobile state (a type of "conventional" epitaxial phase).

It follows that the rigor-state myofibril may possess many alternative distributions of these cluster-like defect regions corresponding to alternative metastable near-degenerate configurations of the system. The observed dynamic behavior of rigor-state muscle fibers (Fig. 1) appears to be consistent with a fluctuational switching between such cluster-like defect states.

In conclusion, protein assemblages such as the helical flagellum and rigor-state myofibril may represent modulated biological structures in which a "built-in" conflict leads to certain novel features. The role of conflict as a physical design principle in biological structures has, however, apparently received only limited attention to date [6,13].

References

1. D.L.D. Caspar: Biophys. J. 37, 5a (1982)
2. F. Oosawa, S. Asakura: Thermodynamics of the Polymerization of Protein (Academic Press, New York 1975)
3. M.S. Mooseker, M. Pratt, D.P. Kiehart, R.E. Stephens: Biophys. J. 17, 173a (1977); A. Fabiatio, F. Fabiato: J. Gen. Physiol. 72, 667 (1987)
4. C.J. Ritz-Gold: Biophys. J. 47, 395a (1985)
5. K. Nishiyama, H. Shimizu: Mathemat. Biosci. 54, 115 (1981)
6. J.A. Krumhansl, D.M. Alexander: In Structure and Dynamics: Nucleic Acids and Proteins, ed. by E. Clementi, R.H. Sarma (Adenine, N.Y. 1983) p. 61
7. A.R. Bishop: In Nonlinear Electrodynamics in Biological Systems, ed. by W.R. Adey, A.F. Lawrence (Plenum, New York 1984) p. 155
8. D.L.D. Caspar: In Principles of Biomolecular Organization, ed. by G.E.W. Wolstenholme, M. O'Connor (Little, Brown & Co., Boston 1965) p. 7
9. P. Bak: Rep. Progr. Phys. 45, 587 (1982)
10. A.R. Bishop: Physica Scripta 20, 409 (1979)
11. B. Horovitz, J.L. Murray, J.A. Krumhansl: Phys. Rev. B 18, 3549 (1978)
12. F. Hardouin, A.M. Levelut, G. Sigaud: J. Chim. Phys. 80, 53 (1983)
13. J.C. Bean: Science 230, 127, (1985)
14. H. Toyoki, S. Yomosa, S. Takeno, S. Homma: Phys. Lett. 97A, 70 (1983)

Is Universality Valid on Hierarchical and Fractal Lattices?

Bambi Hu

Department of Physics, University of Houston-University Park,
Houston, TX 77004, USA

There is a lack of conclusive evidence, theoretical or experimental, to establish the validity of universality in phase transitions on hierarchical and fractal lattices. The critical properties depend on so many details of the lattice that it seems quite unlikely that a complete or finite set of universality criteria exists. It is proposed that a distinction be made between a "strong statement of universality" and a "weak statement of universality."

In recent years, there has been considerable interest in the study of phase transitions on hierarchical [1,2] and fractal [3] lattices. Such lattices are iteratively constructed to be exactly self-similar. They therefore possess scale invariance, but not translational invariance. Due to their inhomogeneity, there is hope that they may serve as a convenient theoretical laboratory for the study of such physical systems as random magnets, polymers, precolation clusters, and interfaces, etc.

Universality is one of the most remarkable and important characteristics of phase transitions. Indeed, it was the fixed-point theory of the renormalization group that provided a precise explication of universality. However, it serves to remember, historically as well as logically, that before the advent of the renormalization group, the validity of universality on Bravais lattices had been checked to a high degree of accuracy. In the case of hierarchical and fractal lattices, universality seems to have been taken for granted [4].

Universality depends on a number of factors, and physical systems can be classified accordingly. On Bravais lattices, the universality criteria are dimensionality and symmetry. On hierarchical and fractal lattices, it is obvious that one would need more than two criteria. Hence, the crucial question is not that of whether a few more criteria are needed, but that of whether a complete or finite set of universality criteria exists. [Of course, if the set is finite but very large, there is also the additional question of how useful these criteria will become in practice.]

Geometric factors such as the order of ramification, fractal dimension, connectivity and lacunarity may play a useful role in the characterization of hierarchical or fractal lattices; however, it is entirely unclear whether they can serve as universality criteria. Logically speaking, even before one talks about universality or universality criteria, one should first ask the question: Is there any conclusive evidence, theoretical or experimental, for universality on hierarchical or fractal lattices? At this point, it seems fair to answer this question in the negative. Moreover, non-rigorous notwithstanding, there are simple reasons to believe that it is unlikely that a complete or finite set of universality criteria exists. There is virtually unlimited freedom in the construction of hierarchical and fractal lattices. Since these lattices are exactly self-similar, any minute change in their construction will persist to all length scales, resulting in different critical exponents. For example [5], consider the following two slightly different hierarchical lattices shown in Fig. 1.

Fig. 1

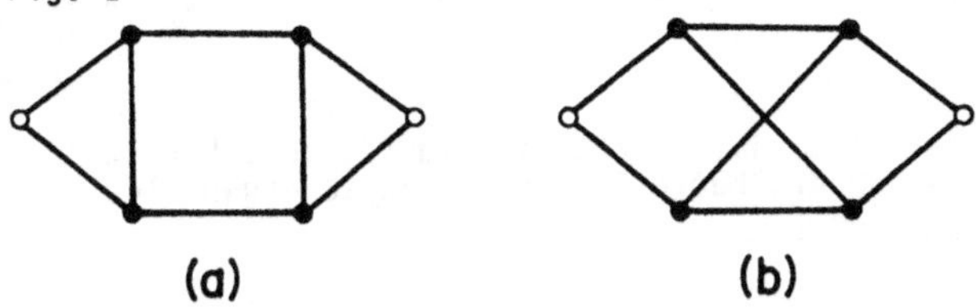

They not only have the same order of ramification, fractal dimension $D = \ln A/\ln b$, and connectivity $Q = \ln C/\ln b$, but also the same aggregation number A, minimum cut C, and the lattice rescaling factor b. However, an exact analytical calculation of the Ising model shows that their thermal eigenvalues λ_t are different [Table 1].

Table 1. Thermal eigenvalues λ_t of the Ising model on the two hierarchical lattices shown in Fig. 1. They have identical A = 8, C = 2, b = 3, and $D = 3(\ln 2/\ln 3)$, $Q = \ln 2/\ln 3$. x^* denotes the fixed point ($x = \exp K$); analytical results are enclosed in curly brackets.

Fig. 1	x^*	λ_t
(a)	3.87513	2.07364
	$\{ [\frac{1}{3}[4 + (73 + 6\sqrt{87})^{1/3} + (73 - 6\sqrt{87})^{1/3}] \}$	
(b)	3.25426	2.15767
	$\{ \frac{1}{4}[3 + \sqrt{17} + (\sqrt{10} + 6\sqrt{17})^{1/2}] \}$	

Of course, one point of view, taking universality for granted, is to say that another universality criteria (e.g., planarity) has to be introduced to distinguish these two lattices. However, a different point of view, not taking universality for granted, is to regard the source of difference as dependence on the detailed structure of the lattice.

There are many peculiar features of phase transitions on hierarchical and fractal lattices. For example, consider these two distinct hierarchical lattices shown in Fig.2.

Fig. 2

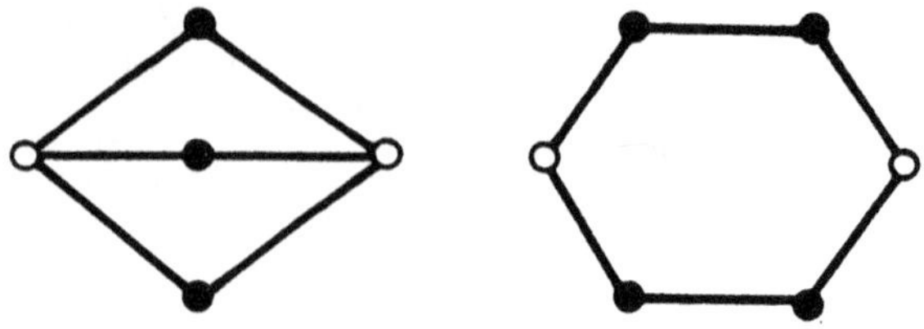

They have the same thermal eigenvalue λ_t but different magnetic eigenvalue λ_h [Table 2].

Table 2. Two hierarchical lattices, illustrated in Fig. 2, that have the same thermal eigenvalue λ_t, but different magnetic eigenvalue λ_h.

Fig. 2	λ_t	λ_h
(a)	1.85410	4.85410
(b)	1.85410	5.78011

Moreover, this example is by no means a singular case. One can, by design, construct reducible hierarchical lattices from their irreducible components such that, although their renormalization-group recursion relations would give the same thermal exponent, their geometry is however different to produce different magnetic exponents. Such examples render an extended universality principle untenable [4].

There are many other peculiar features, such as critical amplitude oscillation [6] on hierarchical lattices, which have no experimental evidence, that should caution people that conventional wisdom of phase transitions should not be taken for granted. One should re-examine all the premises and conclusions of conventional phase transitions before one extrapolates their validity to hierarchical or fractal lattices.

In any attempt to rescue universality on hierarchical or fractal lattices, it is logically more appropriate to first identify the factors which critical properties are independent of. In other words, one should repeat the careful work done on Bravais lattice before universality was established. In the case of hierarchical lattices, one such factor identified is spin. We have studied [4] the spin-1 Blume-Emery-Griffiths model [7] on a family of generalized diamond hierarchical lattices with N rungs shown in Fig. 3.

Fig. 3

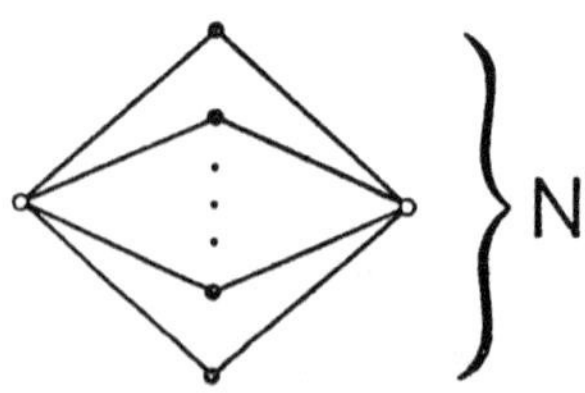

As is evident from Table 3, the thermal eigenvalues at the Ising fixed point are the same as those of the spin-1/2 Ising model.

Table 3. Thermal eigenvalues of the spin-1/2 Ising model and the Spin-1 Blume-Emery-Griffiths model on a family of generalized diamond lattices with N rungs shown in Fig. 3.

N	Spin-1/2	Spin-1
2	1.67857	1.67857
3	1.85410	1.85410
4	1.91737	1.91737
5	1.94695	1.94695
6	1.96310	1.96310
7	1.97286	1.97286
8	1.97921	1.97921
9	1.98357	1.98357
10	1.98668	1.98668

Therefore it seems that the spin-independent aspect of universality is retained for hierarchical lattices. In the case of fractal lattices, since the Midgal-Kadanoff renormalization group is only approximate, one probably has to resort to series-expansions methods. The high accuracy attainable by series expansions will help to settle many of the important questions clouding phase transitions on hierarchical and fractal lattices.

Recently, Hu [4] proposed that a distinction be made between a "weak statement of universality" and a "strong statement of universality." In making a "weak statement of universality," one states what factors critical properties are <u>independent</u> of; whereas in making a "strong statement of universality," one states what factors critical properties are <u>only dependent</u> on. It is always safer to start with a weak statement and be wary of the pitfalls in making the transition to a strong statement.

In a broader context, the fundamental question is that of what the necessary and sufficient conditions for universality are. Translational invariance does seem to play an essential role. Universality as an <u>exact</u> principle may however be valid only for a certain class of "high symmetry" systems under appropriate conditions. It is not inconceivable, especially for "low-symmetry" or inhomogeneous systems, that universality is only approximate, restricted, or even invalid. In any case, it will be very desirable to have a new generation of precision measurements of the critical exponents so that more insights into universality, and critical phenomena in general, can be gained.

References

1. A.N. Berker and S. Ostlund, J. Phys. C 12, 4961 (1979)

2. R.B. Griffiths and M. Kaufman, Phys. Rev. B 26, 5022 (1982), and references therein

3. Y. Gefen, A. Aharony, and B.B. Mandelbrot, J. Phys. A 17, 1277 (1984), and references therein; A. Aharony, "Fractals in Statistical Physics," paper presented at the International Conference on Collective Phenomena, Tel Aviv University (May 1984)

4. B. Hu, Phys. Rev. Lett. 55, 2311 (1986); B. Hu, Phys. Rev. B 33, 6503 (1986); Y. Wu and B. Hu, "Comments on Phase Transitions on Fractals," University of Houston preprint (1986)

5. I will use hierarchical lattices for illustrative purposes since the renormalization group is exact there. However, the same comments should also be applicable to fractal lattices. For a more detailed discussion of phase transitions on fractals and discrepancies with previous works, the reader is referred to Wu and Hu in Ref. 4

6. B. Derrida, C. Itzykson, and J.M. Luck, Commun. Math. Phys. 94, 115 (1984)

7. M. Blume, V.J. Emery, and R.B. Griffiths, Phys. Rev. A 4, 1071 (1971)

Part IX

Summary

Lessons from the Past and Prospects for the Future?

G. Baym

Theoretical Division, Los Alamos National Laboratory,
Los Alamos, NM 87545, USA and
Department of Physics, University of Illinois at Urbana-Champaign,
1110 West Green Street, Urbana, IL 61801, USA

1. INTRODUCTION

I must first of all point out that I am here under somewhat false pretenses, being a truly linear physicist who has not done serious work on any of the problems discussed this week. In fact, when I was a graduate student of Julian Schwinger's, we were firmly taught that all soluble problems could be reduced to the harmonic oscillator, the quintessential linear problem. [Of course, he was very skilled at transforming all sorts of problems, such as the hydrogen atom and angular momentum, to the harmonic oscillator.] I will thus not treat the subject on a state-by-state basis, diving into the various controversies we have heard, but only give an incomplete "statistical" thermodynamic overview of the meeting, trying to convey my impressions of how the conference answered its implicit assignment: what are the lessons from the past and prospects for the future?

The first lesson learned from all the wonderful talks this week is the enormous progress that has been made in entering the world of real nonlinearities and their role in condensed matter physics. The subject of nonlinearity in condensed matter is vital and growing, and taking shape as a major subfield. One cannot help but be impressed by the wide range of people from very different disciplines that have come together this week, ranging from condensed matter theorists and experimentalists, chemists, real materials people, mathematicians, and even a few high-energy and nuclear types. Equally impressive has been the wide range of problems, systems and materials considered: from anisotropic magnetic compounds, electronic materials such as polyacetylene, PDA, and other polymers, heavy fermion compounds, ferroelectrics, spin glasses, quasicrystals, charge-density wave materials, quantum-Hall materials, and other structured materials, to biological materials such as rhodopsin and DNA, to artificial structures, including electrical circuits and quasicrystalline superconducting loop arrays, and finally to macroscopic quantum systems.

2. LESSONS FROM THE PAST AND PRESENT

What are the major lessons we, or at least I, have learned? One is clearly the recognition of the important role that simplified pictures have played as starting points for understanding realistic systems. In its first phase, from 1927-33, the quantum theory of solids was concerned with developing the most elementary and qualitative descriptions, e.g., the free electron model of metals by Pauli and Sommerfeld, the nature of energy bands in solids by Bloch, Peierls, and Wilson, the origin of positive Hall coefficients by Peierls, the quantum basis of the coupling of elementary magnetic moments in ferromagnets by Heisenberg, etc. Indeed, only after 1933, with the introduction of the Wigner-Seitz method, were solid-state physicists first able to confront quantitatively the quantum theory of condensed matter systems with experiment on real solids [1]. We have barely stepped over this threshold now with the nonlinear systems discussed at this meeting.

More generally, advances in understanding strongly interacting systems have tended to use linear models as guides. Examples are Landau's theory of superfluid He and his Fermi liquid theory; both these theories of strongly interacting systems begin with a decomposition of states into quasiparticle "elementary excitations" – the linearization – and then include interactions as tractable nonlinearities.

The problem now is that we are beginning to face intrinsically nonlinear strongly-coupled systems, such as spin glasses and those with a quantized Hall effect, for which the starting simplifications are less obvious. In highly-structured materials, especially ones with randomness, the nature of the states and spectra remains a difficult problem, e.g., the role of itinerant versus localized modes. How does one develop systematic approaches to solving such nonlinear problems? Here we must turn, as a guide to understanding these systems, to a wider variety of starting points,including exactly soluble models, numerical simulations, and occasional inspired guesses, such as Laughlin's wave function for the fractionally-quantized Hall effect.

Historically, the Ising, Heisenberg and other lattice models, both classical and quantum, have provided great insight into ferromagnetism and other cooperative phenomena, as more recently have lattice gauge theories, and model theories with solitons, such as sine-Gordon, into intrinsically nonlinear systems. Important connections, as we have heard, are also now being made between nonlinear condensed matter systems and relativistic field theories, such as those between polyacetylene and the Gross-Neveu model, and magnetic models and conformally invariant field theories. Controversy over the utility of soluble models does not seem to be past us. The questions raised this week on the usefulness for magnetic systems of the Bethe Ansatz versus the elementary excitation, or spin wave, picture were a strong echo of Bloch's immediate reaction to Bethe's paper containing his Ansatz, shortly after Bloch (and Slater) invented spin waves, and Bloch showed that they would destroy spontaneous magnetization in one and two dimensions. As Bloch wrote to Peierls from Copenhagen on 6 November 1931, "It appears to me that Bethe's tedious algebraic manipulation [Ixereien] is somewhat academic in character, in particular because it does not sufficiently discuss the neighborhood of the lowest eigenvalues. I believe that in this regime, however, my calculations are reasonable, since they neglect only the exclusion of spins on the same site and this cannot play a role in a very dilute spin gas." [2]

Heisenberg, incidentally, did not conceive of his model in any way as an exactly soluble toy; rather he felt that he was solving ferromagnetism as a natural extension of his work on spin alignment, via exchange, in the helium atom. Only some two years after he began his work on ferromagnetism did he appear to worry about the connection of his model with Ising's. As he wrote then (July 1928) to Pauli, "I'd like very much if Weyl could try this problem [the Heisenberg model]. I've completely given it up. The whole question seems important to me on account of the similarity between my model and Ising's. My present view is that Ising should have obtained ferromagnetism if he had assumed sufficiently many neighbors (perhaps $z \geq 8$)... . That Ising uses this ['wild spatial'] model as an argument against ferromagnetism seems to me an indication that he did not understand in perspective his own work." [3]

The approach to nonlinear systems by means of controlled numerical simulations in lattice models, as discussed often this week, is becoming a significant tool, especially with the advent of large computer flop rates enabling high statistics Monte Carlo calculations. Particularly, we are beginning to have available exact and informative studies of small systems, such as magnetic, and those with charge-density waves, and fractionally-quantized Hall effect.

Once simple models are understood it is necessary to face the problem of their relations to real systems. As several examples this week made clear – for instance, the failure of pure sine-Gordon models of magnetic systems to take into account out-of-plane degrees of freedom – one must avoid the temptation to stretch the physics to make it fit the secure models; rather, failure of the simple models is a signal that more interesting physics is waiting to be dealt with. Still unresolved is the development of a convergent approach to polyacetylene, reconciling the momentum space versus real space, or solid-state versus chemical points of view, the question of "correlation versus dimerization gaps." To what extent do these approaches account for all the relevant degrees of freedom? The arguments are reminiscent of the ancient controversy that arose in the theory of ferromagnetism of Heisenberg's Heitler-London method versus Bloch's tight-binding Bloch-wave approach. Slater finally brought the two points of view into harmony in his 1930 paper on "Cohesion in monovalent metals," [4] where he discusses "the relations of the methods of Heisenberg and of Bloch," and shows that, as different bases to build perturbation expansions on, "they are essentially equivalent in their results when properly handled."

A further lesson is the importance of dynamical studies, both experimental and theoretical, beyond thermodynamic ones. We have seen many examples of the crucial role of dynamical response in elucidating the properties of systems, e.g., quantum spin chains, charge density waves, polymers, random field systems, and non-equilibrium systems such as spin glasses, whose states, as in many real materials, depend on the past history; in the latter, for instance, dynamical studies may be only way to probe the transition line in magnetic field, temperature plane. Systematic studies of finite frequency properties should prove invaluable as well in heavy fermion systems, quantum Hall systems, and quasicrystals, where one is only beginning to pin down the states.

What is the relation of the dynamics of real systems to theorists' integrable systems? For example, dynamics of spin-chains, as we have seen, appear to deviate substantially from simple kink-antikink scattering in sine-Gordon models. How does one deal with dynamical systems with a large or infinite number of crucial degrees of freedom? A particular challenge for the theorists is to develop numerical simulation techniques for predicting dynamics. How can one use Monte Carlo methods to go beyond study of elementary lattice thermodynamics and low-lying excitations, to calculate scattering vertices, transport coefficients, correlation functions, and further dynamical response?

Another recurring theme this week has been the role of defects and impurities – dirt physics. Here we should not heed Pauli's advice to Peierls in 1931, "I consider it harmful when younger physicists become accustomed to order-of-magnitude physics. The residual resistance is a dirt effect and one shouldn't wallow in dirt," or later, "One shouldn't work on semiconductors, that's a filthy mess (Schweinerei)...." [5]. What are the effects of defects and impurities, e.g., in magnetic chains on solitons and diffusive behavior, in single crystal polymers, in charge-density-wave dynamics, in pinning of discommensurations, in ferroelectrics, etc.? A closely related and recurrent question has been the role of noise, ranging from problems of small systems, charge density wave motion, spin glasses, pattern growth, etc., to large systems exhibiting quantum coherence and interference.

One important lesson we have not yet learned is how generally to recognize the "smoking guns" of nonlinearity. What behavior is truly nonlinear, as for example that seen in the solitons in magnetic chains and in 3He-A, and in the existence of spinless charge carriers in polyacetylene? How does one distinguish dirt effects from nonlinearities, if possible; impurity problems, as we learned in the Kondo effect and can expect to see in heavy fermion systems, can themselves become highly nonlinear.

3. PROSPECTS FOR THE FUTURE

Although it is difficult to predict the course of the field of nonlinearities in condensed matter in detail, its future prospects, as brought out at this meeting, certainly appear very bright. The field, in possession of both very good theoretical tools and experimental techniques, e.g., resonance, and fabrication methods, is clearly emerging as one of the forefront subjects in condensed matter physics. Improvements in experimental techniques are driving the development of more accurate theory than we have at present. In addition to its more predictable developments, the field has and should continue to yield fascinating surprises, such as the integer and fractionally-quantized Hall effects, and heavy fermion compounds - two as intriguing as any in physics.

It is an inexpensive area, in contrast to large-scale physics, and very important, both for its intrinsic physics interest and its possible applications. Overall the field is leading to a new sophisticated level of material science, on many scales, macro, meso, micro: including artificial and fabricated structures, and study of intrinsically inhomogeneous systems. To mention two examples discussed, the development of organic electronic materials as excellent nonlinear optical materials, and the very novel uses of superconductivity, from networks of Josephson junctions, to testing of quantum coherence on large scales.

Lastly, one of the pleasures of being the final speaker is to be able to acknowledge the hard work of the organizers of the meeting; I know that I speak for all of us in thanking Alan Bishop, David Campbell, Pradeep Kumar, and Steve Trullinger for all their efforts in giving us the opportunity to be exposed to and participate in such a fine overview of the present state of nonlinearities in condensed matter. I would also like to express my thanks to Doug Scalapino for conversations which helped to shape this summary.

* Supported in part by NSF Grant DMR84-15063.

REFERENCES

1. L. Hoddeson, G. Baym, and M. Eckert: Rev. Mod. Phys., Jan 1987.
2. Letter privately communicated by Felix Bloch.
3. Heisenberg to Pauli, 31 July 1928, in A. Hermann, K. von Meyenn, and V. F. Weisskopf (eds.), Wolfgang Pauli. Scientific Correspondence with Bohr, Einstein, Heisenberg, a.o., V. I: 1919-29 (Springer, Berlin, Heidelberg 1979).
4. J. C. Slater: Phys. Rev. 35, 509 (1930).
5. Pauli to Peierls in Ann Arbor, 1 July 1931, Pauli to Peierls, 29 Sept. 1931, in K. von Meyenn, A. Hermann and V. F. Weisskopf (eds.), Wolfgang Pauli. Scientific Correspondence with Bohr, Einstein, Heisenberg, a. o., V. II: 1930-1939 (Springer, Berlin, Heidelberg 1985).

Index of Contributors

Springer Series in Synergetics

Editor: H. Haken

New Volumes

Volume 33

Selforganization by Nonlinear Irreversible Processes

Proceedings of the Third International Conference Kühlungsborn, GDR, March 18-22, 1985

Editors: W. Ebeling, H. Ulbricht

1986. 78 figures. IX, 221 pages. ISBN 3-540-16255-0

Volume 32

Dimensions and Entropies in Chaotic Systems

Quantification of Complex Behavior

Proceedings of an International Workshop at the Pecos River Ranch, New Mexico, September 11-16, 1985

Editor: G. Mayer-Kress

1986. 139 figures. IX, 257 pages. ISBN 3-540-16254-2

Volume 31

Complex Systems - Operational Approaches in Neurobiology, Physics, and Computers

Proceedings of the International Symposium on Synergetics at Schloß Elmau, Bavaria, May 6-11, 1985

Editor: H. Haken

1985. 208 figures. X, 365 pages. ISBN 3-540-15923-1

Volume 30

Dynamical Problems in Soliton Systems

Proceedings of the Seventh Kyoto Summer Institute, Kyoto, Japan, August 27-31, 1984

Editor: S. Takeno

1985. 102 figures. XIII, 281 pages. ISBN 3-540-15372-1

Volume 29

Temporal Order

Proceedings of a Symposium on Oscillations in Heterogeneous Chemical and Biological Systems, University of Bremen, September 17-22, 1984

Editors: L. Rensing, N. I. Jaeger

1985. 201 figures. IX, 325 pages. ISBN 3-540-15274-1

Volume 25

J. S. Nicolis

Dynamics of Hierarchical Systems

An Evolutionary Approach

1986. 145 figures. XV, 397 pages. ISBN 3-540-13323-2

Volume 13

C. W. Gardiner

Handbook of Stochastic Methods

for Physics, Chemistry and the Natural Sciences

2nd edition. 1985. 29 figures. XIX, 442 pages. ISBN 3-540-15607-0

Volume 8

Stochastic Nonlinear Systems in Physics, Chemistry, and Biology

Proceedings of the Workshop, Bielefeld, Federal Republic of Germany, October 5-11, 1980

Editors: L. Arnold, R. Lefever

1st edition. 1981. 2nd printing 1985. 48 figures. VIII, 237 pages. ISBN 3-540-10713-4

Springer-Verlag
Berlin Heidelberg New York
London Paris Tokyo

Springer

www.ingramcontent.com/pod-product-compliance
Lightning Source LLC
Chambersburg PA
CBHW071412230926
51846CB00001B/62
9783642830358

This powerful book draws back the curtain on the shadowy world of private nature conservation. African farmers along the Orange River have managed, against all odds, to maintain access to their ancestral lands through a century and a half of colonialism, genocide, and apartheid – only to find their access threated by elite conservation projects. Moore and Lenggenhager reveal the long history of human occupation in the very place that capitalist investors now insist must be 'rewilded' to save the planet. *Space is the Ultimate Luxury* challenges the moralizing claims of the men who fly private planes and make millions from ecologically destructive industries but who nevertheless insist that it is African pastoralists who threaten our planet's future.

– Meredith McKittrick: Associate Professor of History at Georgetown University and author of *Green Lands for White Men: Desert Dystopias and the Environmental Origins of Apartheid.*

This book analyses the complex intersections of capital, conservation, land, and identity in one of Namibia's most expansive, ecologically and historically rich landscapes. This compelling narrative explores the tensions between individuals seeking large parcels of land for exclusive use, conservationists claiming to protect biodiversity, and local communities fighting to reclaim their ancestral land rights. It provides a thought-provoking analysis of this modern-day clash, and it is an important book for Namibia's land question and conservation agenda.

– Romie Nghitevelekwa: Senior Lecturer in Sociology at the University of Namibia and author of *Securing Land Rights: Communal Land Reform in Namibia.*

This is a book of history and anthropology, the past and the present. Effortlessly, the authors draw on academic fields as diverse as geology and biology, moulding reflections on deep-time and corporate green-washing into a single highly readable whole. The absences of diatribes and the cool understated approach of the text makes it all the more powerfully devastating. In fifteen short and succinct chapters the authors draw attention to the true inhabitants of the deep south of Namibia, delivering a volley of sucker-punches which make a mockery of the environmental claims of the mega-rich, be they mining executives, gap-year youth, corporate traders, or hunters.

– Jan-Bart Gewald: Professor of History at Leiden University and author of *Herero Heroes: A Socio-Political History of the Herero of Namibia*